中国植物病理学会
2019 年学术年会论文集

◎ 彭友良 王文明 陈学伟 主编

Proceedings of the Annual Meeting of Chinese Society for Plant Pathology (2019)

中国农业科学技术出版社

图书在版编目（CIP）数据

中国植物病理学会2019年学术年会论文集 / 彭友良，王文明，陈学伟主编 .—北京：中国农业科学技术出版社，2019. 6

ISBN 978-7-5116-4266-0

Ⅰ. ①中… Ⅱ. ①彭…②王…③陈… Ⅲ. ①植物病理学-学术会议-文集 Ⅳ. ①S432. 1-53

中国版本图书馆CIP数据核字（2019）第120732号

责任编辑 姚 欢 邹菊华
责任校对 贾海霞

出 版 者 中国农业科学技术出版社
北京市中关村南大街12号 邮编：100081
电 话 (010)82106636(发行部) (010)82106631(编辑室)
(010)82109703(读者服务部)
传 真 (010) 82106631
网 址 http://www.castp.cn
经 销 者 各地新华书店
印 刷 者 北京富泰印刷责任有限公司
开 本 889 mm×1 194 mm 1/16
印 张 37. 75
字 数 1000千字
版 次 2019年6月第1版 2019年6月第1次印刷
定 价 100. 00元

《中国植物病理学会 2019 年学术年会论文集》

编辑委员会

前　言

经中国植物病理学会第十一届理事会研究决定，“中国植物病理学会2019年学术年会暨学会成立90周年庆祝大会”将于2019年7月20—24日在成都召开。会议期间将分大会场、分会场及墙报形式交流我国植物病理学理论研究与实践的主要进展，以促进我国植物病理学科发展和科技创新。

会议通知发出后，全国各地植物病理学科技工作者投稿踊跃，为了便于交流，会议论文编辑组对收到的论文和摘要进行了编辑，并委托中国农业科学技术出版社出版。本论文集收录论文及摘要共501篇，其中真菌及真菌病害202篇、卵菌及卵菌病害11篇、病毒及病毒病害81篇、细菌及细菌病害35篇、线虫及线虫病害4篇、植物抗病性68篇、病害防治81篇，以及其他19篇。这些论文及摘要基本反映了近年来我国植物病理学科技工作者在植物病理学各个分支学科基础理论、应用基础研究与病害防治实践等方面取得的研究成果。

由于本论文集论文数量多，编辑工作量大，时间仓促，在编辑过程中，本着尊重作者意愿和文责自负的原则，对论文内容一般未做改动，仅对某些论文的编辑体例上和个别文字做了一些处理和修改，以保持作者的写作风貌。因此，论文集中如果存在不妥之处，诚请读者和论文作者谅解。另外，在本论文集发表的摘要不影响作者在其他学术刊物上发表全文。

本次会员代表大会及学术年会的召开，得到了四川农业大学、四川省植物病理学会等承办单位的鼎力支持。在大会筹办和论文集编辑出版期间，中国植物病理学会和上述单位的众多专家和工作人员，为本次大会的召开和论文集的出版，付出了辛勤劳动。在此，我们表示衷心的感谢！

最后，谨以此论文集庆贺“中国植物病理学会2019年学术年会暨学会成立90周年庆祝大会”胜利召开，祝大会圆满成功！

编　者

2019年7月

目 录

第一部分 真 菌

第二部分 卵 菌

第三部分　病　毒

第四部分　细　菌

第五部分　线　虫

第六部分　抗病性

第七部分　病害防治

第八部分 其 他

第一部分　真　菌

A *Cytospora chrysosperma* virulence effector CCG_07874 localizes to the plant nucleus to suppress the plant immune response*

Han Zhu [1**], Xiong Dianguang [1***], Xu Zhiye[1], Liu Tingli [2], Tian Chengming [1***]

(1. *The Key Laboratory for Silviculture and Conservation of Ministry of Education*, *College of Forestry*, *Beijing Forestry University*, *Beijing* 100083, *China*; 2. *Provincial Key Laboratory of Agrobiology*, *Jiangsu Academy of Agricultural Sciences*, *Nanjing* 210095, *China*)

Abstract: Plant pathogens secrete a large number of virulence-related effectors into host cells during infection to regulate plant immunity and promote parasite colonization. However, the function of effector in *Cytospora chrysosperma*, the causal agent of canker disease damaging a wide range of woody plants and resulting in significant annual losses in economy and ecology, remains largely enigmatic. In this study, we identified a *C. chrysosperma* effector CCG_07874 which was highly expressed during the early infection of host. Sequence analysis revealed that CCG_07874 contained 290 amino acids with an N terminal Signal peptide (1-18 aa) and was rich in cysteines (four). Remarkably, CCG_07874 contained a CAP superfamily domain that was conserved in fungi and showed highly conserved regions at cysteines sites in the C terminus. Deletion of CCG_07874 using the split-marker method showed that CCG_07874 contributed no obvious differences to the vegetative growth, but was essential for ROS tolerance and the virulence of *C. chrysosperma*. Expression of CCG_07874 in *Nicotiana benthamiana* indicated that CCG_07874 acted as an immune inhibitor which could promote the infection of *Botrytis cinerea* by suppressing of callose deposition and cell death triggered by BAX and INF1. Expression of GFP-labeled CCG_07874 in *N. benthamiana* revealed that it localized to the plant nucleus and cytosol, while only the plant nucleus location were essential for maintaining its full immune inhibiting activity. Overall, our investigation on a *C. chrysosperma* virulence effector CCG_07874 demonstrates that CCG_07874 acts in the nucleus to suppress the plant immune response.

Key words: *Cytospora chrysosperma*; virulence effector; subcellular localization; plant immunity

---

* Funding: This work was supported by funding from the National Key Research and Development Program (2017YFD0600100), National Natural Science Foundation of China (31800540) and outstanding Youth Fund of Jiangsu Province (BK20160016)

** First author: Han zhu; E-mail: hanandzhu@163.com

*** Corresponding authors: Tian Chengming; E-mail: chengmt@bjfu.edu.cn; Xiong Dianguang; E-mail: xiongdianguang@126.com

A novel septin ring colocalization protein FgCsp1 plays important role in growth, pathogenesis, and differentiation through Mgv1 MAP kinase pathway in *Fusarium graminearum*

Chen Daipeng[1,2*], Wu Chunlan[1,2], Yin Jinrong[1], Zheng Wenhui[3], Wang Yulin[1], Ma Jiwen[1], Jiang Cong[1], Liu Huiquan[1], Xu Jinrong[2**]

(1. *State Key Laboratory of Crop Stress Biology for Arid Areas*, *College of Plant Protection*, *Northwest A&F University*, *Yangling*, *Shaanxi* 712100, *China*; 2. *Department of Botany and Plant Pathology*, *Purdue University*, *West Lafayette*, *IN* 47907, *USA*; 3. *State Key Laboratory of Ecological Pest Control for Fujian and Taiwan Crops*, *College of Plant Protection*, *Fujian Agriculture and Forestry University*, *Fuzhou* 350002, *China*)

Abstract: Fusarium head blight (FHB) caused by *Fusarium graminearum* is a devastating disease of wheat, barley and other cereals worldwide. To develop an efficient approach for tagging growth associate genes in *F. graminearum*, we tested the *mimp*1/*impalaE* transposon tagging system. A transformant of the *nia*1 strain with efficient *mimp*1/*impalaE* transposition was generated and used to collect 522 insertion mutants. Among them, 27 had growth or colony morphology defects. Mutation sites of 36 mutants was identified by Blast via flanking sequences generated by Hi Tail-PCR. Results from mutation site analysis indicate random distribution of mutation sites in the *F. graminearum* genome. Whereas 22 of the insertion sites were in the non-coding regions, 14 mutants had *mimp*1 inserted to the promoter or 5′-UTR regions. In one of the mutants with pleiotropic defects, the insertion site is 349-bp upstream from the start codon of a gene named *FgCSP*1 (Cytoplasm Septal localization Protein) in this study. Although FgCsp1 orthologs are conserved in filamentous ascomycete, none of them have been functionally characterized. The deletion mutant of *FgCSP*1 generated in this study has similar phenotypes with that of insertion mutant. It reduced 40% in growth, rarely produce aerial hyphae, and the colony edge is irregular. Furthermore, the *Fgcsp*1 deletion mutant with hyperbranched hyphae and the hyphal tips were swollen. It also reduced in conidiation and defect in conidia morphology and germination. The deletion mutant *Fgcsp*1 was defect in pathogenesis and reduced 50% in DON production. Whereas the deletion mutant *Fgcsp*1 can produce perithecium, defect in ascus development and ascospore formation was observed. Additionally, deletion of *FgCSP*1 also affected responses to various cell wall and membrane stresses. The activation of the Mgv1 MAP kinase that is important for growth, pathogenesis, and environmental responses was reduced in the *Fgcsp*1 mutant and express the *MKK*2, upstream of Mgv1, dominant active gene in *Fgcsp*1 partially recovery its defects. The *Fgcsp*1 deletion mutant was also defect in septation. Interestingly, FgCdc3, key component of septin which is important for nuclear division, morphogenesis and pathogenesis was identified by affinity purification assay and found to be co-localized with

* First author: Chen Daipeng; PhD, major in plant pathology; E-mail: chendp@ nwsafu. edu. cn

** Corresponding author: Xu Jinrong; Professor of fungal biology; E-mail: Jinrong@ purdue. edu

FgCsp1 in septin ring in living cells. Our results indicate that FgCsp1 is a conserved septal protein associate with septin in filamentous ascomycetes and it plays important role in hyphal growth, differentiation, and pathogenesis via Mgv1 MAP kinase pathway.

Key words: *Mimp*1; Septin ring; Polarity growth; Mgv1; FgCdc3; *Gibberella zeae*

A pathogenic fungus *Alternaria mali* was firstly identified from tea leaf spot disease in Kaiyang, Guizhou Province, China*

An Xiaoli[1,2**], Wang Xue[2], Dharmasena Dissanayake-saman-pradeep[2], Ren Yafeng[2], Wang Delu[1], Song Baoan[2], Chen Zhuo[2***]

(1. *College of Forestry*, *Guizhou University*, *Guiyang* 550025, *China*;
2. *State Key Laboratory Breeding Base of Green Pesticide and Agricultural Bioengineering*, *Guizhou University*, *Guiyang* 550025, *China*)

Abstract: Leaf spot disease was an important disease of tea (*Camellia sinensis*) in Kaiyang county, Guizhou Province, which mainly damaged young leaves and shoot of tea and led to a huge loss of the production of tea. The spots initially represent brown and round, and then the diameter of the spot was 5-8 mm during later period, with the color of the center in the spot changing white. Tea leaf spot disease always occurs in early spring and the region with 1300 m of altitude. During 2016, disease incidence of leaves was estimated at 84% to 92%, depending on the field. In order to identify the pathogens of this disease, we isolated and identified the pathogens from tea leaf spot disease in this region. The gene of the internal transcribed spacer (ITS), parts of the 18S nrDNA (SSU), translation elongation factor 1-alpha (*EF*1), the *Alternaria* allergen a1 (*Alt*), glyceraldehyde-3-phosphate dehydrogenase (*GDP*), 28S nrDNA (LSU), *anonymous region* (OPA1, OPA2) and *endopolygalacturonase* (*endo*) of the isolates were amplified, sequenced and deposited in Genbank. According to the Koch's postulate, the pathogenicity test was conducted on tea leaves using the methods of the puncture, cut and unwound. The result indicated that the isolates on PDA, OA and MEA medium represented initially round form, and white mycelium on PDA and MEA, and black brown mycelium on OA. The reverse sides of the isolates firstly displayed light yellow on PDA and MEA, and then rapidly became dark brown from the center on MEA. Conidiophores represent dark brown, geniculate, lenth: 11.98 ±9.14 μm, width: 4.47±0.64 μm, ranging from 4 to 39×3 to 5 μm length/width. Brown conidia, narrow ovoid, lenth: 22.94±4.48 μm, width: 11.09±1.68; 14 to 34×7 to 14 μm length/width, with 4 to 8 transverse septa and with conspicuously ornamented walls. Maximum parsimony phylogenetic analysis based on concatenated sequences of combined *OPA*1 (1-610), *OPA*2 (611-1200) and *endo* (1201-1658) indicated that the strain AXLKY_2019_010 was identical to reference strains *Alternaria mali* strain EGS38-029, and the clade was supported by 100% bootstrap values. The pathogenicity test indicated that the isolates can induce leaf spot on leaves of tea using the methods of puncture and cut. To our knowledge, this is the first report of *A. mali* causing leaf spot on tea plants in China.

Key words: *Alternaria mali*; identification; pathogenicity test

---

* Funding: This work was supported by National Key Research Development Program of China (2017YFD0200308) and its Post-subsidy project (2018-5262), and the Major Science and Technology Projects inGuizhou Province (No. 2012-6012)

** First author: An Xiaoli; E-mail: 2489221766@qq.com

*** Corresponding author: Chen Zhuo; E-mail: gychenzhuo@aliyun.com

A pathogenic fungus *Didymella bellidis* was firstly identified from tea leaf spot disease in Yuqing county, Guizhou Province, China*

Yin Qiaoxiu[1**], Ren Yafeng[1], Li Dongxue[1], Wang Xue[1],
Dharmasena Dissanayake-saman-pradeep[1],
Jiang Shilong[1,2], Wu Xian[1], Song Baoan[1], Chen Zhuo[1***]
(1. *State Key Laboratory Breeding Base of Green Pesticide and Agricultural Bioengineering, Guizhou University, Guiyang* 550025, *China*; 2. *College of Agriculture, Guizhou University, Guiyang* 550025, *China*)

Abstract: Tea leaf spot was an important disease of tea plant (*Camellia sinensis*) in Yuqing county, Guizhou province. The disease mainly damaged the young leaves and shoots tea, and then led to a huge loss of tea leaves. Early symptoms were light brown spots, which was gradually increased in size and developed into brown, scattered, elliptical or irregular lesions. The lesions later became greyish at the central, and then were surrounded by an apparent bright yellow halo. The adjacent lesions have been merged to the larger lesion, which was covered the margin of the leaves. In this study, we isolated and identified some fungi from tea leaves samples with the symptoms of leaf spot, and then conducted the pathogenicity test using Koch's rule. The colony was white to pale yellow and produced masses of flocculent aerial mycelium, with an average growth rate of 12.9±0.062 mm/d on PDA after 7 days of post-inoculation in the dark. The colony displayed pale yellow on reverse side. After 22 days of post-inoculation, the colonies represented white to rosy buff. Pycnidia produced on PDA were usually globose to subglobose, scattered or gregarious, olivaceous to olivaceous black. The unicellular and hyaline conidia represent ellipsoidal or obovate, (4.09-9.04) μm× (1.77-4.06) μm, which usually became rounded in both ends. 28S nrDNA (LSU), internal transcribed spacer (ITS), *β-tubulin* 2 (*β-TUB*2) was amplified with the primer pair LR0R/LR7, V9G/ITS4, Btub2Fd/Btub4Rd for the isolates, and then the sequences were deposited in GenBank. Maximum parsimony phylogenetic analysis based on concatenated sequences of combined *ITS* (1-493), *LSU* (494-1897), *β-TUB*2 (1898-2254) indicated that the isolate GZYQ-5-2-b was identical to reference strains *Didymella bellidis* PD 94/886, and the clade was supported by 100% bootstrap values. To fulfill Koch's postulates, the tea leaves of two tea varieties (Moss tea and Fuding Dabai cha) were inoculated with mycelial plugs on PDA or the spore suspension (1×10^6 spores per mL) using the methods of unwound and wound. The symptoms for puncture- or cut-treatment developed brown spots, which was similar to that of naturally infected leaves in the field condition. The symptom was not observed for the unwound-treatment. Our results will contribute to the field management of the disease and the study of pathogenesis in the future.

Key words: *Didymella bellidis*; identification; pathogenicity test; leaf spot

---

* Funding: This work was supported by National Key Research Development Program of China (2017YFD0200308) and its Post-subsidy project (2018-5262), and the Major Science and Technology Projects in Guizhou Province (No. 2012-6012)

** First author: Yin Qiaoxiu; E-mail: yinqiaoxiu@ aliyun. com

*** Corresponding author: Chen Zhuo; E-mail: gychenzhuo@ aliyun. com

A pathogenic fungus *Pestalotiopsis trachicarpicola* was firstly identified from tea leaf spot disease in Yuqing county, Guizhou Province, China*

Yin Qiaoxiu[1**], Ren Yafeng[1], Li Dongxue[1], Wang Xue[1], Wu Xian[1], Jiang Shilong[1,2], Wang Yong[1,2], Wu Xian[1], Song Baoan[1], Chen Zhuo[1***]

(1. *State Key Laboratory Breeding Base of Green Pesticide and Agricultural Bioengineering*, *Guizhou University*, *Guiyang* 550025, *China*; 2. *College of Agriculture*, *Guizhou University*, *Guiyang* 550025, *China*)

Abstract: Leaf spot disease caused by pathogenic fungus seriously damaged the young leaves and shoots tea, and then affects the yield and quality of tea leaves in Yuqing county, Guizhou province. It was an important disease of tea plant (*Camellia sinensis*) in this region. The symptom of the disease represents the characteristic as follow: Purple–brown or reddish–brown dots appeared on leaves in the initial stage, and then the lesions developed into round or amorphous dark spots. The haloed yellow lesions were around the radial edges. In this study, we isolated and identified some fungi from tea leaves samples with the typical symptoms, and then conducted the pathogenicity test using Koch's rule. The colonies of these isolates represent whitish to pale yellow, dense aerial mycelium, circular on PDA, with fimbriate edge, and produce black fruiting bodies, with developing in concentric circles. The colonies display yellow on reverse side. The conidiomata was globose, black, semi–immersed on PDA. Its conidiogenous cells were fusiform, hyaline, short and thin walled. Conidia consist of five cells [per size (17.49–30.16) μm× (4.74–7.4) μm]. Basal cell represents conic to acute, hyaline, thin–walled, verruculose, 2.66 to 7.70 μm. The apical cell represents conic to subcylindrical, hyaline, verruculose, 2.99 to 6.69 μm. The middle three cells represent olive color, from the top the second cell was 3.52 to 7.02 μm, the third cell was 3.37 to 6.82 μm, the fourth cell was 3.11 to 7.26 μm, two to three apical appendages, length of 6.69 to 14.85 μm, and a basal appendage, length of 2.66 to 7.70 μm. The primers ITS4/ITS5 for internal transcribed spacer (ITS) region, T1/BT2B for *β–tubulin gene* (*TUB*), and EF1–728F/EF–2 for *translation elongation factor* 1–*alpha* (*TEF*–1*α*) was used to amplify, the sequences were deposited in GenBank. The phylogenetic tree was constructed by PAUP v 4.0b10 software coupled with the Maximum Parsimony method, and the isolates were identical to reference strains *Pestalotiopsis trachicarpicola* MFLUCC 12–0266, and the clade was supported by 100% bootstrap values. To fulfill Koch's postulates, the leaves of the tea plant varieties of Moss tea and Fuding Dabaicha were inoculated with mycelial plugs on PDA or the spore suspension (1×10^6 spores per mL) using the methods of unwound, puncture or cut. Unwound treatment cannot induce the formation of the lesion. However,

* Funding: This work was supported by National Key Research Development Program of China (2017YFD0200308) and its Post–subsidy project (2018–5262), and the Major Science and Technology Projects inGuizhou Province (No. 2012–6012)

** First author: Yin Qiaoxiu; E-mail: yinqiaoxiu@ aliyun. com

*** Corresponding author: Chen Zhuo; E-mail: gychenzhuo@ aliyun. com

wound treatment can lead to the brown spot after 2 or 11 days of post-inoculation for inoculated methods of mycelial plugs or the spore suspension. To our knowledge, this is the first report of *P. trachicarpicola* causing leaf spot on tea plants in China.

Key words: *Pestalotiopsis trachicarpicola*; identification; pathogenicity test; leaf spot

A pathogenic fungus *Pseudopestalotiopsis theae* was firstly identified from tea leaf spot disease in Kaiyang, Guizhou Province, China*

Ren Yafeng[1**], An Xiaoli[2], Li Dongxue[1], Bao Xingtao[1],
Jiang Shilong[2,3], Wang Delu[2], Song Baoan[1], Chen Zhuo[1***]
(1. *State Key Laboratory Breeding Base of Green Pesticide and Agricultural Bioengineering, Guizhou University, Guiyang* 550025, *China*; 2. *College of Forestry, Guizhou University, Guiyang* 550025, *China*; 3. *College of Agriculture, Guizhou University, Guiyang* 550025, *China*)

Abstract: Tea leaf spot disease mainly damaged young leaves and shoot of tea (*Camellia sinensis*) and led to a huge loss of the production of tea in Kaiyang county, Guizhou Province. In order to identify the causal agent of this disease, we isolated and identified the pathogen. The morphologies of the colony on PDA were undulating, with regular margin and the aerial mycelia represented woolly-cottony and pure white. Globose conidiomata containing black conidial masses were formed on the aerial mycelia. Conidiogenous cells were in clusters, simple filiform, hyaline and smooth-walled. Conidia represented fusiform, 4-septate, measured 24.99±2.10 μm (20.25-28.52 μm) × 6.00±0.70 μm (4.83-8.56 μm), with 3-4 apical appendages, and the three median cells were brown, whereas basal cells were hyaline. The genes of translation elongation factor 1-alpha (*EF*1), the internal transcribed spacer (ITS), 28S nrDNA (*LSU*) and beta-tubulin (*TUB*) were amplified and sequenced. Maximum parsimony phylogenetic analysis based on concatenated sequences of combined *EF*1 (1-356), ITS (357-969), *LSU* (970-1895) and *TUB* (1896-2426) indicated that the strain GZ-KY2018AXL4D was identical to reference strain *Pesudopestalotiopsis theae* MFLUCC12-0055, and the clade was supported by 97.3% bootstrap values. To fulfill the Koch's postulate, wounded tea leaves were inoculated with actively growing mycelia plugs on PDA. The light brown spots were formed on inoculation site after 1-2 d of post-inoculation. To our knowledge, this is the first report of *Ps. theae* causing tea leaf spot in this region, and this work will help the management of disease in the field.

Key words: *Pseudopestalotiopsis theae*; identification; pathogenicity test

---

* Funding: This work was supported by National Key Research Development Program of China (2017YFD0200308) and its Post-subsidy project (2018-5262), and the Major Science and Technology Projects inGuizhou Province (No. 2012-6012)

** First author: Ren Yafeng; E-mail: renyafeng@ aliyun. com

*** Corresponding author: Chen Zhuo; E-mail: gychenzhuo@ aliyun. com

A qPCR system for *Plasmodiophora brassicae* detection in multiple samples*

Guan Gege[1**], Xing Manzhu[1**], Piao Zhongyun[2], Liang Yue[1***]

(1. *College of Plant Protection*, *Shenyang Agricultural University*, *Shenyang* 110866, *China*;
2. *College of Horticulture*, *Shenyang Agricultural University*, *Shenyang* 110866, *China*)

Abstract: Clubroot is a soil-borne disease caused by *Plasmodiophora brassicae*, which mainly attack the cruciferous plants and causes serious economic losses in the world. The pathogen produces numerous resting spores that can survive and keep their pathogenicity in soil for seven and more years. Such survival characteristics of resting spores led to the difficult management of clubroot. In this study, a detection system of *P. brassicae* in multiple samples was established using the qPCR approach. A plasmid containing the qPCR target DNA sequence was constructed and used to generate a standard curve. The established qPCR system was used to measure the abundance of resting spores in samples containing resting spores, including the purified resting spores, soils, and roots and seeds from *Brassica napus*, *B. rapa* and *B. juncea*, respectively. The sensitivity levels were analyzed with the baseline concentrations, including 1 000 copies/μL plasmid, 10 resting spores and 1 000 spores/g soil, 1 000 spores/g root and 1 000 spores/g seeds. The variation of the spore concentrations in the multiple samples was evaluated. A level of 1×10^5 spores showed the minimum variation with the positive correlation to a certain disease index. Herein, a quantitative system of *P. brassicae* in the multiple contaminated samples was established, which will facilitate the accurate detection and timely monitoring of clubroot. This study also provides a technical approach and theoretical reference for the early forecast and epidemic in the sustainable management of clubroot.

Key words: Clubroot; *Plasmodiophora brassicae*; quantitative detection; Liner regression

* 基金项目：辽宁省“兴辽英才计划”项目（XLYC1807242）；国家油菜产业技术体系（CARS-12）；沈阳农业大学引进人才科研启动费项目（20153040）

** 第一作者：关格格，硕士研究生，研究方向为植物保护学；E-mail：guangege1995@163.com
邢曼竹，硕士研究生，研究方向为植物病理学；E-mail：fohao19@126.com

*** 通信作者：梁月，教授，博导，研究方向为植物病理学与真菌学；E-mail：yliang@syau.edu.cn

An effector protein of the wheat stripe rust fungus targets chloroplasts to suppress chloroplast-mediated host immunity*

Xu Qiang, Tang Chunlei, Zhao Jinren, Sun Shutian,
Wang Xiaodong, Kang Zhensheng*, Wang Xiaojie*
(*State Key Laboratory of Crop Stress Biology for Arid Areas and College of Plant Protection, Northwest A&F University, Yangling, Shaanxi* 712100, *China*)

Abstract: Chloroplasts are important for photosynthesis and plant immunity against microbial pathogens. However, there is only limited knowledge on fungal effectors targeting chloroplasts to promote infection of plant pathogenic fungi. Here, we identified and characterized a novel haustorium-specific protein from *Puccinia striiformis* f. sp. *tritici* (*Pst*) that was secreted and translocated into chloroplasts after the cleavage at the predicted transit peptide. Transient expression of *Pst*_12806 inhibited BAX-induced cell death in *Nicotiana* and reduced Pseudomonas-induced hypersensitive response in wheat. It suppressed plant basal immunity by reducing callose deposition and the expression of defense-related genes. *Pst*_12806 was highly expressed during infection and the growth and development of *Pst* were compromised in *Pst*_12806 knockdown plants by HIGS, likely due to increased ROS accumulation. Pst_12806 interacted with the TaISP protein of the Cyt b6/f complex and the binding of Pst_12806 with its C-terminal Rieske domain disrupted the chloroplast functions. Expression of *Pst*_12806 in plants decreased the electron transport rate (ETR), reduced photosynthesis, and attenuated the production of chloroplast-derived ROS. Silencing *TaISP* by VIGS in a susceptible wheat cultivar reduced fungal growth and uredinium development, suggesting an increase in resistance against *Pst* infection. Taken together, our results showed that Pst_12806 is translocated into chloroplasts to perturb photosynthesis to avoid triggering cell death and support pathogen survival on live plants, indicating the importance of interfering chloroplast functions during the infection of *Pst* and other biotrophic pathogens.

Key words: *Puccinia striiformis* f. sp. *tritici* (*Pst*); chloroplast; effector; photosynthesis; plant immunity

Analysis of host specificity andpathogenicity of Cas5 isoform cassiicolin in *Corynespora cassiicola* from *Hevea brasiliensis*[*]

Li Boxun[**], Liu Xianbao, Shi Tao, Feng Yanli, Huang Guixiu[***]

(*Environment and Plant Protection Institute*, *Chinese Academy of Tropical Agricultural Sciences*, *Haikou*, *Hainan* 571101, *China*)

Abstract: *Corynespora* leaf fall disease (CLFD) caused by the fungus *Corynespora cassiicola* can lead to massive defoliation and death of infected rubber trees (*Hevea brasiliensis*). Cassiicolin is a new type of structural toxin protein obtained by pro-cassiicolin splicing and encoding up to six distinct protein isoforms. Some studies found that among the six distinct protein isoforms cassiicolin, only the Cas5 isoform cassiicolin had obvious host specificity and was specific and advantage population on rubber tree in China. The objectives of this study are to use Liquid Chromatography-Electrospray Ionization Mass Spectrometry (LC-ESI-MS) and Nuclear Magnetic Resonance (NMR) methods to analyze the structure of the Cas5 isoform cassiicolin protein. Meanwhile we will inoculate host leaves with pure toxins to analyze the host specificity and host cell lethal effect of Cas5 isoform cassiicolin, and further reveal the relationships between the structural differences of Cas5 isoform cassiicolin protein and host specialization. In addition, we plan to adopt the transcriptome sequencing techniques to analyze the regulatory genes and the regulatory networks that are associated with the pathogenicity and host specificity of Cas5 isoform cassiicolin gene, Which will reveal the roles of Cas5 toxin protein in the pathogenicity of *C. cassiicola*. The results of this research will significantly contribute to the developments of new disease control strategies for reducing *Corynespora* leaf fall disease of *Hevea brasiliensis*.

Key words: *Hevea brasiliensis*; *Corynespora cassiicola*; Cas5 isoform cassiicolin; pathogenicity

* 基金项目：国家重点研发计划项目（2017YFC1200600）；海南省自然科学基金面上项目（317233）；农业部现代农业人才支撑计划项目（0316001）

** 第一作者：李博勋，硕士，助理研究员，研究方向：热带作物多主棒孢病害监控研究；E-mail：diyningxiang@126.com

*** 通信作者：黄贵修；E-mail：hgxiu@vip.163.com

橡胶树多主棒孢病菌 Cassiicolin 基因条形码数据库构建及分子检测技术*

李博勋**，刘先宝，冯艳丽，李　希，黄贵修***

（中国热带农业科学院环境与植物保护研究所，农业部热带作物有害生物综合治理重点实验室，海南省热带农业有害生物监测与控制重点实验室海南，海口　571101）

摘　要：多主棒孢（*Corynespora cassiicola*）是为害我国主要热带作物的植物病原真菌，其寄主范围广、形态差异显著、症状类型多样，且含有一种寄主专化性毒素 Cassiicolin，存在 6 种毒素类型。本研究利用已公布的 Cassiicolin 毒素基因（Cas1—Cas6 型），构建了含 287 个菌株的国内橡胶树和部分热作多主棒孢 Cassiicolin 基因条形码数据库，建立了一套特异性强、灵敏度高的分子检测技术，可检测 100pg/μL 的目标基因组 DNA，系统分析了我国橡胶树、木薯、番木瓜、瓜菜等主要热作的 919 株多主棒孢的毒素类型，发现国内仅存在 Cas2 和 Cas5 两种毒素类型，其中 Cas5 型的菌株占 94.8%，为橡胶树多主棒孢的优势种群和特有毒素类型。而 Cas2 型是橡胶树和其他作物多主棒孢共有的毒素型。系统发育树分析发现，不同毒素类型的多主棒孢菌株与寄主来源密切相关，但与地理来源没有明显的相关性，且 Cas5 型多主棒孢具有明显的寄主专化性。为了进一步明确我国橡胶树多主棒孢的遗传种群结构，本研究利用核糖体 DNA 内转录间隔区（ITS）、翻译延长因子（TEF）和 β 微管蛋白基因，对来自橡胶树、木薯、番木瓜和瓜菜等多种热带作物的 71 株多主棒孢进行种群多样性分析。结果发现，利用最大似然法，在相似系数为 0.97 时可将供试菌株划分为两大遗传类群，类群 Ⅰ 为国内和部分境外国家的橡胶树多主棒孢；类群 Ⅱ 为其他寄主多主棒孢，遗传类群与寄主来源具有显著的相关性。致病力测定表明，多主棒孢种内致病力分化明显，类群 Ⅰ 的橡胶树多主棒孢仅能侵染橡胶树，不能侵染其他寄主，具有明显的寄主专化性，且不同地理来源的菌株致病力差异不显著。类群 Ⅱ 的多主棒孢在不同寄主间可以相互侵染，但菌株在致病力和发病症状上存在一定差异，表现出对其原寄主的高度致病力。通过构建多主棒孢 Cassiicolin 基因条形码数据库、多基因序列聚类以及致病力分化分析，明确我国主要热作多主棒孢病菌的种群结构和优势种群情况，这为弄清我国发掘、保存多主棒孢菌种资源以及制订病害的防治策略具有重要的指导意义。

关键词：橡胶树；多主棒孢（*Corynespora cassiicola*）；基因条形码；种群结构；致病力分化

---

* 基金项目：国家重点研发计划项目（2017YFC1200600）；海南省自然科学基金面上项目（317233）；农业部现代农业人才支撑计划项目（0316001）

** 第一作者：李博勋，硕士，助理研究员，研究方向：热带作物多主棒孢病害监控研究；E-mail：diyningxiang@126.com

*** 通信作者：黄贵修；E-mail：hgxiu@vip.163.com

Biological characteristics of *Ustilaginoidea virens* from different regions in Liaoning province*

Li Xinyang**, Wei Songhong***, Li Shuai, Wang Haining,
Zhu Lijun, Zhang Zhaoru, Liu Wei
(*College of Plant Protection*, *Shenyang Agricultural University*, *Liaoning* 110866, *China*)

Abstract: In order to reveal the relationship between the mycelial growth of *Ustilaginoidea virens* and the environmental conditions from different regions in Liaoning Province, and understand its infection mechanism. The results would provide a scientific basis for forecasting, control and resistance breeding of the rice false smut. The pathogens were isolated and purified by Streptomycin washing. Biological characteristics of *Ustilaginoidea virens* were determined by using the crossing method. The results showed that the mycelia of the pathogen grew better on the medium OA、XBZ、PSA and PDA. They could grow at temperature of 10℃ to 30℃, best at 25-30 ℃. They grew well at pH varied from 3 to 12, best at 4-7. They could utilize many substances as carbon and nitrogen sources. Among carbon sources tested, starch and sucrose were found to be the best ones for the growth of *Ustilaginoidea virens*. Yeast extract and peptone were the most suitable nitrogen sources. Darkness was good for mycelial growth. The lethal temperature for the mycelium was 55 ℃. Based on the studies about biological characteristics of *Ustilaginoidea virens*, apart from chemical and biological controls, the growth of *Ustilaginoidea virens* can be inhibited by changing cultivation environment, reasonable fertilization, irrigation, planting density and rotation, so as to reduce the disease damage.

Key words: Liaoning Province; Rice False Smut; *Ustilaginoidea virens*; Biological Characteristics

* Funding: China Agriculture Research System (CARS-01) and Liaoning BaiQianWan Talents Program

** First author: Li Xinyang, Master student; E-mail: 17309879162@163.com

*** Corresponding author: Wei Songhong, professor; E-mail: songhongw125@163.com

Chaetosphaeronema carmichaeli sp. nov., causing stem spots of *Aconitum carmichaeli* in China[*]

Wang Yan[**], Jing Lin, Zhu Tiantian, Zeng Cuiyun, Chen Honggang
(*Gansu University of Chinese Medicine*, *Lanzhou* 730000, *China*)

Abstract: A new kind of stem spot disease was detected on the stems of *Aconitum carmichaeli* (Chinese Wutou) in Gansu Province, China, in 2010. A *Chaetosphaeronema*-like fungus was isolated and completion of Koch's postulates confirmed that the fungus was the causal agent of the stem spot disease. The morphology and molecular methods were combined to identify the pathogen. The conidiomata were thickwalled, 250-600 μm wide and 150-500 μm high. The conidia were 0-1 septate, 7.7 (-11.4)-19.2μm × 2.0 (-2.8) × 5.1μm. A multi-locus phylogenetic analysis of five loci, LSU, ITS, tef1-α, RPB2 and BTUB demonstrated that this *Chaetosphaeronema*-like fungus is closely related to *C. hispidulum* and *C. achilleae*. However, on the basis of larger pycnidia with no setae this pathogen is viewed as a new taxon of *Chaetosphaeronema*.

Key words: *Aconitum carmichaeli*; Pathogen identification; Multi-locus phylogeny; Taxonomy

---

* Funding: This study has been financially supported by the National Natural Science Foundation Committee of China (No. 31460013), Gansu Natural Science Foundation Committee (17JR5RA164), Key Laboratory of Traditional Chinese Medicine Quality and Standard Project (ZYZL18-001), The Fourth Census of Chinese Medicine Resources in Gansu, China

** First author: Wang Yan, gswangyan101@163.com

Cladosporium cladosporioides was firstly identified from tea leaf spot disease in Yuqing, Guizhou Province, China*

Wang Xue[1**], Yin Qiaoxiu[1], Li Dongxue[1], Ren Yafeng[1], Wu Xian[1], Jiang Shilong[1,2], Wang Yong[1,2], Song Baoan[1], Chen Zhuo[1***]

(1. *State Key Laboratory Breeding Base of Green Pesticide and Agricultural Bioengineering*, *Guizhou University*, *Guiyang* 550025, *China*; 2. *College of Agriculture*, *Guizhou University*, *Guiyang* 550025, *China*)

Abstract: Tea leaf spot disease was an important disease in Yuqing county, Guizhou province. Early symptoms were small grey dark elliptical lesions. The lesion was gradually enlarged with irregular shape and the center of lesion appeared to be necrotic. Disease incidence of leaves was estimated at 46% to 52%, depending on the field. This disease can seriously decrease the production and quality of tea leaves. In this study, we isolated and identified some fungi from tea samples with the symptoms of leaf spot, and then conducted the pathogenicity test. For example, *Nigrospora oryzae*, *Nigrospora sphaerica*, et al were identified from these leaves samples. In addition, we further identified some isolates from the samples of tea leaves, with the similar morphological characteristics. The colony of GZYQ-08-01 on PDA initially was grey olivaceous, and then changed to be grey or olivaceous black, which showed a velvety texture. The conidia were numerous, elliptical, limoniform, branched chains, aseptate, and measured 5.66±3.16 (3.44 to 26.5) μm× 3.21±0.61 (2.34 to 5.02) μm ($n=50$). The internal transcribed spacer (ITS) regions, the translation elongation factor (*TEF*-1α) and actin (*ACT*) sequences were amplified and sequenced from the isolates, and then deposited in Genbank (Accession Nos. MK852271, MK852273, MK852272, and for GZYQ-08-01). The phylogenetic tree was constructed by PAUP v 4.0b10 software coupled with the Maximum Parsimony method. The isolate was identical to reference strains *Cladosporium cladosporioides CBS*15038, and the clade was supported by 100% bootstrap values. The mycelial plugs on PDA or the spore suspension (1×10^6 spores per mL) were inoculated to the tea leaves using the methods of unwound, puncture or cut. The lesions were not observed on tea leaves after 7 days post-inoculation. Some literatures reported that the fungus caused blossom blight of strawberry, leaf spot of *Alstroemeria aurea*, raceme blight of macadamia nuts. Because *C. cladosporioides* was involved with the formation of leaf spot in Yuqing county, Guizhou Province, it is worthy to further study the molecular mechanism of tea leaf spot in the region.

Key words: *Cladosporium cladosporioides*; identification; pathogenicity test; leaf spot; pathogenicity mechanism

---

* Funding: This work was supported by National Key Research Development Program of China (2017YFD0200308) and its Post-subsidy project (2018-5262), and the Major Science and Technology Projects inGuizhou Province (No. 2012-6012)

** First author: Wang Xue; E-mail: gdwangxue@ aliyun. com

*** Corresponding author: Chen Zhuo; E-mail: gychenzhuo@ aliyun. com

MoSnt2-mediated deacetylation of histone H3 regulates autophagy-dependent plant infection by the rice blast fungus

He Min[1,2#], Xu Youping[1], Chen Jinhua[1,#], Luo Yuan[1,#], Lv Yang[1], Su Jia[1], Michael J. Kershaw[2], Li Weitao[1], Wang Jing[1], Yin Junjie[1], Zhu Xiaobo[1], Liu Xiaohong[3], Mawsheng Chern[5], Ma Bingtian[1], Wang Jichun[1], Qin Peng[1], Chen Weilan[1], Wang Yuping[1], Wang Wenming[1], Ren Zhenglong[1], Wu Xianjun[1], Li Ping[1], Li Shigui[1], Peng Youliang[4], Lin Fucheng[3], Nicholas J. Talbot[2], Chen Xuewei[1*]

(1. *State Key Laboratory of Exploration and Utilization of Crop Genetic Resource in Southwest China* (*In preparation*), *State Key Laboratory of Hybrid Rice*, *Rice Research Institute*, *Sichuan Agricultural University at Wenjiang*, *Chengdu*, *Sichuan* 611130, *China*; 2. *School of Biosciences*, *University of Exeter*, *Geoffrey Pope Building*, *Stocker Road*, *Exeter*, *United Kingdom*; 3. *State Key Laboratory for Rice Biology*, *Biotechnology Institute*, *Zhejiang University*, *Hangzhou*, *China*; 4. *State Key Laboratory of Agrobiotechnology and MOA*, *Key Laboratory of Plant Pathology*, *China Agricultural University*, *Beijing*, *China*; 5. *Department of Plant Pathology*, *University of California*, *Davis*, *California*, *USA*)

Abstract: Autophagy is essential for pathogenicity of many fungal pathogens including *Magnaporthe oryzae*, the causal agent of rice blast disease. However, the regulatory mechanism underlying pathogenicity-associated autophagy is poorly understood. We found that the histone protein H3 deacetylation regulator MoSnt2 is necessary for autophagy and plant infection by *M. oryzae*. *MoSNT2* deletion mutants are compromised in autophagic homeostasis, autophagy-dependent fungal cell death and pathogenicity. These mutants are defective in infection structure development, conidiation, oxidative stress tolerance and cell wall integrity. MoSnt2 recognizes acetylated-H3 through its plant homeodomain (PHD) and recruits the histone deacetylase complex via Egl-27 and MTA1 homology 2 (ELM2) domain, thereby resulting in deacetylation of H3. MoSnt2 binds to promoters of autophagy genes *MoATG*6, 15, 16, and 22 to modulate their expression. Besides, *MoSNT*2 expression is positively regulated by MoTor signaling which is necessary for autophagy and rice infection. Our study uncovers a direct link between MoSnt2 and MoTor signaling and defines a novel histone deacetylation mechanism whereby *MoSNT*2 regulates infection-associated autophagy and plant infection in the rice blast fungus.

Key words: Autophagy; *Magnaporthe oryzae*; MoSnt2; MoTor signaling; Pathogenicity

Colletotrichum siamense, a new leaf pathogen of *Sterculia nobilis* Smith recorded in China

Zhang Yaowen*, He Yonglin, Li Qiqin, Lin Wei, Yuan Gaoqing**
(*College of Agriculture*, *Guangxi University*, *Nanning* 530004, *China*)

Abstract: *Sterculia nobilis* Smith is a tropical woody plant, which is used as garden trees in Guangxi, Guangdong, Fujian, Yunnan and Taiwan of China, Vietnam, India, Indonesia, Malaysia and Japan, etc. The fruit of *S. nobilis* contains various nutrient components and are commonly used in traditional Chinese medicine for treating gastroenteric disorder and bloody flux, and also can be used as food raw materials in Southern China. In August 2018, a leaf spot was observed on *S. nobilis* in Nanning, China. The symptoms of the leaf spots on the old leaves were nearly round or an irregular shape. The new leaf lesions were yellowish-brown, and expanded rapidly from the leaf margin to center, forming a large area of dead spots.

Four isolates (ZBPP07, ZBPP08, ZBPP10 and ZBPP12) wereisolated from infected leaves. The mycelial discs (5 mm in diameter) and conidia suspension (1×10^6 spores/mL) of isolates was used for the pathogenicity test. Inoculated leave exhibited symptoms similar to that of natural infection. The same fungus was re-isolated from the lesions in inoculating leave. The colony of isolates was yellowish to pinkish with dense whitish-grey aerial mycelium and a few bright orange conidial masses near the inoculum point 5 days after culture. The average mycelial growth rate of four isolates on PDA was 16.23-17.48 mm/d at 28 ℃. Conidia were all hyaline, guttulate, one-celled and slightly rounded ends. The average conidial size of four isolates was (15.79-17.98) × (4.68-5.21) μm. Appressoria shape varied from ovoid, clavate, inverted trapezoid or slightly irregular to irregular. The average appressoria size of four isolates was (6.76-7.87) × (5.13-6.12) μm. Six gene (ITS, ACT, GAPDH, CAL, CHS-1 and TUB2) were used to identify the four isolates, and the isolates clustered unambiguously with *C. siamense* strains in Phylogenetic tree. According to the morphological and phylogenetic identification, the pathogen caused leaf spot of *S. nobilis* was *C. siamense*. It was first report the leaf spot of *S. nobilis* caused by *C. siamense* in China.

* First author: Zhang Yaowen; E-mail: 2297887061@ qq. com
** Corresponding author: Yuan Gaoqing; ygqtdc@ sina. com

Competition of different genotype infection in *Phytophthora infestans*

Liu Yuchan[1,2 *], Duan Guohua[1,2], Xie Yekun[1,2], Yang Lina[1,2], Zhan Jiasui[1,2 **]

(1. *State Key Laboratory of Ecological Pest Control for Fujian and Taiwan Crops, Fujian Agriculture and Forestry University*, *Fuzhou* 350002, *China*; 2. *Fujian Key Laboratory of Plant Virology*, *Institute of Plant Virology*, *Fujian Agricultural and Forestry University*, *Fuzhou* 350002, *China*)

Abstract: It is common that several genotypes of same pathogen species infect host at the same time, which also called co-infection. In this study, we examined the competition of different genetic similarity of the genotypes in co-infection of *Phytophthora infestans*. The six co-infections consisted of two different genetic similarity genotypes, genotyping the single sporangium of co-infection to study the competition of the different genotypes. The results showed that only one genotype dominated in four of co-infections of total six two-genotypes infections. These four co-infections consisted of two lower genetic similarities genotypes. We reasonably postulated that the one genotype can restrain the other one in the two genotypes infection, suggesting the competition existed in the combination of lower genetic similarity of the genotype co-infection.

Key words: P*hytophthora infestans*; mixed-genotype infection; potato late blight

---

* First author: Liu Yuchan, master student, major in molecular plant pathology; E-mail: 693339612@ qq. com

** Corresponding author: Zhan Jiasui, Professor; E-mail: Jiasui. zhan@ fafu. edu. cn

Crinum asiaticum Anthracnose Caused by *Colletotrichum fructicola* in China

Shen Yanan[1], Xiao Dan[1], Li Shuang[1], Lin Yinfu[1], Pan Limei[4],
Chen Baoshan[2,3*], Wen Ronghui[1,3*]

(1. *College of Life Science and Technology*, *Guangxi University*, *Nanning* 530004, *China*; 2. *College of Agricultural*, *Guangxi University*, *Nanning* 530004, *China*3. *State Key Laboratory for Conservation and Utilization of Subtropical Agro-bioresources*, *Guangxi University*, *Nanning* 530004, *China*; 4. *Tuangxi Botanical Garden of Medicinal Plants*, *Nanning* 530004, *China*)

Abstract: *Crinum asiaticum* is (Genus: Lycoris) an ornamental medicinal plant grown in gardens in tropical and subtropical countries. The ornamental and medicinal value of *C. asiaticum* has been significantly affected. In September 2018, *C. asiaticum* plant shaving red spots or irregular light-brown necrotic lesions, symptoms of anthracnose were observed by chance in the Nanning Botanical Garden of Medicinal Plants, Guangxi Province, China. The feature of leaves had red spots or irregular light-brown necrotic lesions. Leaves withered and dropped when they was severely infected. The causal agent was isolated from small pieces of the symptomatic disinfested leaf and plated on potato dextrose agar (PDA) and incubated at 28 ℃. In order to obtain pure cultures, hyphal tips were transferred onto fresh PDA after three days and incubated at 28℃. Colony morphology was observed daily. Colonies went from white to gray with age, occasionally producing orange conidial masses. Conidia were (15.76±1.4) μm× (6.10±1.4) μm, hyaline, unicellular, cylindrical or fusiform. The morphological characteristics of the fungus were consistent with previous descriptions of the *Colletotrichum fructicola* species complex. Using the method of Multilocus Sequence Analysis identification was confirmed by amplification and sequencing of five DNA regions corresponding to the internal transcribed spacer (ITS), actin (ACT), β-tubulin (TUB2), chitin (CHS-1), and glyceraldehyde 3-phosphate dehydrogenase (GAPDH). The ACT sequences had 100% sequence similarity with *Colletotrichum fructicola* (KR262556.1), ITS and GAPDH sequences had >99% (KP748219.1 and MH463893.1), and CHS-1 and THB2 sequences had >98% similarity (LC469131.1 and MF111081.1). Sequences were deposited in the GenBank. To verify pathogenicity, leaves from 2-month-old *C. asiaticum* plants were damaged with a sterile tip after their surfaces had been wiped with 75% ethanol and wiped with sterilized distilled water three times. Spore suspensions were prepared from 7-day-old PDA cultures and then inoculated using the wound/drop inoculation method. Control plants were established using sterile water during the inoculation process, All plant were placed in a growth chamber with a temperature of 28 ± 2 ℃ and a humidity of 90%. After five days, anthracnose symptoms identical to those observed in the field, developed on the inoculated leaves; and no symptoms on the control leaves. Simultaneously, similar results were obtained by inoculating mycelium and PDA as a control, thus confirming Koch's postulates. To our knowledge, this is the first report of *C. fructicola* causing anthracnose of *C. asiaticum* in China.

Key words: *Colletotrichum fructicola*; Crinum asiaticum

---

* 通信作者：温荣辉，教授，主要从事植物分子病毒学研究；E-mail：wenrh@gxu.edu.cn

Development of a loop-mediated isothermal amplification assay for detection of *Phytophthora parasitica*[*]

Wu Na[**], Li Shujun, Kang Yebin[***]

(*College of Forestry, Henan University of Science and Technology, Luoyang* 471003, *China*)

Abstract: In order to establish a rapid detection method for *Phytophthora parasitica*-mediated isothermal amplification visualization, four primers (2 Inner primers、2 external primers) were designed to test the DNA fragment of *P. parasitica*-specific transcribed spacer (ITS) sequence, optimize LAMP reaction conditions and reaction system, and verify the specificity and sensitivity. The results show that the optimum reaction temperature is 64. 3℃, and the reaction time is 1 h. The optimal reaction system is: 10×ThermoPol Buffer, 7 mmol/L $MgSO_4$, 1. 4 mmol/L dNTP Mix, 1. 6 μm for FIP/BIP, 0. 2 μm for F3/B3, 1 m betaine, 320 U/mL Bst DNA polymerase, 200 μmol/L HNB, 1 μL of template DNA, sterilized with water to 25 μL. Specific detection only the LAMP product of the *P. parasitica* strain is positive (blue), and the electrophoresis results can produce ladder-shaped bands; the sensitivity verification can reach 100 fg at the DNA level, which is 1 000 times the detection sensitivity of the common PCR method. The detection of suspected tobacco plants and soil samples was consistent with the results of the traditional tissue isolation method for *P. parasitica*. For the tobacco plants that were artificially inoculated with the target bacteria, *P. parasitica* was detected on 2 days.

The LAMP detection system of *P. parasitica* established in this study can eliminate the interference of host plants and other non-target bacteria, and quickly and accurately detect *P. parasitica* from tobacco pathogenic tissues and field soils. This study is great significance for the development of field monitoring and guidance for tobacco blackleg disease.

Key words: *P. parasitica*; tobacco black shank; LAMP; HNB

---

* 基金项目：河南省烟草公司科学研究与技术开发重点项目（HYKJ201610）；河南省烟草公司洛阳市公司科学研究与技术开发项目（LYKJ201804）

** 第一作者：吴娜，在读硕士研究生，主要从事植物免疫学研究

*** 通信作者：康业斌，教授，博士研究生导师，主要从事植物免疫学研究；E-mail：kangyb999@163.com

Development of rice conidiation media for *Ustilaginoidea virens*[*]

Wang Yufu[**], Xie Songlin, Liu Yi, Qu Jinsong,
Lin Yang, Huang Junbin, Yin Weixiao[***], Luo Chaoxi

(*Department of Plant Pathology, College of Plant Science and Technology and the Key Lab of Crop Disease Monitoring and Safety Control in Hubei Province, Huazhong Agricultural University, Wuhan* 430070, *China*)

Abstract: Rice false smut, caused by the ascomycete *Ustilaginoidea virens*, is a serious disease of rice worldwide. Conidia are very important infectious propagules of *U. virens*, but the ability of pathogenic isolates to produce conidia frequently decreases in culture, which impacts pathogenicity testing. Here, we developed tissue media amended with rice leaves or panicles that stimulate conidiation by *U. virens*. Generally, rice leaf media more effectively increased conidiation than rice panicle media, and certain non-filtered tissue media were better than their filtered counterparts. Among the tested media, the Indica rice leaf medium with 0.06 g/mL of Wanxian 98 leaves was most efficient at inducing conidiation by *U. virens*, and it was able to induce conidiation in conidiation-defective isolates. Although the conidia induced in rice tissue media were smaller, they were able to germinate on potato sucrose agar medium and infect rice normally. This method provides a foundation for the production of conidia by *U. virens* that will be widely applicable in its pathogenicity testing as well as in genetic analyses for false smut resistance in rice cultivars.

Key words: Rice false smut; *Ustilaginoidea virens*; Conidiation media; Rice tissues; Pathogenicity

---

* Funding: This work was supported by the National Key Research and Development Program of China (2016YFD0300700), the Fundamental Research Funds for the Central Universities (2662017JC003) and the National Natural Science Foundation of China (No. 31701736)

** First author: Wang Yufu, Doctor; E-mail: 635531255@qq.com

*** Corresponding author: Yin Weixiao; E-mail: wxyin@mail.hzau.edu.cn

Discovery and mapping of southern rust resistance gene in maize*

Lu Lu[1], Xu Zhengnan[2], Sun Suli[1], Zhu Zhengdong[1],
Weng Jianfeng[2], Duan Canxing[1**]
(1. *Institute of Crop Sciences, Chinese Academy of Agricultural Sciences, National Key Facility for Crop Gene Resources and Genetic Improvement, Beijing* 100081, *China*;
2. *Institute of Crop Science, Chinese Academy of Agricultural Sciences, National, Engineering Laboratory for Crop Molecular Breeding, Beijing* 100081, *China*)

Abstract: Southern corn rust is a highly contagious disease caused by *Puccinia polysora* Underw., disseminated by airflow, which can reduce the yield of maize (*Zea mays* L.). In the study, using recombinant inbreeding line (RILs) derived from a cross between Qi319 (resistant parent) and Ye478 (susceptible parent), a high-density genetic map was constructed. Five quantitative trait loci (QTLs), associated with resistance to southern corn rust, were located on chromosomes 3, 5, 6, 9, 10, respectively. Each QTL could explain the total phenotypic variation of 2.84%-24.15%, with a total of 56.11% of phenotypic variation. In 2017, chromosome segment substitution lines (CSSL), constructed with Qi319 as non-recurrent parent and Ye478 as a recurrent parent, were used to further verify the rust resistance gene. The CL183 of CSSL showed resistance to southern rust, which was clearly different from other CSSLs. Further localization of the rust resistance gene showed that the resistance gene was linked markers InDel321-InDel323. It can provide a reference for breeding resistant varieties.

Key words: maize; southern corn rust; QTL; chromosome segment substitution lines; recombinant inbreeding lines

---

* Funding: the Program of Protection of Crop Germplasm Resources (2018NWB036-12); the Scientific Innovation Program of the Chinese Academy of Agricultural Sciences; the National Infrastructure for Crop Germplasm Resources (NICGR2016-008)

** Corresponding author: Duan Canxing; E-mail: duancanxing@caas.cn

Draft genome resource of the tea leaf spot pathogen *Didymella segeticola* *

Li Dongxue**, Ren Yafeng, Wang Xue, Yin Qiaoxiu, Bao Xingtao, Dharmasena Dissanayake-saman-pradeep, Wu Xian, Song Baoan, Chen Zhuo***

(*State Key Laboratory Breeding Base of Green Pesticide and Agricultural Bioengineering, Guizhou University*, *Guiyang* 550025, *China*; 2. *College of Agriculture*, *Guizhou University*, *Guiyang* 550025, *China*)

Abstract: The fungal pathogen *Didymella segeticola* (Basionym: *Phoma segeticola*) causes leaf spot on tea (*Camellia sinensis*), which leads to a loss in tea leaf production in Guizhou Province, China. *D. segeticola* isolate GZSQ-4 was sequenced on Illumina Hi-Seq platform, generating 53, 530, 422 raw reads. The Illumina reads were trimmed, corrected and filtered. A total of 51, 270, 673 clean reads passed the quality control criteria and were used in subsequent analyses of *D. segeticola* isolate GZSQ-4. Meanwhile, a Pacific Biosciences (PacBio) RS library was prepared using G-tubes and sequenced using single-molecule real-time cells, generating PacBio reads. PacBio reads were further filtered, generating PacBio subreads, with 1 489 032 total reads. The largest length was 63 621 bp, and the N50, N90 and average lengths were 10 453 bp, 5 327 bp and 7 939 bp, respectively. The genome size was ~36. 7 Mbp based on read sequence data being processed with the condition of 21-mers using the K-mer statistical method. The whole-genome assembly was ~33. 4 Mbp with a scaffold N50 value of ~2. 3 Mbp. In total, 10 893 genes were predicted using the Non-redundant, Gene Ontology, Clusters of Orthologous Groups, Kyoto Encyclopedia of Genes and Genomes, and SWISS-PROT databases. The genome sequence has been deposited in the NCBI Sequencing Read Archive database and can be accessed with the gene accession SDAV00000000 (BioProject: PRJNA516041; BioSample: SAMN10768991; SRA accession: SRR8492863). The whole-genome sequence of *D. segeticola* will provide a resource for future research on host-pathogen interactions, determination of trait-specific genes, pathogen evolution and plant-host adaptation mechanisms.

Key words: whole genome sequences; fungal pathogen; *Didymella segeticola*; leaf spot

---

* Funding: This work was supported by National Key Research Development Program of China (2017YFD0200308) and its Post-subsidy project (2018-5262), and the Major Science and Technology Projects inGuizhou Province (No. 2012-6012)

** First author: Li Dongxue; E-mail: gydxli@ aliyun. com

*** Corresponding author: Chen Zhuo; E-mail: gychenzhuo@ aliyun. com

Field Control Effect of Biocontrol Bacteria *Pseudomonas guariconensis* Strain ST4 on Sugarcane Smut

Lin Nuoqiao, Zhang Lianhui*

(*Guangdong Province Key Lab oratory of Microbial Signals and Disease Control, Integrative Microbiology Research Centre, South China Agricultural University, Guangzhou* 510642, *China*)

Abstract: Sugarcane is an important cash crop in China, widely planted in Guangxi, Guangdong and Hainan province. Sugarcane smut, a fungal disease caused by *Sporisorium scitamineum*, is one of the vital diseases affecting quality and yield of sugarcane. Infection of *S. scitamineum* depends on the mating of bipolar sporidia to form a dikaryon and develops hyphae to penetrate the meristematic tissue of sugarcane. *Pseudomonas* sp. strain ST4, isolated from rhizosphere in Shantou city of Guangdong Province, was known to produce bioactive compounds to inhibit the mating process of *S. scitamineum* and *Ustilago maydis*. A bioactive inhibitor produced by strain ST4 has been isolated and identified as an indole derivative. The trial in greenhouse showed that control efficiency of this inhibitor on maize smut is up to 94%. In order to verify the field control effect of strain ST4 on sugarcane smut, sugarcane buds was soaked in strain ST4 solution prior to plantation in the field where the disease was serious the previous year. The microbial agent was applied at the sowing stage and the tillering stage in order to compare the control effect of primary infection and re-infection of *S. scitamineum*. The results showed that strain ST4 had better biocontrol effect on the re-infection of *S. scitamineum*. The biennial root of sugarcane was also treated with strain ST4. The results showed that the control effect of strain ST4 on sugarcane root was better than that of newly planted sugarcane, with biocontrol efficiency up to 76%. Optimization of treatment procedures might further improve the biocontrol efficiency.

Key words: Sugarcane smut; *Sporisorium scitamineum*; *Pseudomonas guariconensis* Strain ST4; Field trial

* Corresponding author: Zhang Lianhui

First report of *Fusarium solani* and *Fusarium proliferatum* causing root rot of *Clivia miniata* in China*

SunYue, Wang Rui, Wang Fengting, Pan Hongyu, Liu Jinliang**

(*Department of Plant Protection, College of Plant Sciences, Jilin University, Changchun* 130062, *China*)

Abstract: *Clivia miniata* is a famous and precious indoor ornamental flower which has been reported to have extensional and commercial value. During the summer of 2018, root rot disease was observed on *C. miniata* during the flowing phase in flower nurseries in Changchun city, Jilin Province, China. Early symptoms occurred on succulent roots as small brown irregular necrotic spots with a diameter ranging from <1 up to 10 mm initially. The necrosis gradually expanded to the whole root, later developed to the root cortex rot, usually accompanied by yellow leaves. Severely infected plants were broken off from the roots. The causal agent was isolated on potato dextrose agar (PDA) and incubated at 25℃ in the dark for 7 days. Colonies with different morphological characteristics were further purified. All these isolates were first identified by morphological characteristics as *Fusarium solani* and *Fusarium proliferatum* respectively. The internal transcribed spacers (ITS1 and ITS2) and the 5.8S gene were amplified and sequenced. The sequence with 539 nt and 529 nt were deposited in GenBank (accession No. MK642620, MK644026 respectively). BLAST analysis of the sequences showed 100% identity with *F. solani* (accession No. FJ459978.1) and *F. proliferatum* (accession No. MG251448.1), respectively. Based on morphological and molecular analysis, the isolated fungal strains were identified as *F. solani* and *F. proliferatum*. Fungal pathogenicity was confirmed through inoculation by injecting conidial suspension into roots of healthy plants to fulfill Koch's postulates. One month after inoculation, only conidial suspension-inoculated plants (either strain or both) showed root rot symptoms similar to those found in plants used as the inoculum source, whereas the control plants remained no symptoms. To confirm the identity of causal pathogen on symptomatic inoculated plants, *F. solani* and *F. proliferatum* were successfully re-isolated separately or together from the corresponding inoculated roots, thus completing Koch's postulates. To the best of our knowledge, this is the first report of *F. solani*-and *F. proliferatum*-causing root rot of *C. miniata* in China, as well as in the world.

Key words: *Fusarium solani*; *Fusarium proliferatum*; Root Rot; *Clivia miniata*

* Funding: This research was supported by The National Key Research and Development Program of China (2016YFC0501202, 2017YFD0201802)

** Corresponding author: Liu Jinliang; E-mail: jlliu@jlu.edu.cn

Function on the transcriptional factor *SsFKH*1 to sclerotia and cushion development in *Sclerotinia sclerotiorum* *

Cong jie, Liu Jinliang, Zhang Yanhua, Zhang Xianghui, Pan Hongyu**
(*College of Plant Sciences*, *Jilin University*, *Changchun* 130062, *China*)

Abstract: *Sclerotinia sclerotiorum*, the white mold or stem rot fungus, is a necrotrophic pathogen known to infect more than 400 plant species. This disease causes significant yield losses and economic damage to many important crops. On the basis of some research previously, a transcriptional factor SsFkh1 gene from *S. sclerotiorum* was cloned and the function of *SsFkh*1 was charactrerized by gene knockout. The *SsFkh*1 deletion mutants were defective in compound sclerotia and infection cushion formation, which will influence virulence development of *S. sclerotiorum.*

In this study, the *SsFkh*1 deletion mutants and wild type strains were used to analyse differentially expressed genes (DEGs) by RNA Sequencing. The partial candidate genes of compound sclerotia and infection cushion formation were functional analyzed. The RNA - seq results revealed 2005 DEGs ($Log2FC>1$ or <-1, $FDR<0.01$) enriched in 5586 GO terms. KEGG analysis indicated that the up-regulated genes and down-regulated genes were significantly enriched in 12 pathways and 6 pathways ($P<0.05$), respectively.

Based on these results, some valuable genes of *S. sclerotiorum* were deeply excavated and the expression levels of DEGs were verified by fluorescence quantitative PCR which showed consistent results with the sequencing analysis outcomes. The role of these genes were under investigation.

* Funding: This work was financially supported by the National Natural Science Foundation of China (31772108; 31471730)

** Corresponding author: Pan Hongyu; E-mail: panhongyu@ jlu. edu. cn

Functional research on a secreted protein SsCFEM1 in *Sclerotinia sclerotiorum* *

Ji Xu, Liu Jinliang, Zhang Yanhua, Zhang Xianghui, Pan Hongyu**

(*College of Plant Sciences*, *Jilin University*, *Changchun* 130062, *China*)

Abstract: *Sclerotinia sclerotiorum* (Lib.) de Bary is a broad host range plant pathogen and causes lots of loss in the worldwide scale every year. *Sclerotinia sclerotiorum* can form the infection structure, named comprounded appressorium, in order to adhere and penetrate the host plant during pathogenic processes. In addition, *Sclerotinia sclerotiorum* is able to synthesize oxalic acid and cell wall degrading enzymes, which play important role in resisting host defense reaction and accelerating pathogenicity.

CFEM (common in several fungal extracellular membrane proteins) domain is unique in fungi and contains eight conserved cysteine residues. In*Magnaporthe oryzae*, the CFEM domain of seven-transmembrane protein Pth11, a significant G-protein-coupled receptor, is necessary for proper development of appressoria, appressorialike structures and pathogenicity. However, the underlying mechanisms by which CFEM domain-containing proteins act remain largely unknown, especially in broad-hostrange necrotrophic pathogen *S. Sclerotiorum.*

The transcript level of SsCFEM1 increased highly at 24 hours post inoculation (hpi) and then maintained at a high level expression during late stages (24~48 hpi), indicating that SsCFEM1 may be involved in infection of *S. sclerotiorum.* In order to futher investigate the role of SsCFEM1 in *S. Sclerotiorum*, the knock-out mutant was obtained by PEG-mediated protoplast transformation. The knock-out mutant showed no significant changes on mycelial growth, colony morphology and sclerotial development compared with wild type strain. Although the knock-out mutant can form normal comprounded appressorium, the number of it is significantly reduced. Virulence of knock-out mutants was significant declined on detached leaves on *Nicotiana benthamiana*, suggesting that SsCFEM1 is involved in the virulence of *S. sclerotiorum.* Therefore, our data suggested that SsCFEM1 contributes to virulence and comprounded appressorium production.

Key words: *Sclerotinia sclerotiorum*; secreted protein; SsCFEM1

---

* Funding: National Natural Science Foundation of China (31772108; 31471730)

** Corresponding author: Pan Hongyu; E-mail: panhongyu@jlu.edu.cn

Functional study on genes encoding photoresponsive system components of the sugarcane smut fungus *Sporisorium scitamineum*

Cui Guobing

(*South China Agricultural University*, *Guangzhou* 510642, *China*)

Abstract: *Sporisorium scitamineum* causing sugarcane smut is one of the most important fungal pathogens in sugarcane production. Photo-responsive system has been reported in its closely related smut fungus *Ustilago maydis*, but has not been characterized in *S. scitamineum*. We identified 11 candidate genes orthologous to *U. maydis* photo-responsive system in *S. scitamineum*, 8 out of which displayed differential expression in response to light (photo exposure). By homologous recombination we generated single deletion mutant of these genes individually and we will assess the mutants in aspects of sexual-mating, filamentous growth, and pathogenicity. Our study may reveal the potential role of photo response in sugarcane smut disease establishment.

Key words: *Sporisorium scitamineum*; sugarcane; encoding photoresponsive system

Genetic diversity of strawberry anthracnose in Zhejiang, China

Chen Xiangyang*, Wu Jianyan, Zhang Chuanqing**

(*Department of Plant Pathology, Zhejiang Agriculture and Forest University, Lin'an* 311300, *China*)

Abstract: Strawberry anthracnose is a serious fungal disease, which can damage plants throughout the development period of strawberries, mainly infecting strawberry roots and stolons. In this study, a total of 91 isolates from different regions of Zhejiang province of China were collected, and the phylogenetic analysis of multiple genes (actin, internal transcribed spacer, calmodulin, glyceraldehyde - 3 - phosphate dehydrogenase, and chitin synthase) combined with morphological characteristics showed that *Colletotrichum* causing strawberry disease belonged to *C. gloeosporioides* species complex, including of 48 isolates of *C. fructicola*, 21 isolates of *C. siamense*, 13 isolates of *C. gloeosporioides* and 9 isolates of *C. anenigma*. *C. fructicola* is the dominant species of strawberry anthrax in Zhejiang Province. *C. siamense*, mainly distributed in the central and eastern regions of Zhejiang Province (Hangzhou, Jinhua, Shaoxing, Ningbo and Taizhou), is first reported as a pathogen on strawberries. Our results indicate the four kinds of *C. gloeosporioides* species cause strawberry anthracnose in Zhejiang Province, which will help the exploration of epidemic of strawberry anthracnose and disease control in the future.

Key words: Strawberry anthracnose; *Colletotrichum gloeosporioides*; Phylogenetic analysis; Morphological characterization; Disease distribution.

* 第一作者：陈向阳，研究生，学生，植物病理学；E-mail：308892049@qq.com

** 通信作者：张传清，博士，教授，植物病理学；E-mail：cqzhang@zafu.edu.cn

Histone demethylase BcJar1 regulating pathogenic development and virulence of the gray mold fungus *Botrytis cinerea*

Hou Jie[1,2], Feng Huiqiang[1], Chang Haowu[3], Liu Yue[1], Li Guihua[1], Yang Song[1], Sun Chenhao[1], Zhang Mingzhe[1], Yuan Ye[1], Sun Jiao[1], Zhang Hao[3], Qin Qingming[1*]

(1. *College of Plant Sciences*, *Jilin University*, *Changchun* 130062, *China*;

2. *College of Forestry*, *BeiHua University*, *Jilin* 132013, *China*;

3. *College of Computer Science and Technology*, *Jilin University*, *Changchun* 130062, *China*)

Abstract: Histone 3 Lysine 4 (H3K4) demethylation is ubiquitous in organisms; however, the roles of H3K4 demethylase JARID1 (Jar1) /KDM5 in the development and pathogenesis of fungal pathogens remain largely unexplored. Here, we demonstrate that Jar1/KDM5 in *Botrytis cinerea*, the gray mold fungus, plays crucial roles in these processes. The *BcJAR*1 gene was deleted in *B. cinerea* wild-type strain B05. 10 and was reintroduced into the *BcJAR*1 deletion mutant to generate the complemented strains via *Agrobacterium tumefaciens*-mediated transformation (ATMT) approach. The roles of *BcJAR*1 in the fungus development and pathogenesis were investigated using approaches of genetics, molecular/cell biology, pathogenicity and transcriptomic profiling. Our findings demonstrate that BcJar1 is a H3K4 demethylase and is functionally equivalent to ScJhd2, the sole H3K4 demethylase in *Saccharomyces cerevisiae* that belongs to the JARID-family. BcJar1 regulates H3K4me3 and H3K4me2/me3 methylation levels during vegetative and pathogenic development, respectively. Loss of *BcJAR*1 impairs conidiation, appressorium formation and stress adaptation; abolishes infection cushion (IC) formation and virulence, but promotes sclerotium production, of the Δ*Bcjar*1 mutant. BcJar1 mediates reactive oxygen species (ROS) production and proper accumulation and assembly of Sep4, a virulence determinant that initiates infection structure (IFS) formation and host penetration. Exogenous cAMP partially restored the ability of appressorium formation, but failed to rescue IC formation of the Δ*Bcjar*1 mutants, suggesting that cAMP can trigger processes that bypass the defect resulting from the absence of the histone demethylase. BcJar1 orchestrates global expression of genes related to ROS production, stress response, carbohydrate transmembrane transport, secondary metabolites, etc., which are required for conidiation, IFS formation, host penetration, and virulence of the gray mold fungus. Our work systematically elucidates BcJar1 functions and provides novel insights into Jar1/KDM5-mediated H3K4 demethylation in regulating fungal development and pathogenesis.

Key words: *Botrytis cinerea*; H3K4 demethylase JARID1 (Jar1) /KDM5; stress adaptation; infection structures; development and pathogenesis; septin protein Sep4; reactive oxygen species; cAMP signaling

---

* Corresponding author: Qin Qingming; E-mail: qmqin@ jlu. edu. cn

Identification and characterization of pestalotioid fungi causing leaf spots on mango in southern China*

Shu Juan[1,2]**, Yu Zhihe[2], Sun Wenxiu[2], Zhao Jiang[2], Li Qili[1]***,
Tangt Lihua[1], Guo Tangxun[1], Huang Suiping[1], Mo Jianyou[1]***
(1. *Institute of Plant Protection, Guangxi Academy of Agricultural Sciences and Guangxi Key Laboratory of Biology for Crop Diseases and Insect Pests, Nanning, Guangxi*, 530007, *China*; 2. *College of Life Sciences, Yangtze University, Jingzhou, Hubei*, 434025, *China*)

Abstract: During 2016 and 2017, 21 isolates of pestalotioid fungi associated with leaf spots on mango leaves were collected from seven provinces in southern China: Guangxi, Hainan, Yunnan, Sichuan, Guangdong, Guizhou and Fujian. All the 21 isolates were subjected to morphological characterization and DNA sequence analysis. The morphological data were combined with analyses of concatenated sequences of the ITS (internal transcribed spacer), TEF1-α (translation elongation factor) and TUB2 (β-tubulin) for higher resolution of the species identity of these isolates. The results showed that these isolates belong to *Neopestalotiopsis clavispora*, *Pestalotiopsis adusta*, *P. anacardiacearum*, *P. asiatica*, *P. photinicola*, *P. saprophyta*, *P. trachicarpicola* and *Pseudopestalotiopsis ampullacea*. Pathogenicity test results showed that all the isolates could cause symptoms on detached mango leaves (cv. Tainong). To our knowledge, this is the first description of *N. clavispora*, *P. adusta*, *P. asiatica*, *P. photinicola*, *P. saprophyta*, *P. trachicarpicola* and *Ps. ampullacea* as causal agents for leaf spots on mango in China.

Key words: Mango, leaf spots; *Neopestalotiopsisis*; *Pestalotiopsis*; *Pseudopestalotiopsis*; morphology, phylogeny

1 Background

As a fruit-producing tree species in the Anacardiaceae, mango (*Mangifera indica* L.) is an important tropical and subtropical crop (Okigbo, *et al.*, 2003). In China, mango is cultivated mainly in southern provinces such as Guangxi, Hainan, Yunnan, Sichuan, Guangdong, Guizhou and Fujian. In 2014, the total planted area of this fruit in mainland China was almost 173 300 ha, and the production amounted to almost 1 463 000 tons (Nan, *et al.*, 2017).

Leaf spots on mango are known to be caused by fungi species in a few genera such as *Pestalotia*, *Botryodiplodia*, *Fusariella* and *Macrophoma* (Okigbo, *et al.*, 2003). Currently, there are more than 220 established pestalotioid species found on various hosts worldwide (Ge, *et al.*, 2009), but only a few species in this genus have been reported on mango. *P. glandicola* was the first reported causing mango leaf spot in India by Ullasa *et al.*, (1985). *P. mangiferae* was reported causing mango leaf spots in Taiwan, China (Yao *et al.*, 2007). Other recent reports on new species associated with mango in the genus pestalotioid

* 项目基金：国家自然科学基金（31600029，31560526）；广西自然科学基金（2016GXNSFCB380004）
** 第一作者：舒娟，硕士研究生，研究方向为植物真菌病害；E-mail：201872522@yangtzeu.edu.cn
*** 通信作者：李其利；E-mail：liqili@gxaas.net；65615384@qq.com
莫贱友；E-mail：mojianyou@gxaas.net

include: *P. anacardiacearum* in Yunnan, China, and *P. samarangensis* in Thailand by Maharachchikumbura (2013); and *P. uvicola* and *P. clavispora* were reported in Italy by Ismail (2013).

Pestalotioid fungi are major causal agents for mango leaf spots in China. However, there are limited reports to characterize and identify possibly different species. The genus *Pestalotia* was described as having fusiform conidia composed of six cells, and appendages in the apical and basal extremes (De Notaris 1839). Just over a hundred years later, the genus was divided into three genera, based on the number of cells comprising the conidia (Steyaert, 1955). Specifically, conidia of the genera *Pestalotia*, *Pestalotiopsis*, and *Truncatella* possess, respectively, 6, 5, and 4 cells (Steyaert, 1955). Even after more than 60 years, this classification system is still in current use (Maharachchikumbura *et al.*, 2011). However, there are controversies about generic divisions among pestalotioid fungi (Maharachchikumbura, *et al.*, 2014). Sutton (1980) used electron microscopy to examine the development of the cell wall for two species of pestalotioid and the species *Pestalotia pezizoides*, and the findings further supported Steyaert's classification (Griffiths and Swart, 1974).

Until the 1990s, the taxonomy of pestalotioid and related genera was based on stable conidial characteristics such as the pigmentation of the three medium cells of the conidia is concolorous in pestalotioid and versicolorous in *Neopestalotiopsis* (Maharachchikumbura, *et al.*, 2011; 2012). However, using conidial characteristics for species identification for pestalotioid genera has been controversial due to the large variability of other morphological characteristics such as colony color, texture, and shape and conidial characteristics when grown on culture media (Egger, 1995; Hu, *et al.*, 2007). Recently, pestalotioid was redefined by analyzing the sequence of the 18S together with the previous multiple-gene analyses by Maharachchikumbura *et al.*, (2014), clearly showing three highly diverse lineages. Thus, two new genera: *Neopestalotiopsis* and *Pseudopestalotiopsis* were introduced (Maharachchikumbura, *et al.*, 2014).

Molecular techniques have become the norm for species differentiation and identification especially for fungi diversity. Maharachchikumbura *et al.*, (2012) evaluated10 groups of primers, seeking the best regions and/or genes on which to perform phylogenetic analyses of multiple, replicable, and reliable genes. They found that the internal transcribed spacer (ITS) region, the β-tubulin (TUB2) and translation elongation factor (TEF1-α) genes provided sharper definition of species limits.

The purpose of this study was to identify the species of pestalotioid fungi which cause mango leaf spots in China using morphological observation, molecular identification and pathogenicity testing of the isolates on detached mango leaves.

2 Results

2.1 Fungal isolation

Representative samples of leaf spots on mango were collected to represent different stages of the disease. Early disease foliar symptoms on leaves were small yellow-to-brown lesions. Later, these spots expanded with uneven borders, and turned white to gray and coalesced to form larger gray patches. After lesions had matured, small blackacervuli were sometimes observed. Samples were collected from seven southern provinces in China: Guangxi, Hainan, Yunnan, Sichuan, Guangdong, Guizhou and Fujian, with at least one sample from each site. These were stored in coolers until processed in the lab and isolations made. Based on colony characteristics, conidial morphology and ITS sequence analysis, 21 pestalotioid isolates were obtained (Table 1). The 21 isolates included six from Hainan, nine from Yunnan, one from Guangdong, two from Guangxi, and three from Fujian province.

Table 1　Pestalotiod isolates collected from Mango leaves in southern China, with species, geographic original and GenBank accession details

Isolate	Species	Geographic area	Source	GenBank accession numbersa		
				ITS	TEF1-α	TUB2
FJ2-G-2	*Neopestalotiopsis clavispora*	Xiamen, Fujian	Leaf	MK228984	MK512478	MK360925
HN40-1	*N. clavispora*	Jiangchang, Hainan	Leaf	MK228985	MK512479	MK360926
HN51	*N. clavispora*	Dongyun, Hainan	Leaf	MK228986	MK512480	MK360927
YN27-2-2	*N. clavispora*	Ganzuang, Yuanjiang	Leaf	MK228987	MK512481	MK360928
YB27-3	*Pestalotiopsis adusta*	Yuanjiang, Yunan	Leaf	MK228988	MK512482	MK360929
YB56-1	*P. adusta*	Xishuangbanna, Yunnan	Leaf	MK228989	MK512483	MK360930
FY10-12	*P. anacardiacearum*	Pumei, Fujian	Leaf	MK228990	MK512484	MK360931
HN37-4	*P. anacardiacearum*	Jiangchang, Hainan	Leaf	MK228991	MK512485	MK360932
YB41-2	*P. anacardiacearum*	Yuanyang, Yunnan	Leaf	MK228992	MK512486	MK360933
YN30-1	*P. asiatica*	Ganzuang, Yuanjiang	Leaf	MK228993	MK512487	MK360934
HN44-1	*P. asiatica*	Jiangchang, Hainan	Leaf	MK228994	MK512488	MK360935
YN54-2	*P. asiatica*	Xishuangbanna, Yunnan	Leaf	MK228995	MK512489	MK360936
FY7-3	*P. photinicola*	Nanshan, Fujian	Leaf	MK228996	MK512490	MK360937
YB28-2	*P. photinicola*	Yuanjiang, Yunan	Leaf	MK228997	MK512491	MK360938
GD22-1	*P. saprophyta*	Leizhou, Guangdong	Leaf	MK228998	MK512492	MK360939
YN44-2-1	*P. saprophyta*	Xishuangbanna, Yunnan	Leaf	MK228999	MK512493	MK360940
HN56-2	*P. trachicarpicola*	Sanya, Hainan	Leaf	MK229000	MK512494	MK360941
GX17-1	*Pseudopestalotiopsis ampullacea*	Tianyang, Guangxi	Leaf	MK229001	MK512495	MK360942
GX23-1	*Ps. ampullacea*	Tiandong, Guangxi	Leaf	MK229002	MK512496	MK360943
HN41-1	*Ps. ampullacea*	Jiangchang, Hainan	Leaf	MK229003	MK512497	MK360944
YB36-2	*Ps. ampullacea*	Honghe, Yunnan	Leaf	MK229004	MK512498	MK360945

[a]ITS= internal transcribed spacer; TEF1-α=translation elongation factor; TUB2=β-tubulin.

2.2 Morphological and cultural characterization

After 7 daysat 28℃, the colonies on PDA grew up to 7–9 cm diam with wavy edges, off–white color, dense aerial growth, and a reverse–side yellowish color. After two to three weeks, black and gregarious fruiting bodies were obvious (Fig. 1). There were considerable variations in mycelial growth rate among 21 isolates in this study (Table 1). The average mycelial growth rates ranged from 2.1 to 7.1 mm/d. Conidia were all five–celled, with transparent end cells and the middle three cells were a dark purplish–brown. All the individual conidia had two or three chaetae. The average conidial size for the 21 isolates ranged from (28.5–15.3) μm× (8.5–4.5) μm. There were significant differences between conidial size and growth rates among the isolates (Table 1).

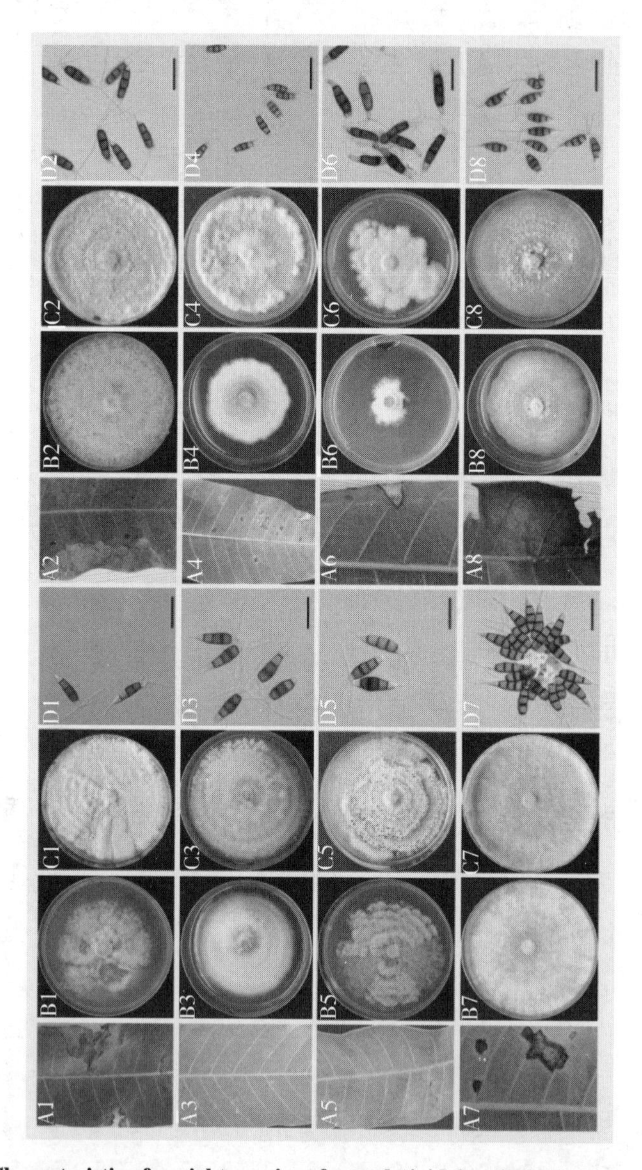

Fig. 1 Characteristics for eight species of pestalotioid fungi obtained from mango

Each species is represented by four pictures (A, B, C and D). A shows symptoms of mango leaf spots, B has colony growth on PDA after 7 days at 28℃, C is growth after 14 days at 28℃, D shows conidial features. Plates 1 and 8 refer to *Pestalotiopsis saprophyta*, *Pseudopestalotiopsis ampullacea*, *Neopestalotiopsis clavispora*, *P. trachicarpicola*, *P. asiatica*, *P. anacardiacearum*, *P. adusta*, and *P. photinicola*.

Table 2　Morphology (conidia size, mycelial growth) and pathogenicity (lesion diameter) of pestalotioid isolates collected in this study

Species	Isolate	Conidial size (μm)[a]	Mycelial growth (mm/d)[b]	Lesion diameters on mango leaves (mm)[c]
Neopestalotiopsis clavispora	FJ2-G-2	19.4 ± 0.2× 6.0 ± 0.1	6.5 a	7.8 ab
N. clavispora	HN40-1	22.4 ± 0.2× 5.8 ± 0.1	5.6 bcdefg	26.4 a
N. clavispora	HN51	23.6 ± 0.3× 7.0 ± 0.1	5.6 bcdefg	13.3 ab
N. clavispora	YN27-2-2	28.5 ± 0.3× 6.9 ± 0.1	6.1 abcd	11.7 ab
Pestalotiopsis adusta	YB27-3	18.7 ± 0.8× 7.2 ± 0.3	4.7 bcdefgh	7.0 ab
P. adusta	YB56-1	21.4 ± 0.2× 4.5 ± 0.1	6.6 ab	4.6 ab
P. anacardiacearum	FY10-12	25.1 ± 0.3× 7.2 ± 0.1	2.9 efgh	5.7 ab
P. anacardiacearum	HN37-4	21.3 ± 0.3× 7.1 ± 0.1	2.7 fgh	2.6 b
P. anacardiacearum	YB41-2	27.7 ± 0.3× 8.5 ± 0.1	2.7 fgh	4.1 b
P. asiatica	YN30-1	25.2 ± 0.2× 6.1 ± 0.1	6.5 ab	2.4 b
P. asiatica	HN44-1	15.3 ± 0.3× 6.3 ± 0.1	4.8 bcdefgh	8.0 ab
P. asiatica	YN54-2	18.7 ± 0.3× 7.1 ± 0.1	6.7 ab	4.1 b
P. photinicola	FY7-3	20.6 ± 0.2× 5.5 ± 0.1	6.3 abc	8.1 ab
P. photinicola	YB28-2	18.6 ± 0.2× 6.4 ± 0.1	3.2 cdefgh	4.5 ab
P. saprophyta	GD22-1	22.9 ± 0.2× 6.0 ± 0.1	5.5 bcdefg	10.0 ab
P. saprophyta	YN44-2-1	24.3 ± 0.2× 6.3 ± 0.1	6.5 ab	9.4 ab
P. trachicarpicola	HN56-2	18.4 ± 0.2× 4.9 ± 0.1	6.0 abcdef	3.4 b
Pseudopestalotiopsis ampullacea	GX17-1	25.8 ± 0.2× 6.5 ± 0.1	7.1 ab	4.2 b
Ps. ampullacea	GX23-1	21.5 ± 0.2× 5.8 ± 0.3	7.0 ab	11.4 ab
Ps. ampullacea	HN41-1	27.9 ± 0.7× 6.5 ± 0.3	2.9 defgh	10.3 ab
Ps. ampullacea	YB36-2	24.5 ± 0.4× 6.6 ± 0.1	5.3 bcdefgh	15.6 ab

[a]The length and width of 100 conidia per isolate on PDA were measured after 14-21 days at 28℃.

[b]The colony diameter (mm) was measured in two perpendicular directions. The colony diameter data were used to calculate the mycelial growth rate (mm/day).

[c]The virulence of the isolate was evaluated by measuring lesion length at 5 days post inoculation in two perpendicular directions on leave

2.3 Phylogenetic analysis

The sequences of ITS, tub2, tef1-α for all the isolates in this study were compared with sequences in GenBank, and matches with high similarity (≥99%) were selected. The results showed that isolates FY10-12, YB41-2 and HN37-4 were highly similar to *P. anacardiacearum*; YB27-3 and YB56-1 with *P. adusta*; HN44-1, YN54-2 and YN30-1 with *P. asiatica*; YB28-2 and FY7-3 with *P. photinicola*; HN56-2 with *P. trachicarpicola*; GD22-1 and YN44-2-1 with *P. saprophyta*; HN51, HN40-1, YN27-2-2 and FJ2-G-2 with *N. clavispora*; and GX17-1, HN41-1, GX23-1 and YB36-2 with *Ps. ampullacea*.

Thesequences of the ITS region, β-tubulin, tef1 genes for all 21 isolates were submitted to GenBank (Table 1). The sequences of pestalotioid fungi species formerly reported by others were also used in this study for comparison. Combined ITS, tub2 and tef1-α sequence data were use in Neighbor-Joining analyses, and the relationships among our 21 isolates and the selected ex-type or ex-epitype species from the GenBank are shown in Fig. 2.

2.4 Pathogenicity and virulence on mangoleaves

In this study, 21 pestalotiod isolates were inoculated onto artificially wounded mango leaves, and all the isolates produced symptoms, while the mock-inoculated PDA plugs did not. The virulence (lesion sizes) for these isolates showed significant differences, with lesion diameters ranging from 2.4 to 26.4 mm (Table2). The two largest lesions diameters were 26.4 mm for HN40-1 from Hainan, 15.6 mm for YB36-2 from Yunnan. The two smallest lesions diameters were 2.4 mm for YN30-1 from Yunnan, and 2.6 mm for HN37-4 from Hainan. There were no obvious statistically significant relationships observed between the virulence level and the fungi species tested, nor between virulence and origin.

3 Discussion

Based on molecular and morphological characterization along with virulence assays, we found that 21 pestalotiod isolates from seven provinces in China from mango leaf spots involve a variety of species from the genera *Pestalotiopsis*, *Neopestalotiopsis* and *Pseudopestalotiopsis*. Except for *P. anacardiacearum*, the other seven fungi species including *P. adusta*, *P. asiatica*, *P. saprophyta*, *P. trachicarpicola*, *P. photinicola*, *N. clavispora*, *Ps. ampullacea* are the first reports on mango in China.

P. asiatica and *P. saprophyta* were reported on other host by Maharachchikumbura *et al.* (2012). Zhang at al. (2012) first reported *P. trachicarpicola* on palm (*Trachycarpus fortunei* (Hook.) H. Wendl.), and later Jayawardena *et al.* (2015) found it on grapevine (*Vitis vinifera* L.). Other reported pestalotioid species include *P. photinicola* on moor besom (*Photinia serrulata* Lindl.) (Chen, at al., 2017), *P. adusta* from an unidentified tree in Hainan province by Li *et al.* (2008), and on *Clerodendrum canescens* Wall. (Xu, *et al.*, 2016).

Neopestalotiopsis clavispora was reported on strawberry (*Fragaria* × *ananassa* Duch.) in Spain (Chamorro, *et al.*, 2016), on *Rosa chinensis* in China (Feng, *et al.*, 2014), on blueberry in Uruguay (Gonzάlez, *et al.*, 2012), on Highbush Blueberry in Anhui Province in China by Chen *et al.* (2016). As a whole, this species has been described as a pathogen for a variety of shrubby hosts.

Traditionally, only morphological data was available, and it was difficult to identify isolates to a species level, especially for isolates from *Neopestalotiopsis*, *Pestalotiopsis* and *Pseudopestalotiopsis* that showed variability in their morphological characteristics (Keith, *et al.*, 2006). In this study, colony

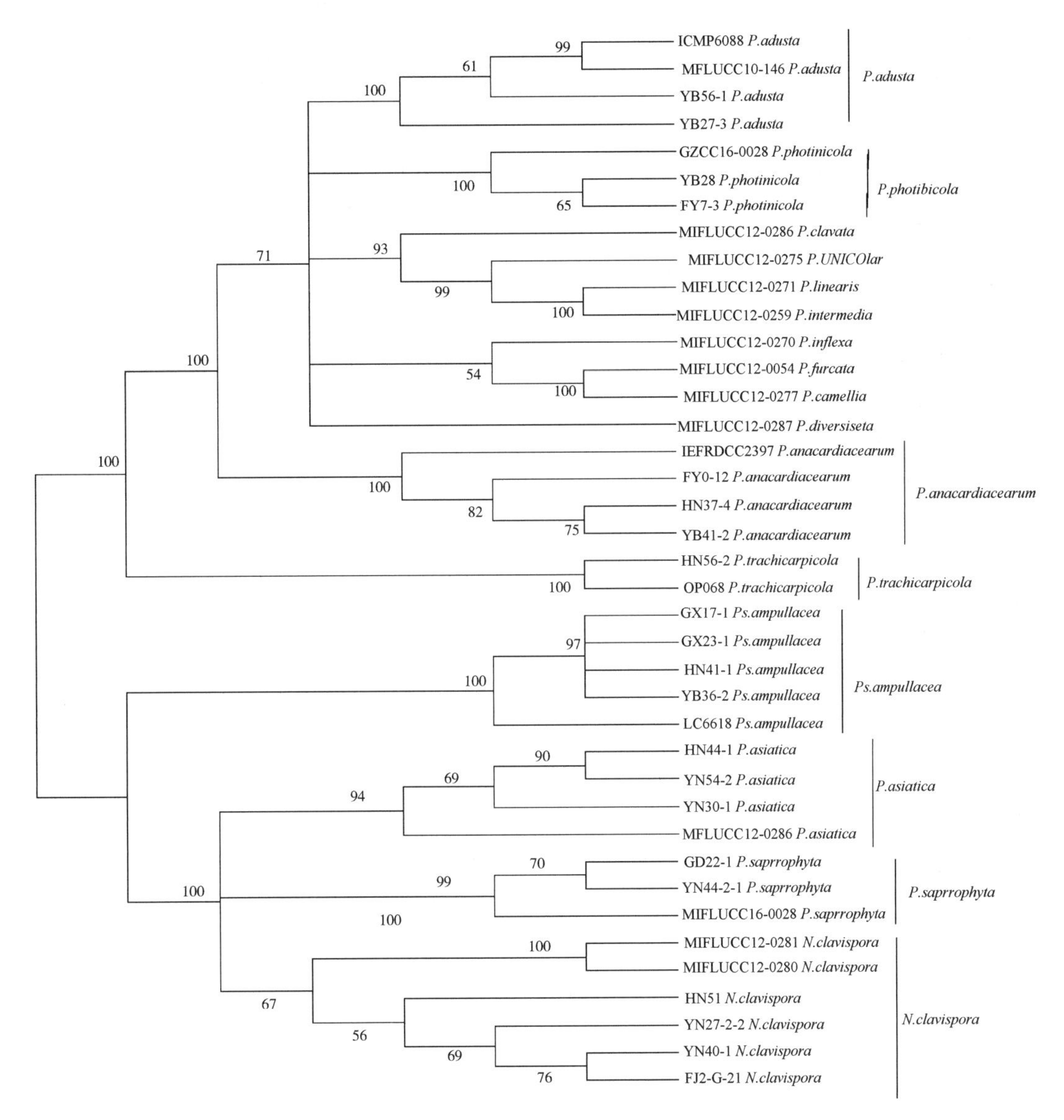

Fig. 2 Phylogenetic tree based on combined ITS, tub2, and tef1-α genes for 21 pestalotioid isolates from mango plus 18 ex-type or ex-epitype pestalotioid isolates from GenBank

Bootstrap values >50% are given at the nodes.

characteristics of the single-spore isolates were not stable when sub-cultured because the isolates varied in one or more traits. This phenomenon was previously reported by Hu *et al.* (2007) that the cultural characteristics (eg. color, growth rate, and texture) of the colony for pestalotioid were flexible upon sub-culturing (Jeewon, *et al.*, 2003, Tejesvi, *et al.*, 2007; Maharachchikumbura, *et al.*, 2011). On the other hand, conidial characteristics were relatively stable, especially the length, width and color of the conidia. These characteristics were more effective for resolving to genus, but not always dependable for resolving to species (Maharachchikumbura, *et al.*, 2011).

Currently, phylogenetic analysis using sequence data can reveal the genetic relationships at a species level based on available reference sequences from type species. In our study, all pestalotioid isolates were assigned to one of these taxa: *P. anacardiacearum*, *P. adusta*, *P. asiatica*, *P. saprophyta*, *P. trachicarpicola*, *P. photinicola*, *N. clavispora* and *Ps. ampullacea*. Species differences were evident in the sequence data for the genomic regions used in this study. However previous research encountered some issues regarding how many nucleotide differences were sufficient to place different isolates as separate species. Maharachchikumbura *et al.* (2012; 2014) were able to differentiated two species by only one nucleotide (ITS), but they had done extensive work to reveal this resolving site. In 2014, they defined two species: *N. honoluluana* and *N. zimbabwana* based on only six nucleotides differences (ITS, TUB2, TEF1-α, ACT, GAPDH, LSU). In contrast, a study in 2012, one of the *N. foedans* isolate with six nucleotides differences (ITS, TUB2, TEF1-α, ACT, GAPDH, LSU) was still considered to be the same species as the other two *N. foedans* isolates evaluated. Liu *et al.* (2018) used only one genomic region (ITS) to define a new species: *P. mangiferae*. In this study, three genes (ITS, TUB2, TEF1-α) were used, and the results indicated that this set of genes can also resolve pestalotioid species, at least for the taxa studied here.

All isolates recovered from mango leaf spots lesions were pathogenic in the detached mango leaf test. Pathogenicity results did not seem to have any obvious relationship togeographical location. The results might not reflect the real virulence potential of these isolates in the field, so further research should be done in the field to confirm the lab results.

This research revealed the species diversity within the genera *Pestalotioipsis*, *Neopestalotiopsis* and *Pseudopestalotiopsis* associated with mango leaf spots in China. These findings will allow for development of strategies to deal with epidemics of these pathogens after more research on the life cycles of each of the species. In future studies, more isolates should be examined as well as other research on disease prevention and management.

4 Methods

4.1 Fungal isolation

During 2016 and 2017, mango leaves with leaf spots were collected from sevenmajor mango growing areas in China (Guangxi, Hainan, Yunnan, Sichuan, Guangdong, Guizhou and Fujian provinces, with at least one sample from each site). Tissues were cut to 5 mm × 5 mm pieces and soaked in 75% ethanol for 10 seconds and 5% hypochlorite for 1 min, followed by three washes with autoclaved water for 3 min. The samples were placed onto potato dextrose agar (PDA), and incubated at 28℃ under 12/12 h light/darkness for 3-5 days. After colonies grew out, the hyphal margins were chosen for sub-culturing onto fresh PDA and incubated at 28℃ for 14 to 21 days. Isolates were purified by single-spore transfers (Liu, *et al.*, 2003), and stored for further study.

4.2 Morphological identification

Single-spore isolates were sub-cultured on PDA at 28℃ and incubated up to 7 days, and then 5-mm-diam plugs from colony margins were placed in the center of each 90-mm-diam plate, with five replications per isolate. The color and culture diameters (two perpendicular directions) of each colony were recorded after 3, 7, and 14 days at 28℃, and hyphal growth rates (mm/d) calculated. Measurements were made of conidia were produced on PDA after two to three weeks at 25℃. For those isolates that did not produce conidia on regular PDA, sterilized pine needles were added to PDA to induce conidial production following Margues *et al.* (2013).

4.3 DNA extraction, PCR, and sequencing

The 21 isolates were cultured on PDA overlaid with cellophane and incubated at 28℃ for 7 days. Fungi mycelia (100 mg) were scraped from the surface, and genomic DNA was extracted using a DNA Kit [Tiangen Biotech (Beijing) Co., Ltd] and diluted to 1 ng/μL. The ITS region was amplified using primer pair ITS4 / ITS5 (White, *et al.*, 1990), the β-tubulin gene using BT2a / BT2b (Glass and Donaldson 1995; O'Donnell and Cigelnik, 1997), and tef1 using EF1-526F /EF1-1567R (Rehner, 2001) (Table 3).

Table 3 Primers used for PCR amplification and DNA sequencing

Genomic region[a]	Primer	Primer sequences
ITS	ITS4	TCCTCCGCTTATTGATATGC
	ITS5	GGAAGTAAAAGTCGTAACAAGG
TEF1-α	EF1-526F	GTCGTYGTYATY GGHCAYGT
	EF1-1567R	ACHGTRCCRATACCACCRATCTT
TUB2	Bt2a	GGTAACCAAATCGGTGCTGCTTTC
	Bt2b	CCCTCAGTGTAGTGACCCTTGGC

aITS= internal transcribed spacer; TEF1-α = translatinelongation factor; TUB2=β-tubulin.

PCR was done ina 25 μL reaction volume containing 11.5 μL of sterilized distilled water, 10.5 μL 2× Ex Taq Master Mix (Shanghai Sangon Biotech Co., Ltd), 2.0 μL primer (R+F), and 1 μL template DNA. The PCR profiles were set as follows: an initial denaturing step of 3 min at 95℃, followed by 35 cycles of denaturation at 95℃ for 30 s, annealing at 55℃ for 30 s (for ITS), and elongation at 72℃ for 30 s, a final extension was performed at 72℃ for 10 min. The annealing temperatures for tef1-α and tub2 were 58℃ and 61℃, respectively. The PCR products were sent to Nanning Guotuo Company for sequencing with both primers.

4.4 Phylogenetic analyses

DNAMAN (version 5.2.2; Lynnon Biosoft) was used to edited and assembled sequences and produce consensus sequences, which were compared by BLAST (Altschul *et al.* 1990) against the NCBI NR database. Sequences from extype or ex-epitype isolates of pestalotioid fungi species from GenBank were also selected for phylogenetic analyses (Table 4). Sequence alignments for each locus and the combined loci were analyzed with ClustalX (v. 1.83). Phylogenetic trees were constructed with MEGA4 (version 4.0) using the Neighbor-Joining (NJ) method.

4.5 Pathogenicity and virulence on mango leaves

All the 21 isolates were used for pathogenicity and virulence tests on detached mango leaves (cv. Tainong) in controlled conditions. The leaves were disinfected by immersion in 75% alcohol for 10 sec, then in 1% NaClO for 1 min, and finally rinsed three times in sterilized distilled water for 3 min. The fresh young leaves (collected from Tiandong, Guangxi province) were inoculated using the following procedure: multiple punctures were made with a sterilized toothpick in a 5-mm-diameter circle, followed by inoculation with a 5-mm-diameter mycelial plug from a 7-day-old PDA culture. Leaves were then placed in transparent plastic boxes lined with water saturated filter paper to maintain high humidity. For each isolate, there were ten replicate leaves. The boxes were then sealed and incubated at 28℃ under 12/12 hours of light/darkness for 5 days. Virulence was assessed by measuring lesion length at 5 days post inoculation (DPI) in two perpendicular directions on each leaf. The data were subjected to analysis of variance, and when significant treatment effects were found, the mean values were compared by LSD (P=0.05) using Data Processing System software (DPS 3.0, Tang, *et al.*, 2013).

Table 4 Sequences from pestalotioid fungi obtained from GenBank and used as references in this study

Species	Culture accession No.	Host	Location	GenBank accession[a]		
				ITS	TUB2	TEF1-α
Neopestalotiopsis clavispora	MFLUCC 12-0281	*Magnolia* sp.	China	JX398979	JX399014	JX399045
N. clavispora	MFLUCC12-0280	*Magnolia* sp.	China	JX398978	JX399013	JX399044
Pestalotiopsis adusta	ICMP 6088	On refrigerator door	Fiji	JX399006	JX399037	JX399070
P. adusta	MFLUCC10-146	*Syzygium* sp.	Thailand	JX399007	JX399038	JX399071
P. anacardiacearum	IFRDCC 2397	*Mangifera indica*	China	KC247154	KC247155	KC247156
P. asiatica	MFLUCC 12-0286	Unidentified tree	China	JX398983	JX399018	JX399049
P. camellia	MFLUCC12-0277	*Camellia japonica*	China	JX399010	JX399041	JX399074
P. clavata	MFLUCC12-0268	*Buxus* sp.	China	JX398990	JX399025	JX399056
P. diversiseta	MFLUCC12-0287	*Rhododendron* sp.	China	JX399009	JX399040	JX399073
P. furcata	MFLUCC12-0054	*C. sinensis*	Thailand	JQ683724	JQ683708	JQ683740
P. inflexa	MFLUCC12-0270	Unidentified tree	China	JX399008	JX399039	JX399072
P. intermedia	MFLUCC12-0259	Unidentified tree	China	JX398993	JX399028	JX399059
P. linearis	MFLUCC12-0271	*Trachelospermum* sp.	China	JX398992	JX399027	JX399058
P. photinicola	GZCC 16-0028	*Punica granatum*	China	KY092404	KY047663	KY047662
P. saprophyta	MFLUCC 12-0282	*Magnolia* sp.	China	JX398982	JX399017	JX399048
P. trachicarpicola	OP068	*Trachycarpus fortunei*	China	JQ845947	JQ845945	JQ845946
P. unicolor	MFLUCC12-0275	Unidentified tree	China	JX398998	JX399029	JX399063
Pseudopestalotiopsis ampullacea	LC6618	*C. japonica*	China	KX895025	KX895358	KX895244

[a]ITS= internal transcribed spacer; TEF1-α =elongation factor; TUB2=β-tubulin.

References

Baldwin E A, Burns J K, Kazokas W, *et al.* Effect of two edible coatings with different permeability characteristics on mango (*Mangifera indica* L.) ripening during storage [J]. Postharvest Biology and Technology, 1999, 17: 215-226.

Chamorro M, Aguado A, De los Santos B. First report of root and crown rot caused by *Pestalotiopsis clavispora* (*Neopestalotiopsis clavispora*) on strawberry in Spain [J]. Plant Disease, 2016, 100: 1495.

Chen Y Y, Maharachchikumbura S S N, Liu J K, *et al.*. Fungi from Asian Karst formations I. *Pestalotiopsis photinicola* [J]. Mycosphere, 2017, 8: 103-110.

Chen Y, Zeng L, Shu N, *et al. Pestalotiopsis*-like species causing gray blight disease on *Camellia sinensis* in China [J]. Plant Disease, 2018, 10: 98-106.

Chen Y, Zhang A F, Yang X, *et al.* First report of *Pestalotiopsis clavispora* causing twig blight on highbush blueberry (*Vaccinium corymbosum*) in Anhui province of China [J]. Plant Disease, 2016, 100: 859.

De Notaris G. Micromycetes italiei Dec II [J]. Mere R Acad Sci Torino II, 1839, 3: 80-81.

Egger K N. Molecular analysis of ectomycorrhizal fungal communities [J]. Canadian Journal of Botany, 1995, 73: 1415-1422.

Feng Y R, Liu B S, Sun B B. First report of leaf blotch caused by *Pestalotiopsis clavispora* on *Rosa chinensis* in China [J]. Plant Disease, 2014, 98: 1009.

Ge Q X, Chen Y X, Xu T. Flora fungorum sinicorum. Vol. 38, *Pestalotiopsis* [M]. Science Press, Beijing, 2009.

Glass N L, Donaldson G C. Development of primer sets designed for use with the PCR to amplify conserved genes from filamentous ascomycetes [J]. Applied and Environmental Microbiology, 1995, 61: 1323-1330.

González P, Alaniz S, Montelongo M J, *et al.* First report of *Pestalotiopsis clavispora* causing dieback on blueberry in Uruguay [J]. Plant Disease, 2012, 96: 914.

Griffiths D A, Swart H J. Conidial structure in two species of *Pestalotiopsis* [J]. Transactions of the British Mycological Society, 1974, 62: 295-304.

Hu H, Jeewon R, Zhou D, *et al.* Phylogenetic diversity of endophytic *Pestalotiopsis* species in *Pinus armandii* and *Ribes* spp.: evidence from rDNA and β-tubulin gene phylogenies [J]. Fungal Diversity, 2007, 24: 1-22.

Ismail A M, Cirvilleri G, Polizzi G. Characterisation and pathogenicity of *Pestalotiopsis uvicola* and *Pestalotiopsis clavispora* causing grey leaf spot of mango (*Mangifera indica* L.) in Italy [J]. European Journal of Plant Pathology, 2013, 135: 619-625.

Jayawardena R S, Zhang W, Liu M, *et al.* Identification and characterization of *Pestalotiopsis*-like fungi related to grapevine diseases in China [J]. Fungal Biology, 2015, 119: 348-361.

Jeewon R, Liew E C, Simpson J A, *et al.* Phylogenetic significance of morphological characters in the taxonomy of *Pestalotiopsis* species [J]. Molecular Phylogenetics and Evolution, 2003, 27: 372-383.

Keith L M, Velasquez M E, Zee F T. Identification and characterization of *Pestalotiopsis* spp. causing scab disease of guava, *Psidium guajava*, in Hawaii [J]. Plant Disease, 2006, 90: 16-23.

Ko Y, Yao K S, Chen C Y, *et al.* First report of gray leaf spot of mango (*Mangifera indica*) caused by *Pestalotiopsis mangiferae* in Taiwan [J]. Plant Disease, 2007, 91: 1684.

Li E, Jiang L, Guo L, *et al. Pestalachlorides* A-C, antifungal metabolites from the plant endophytic fungus *Pestalotiopsis adusta* [J]. Bioorganic & Medicinal Chemistry, 2008, 16: 7894-7899.

Li R, Huang G, Su M, *et al.* Status and developmental strategies of mango industry in China [J]. Journal of Southern Agriculture, 2013, 44: 875-878.

Liu A R, Zhang R Y, Xie X C. Isolation and identification of pathogen of banana crown rot [J]. Journal of South China University of Tropical Agriculture, 2003, 9: 1-5.

Liu F, Hou L, Raza M, *et al. Pestalotiopsis* and allied genera from *Camellia*, with description of 11 new species from China [J]. Scientific Reports, 2017, 7: 866-885.

Liu Y J, Huang B. *Pestalotiopsis mangiferae* causes leaf spot of Chinese hickory (*Carya cathayensis*) in China [J]. Plant Disease, 2018, 102: 674.

Maharachchikumbura S S, Guo L D, Cai L, *et al.* A multi-locus backbone tree for *Pestalotiopsis*, with a polyphasic characterization of 14 new species [J]. Fungal Diversity, 2012, 56: 95-129.

Maharachchikumbura S S, Guo L D, Chukeatirote E, *et al. Pestalotiopsis*—morphology, phylogeny, biochemistry and diversity [J]. Fungal Diversity, 2011, 50: 167-187.

Maharachchikumbura S S, Guo L D, Chukeatirote E, *et al.* Destructive new disease of *Syzygium samarangense* in Thailand caused by the new species *Pestalotiopsis samarangensis* [J]. Tropical Plant Pathology, 2013, 38: 227-235.

Maharachchikumbura S S, Hyde K D, Groenewald J Z, *et al. Pestalotiopsis* revisited [J]. Studies in Mycology, 2014, 79: 121-186.

Maharachchikumbura S S, Zhang Y, Wang Y, *et al. Pestalotiopsis anacardiacearum* sp. nov. (*Amphisphaeriaceae*) has an intricate relationship with *Penicillaria jocosatrix*, the mango tip borer [J]. Phytotaxa, 2013, 99: 49-57.

Mo J, Zhao G, Li Q, *et al.* Identification and characterization of *Colletotrichum* species associated with mango anthracnose in Guangxi, China [J]. Plant Disease, 2018, 102: 1283-1289.

Nan N, Fu Z J., Xu J C. Mango industrialized development in China [J]. Journal of Yunnan Agricultural University, 2017, 11: 80-84.

Okigbo R N, Osuinde M I. Fungal leaf spot diseases of mango (*Mangifera indica* L.) in southeastern Nigeria and biological control with *Bacillus subtilis* [J]. Plant Protection Science-prague, 2003, 39: 70-78.

Reddy M S, Murali T S, Suryanarayanan T S, *et al. Pestalotiopsis* species occur as generalist endophytes in trees of western Ghats forests of southern India [J]. Fungal Ecology, 2016, 24: 70-75.

Solarte F, Muñoz C G, Maharachchikumbura S S, *et al.* Diversity of *Neopestalotiopsis* and *Pestalotiopsis* spp., causal agents of *Guava scab* in Colombia [J]. Plant Disease, 2018, 102: 49-59.

Steyaert R L. *Pestalotia*, *Pestalotiopsis* et *Truncatella*. Bulletin du Jardin botanique de l'Etat, Bruxelles/Bulletin van den Rijksplantentuin [J]. Brussel, 1955: 191-199.

Sutton B C. The *Coelomycetes*: Fungi Imperfecti with pycnidia, acervuli and stomata [M]. England: Commonwealth Mycological Institute, 1980.

Tang Q Y, Zhang C X. Data Processing System (DPS) software with experimental design, statistical analysis and data mining developed for use in entomological research [J]. Insect Science, 2013, 20: 254-260.

Tsai I, Maharachchikumbura S S, Hyde K D, *et al.* Molecular phylogeny, morphology and pathogenicity of *Pseudopestalotiopsis* species on Ixora in Taiwan [J]. Mycological Progress, 2018, 17: 1-12.

Ullasa B A, Rawal R D. Occurrence of a new post-harvest disease of mango due to *Pestalotiopsis glandicola* [J]. Acta Horticulturae, 1985, 231: 540-543.

Xu M F, Jia O Y, Wang S J, *et al.* A new bioactive diterpenoid from *Pestalotiopsis adusta*, an endophytic fungus from *Clerodendrum canescens* [J]. Natural Product Research, 2016, 30: 2642-2647.

Zhang Y, Maharachchikumbura S S, McKenzie E H, *et al.* A novel species of *Pestalotiopsis* causing leaf spots of *Trachycarpus fortunei* [J]. Cryptogamie, Mycologie, 2012, 33: 311-318.

Identification and mapping of a new blast resistance gene *Pi67* (t) in *Oryza glaberrima* *

Dong Liying[1**], Liu Shufang[1], Xu Peng[2], Li Xundong[1], Zhou Jiawu[2], Li Jing[2], Deng Wei[2], Tao Dayun[2], Yang Qinzhong[1***]

(1. *Agricultural Environment and Resources Institute*, *Yunnan Academy of Agricultural Sciences*, *Kunming* 650205, *China*; 2. *Food Crops Institute*, *Yunnan Academy of Agricultural Sciences*, *Kunming* 650205, *China*)

Abstract: *Oryza glaberrima* with strong resistance to biotic and abiotic stress was regarded as an excellent gene pool for Asian cultivated rice improvement. Mining and utilization of favorable genes of *O. glaberrima* would be important for broadening genetic basis of modern rice. Several *O. glaberrima* accessions showed highly resistance against *Magnaporthe oryzae* in natural blast nursery, indicating that *O. glaberrima* could possess new blast resistance genes. To identify new blast resistant genes from *O. glaberrima*, *O. sativa* subsp. *japonica* cultivar Dianjingyou 1 was crossed with accessions of *O. glaberrima*, and then successively backcrossed with recurrent parent Dianjingyou 1 to breed introgression lines. A set of BC_5F_4 introgression lines (ILs) was evaluated for blast resistance in natural blast nursery. The resistant ILs in natural blast nursery were inoculated again with strong virulent to monogenic lines in greenhouse. The resistant IL106 was selected for genetic and molecular analysis. Using a BC_6F_2 population derived from a cross of IL106/ Dianjingyou 1, a new dominant blast resistant gene, designated as *Pi67* (t), was identified and preliminarily mapped on chromosome 6 of rice flanked by SSR marker RM30 and RM345, and co-segregated with STS67-5. It was found that *Pi67* (t) conferred a broad-spectrum resistance to *M. oryzae* strains collected from different rice-growing regions. The new identified blast resistance gene *Pi67* (t) would be a promising gene for disease-resistant breeding.

Key words: *Magnaporthe oryzae*; resistance gene; *Pi67* (t); mapping; *Oryza glaberrima*

---

* Funding: This project was supported by National Natural Science Foundation of China (31860524)

** First author: Dong Liying, Master, major in plant pathology; E-mail: dliying70@ 163. com

*** Corresponding author: Yang Qinzhong, major in plant pathology; E-mail: qzhyang@ 163. com

Identification of physiological races and avirulence genes of *Pyricularia oryzae* in Liaoning province in 2017[*]

Liu Wei[**], Wei Songhong[***], Zhu Lijun, Wang Haining, Zhang Zhaoru, Li Xinyang
(*College of Plant Protection*, *Shenyang Agricultural University*, *Liaoning* 110866, *China*)

Abstract: In order to understand the dynamics of physiological races and the composition of avirulence genes of *Pyricularia oryzae* and provide a theoretical basis for rational distribution of rice varieties and breeding of blast-resistant cultivars in Liaoning province, 7 identified rice varieties were used to identify the physiological race of 75 isolates of *Pyricularia oryzae* collected in 2017 and the primers were designed to amplify the target sequences based on the sequences of 10 Avr-genes which had been cloned. The results showed that the tested isolates were divided into 6 groups of 32 physiological races, ZA and ZB were the predominant groups, accounting for 42.67% and 28.00%, respectively, ZC, ZD, ZE and ZF were 9.33%, 5.33%, 4.00% and 10.67%. ZG was not detected. All the test isolates carried the avirulence genes *Avr-Pi9* and *ACE1*. Most carried the avirulence genes *Avr-Pita*, *Avr-Piz-t*, *Avr-Pik*, *PWL2 and Avr-Pib*, carrying frequencies of 92%, 94.67%, 93.33%, 93.33% and 49.33%. The avirulence genes *Avr-Pii*, *Avr-CO39* and *Avr-Pia* were not detected. In summary, the composition and structure of the physiological races and avirulence genes of *Pyricularia oryzae* are complex in Liaoning province. In the rice variety layout, the rice cultivars carrying resistant genes *Pita*, *Pi9*, *Pizt* and *Pik* should be generalized in Liaoning province.

Key words: Liaoning Province; *Pyricularia oryzae*; physiological races; avirulence gene

---

* Funding: The National Key R&D Program of China (2018YFD0300307); China Agriculture Research System (CARS-01)
** First author: Liu Wei, Master student; E-mail: lwlwliuwei@126.com
*** Corresponding author: Wei Songhong, professor; E-mail: songhongw125@163.com

Identification of the pathogen *Alternaria longipes* causing *Scaevola taccada* leaf spot*

Wang Yi[1,2**], Zhao Chao[1***], Hu Meijiao[2], Li Min[2], Gao Zhaoyin[2], Li Chunxia[2], Hong Xiaoyu[2], Zhou Ziqian[2,3], Zhang Wu[4]

(1. *Department of microbiology*, *College of Tropical Agriculture and Forestry*, *Hainan University*, *Haikou* 570228, *China*; 2. *Institute of Environment and Plant Protection*, *Chinese Academy of Tropical Agricultural Sciences*, *Haikou* 571101, *China*; 3. *College of Plant Science and Technology*, *Huazhong Agriculture University*, *Wuhan* 430070; 4. *Lingnan Normal University*, *Zhanjiang* 524048, *China*)

Abstract: *Scaevola taccada* is a main bushy shrub distributed widely in the coastal strands of Pacific and Indian Oceans. It usually used as windbreak and sand fixation plant due to its strong salt and drought tolerance, high photosynthetic capacity and water use efficiency. In May and October 2018, a leaf spot disease was observed on *S. taccada* during the disease survey of the native plants on the Xisha Islands, China. In all cases, a similar fungus was consistently isolated from symptomatic leaves tissue. One representative single-spore isolate 35 was selected randomly, used for pathogenicity tests, and identified with morphological and molecular methods. In addition, the same fungal cultures were successfully re-isolated from symptomatic plants and confirmed by morphological characters, which completed Koch's postulates. DNA was extracted from the mycelium of 10-days-old culture, and the internal transcribed spacer (ITS) rDNA regions were amplified with the primer pairs ITS1 and ITS4, and the glyceraldehyde-3-phosphate dehydrogenase (*gapdh*) gene regions were amplified with GPD1 and GPD2 primers. The sequences were compared the GenBank nucleotide database by using a BLAST alignment, which revealed that 35 had 99% to 100% identity with *A. longipes* for the ITS and *gapdh* regions. The fungus was identified as *A. longipes*. To our knowledge, this is the first report of *A. longipes* causing leaf spot on *S. taccada* in China.

Key words: *Scaevola taccada*; leaf spot; *Alternaria longipes*; Identification

---

* Funding: This research was supported by Financial Fund of the Ministry of Agriculture and Rural Affairs, P. R. China (No. NFZX2018)

** First author: Wang Yi, Master's degree

*** Corresponding author: Zhao Chao, Ph. D

Identification of the pathogen of *Stenotaphrum helferi* leaf spot

Zheng Jinlong[1], Liu Wenbo[3], Zhan Min[3], He Chunping[1], Yi Kexian[1*],
Xi Jingen[1], Huang Xing[1], Wu Weihuai[1], Gao Jianming[2], Zhang Shiqing[2],
Chen Helong[2], Liang Yanqiong[1], Tbokie Jr Thomas[4]

(1. *Environment and Plant Protection Institute*, *CATAS*, *Key Laboratory of Integrated Pest Management on Tropical Crops*, *Ministry of Agriculture*, *P. R. China*, *Haikou*, *Hainan* 571101, *China*; 2. *Institute of Tropical Bioscience and Biotechnology*, *CATAS*, *Key Laboratory of Tropical Crop Biotechnology*, *Ministry of Agriculture Haikou* 571101, *China*; 3. *College of plant protection*, *Hainan University Haikou* 570228, *China*; 4. *College of Plant Protection*, *Nanjing Agricultural University*, *Nanjing* 210095, *China*)

Abstract: Pathogens of 61 *Stenotaphrum helferi* leaf spot obtained from six provinces in China (Hainan, Guangdong, Guangxi, Yunnan, Fujian and Jiangsu), the United States of America and South Africa were isolated, purified, characterized base on morphology and pathogenicity analyzed by using rDNA internal transcribed spacer region (rDNA - ITS), Glyceraldehyde - 3 - phosphate dehydrogenase (GAPDH) and β - Tubulin (TUB2) sequences. The pathogenicity assays were conducted in accordance with Koch's law, and the observed morphological characteristics of the isolates confirmed the pathogen as *Culvularia lunata*. Results of the molecular analyses showed 99% similarity with *C. lunata*. Our research findings provide insights to support future studies on the epidemiology, resistance mechanism and control strategy of the leaf spot disease of bluestocked grass.

Key words: *Stenotaphrum helferi*; Leaf spot; *Curvularia lunata*; pathogenic fungus; pathogenicity assay

* 通信作者：易克贤，博士，研究员，研究方向：热带作物真菌病害及其抗性育种；E-mail：yikexian@ 126. com

Infection mechanism of plant pathogenic *Diaporthe actinidiae* to kiwifruit deciphered by genome and transcriptome analysis*

Li Li**, Deng Lei, Chen Lianfu, Pan Hui, Wang Zupeng, Zhong Caihong***

(1. *CAS Key Laboratory of Plant Germplasm Enhancement and Specialty Agriculture, Wuhan Botanical Garden, Wuhan* 430074, *China*; 2. *Innovation Academy for Seed Design, Chinese Academy of Sciences, Wuhan* 430074, *China*; 3. *CAS Engineering Laboratory for Kiwifruit Industrial Technology, Chinese Academy of Sciences, Wuhan* 430074, *China*)

Abstract: *Diaporthe* species are fungal pathogens that devastate crop plants and fruits worldwide. *Diaporthe actinidiae* strain, the pathogens caused kiwifruit stem-end rot, has been regarded as one of the most serious threat to kiwifruit production in the world. In the paper, we reported the whole genome and comparative transcriptome analyses of *D. actinidiae* infecting kiwifruit. The genome size of *D. actinidiae* strain was 58. 30 Mb, 17, 176 protein-encoding genes were predicted. Comparative genomics with 30 other fungi showed 18. 94% of the proteins were unique to *D. actinidiae*, and the strain have highest level of CAZYme genes, especially Glycoside Hydrolases and Auxiliary activities enzymes. 3140 pathogenicity-related genes were identified in *D. actinidiae* genome, including 135 lethal genes, 85 hyper-virulence genes and 9 genes resistance to chemical. RNA-Seq analysis was conducted to compare the expression profiles among *D. actinidiae* mycelium and infected flesh tissue of *D. actinidiae* infecting three suspectable cultivars. In the differentially expressed genes (DEGs), 340 up-regulated and 1161 down-regulated genes could be assigned into functional categories of the gene ontology (GO): biological process, molecular function, and cellular component. The function of 23 significant DEGs genes involving in Pathogen Host Interactions (PHI) progress among *D. actinidiae* and kiwifruit was verified. This study build the foundations for further characterize the underlying infecting mechanisms of this important fungal pathogen to kiwifruit.

Key words: *Diaporthe* spp. ; kiwifruit rot disease; fungal genomes; pathogenicity-related genes

---

* 基金项目：国家自然科学基金青年科学基金项目（31701974）；湖北省自然科学基金面上项目（2017CFB443）；湖北省农业科技创新行动项目“特色水果生态高效栽培与采后处理”项目；武汉市科技局前资助科技计划（2018020401011307）

** 第一作者：李黎，副研究员，主要从事猕猴桃病害研究；E-mail：lili@ wbgcas. cn

*** 通信作者：钟彩虹，研究员，主要从事猕猴桃资源挖掘、育种、病害等相关研究；E-mail：zhongch1969@ 163. com

Leaf blight caused by *Curvularia coicis* on Chinese pearl barley (*Coix chinensis*) in Fujian Province, China*

Dai Yuli**, Gan Lin, Ruan Hongchun, Shi Niuniu,
Du Yixin, Chen Furu, Yang Xiujuan***

(*Fujian Key Laboratory for Monitoring and Integrated Management of Crop Pests*, *Institute of Plant Protection*, *Fujian Academy of Agricultural Sciences*, *Fuzhou* 350013, *China*)

Abstract: Chinese pearl barley (*Coix chinensis* Tod.) is grown worldwide as a cereal food and medicinal plant. During 2016 and 2017, a severe leaf blight outbreak was observed in Nanjing prefecture, Fujian Province, China. The disease outbreak affected almost 100% of the Chinese pearl barley plants under field conditions. The pathogen was isolated from infected leaves showing typical symptoms of dark brown spots surrounded by yellow halos, and was identified as *Curvularia coicis* Castellani based on morphological characteristics and sequencing of the ribosomal DNA internal transcribed spacer (rDNA-ITS). The optimal temperature for mycelial growth of *C. coicis* on potato dextrose agar (PDA) ranged from 25℃ to 28℃. Pathogenicity assays were performed using Chinese pearl barley 'Longyi 1' at the seven-to eight-leaf stage. Overall, *C. coicis* had a high virulence on these inoculated plants, even in the absence of wounds, and the pathogen was successfully reisolated. The results from this study identified the causal organism of leaf blight on Chinese pearl barley.

Key words: *Coix* leaf blight; *Curvularia coicis*; Chinese pearl barley; pathogenicity; internal transcribed spacer

---

* 基金项目：福建省属公益类科研院所专项（2017R1024 - 3；2018R1025 - 1）；国家重点研发计划项目（2018YFD0200700）；福建省农业科学院青年科技英才百人计划项目（YC2016-4）；福建省农业科学院植物保护创新团队（STIT2017-1-8）

** 第一作者：代玉立，助理研究员，博士，研究方向：真菌学及植物真菌病害；E-mail：dai841225@126.com

*** 通信作者：杨秀娟，研究员，研究方向：植物病理学；E-mail：yxjzb@126.com

Multi-gene sequence analysis of brown spot pathogen of coffee*

Gbokie Jr Thomas[1,2**], Wu Weihuai[1**], Zhu Mengfeng[1], He Chunping[1], Liang Yanqiong[1], Huang Xing[1], Lu Ying[1], Wang Yuanchao[2], Yi Kexian[1***]

(1. *Key Laboratory of Integrated Pest Management on Tropical Groups*, *Ministry of Agriculture and Rural Affairs*, *P. R. China*; *Hainan Key Laboratory for Monitoring and Control of Tropical Agricultural Pests*; *and Environment and Plant Protection Institute*, *Chinese Academy of Tropical Agricultural Sciences*), *Haikou*, *Hainan* 571101, *China*; 2. *College of Plant Protection*, *Nanjing Agricultural University*, *Nanjing* 210095, *China*)

Abstract: Coffee leaf brown eyespot is a fungal disease commonly found in coffee nurseries and established coffee fields. In order to establish a rapid detection and monitoring technique for this pathogen, three conserved gene sequences, including ITS, Histone *H*3, and β-tubulin were analyzed. Using the ITS and Histone H3, as well as the universal primers of β-tubulin gene, the genomic DNA of coffee leaf brown eyespot pathogen initially identified based on the morphological characteristics, was amplified and followed by subsequent amplifications of the individually band. After cloning and sequencing, we obtained 451bp, 410bp and 405bp bands, respectively, which were separately subjected to nBlast in the NCBI database. The results showed that the sequence similarity between the ITS sequence of the coffee leaf brown eyespot pathogen and different strains of *Cercospora* spp. in the NCBI database was as high as 100%, while the coffee brown eyespot pathogen β-tubulin gene sequence had the highest sequence homology similarity of 99. 75% with the different strains of *Cercospora* spp. The sequence similarity between the Histone *H*3 sequence of this pathogen and the other strains of *Cercospora* spp. was 99. 74%. Further comparative analysis of histone *H*3 gene and β-tubulin gene with homologous sequences of the genus *Cercospora* were performed and the results showed that there were four single nucleotide variation sites between the Histone *H*3 sequence and other sequences of genus *Cercospora*. In the phylogenetic tree constructed based on the *H*3 sequence, coffee leaf brown eyespot pathogen was clearly distinct other *Cercospora* spp., forming an independent branch. Our finding therefore indicates that the Histone *H*3 gene sequence can be used as a molecular marker for the identification of the pathogen of coffee brown eyespot disease.

Key words: Coffee spores; Coffee brown spot; Molecular identification; Biological characteristics

* 基金项目：国家重点研发项目“特色经济作物化肥农药减施技术集成研究与示范”（2018YFD0201100）；中国热带农业科学院基本科研业务费专项资金（1630042017021）

** 第一作者：Gbokie Jr Thomas，在读博士，研究方向：植物病理；E-mail：2017202057@ njau. edu. cn
吴伟怀，副研究员，研究方向：植物病理；E-mail：weihuaiwu2002@ 163. com

*** 通信作者：易克贤，博士，研究员，研究方向：热带作物真菌病害及其抗性育种；E-mail：yikexian@ 126. com

Nemania bipapillata was firstly identified from tea leaf spot disease in Yuqing, Guizhou Province, China*

Wang Xue[1**], Yin Qiaoxiu[1], Li Dongxue[1], Ren Yafeng[1], Wu Xian[1],
Jiang Shilong[1,2], Song Baoan[1], Chen Zhuo[1***]

(1. *State Key Laboratory Breeding Base of Green Pesticide and Agricultural Bioengineering, Guizhou University, Guiyang* 550025, *China*; 2. *College of Agriculture, Guizhou University, Guiyang* 550025, *China*)

Abstract: Tea plant (*Camellia sinensis*) is a major crop in Guizhou province, China. By the end of 2017, the cultivated area of tea plant covered 350 000 hm^2. In November 2018, tea leaf spot was caused a huge loss of the production of tea in Yuqing county in Guizhou province. The symptoms of the infected tea leaves were brown, irregular and the center of lesion appeared to be necrotic. In some tea gardens, disease incidence of leaves was 46% to 52%, and decreased seriously the production of tea leaves. In this study, some fungi from tea disease samples were identified, such as *Phomopsis foeniculi*, *Diaporthe ganias*, *et al.* Meanwhile, some isolates were obtained to our attentation. The colony of GZYQ-03-02 on PDA initially was white, and then the colony was gradually changed to black after 14 days post-inoculation. The genes of DNA-dependent RNA polymerase II second largest subunit (*RPB*2), beta-tubulin 2 (β-*TUB*2) and actin (*ACT*) were amplified and sequenced from the isolates, and then deposited in Genbank (Accession Nos. MK852276, MK852273 and MK852274 for the isolate GZYQ-03-02). Maximum parsimony phylogenetic analysis based on concatenated sequences of combined *ACT* (1-312), β-*TUB*2 (313-506), *RPB*2 (507-847) indicated that the isolate GZYQ-03-02 was identical to reference strains *Nemania bipapillata* 90080610, and the clade was supported by 99% bootstrap values. To fulfill Koch's postulates, unwounded and wounded tea leaves were inoculated with mycelial plugs on PDA, and non-symptoms were observed after 7 days of post-inoculation. *N. bipapillata* was known as synonym of *Nemania diffusa*. *N. diffusa* caused tea stem wood decay in Sri Lanka. Meanwhile, *N. bipapillata* was also known as endophyte, which was isolated from plant hosts of *Lycium chinense*, *Cephalotaxus*, *Vanilla planifolia*, *et al.* To our knowledge, we firstly isolated the fungus from tea leaves. The mechanism of the fungus in the tea leaf spot is worth studying in the future.

Key words: *Nemania bipapillata*; identification; pathogenicity test; leaf spot; pathogenicity mechanism

* Funding: This work was supported by National Key Research Development Program of China (2017YFD0200308) and its Post-subsidy project (2018-5262), and the Major Science and Technology Projects inGuizhou Province (No. 2012-6012)

** First author: Wang Xue; E-mail: gdwangxue@ aliyun. com

*** Corresponding author: Chen Zhuo; E-mail: gychenzhuo@ aliyun. com

Orah Anthracnose caused by *Colletotrichum gloeosporioides* in China

Qing Zhen, Chen Haiyun, Hu Xiaoxuan, Lin Yinfu, Chen Baoshan, Wen Ronghui

(*State Key Laboratory for Conservation and Utilization of Subtropical Agro-bioresources*; *College of Life Science and Technology*, *Guangxi University*; 100 *Daxue Road*, *Nanning*, *Guangxi* 530004, *China*)

Abstract: Orah results from a cross between "Temple" tangor and "Dancy" tangerine which is originated from Israel and then introduced to America, Korea, and China. It has the characteristics of late maturation, high sugar content, high yield, good cold and drought tolerance, and so on. Because of the high economic benefit and broad market prospect, Orah has been popularized in China rapidly, especially in Guangxi province, with a planting area over 66 700 hm^2. In January 2019, a leaf spot disease was observed in an orchard (22. 44 N, 108. 12 E) near the heaven reservoir of the Guangxi Province, China. Initially, leaf tips displayed yellow sunken spots that became irregular light-brown necrotic lesions. Black acervuli developed in concentric rings. Symptoms initially appeared on fruits as brown, round, and sunken spots with salmon-colored spore masses on the lesions. Necrotic lesions gradually expanded to the whole fruits. Disease incidence was 16. 4% (n = 500 plants). In order to isolate the pathogen, symptomatic orah leaves were cut into 0. 5 cm2 pieces, surface disinfestedin 75% ethanol during1 min, washed three times with sterile distilled water (SDW) and subsequently disinfested in 5% sodium hypochlorite during 6 min. This treatment was followed by five rinses in SDW and leaf peices were plated on potato dextrose agar (PDA) prior to incubation at 28±2℃ in darkness. After 2 days, a fungus harboring white aerial mycelium was isolated. Colonies displayed a gray center that turned atrovirens with age. Hyphae were hyaline and septate. Conidia, measuring 17. 00 (13. 93 to 19. 43) μm× 5. 55 (4. 49 to 6. 56) μm (n = 21), were hyaline, aseptate and cylindrical. To confirm the identification, five genes were amplified and sequenced: internal transcribed spacer region (ITS), glutamine synthetase (GS), calmodulin (CAL), β-tubulin (TUB2) and glyceraldehyde-3-phosphate dehydrogenase (GAPDH). ITS, GS, CAL, TUB2 and GAPDH sequences showed100% sequence similarity with those from *C. gloeosporioides* (MK130755. 1, HM575376. 1, MK138600. 1, KX578813. 1, HM575277. 1). Based on sequence analysis and morphological characteristics, the isolate was confirmed to be *C. gloeosporioides* and the gene sequences were deposited in GenBank with the accession numbers MK934477 (ITS), MK920991 (GS), MK920988 (CAL), MK920990 (TUB2), and MK920989 (GAPDH). Mycelium plugs from the isolate Yan01 were inoculated to orah fruits. Reisolation of the pathogen was performed from the edges of fruit lesions. Pathogen morphology corresponded to that of the inoculated fungus and control fruits remained healthy. *C. gloeosporioides*is known to causeolive anthracnose in Tunisia as well as being responsible for reddish-orange spot disease with chlorotic halos onpitahaya plants in Brazil. To our knowledge, this is the first report on the involvement of *C. gloeosporioides* in the anthracnose of orah fruit in China.

Key words: Orah; Anthracnose; *C. gloeosporioides*

Physiological races identification of rice blast fungi using 3 sets differential systems

Wang Jichun[1], Zhu Feng[1], Wang Dongyuan[2], Liu Xiaomei[1],
Jiang Zhaoyuan[1], Ren Jinpin[1], Li Li[1], Tiang Chengli[1], Sun Hui[1]
(1. *Institute of Plant Protection*, *Jilin Academy of Agricultural Sciences* 130033, *China*;
2. *Plant Protection College*, *Shenyang Agricultural University*, *Shenyang* 110866, *China*)

Abstract: Understanding thepathogenicity of *Magnaporthe oryzae* is important in assisting the development of the rice breeding programs for long lasting effective genetic resistance. The first international differentials system of rice blast physiological race was established in IRRI and was modified in China into the following China National Differentials Varieties (CNDV): Tetep, Zhenlong13, Sifeng43, Hejiang18, Dongnong363, Guandong51 and Lijiangxintuanheigu (LTH) (Anonymous, 1980). A set of monogenic lines containing 24 major *R* genes in the blast-susceptible recurrent *japonica* variety LTH developed by the International Rice Institute Research Institute was used to evaluated pathogenic characteristics of blast fungi isolates (Wang, *et al.*, 2013; Telebanco-Yanoria, *et al*, . 2008). Furthermore, BRMG differential system was established for blast fungi virulence analysis, which was composed of 7 monogenicline varieties (Wu, *et al.*, 2017).

In this study, blast fungi isolates collected in Northeast of China were inoculated upon 3 sets differential system varieties. The results data demonstrated that A, B, C, D, E, F and G group race amount were 20, 15, 2, 2, 1, 4 and 9, respectively, using CNDV system. The data by CNDV system could not to guide breeding and cultivars distribution for lacking closed genetic background information to current rice landraces. By BRGM system, 24 race types were identified of 53 isolates, and the BRMG21 belongs to the dominant race type. Compared to the confused and indirectly races information of JIRCAS system which using 24 monogenic differential system, no dominant race type was obvious. At last, the data demonstrated by BRMG system was advanced for breeding and resistant cultivars distributing in Japonica rice planting area in Jilin Province.

Key words: Rice blast; Physiological race; differentials; CNDV; BRMG; JIRCAS

Pithomyces chartarum was firstly identified from tea leaf spot in Yuqing, Guizhou Province*

Yin Qiaoxiu[1**], Wang Xue[1], Ren Yafeng[1], Li Dongxue[1],
Jiang Shilong[1,2], Wu Xian[1], Song Baoan[1], Chen Zhuo[1***]
(1. *State Key Laboratory Breeding Base of Green Pesticide and Agricultural Bioengineering, Guizhou University, Guiyang* 550025, *China*;
2. *College of Agriculture, Guizhou University, Guiyang* 550025, *China*)

Abstract: Leaf spot disease is an important disease of tea (*Camellia sinensis*) in Yuqing county, Guizhou province. Early symptoms were small, round or irregular shape. The lesion gradually enlarged to display the darker lesion margin with a bright yellowish-black halo. The center of lesion appeared to be necrotic. The adjacent lesions have been merged to the larger lesion and covered the margin of the leaves. The disease takes place in tea garden during autumn on altitude from one thousand to one thousand three hundred meters. Disease incidence of leaves was estimated at 74% to 82%, depending on the field. Disease severity on a plant basis was estimated to be 68% to 76%. In order to identify the pathogens of this disease, we isolated and identified some fungi from tea samples with the symptoms of leaf spot, and conducted the pathogenicity test. We found *Phoma segeticola*, *Didymella bellidis* and *Pestalotiopsis trachicarpicola* cause the lesions of tea leaves by Koch's rule. In addition, we also isolated and identifed the isolate of GZYQ4-1-d from the samples of tea leaves. The colony of GZYQ4-1-d on PDA was regular and round, flat growing, sometimes floccose, hyaline. Over time, the colony became grey to dark grey because the pigment of mycelium formatted. Conidia are typically pigmented, muriform, sometimes echinulate, usually with 3 transverse septa and with 0-2 longitudinal septa, measured to (17.16-28.38) μm× (12.57-18.04) μm. The internal transcribed spacer (ITS) regions 1 and 2 was used to amplify and sequenced, and the sequence revealed 100% identity with *P. chartarum* by Blastn software. The pathogenicity test was conducted result indicated that the strains can induce leaf spot on leaves of tea. Tea leaves of the tea plant varieties of Moss tea and Fuding Dabaicha were inoculated with mycelial plugs on PDA or the spore suspension (1×10^6 spores per mL) using the methods of unwound, puncture or cut. The symptom has not been observed on inoculated leaves for three treatments. Nevertheless, some literatures reported that the fungus can cause leaf spot of *Withania somnifera*, cabbage, smooth bromegrass, switchgrass and wheat, and leaf blight of miscanthus giganteus. Therefore, it is worthy to further study the molecular mechanism of tea leaf spot in the region because *P. chartarum* exist in tea leaf spot.

Key words: *Pithomyces chartarum*; identification; pathogenicity test; leaf spot; pathogenicity mechanism

---

* Funding: This work was supported by National Key Research Development Program of China (2017YFD0200308) and its Post-subsidy project (2018-5262), and the Major Science and Technology Projects inGuizhou Province (No. 2012-6012)

** First author: Yin Qiaoxiu; E-mail: yinqiaoxiu@ aliyun. com

*** Corresponding author: Chen Zhuo; E-mail: gychenzhuo@ aliyun. com

Profiling of microRNA-like RNAs associated with sclerotial development in *Sclerotinia sclerotiorum* *

Wang Zehao**, Xia Zihao, Tian Jiamei, Ding Chengsong, Liang Yue***

(*Collage of Plant Protection*, *Shenyang Agricultural University*, *Shenyang* 110866, *China*)

Abstract: *Sclerotinia sclerotiorum* is one of the most common plant fungal pathogens, which causes great economic losses worldwide. Sclerotium is a dormant structure that plays significantly biological and ecological roles in the life cycle of *S. sclerotiorum* and other species of sclerotium-forming fungi. Sclerotial development is regulated by complicated pathways involving gene regulation on post-transcriptional level. MicroRNA-like small RNA (milRNA) is a non-coding small RNA that plays a regulatory role in the post-transcriptional level in fungi. In this study, candidate regulatory milRNAs were investigated in *S. sclerotiorum* at the differential stages of sclerotial development. The profiles of small RNA from sclerotia sampled from three representative developmental stages were investigated by high-throughput sequencing. A total of 275 milRNAs were identified, of which 243 were predicted as novel milRNAs. Among them, 47 milRNAs were differential expression. The target RNAs of these 47 milRNAs were *in silico* analyzed for their putative functions that were mainly on cellular process, metabolic process, organelle part and catalytic activity. In order to validate the sequencing results, the expression levels of 9 milRNAs and their targets were analyzed by qRT-PCR assays. The results showed that Ss-milR-3 was found to be up-regulated as well as the corresponding target RNA showed down-regulated at the development and maturation stages of sclerotial formation. This study will facilitate the better understanding of the milRNA regulation associated with sclerotial development in *S. sclerotiorum* and other sclerotium-forming fungi.

Key words: small RNA; milRNA; sclerotial development; *Sclerotinia sclerotiorum*

* 基金项目：辽宁省“兴辽英才计划”项目（XLYC1807242）；沈阳农业大学引进人才科研启动费项目（20153040）；辽宁省高等学校优秀人才支持计划（LR2015058）

** 第一作者：王泽昊，硕士研究生，研究方向为植物病理学；E-mail：zhwang9415@163.com

*** 通信作者：梁月，教授，博导，研究方向为植物病理学与真菌学；E-mail：yliang@syau.edu.cn

Rapid and visual detection of *Phytophthora infestans* using recombinase polymerase amplification (RPA) combined with lateral flow strips

Lu Xinyu*, Zhang Fan, Yu Jia, Dou Daolong, Shen Danyu**
(*Department of Plant Pathology*, *Nanjing Agricultural University*, *Nanjing* 210095, *China*)

Abstract: Late blight disease, caused by the oomycete pathogen *Phytophthora infestans*, is one of the most dreaded pathogens to potato production, has been considered a constant threat to food security worldwide. Rapid and sensitive diagnosis of *P. infestanse* is crucial for the effective prevention and control of late blight disease. In the present work, we developed a simple, reliable and visual detection assay by combining recombinase polymerase amplification (RPA) and lateral flow strips for the detection of *P. infestans*. The RPA assay targets *Ypt*1 (Ras-related protein) gene and detects as low as 100 fg DNA of *P. infestans*. Moreover, the RPA assay can be performed within 20 min, at a wide range of temperatures (25-45℃) and showed no cross-reactivity with DNA of other related oomycetes or fungal pathogens. Further analysis showed that the RPA could successfully detect *P. infestans* from the experimentally infected potato samples and actual samples from the field. Overall, RPA assay is a useful and convenient tool for detecting *P. infestans*, and it can be applied widely in the quarantine of late blight disease, providing valid data for disease prevention in resource-limited settings.

* First author: Lu Xinyu; E-mail: 2018202009@njau.edu.cn
** Corresponding author: Shen Danyu; E-mail: shendanyu@njau.edu.cn

Rate of seed and soilborne pathogenic infection among selected sunflower varieties*

Addrah Mandela Elorm[1**], Zhang Yuanyuan, Yang Jianfeng, Zhao Jun[***]

(1. *Inner Mongolia Agricultural University*, *Huhhot* 010018, *China*)

Abstract: In determining the rate of infection among sunflower varieties popularly grown in Inner Mongolia, the rates of infection in the confectionary sunflower seeds were significantly higher than that of the oilseeds at 88% and 71% respectively. The confectionary seeds were mainly infected by *Verticillium dahliae*, *Fusarium* spp., *Alternaria* spp. *and Rhizopus*spp., whereas, the oilseeds were infected by *V. dahliae*only. Among tested confectionary varieties, 24 varieties were infected with *V. dahliae*, 8infected with both *V. dahliae* and *Alternaria* spp., 7 infectedwith *V. dahliae* and *Rhizopus*spp. and 6 infected with *V. dahliae*, *Alternaria*spp. *and Rhizopus* spp. However, for oilseed sunflower varieties, 5 varieties werefree of pathogens; the other 12 varieties were infected by *V. dahliae* only. Fungi colonies growing around seed coats on MNP 10 media were subcultured on PDA and subjected to molecular techniques using Verticillium spp., Alternaria spp. and Fusarium spp. specific primers in PCR to confirm the presence of *Verticillium dahliae*, *Alternaria tenuissima*, *Alternariaalternate*, *Alternaria helianthiinificiens*, *Fusarium proliferatum*, *Fusarium oxysporum*and *Fusarium incarnatum* respectively, a non-determined nucleotide sequence (no hit) was also recorded. *Verticillium dahliae* infection pathway and pathogen accumulation index in emerging hypocotyl during germination period of infected seeds were determined using imagery from confocal electronic microscope and qPCR. Images obtained from confocal lens electronic microscope proved the presence of GFP tagged *Verticillium dahliae* on seed coats of harvested seeds from inoculated sunflower plants. At three days after emergence of hypocotyl GFP tagged *V. dahliae* could be seen in the epidermal structures of the embryo but not the phloem and xylem tissues. After the seventh and fourteenth day of germination, fluorescence of the *V. dahliae* tagged pathogen could be seen through the epidermal and endodermal tissues of the whole embryo but fluorescence intensity was higher after the fourteenth day than in that of the other treatment times. This supports the hypothesis *V. dahliae* infection starts right from the onset of seed germination hence making it difficult to control the spread of the disease after planting.

Key words: *Alternaria* spp.; *Verticilllium dahliae*; *Rhizopus* spp.; *Fusarium* spp.; GFP tagged *Verticillium dahliae*

---

* 基金项目：特色油料产业技术体系 CARS-14

** 第一作者：Addrah Mandela Elorm (Ghana)，硕士研究生；E-mail：965603585@qq.com

*** 通信作者：赵君，教授，主要从事向日葵病理研究；E-mail：zhaojun02@hotmail.com

Resynthesis of clubroot disease resistant rapeseed (AAC^rC^r and $A^rA^rC^rC^r$) through hybridization

Zeng Lingyi, Wang Xiuzhen, Xu Li, Yang Huan, Chen Wang, Liu Fan, Yan Ruibin, Ren Li, Chen Kunrong, Fang Xiaoping

(Key Laboratory of Biology and Genetic Improvement of Oil Crops, Ministry of Agriculture; Oil Crops Research Institute, Chinese Academy of Agricultural Sciences, Wuhan 430062, *China)*

Abstract: Clubroot is a prevailing soil - borne disease affecting rapeseed production worldwide. However, very few clubroot resistant rapeseed accessions were available for the breeding of new varieties. Identification and introgression of new clubroot resistant genes from closely related species by distance hybridization is an effective strategy to control this disease. *Brassica napus* L. (AACC) is an amphidiploid species that originated from a spontaneous hybridisation of *B. rapa* L. (AA) and *B. oleracea* L., (CC) and contains the complete diploid chromosome sets of both parental genomes. In this research, 27 resistant/susceptible *B. rape* and 3 resistant *B. oleracea* varieties were identified to create new resources of clubroot disease resistant rapeseed. Then using resistant B. oleracea (C^rC^r) as pattermal plants and resistant/susceptible *B. rapa* ($A^rA^r$and AA) as maternal plants, *B. napus* ($AAC^rC^r$and $A^rA^rC^rC^r$) was artificially resynthesized through hybridization, ovary culture and chromosome doubling. Through backcrossed to maternal plants, detection of disease resistance related molecular markers and disease resistant identification, a total of 10 lines of resynthesized *B. napus* were generated. This study will provide new sources of clubroot resistance materials for the breeding of new rapeseed varieties.

Key words: clubroot; rapeseed; AAC^rC^r; $A^rA^rC^rC^r$; hybridization

Screening of effector candidates from *Sclerotinia sclerotiorum**

Xiao Kunqin, Liu Jinliang, Zhang Yanhua, Zhang Xianghui, Pan Hongyu**
(*College of Plant Sciences*, *Jilin University*, *Changchun* 130062, *China*)

Abstract: *Sclerotinia sclerotiorum* (Lib.) de Bary is a necrotrophic plant pathogenetic fungus with broad host range. It also causes yield losses and quality decline in crop production. Studies on pathogenesis mechanism of *S. sclerotiorum* can provide potential targetfor *Sclerotinia* disease management and offer a reference basis for breeding of resistant plant. The interaction of necrotrophic with theirhosts is more complex than initially thought, and still poorly understood.

We combined bioinformatics approaches and transcriptome analysis to determine the repertoire of *S. sclerotiorum* effector candidates and conducted detailed sequence and expression analyses on selected candidates. We identified 34 *S. sclerotiorum* secreted protein genes (*SsEC*1-34) expressed in planta, many of which have no predicted enzymatic activity, was up-regulated during the infection period and may be involved in the interaction between the fungus and its hosts.

To elucidate the localization and possible functions of *SsECs* in plant cells, we conducted heterologous expression analysis in *N. benthamiana* through agroinfiltration method including full length (FL) or without signal peptide (NS) fusion with green fluorescent protein (GFP). The fluorescence signal of 4 effectors (FL and NS), including SsEC10, 27, 29 and 34 were detected in nucleusand cytoplasm, SsEC25, 26, 28 were also accumulated in cytoplasm. SsEC24 (FL) were expressed inapoplast. Additionally, in 3immune marker genes, the transcript levels of *NbPR*1, *NbPR*2, and *NbPR*4 were down-regulated by SsEC34 expression, whereasthe transcript levels of *NbPR*1 and *NbPR*4 were up-regulated by SsEC24 expression. Interestingly, SsEC24 (FL) could inducenecrosis like cell death, bioinformatics analysis showed that it is anelicitor-like protein. We speculate that it may act as a Pathogen-associated molecular patterns (PAMPs) activated host immune response.

Key words: *Sclerotiniasclerotiorum*; effector candidates; heterologous expression; immune response

---

* Funding: This work was financially supported by the National Natural Science Foundation of China (31772108; 31471730)

** Corresponding author: Pan Hongyu; E-mail: panhongyu@ jlu. edu. cn

Screening of pathogenic defective *Fusarium oxysporum* mutants established by *Agrobacterium tumefaciens*-mediated transformation*

Dong Yanhong**, Han Zeyuan, Meng Xianglong, Wang Shutong***, Cao Keqiang

(*College of Plant Protection*, *Hebei Agricultural University*, *Baoding* 071001, *China*)

Abstract: Apple Replant disease (ARD) occurs frequently in the apple growing regions, which is an important diseases restricting apple production in China. According to our previous research, *Fusarium oxysporum* was confirmed as one of the causal agents of ARD, however, the pathogenic mechanism of the pathogen is still unclear. In order to explore the potential pathogenic genes of this pathogen, a T-DNA insertion mutant library of *F. oxysporum* HS2 was constructed by an optimized *Agrobacterium tumefaciens*-mediated transformation system (ATMT). The robust ATMT system yielded 160-200 transformants/10^6 conidia of *F. oxysporum* that has relatively high conversion efficiency. The genetically stable of the transformants were tested through five generations of successive subculture on hygromycin-free mediums, and the alien gene was detected by PCR. Three thousand and five hundred transformants were involved in the T-DNA insertion mutant library of *F. oxysporum* HS2, and 223 mutants were randomly selected for sporulation test and pathogenicity testing. The most of the tested mutants showed weakened virulence. and one of them displayed obviously attenuated sporulation and deficient pathogenicity, and T-DNA in seven deficient pathogenicity mutants were all single-copy confirmed by Southern Blot. The T-DNA insertion mutant library and three deficient pathogenicity mutants could be served as a valuable resource to identify the pathogenic gene of *F. oxysporum*.

Key words: *Fusarium oxysporum*; *Agrobacterium tumefaciens*; T - DNA; mutant; apple replant disease

---

* Funding: National Key R&D Program of China] under Grant [number 2016YFD0201100]; [China Agriculture Research System] under Grant [number Cars-27]; and [Hebei Natural Science Foundation] under Grant [number c2016204140]

** First author: Dong Yanhong, Ph. D. candidates, Phytopathology; E-mail: 2550558743@ qq. com

*** Corresponding author: Wang Shutong; E-mail: bdstwang@ 163. com

The cell division cycle protein SsCdc28 is essential for sclerotial development and pathogenicity of *Sclerotinia sclerotiorum* *

Zheng Hanyu**, Zhang Bowen, Li Ke, Li Xiangning, Luo Jia, Zhang Yanhua*

(*College of Plant Sciences*, *Jilin University*, *Changchun* 130062, *China*)

Abstract: The necrotrophic fungal plant pathogen *Sclerotinia sclerotiorum* is responsible for substantial global crop losses annually resulting in localized food insecurity and loss of livelihood. The mitogen-activated protein kinase (MAPK) cascades serve as central signaling complexes that are involved in various aspects of sclerotia development and infection. In this study, the putative downstream of MAPK pathway, SsCdc28, was analyzed in *S. sclerotiorum*.

In this study, the *SsCdc*28 gene was cloned from the wild-type *S. sclerotiorum* DNA as a template, and the full length was 1185 bp. NCBI was used to find the upstream and downstream sequences of *SsCdc*28 gene, and specific primers were designed. The wild type *S. sclerotiorum* DNA was used as a template to clone the 1 246 bp fragment upstream and the downstream 1 225 bp fragment of *SsCdc*28 gene. The upstream and downstream fragments were ligated to the knockout vector pXEH to construct knockout vector pXEH-*Cdc*28. Preparation of *S. sclerotiorum* protoplasts, and transformation of the constructed knockout vector pXEH-*Cdc*28 into protoplasts. Finally three positive *SsCdc*28 knockout mutants *SsCdc*28-7, *SsCdc*28-11, *and SsCdc*28-13 were obtained. Comparing the knockout mutant strains and wild-type *S. sclerotiorum*, the results showed that the mycelial growth rate of the knockout transformants became slower; the sclerotia could not be formed; the ability to resist high osmotic stress is enhanced, and the pathogenicity is weakened. Our results demonstrated that SsCdc28 function was essential in the vegetative mycelial growth, sclerotia development and penetration-dependent pathogenicity.

Key words: *Sclerotinia sclerotiorum*; *SsCdc*28; Sclerotia development; Pathogenicity

---

* Funding: This study was financially supported in part by the National Natural Science Foundation of China (31101394, 31772108, 31471730), the National Key Research and Development Program of China (2017YFD0300606)

** Corresponding author: Zhang Yanhua, yh_zhang@ jlu. edu. cn

The dehydrin-like proteins from *Sclerotinia sclerotiorum* are involved in stress response, virulence and apothecial development via functioning as molecular chaperones*

Zhu Genglin, Liu Jinliang, Zhang Xianghui, Zhang Yanhua, Pan Hongyu**

(*College of Plant Sciences*, *Jilin University*, *Changchun* 130062, *China*)

Abstract: *Sclerotinia sclerotiorum* is the major disease damaging crop, apothecia plays an important role in the life cycle and is the main target of sclerotinia control and prevention. A Forkhead box (FOX) family transcription factor (TF), SsFoxE2, has been shown to regulate apothecial development in *S*, *sclerotiorum*. However, the regulatory mechanisms of how SsFoxE2 affect sexual reproduction remain unclear. In this study, two dehydrin-like proteins, designated as SsDhn1 and SsDhn2, were firstly identified as cooperative partners of SsFoxE2. Bioinformatics analysis reveals that SsDhn1 and SsDhn2 belong to typical fungal dehydrin-like proteins family containing repeated conserved asparagine-proline-arginine (DPR) motif. Consistent to their predicted biological function of protection against environmental stresses, *Ssdhn*1 and *Ssdhn*2 knock-out (KO) mutants were impaired in oxidative and osmotic stress resistance, respectively. Furthermore, we demonstrated the expression of *Ssdhn*1 and *Ssdhn*2 were depended on the mitogen-activated protein kinase (MAPK) SsHog1 phosphorylation. Expression analyses indicated that SsDhn1 and SsDhn2 are involved in the cellular stress response. Deletion of *Ssdhn*1 or *Ssdhn*2 resulted in attenuated virulence on common bean leaves. Meanwhile, our results suggested that SsDhn1 and SsDhn2 are involved in resistance to plant defense responses via opposing infection stress and protecting effector in the process of infection. Additionally, we found SsFoxE2 could recruit SsDhn1 and SsDhn2 into nucleus, and they showed a same expression patterns of which the highest expression level was detected in apothecial stage. Finally, we demonstrated that SsDhn1 and SsDhn2 assist SsFoxE2 to form mature sclerotia structure which results in apothecial development. Taken together, our results suggested that SsDhn1 and SsDhn2 function as molecular chaperones to affect stress resistance, virulence and sexual development of *S. sclerotiorum*.

Key words: *Sclerotinia sclerotiorum*; dehydrin - like protein; stress response; virulence; Forkhead box transcription factor; sexual development; molecular chaperones

---

* Funding: National Natural Science Foundation of China (31772108, 31572031, 31471730, 31271991)

** Corresponding author: Pan Hongyu; E-mail: panhongyu@jlu.edu.cn

The mitogen-activated protein kinase gene *CcPmk*1 is required for fungal growth, cell wall integrity and pathogenicity in *Cytospora chrysosperma*

Yu Lu[1], Xiong Dianguang[1], Han Zhu[1], Liang Yingmei[2], Tian Chengming[1]*
(1. *College of Forestry, Beijing Forestry University, Beijing* 100083, *China*;
2. *Museum of Beijing Forestry University, Beijing Forestry University, Beijing* 100083, *China*)

Abstract: *Cytospora chrysosperma*, the causal agent of canker disease in a wide range of woody plants, results in significant annual economic and ecological losses. Mitogen-activated protein kinase (MAPK) cascades are highly conserved signal transduction pathways that play a crucial role in mediating cellular responses to environmental and host signals in plant pathogenic fungi. In this study, we identified an ortholog of the Fus3/Kss1-related MAPK gene, *CcPmk*1, and characterized its functions in *C. chrysosperma*. The expression of *CcPmk*1 was highly induced by inoculation on poplar twigs, and targeted deletion of *CcPmk*1 resulted in the loss of pathogenicity, indicating that *CcPmk*1 is an important regulator of virulence. In addition, *CcPmk*1 deletion mutants (Δ*CcPmk*1) displayed reduced growth and conidiation, decreased fungal biomass production and hyperbranching. Furthermore, our results indicated that *CcPmk*1 deletion mutants exhibited hypersensitivity to cell wall inhibitors and cell wall-degrading enzymes. Correspondingly, the transcription of cell wall biosynthesis-related genes in the Δ*CcPmk*1 strain was downregulated compared to that in the wild-type strain. Moreover, we found that *CcPmk*1 could positively regulate the expression of several candidate effector encoding genes which were highly induced in planta. Hence, we hypothesized that *CcPmk*1 regulates the expression of a series of effectors to promote virulence. Overall, we concluded that the functions of *CcPmk*1 extend to fungal development, cell wall integrity and pathogenicity in *C. chrysosperma*.

Key words: *Cytospora chrysosperma*; MAPK; Fungal development; Cell wall integrity; Pathogenicity

* Corresponding author: Tian Chengming; E-mail: chengmt@bjfu.edu.cn

The *OsMPK*15 negatively regulates *Magnaporthe Oryza* and *Xoo* disease resistance via SA and JA signaling pathway in rice

Hong Yongbo[1,2], Liu Qunen[1,2], Cao Yongrun[1,2], Zhang Yue[1,2], Chen Daibo[1,2], Lou Xiangyang[1,2], Cheng Shihua[1,2], Cao Liyong[1,2*]

(1. *State Key Laboratory of Rice Biology*, *China National Rice Research Institute*, *Hangzhou* 310006, *China*; 2. *Zhejiang Key Laboratory of Super Rice Research*, *China National Rice Research Institute*, *Hangzhou* 310006, *China*)

Abstract: Mitogen-activated protein kinase (MAPK) cascades play central roles in response to biotic and abiotic stresses. However, the mechanisms by which various MAPK members regulate the plant immune response in rice remain elusive. In this article, to characterize the mechanisms, the knock-out and overexpression mutants of *OsMPK*15 were constructed and the disease resistance was investigated under the various fungal and bacterial inoculations. The knock-out mutant of *OsMPK*15 resulted in the constitutive expression of pathogenesis-related (*PR*) genes, increased accumulation of reactive oxygen species (ROS) triggered by the pathogen-associated molecular pattern (PAMP) elicitor chitin, and significantly enhanced the disease resistance to different races of *M. oryzae* and *Xanthomonas oryzae* pv. *oryzae* (*Xoo*), which cause the rice blast and bacterial blight diseases, respectively. On contrary, the expression of *PR* genes and ROS were down-regulated in the *OsMPK*15-overexpressing (OsMPK15-OE) lines. Meanwhile, phytohormones such as salicylic acid (SA) and jasmonic acid (JA) were accumulated in the *mpk*15 mutant lines but decreased in the OsMPK15-OE lines. The expression of SA-and JA-pathway associated genes were significantly upregulated in the *mpk*15 mutant, whereas it was down regulated in the OsMPK15-OE lines. We conclude that *OsMPK*15 may negatively regulate the disease resistance through modulating SA-and JA-mediated signaling pathway.

Key words: *OsMPK*15; *PRs*; SA/JA; ROS; *M. oryzae*; *Xoo*; rice

Transferring blast resistant genes into *Oryza sativa* subsp. *japonica* cultivars with marker-assisted selection strategy*

Liu Shufang[1**], Dong Liying[1], Zhao Guozhen[2], Li Xundong[1], Yang Qinzhong[1***]

(1. *Agricultural Environment and Resources Institute*, *Yunnan Academy of Agricultural Sciences*, *Kunming* 650205, *China*; 2. *Food Crops Institute*, *Yunnan Academy of Agricultural Sciences*, *Kunming* 650205, *China*)

Abstract: Rice blast, caused by the filamentous fungus *Magnaporthe oryzae*, is one of the most serious diseases in rice production worldwide. Utilization of host resistance is the most economic, effective and environment-friendly strategy for its control. Three *japonica* cultivars, Yunjingyou 5, Yunzijing 41 and Chujing 28 are elite rice cultivars with high yield and good quality, but susceptible to rice blast. In order to improve their blast resistance against *M. oryzae*, seven blast resistance genes *Piz*, *Pita*2, *Pikh*, *Pi*5, *Piz-t*, *Pi*9 and *Pi*20 expressing broad-spectrum resistance to rice blast in Yunnan were introgressed into these cultivars via successive backcross and marker-assisted selection strategy. Total of 11 molecular markers closely linked with these 6 blast resistance genes were developed and used in tracking the resistance genes in each backcross progenies, and then the progenies carrying resistance gene was used for next backcross with recurrent parents. Total 17 BC_4F_1 crosses with single blast resistance gene in recurrent parents Yunjingyou 5, Yunzijing 41 and Chujing 28 were obtained, and the pyramiding of blast resistance genes into recurrent parents are ongoing. These resulting materials would be important genetic resources for improvement of modern cultivated rice for resistance to rice blast.

Key words: *Oryza sativa*; blast resistance genes; marker-assisted selection; *Magnaporthe oryzae*

* Funding: This project was supported by Yunnan Academy of Agricultural Sciences (YJM201707)

** First author: Liu Shufang, Master, major in plant pathology; E-mail: lshufang80@163.com

*** Corresponding author: Yang Qinzhong, major in plant pathology; E-mail: qzhyang@163.com

香蕉枯萎病菌 FOC1 和 FOC4 的毒素表达量差异分析*

叶怡婷**，李华平***
（华南农业大学农学院，广州　510642）

摘　要：香蕉枯萎病菌（*Fusarium oxysporum* f. sp. *cubense*，Foc）是典型的土传维管束病原真菌，能引起香蕉枯萎病，是香蕉产业最重要的一个限制因子。根据其对不同品种香蕉的致病性，将其划分为 3 个生理小种，在中国以 1 号小种（FOC1）和 4 号小种（FOC4）的危害最为严重。为了明确 FOC1 和 FOC4 在培养条件和侵染不同植株中分泌毒素含量的差异，笔者利用乙酸乙酯提取法分别提取了 FOC1 和 FOC4 在 Richard 培养基中不同培养时间点的镰刀菌酸，通过液相-质谱仪测定了其含量。结果表明，两个小种间的镰刀菌酸的含量在不同培养时间点存在显著差异。比较香蕉和番茄（*Fusarium oxysporum* f. sp. *lycopersici*）2 个尖孢镰刀菌专化型的镰刀菌酸合成基因簇中的 12 个基因（*FUB*1-*FUB*12）的一致性，笔者发现在香蕉枯萎病菌 FOC1 和 FOC4 中至少存在 10 个镰刀菌酸的合成基因（*FUB*1、*FUB*3、*FUB*4、*FUB*5、*FUB*6、*FUB*7、*FUB*8、*FUB*9、*FUB*11 和 *FUB*12）。对在 Richard 培养液中培养 7 d、14 d、21 d、28 d 的 FOC1 和 FOC4 菌丝进行镰刀菌酸合成基因簇各个基因的表达量进行荧光定量 PCR 测定。结果表明，除了 *FUB*3 和 *FUB*5 的表达量在 FOC1 中高于 FOC4 外，其余 8 个基因的表达量在 FOC4 中均高于 FOC1；特别是 *FUB*1 在 FOC1 中几乎不表达，而在 FOC4 中存在恒定的较高的表达量。这些结果表明 FOC1 和 FOC4 的毒素表达量在培养基中存在明显的差异。进一步的在培养基中的诱导分析和在侵染植株中的含量和表达分析正在进行中。

关键词：香蕉；香蕉枯萎病菌；毒素表达量；镰刀菌酸

* 基金项目：现代农业产业技术体系建设专项（CARS-31-09）
** 第一作者：叶怡婷，硕士研究生，植物病理学；E-mail：2716945735@qq.com
*** 通信作者：李华平，教授；E-mail：huaping@scau.edu.cn

腐烂病菌在苹果树上的时空分布及取样建议*

韩泽园**，董燕红，祁兴华，孟祥龙，曹克强***，王树桐***
（河北农业大学植物保护学院，保定 071001）

摘 要：苹果树腐烂病菌具有潜伏侵染的特性，前期难以发现，后期防治较为被动，且容易复发。本试验通过实时荧光定量的方法检测无病园及发病园树体在不同季节、不同高度、同一枝条上离树体不同距离枝段，以及不同类型组织中的苹果树腐烂病菌含量。根据腐烂菌在树体中的分布特征，建议在对不同季节苹果园腐烂病菌进行动态监测时，可以采集一年生枝条进行检测。无病园监测建议在春季采集中层离树体较近的韧皮部；夏季采集上层枝条木质部；秋季采集上层枝条中部木质部；冬季采集中层离树体较远的木质部。发病园监测建议春季采集上层木质部，其余季节采集下层木质部。在开展普查式检测时，建议采集树体上层一年生枝条，检测枝条离树体较近部位的木质部和韧皮部。

关键词：苹果树腐烂病菌；潜伏侵染；实时定量 PCR；检测；时空分布

---

* 基金项目：国家重点研发计划项目（2016YFD0201100）；国家苹果产业技术体系（CARS-27）；河北省自然科学基金（C2016204140）；新疆生产建设兵团科技发展专项资金（2018AB035）

** 第一作者：韩泽园，在读硕士研究生，植物病理学；E-mail：1109518223@qq.com

*** 通信作者：曹克强，博士，教授，从事植物病害流行与综合防治研究；E-mail：ckq@hebau.edu.cn

王树桐，博士，教授，从事植物病害流行与综合防治研究；E-mail：bdstwang@163.com

玉米小斑病菌（*Bipolaris maydis*）全基因组候选分泌蛋白预测与分析

曾义青*，郝志刚，李健强，罗来鑫**
（中国农业大学植物病理学系，种子病害检验与防控北京市重点实验室，北京　100193）

摘　要：玉米小斑病是玉米叶片上普遍发生的真菌性病害，病原菌为 *Bipolaris maydis*，发病初期时，在下部叶片形成水渍状斑点，随着病害程度加深，最终导致植株枯萎死亡，严重影响玉米的产量。目前关于该病害体系的报道主要集中在病害传播及防控等方面，而对于该菌分泌蛋白相关的研究尚未报道。

本文利用生物信息学手段，对该菌分泌蛋白进行预测，进而明确其分泌蛋白在侵染、操控植物等过程的作用。具体方法如下：首先，从 NCBI 中下载玉米小斑病菌的全基因组序列（ATCC 48331），其次，通过 SignalP-5.0、Phobius、TMHMM Server v. 2.0、PredGPI 生物信息学软件和预测程序对玉米小斑病菌中 12 705条蛋白序列进行预测，再通过对上述蛋白半胱氨酸含量、信号肽长度及冗余性分析，获得 391 个符合条件的候选分泌蛋白，利用 eggNOG 5.0 蛋白质数据库进行基因功能注释，得到 267 个具有功能注释的基因，分析其功能注释大多属于果胶酶类、水解酶类、蛋白酶类等。

综上，本研究利用生物信息学方法预测出玉米小斑病菌的候选分泌蛋白，为进一步研究分泌蛋白在病原菌与玉米互作中的作用奠定了基础，并为其他病原菌分泌蛋白的预测及分析提供了参考。

关键词：玉米小斑病；全基因组；候选分泌蛋白；生物信息学

* 第一作者：曾义青，硕士研究生，主要从事种子病理学和病原菌与植物互作研究；E-mail：15256432971@163.com
** 通信作者：罗来鑫，博士，博士生导师，主要从事种传病害研究；E-mail：luolaixin@cau.edu.cn

甘蔗芽蜡质对甘蔗黑穗病菌冬孢子萌发的影响

刘宗灵[1]，兰仙软[1]，周　赛[1]，李　茹[1,3]，陈保善[1,2,3] *

（1. 广西大学生命科学与技术学院，南宁　530004；2. 广西大学农学院，南宁　530004；
3. 亚热带农业生物资源保护与利用国家重点实验室，南宁　530004）

摘　要：甘蔗黑穗病是由甘蔗黑穗病菌（*Sporisorium scitamineum*）侵染甘蔗芽造成的一种真菌病害。笔者实验室前期研究发现，甘蔗黑穗病菌冬孢子在黑穗病高抗品种中蔗 1 号、6 号和 9 号芽上的萌发率显著低于高感品种 ROC22。本研究探究甘蔗芽蜡质对甘蔗黑穗病菌冬孢子萌发的影响。利用气质联用仪测定了中蔗 1 号、6 号、9 号和 ROC22 芽上的蜡质成分及含量。结果显示，甘蔗芽蜡质的主要成分为十九烷、软脂酸、硬脂酸、二十五烷、二十七烷、二十九烷、二十六烷醇、二十八烷醇和二十八烷醛。4 个甘蔗品种芽的蜡质总量无显著差别。其中中蔗 1 号、6 号和 9 号的高级醇含量显著低于 ROC22 的高级醇含量。用二十六烷醇、二十八烷醇，二十八烷醛和二十七烷标准品对甘蔗黑穗病菌冬孢子进行萌发实验，发现经二十六烷醇、二十八烷醇和二十八烷醛处理的甘蔗黑穗病菌冬孢子的萌发率高于对照。以上研究结果说明甘蔗芽的蜡质成分可能参与诱导甘蔗黑穗病菌冬孢子的萌发。

关键词：蜡质；甘蔗黑穗病；孢子萌发

* 通信作者：陈保善；E-mail：chenyaoj@ gxu. edu. cn

甘蔗梢腐病病原菌鉴定及其突变体库构建*

蒙姣荣[1,2]，黄海娟[3]，李杨秀[1]，李雨珈[1]，杨惠贞[3]，李界秋[1]，陈保善[1,2**]
（1. 广西大学农学院，南宁　530004；2. 亚热带农业生物资源保护与利用国家重点实验室，南宁　530004；3. 广西大学生命科学与技术学院，南宁　530004）

摘　要：梢腐病是甘蔗的主要真菌病害之一。2016—2018 年，从广西扶绥县采集具有典型的甘蔗梢腐病样本 83 份，通过常规组织分离共获得镰刀菌菌株 131 个，经叶片离体接种证实 94 个菌株具有致病性。翻译延伸因子 Alpha1 亚基（translation elongation factor alpha 1，TEF-1）序列聚类分析表明，这些具有致病性的镰刀菌菌株分别属于为甘蔗镰孢菌（*Fusarium sacchari*）（69 个菌株）、4 个菌株为尖孢镰刀菌（*F. oxysporum*）（4 个菌株）、层出镰刀菌（*F. proliferatum*）（3 个菌株）、藤仓镰孢菌（*F. fujikuroi*）（1 个菌株）和 17 个尚未鉴定到种镰刀菌菌株。这是 *F. sacchari* 在我国引起甘蔗梢腐病的首次报道。

利用根癌农杆菌介导的 T-DNA 插入技术对强致病力菌株 FF001 进行遗传转化，获得 3018 个突变株。人工离体接种的方法接种甘蔗叶片，获得致病力明显变化的突变体 27 株，其中致病力明显变弱的 21 株，致病力增强的 6 个菌株。通过 hiTAIL-PCR 扩增，获得了 23 个突变株 T-DNA 插入的基因组位点及其侧翼序列，相应的基因分别编码寡肽转运蛋白（oligopeptide transporter，*OPT*）、核类 VCP 蛋白基因（nuclear VCP-like protein，*NVL*）、2-脱氢泛解酸 2-还原酶（2-dehydropantoate 2-reductase）等已知功能及一些未知功能蛋白。

采用同源双交换的方法构建了 *OPT* 和 *NVL* 的基因缺失突变株，离体接种和盆栽活体接种均显示 Δ*OPT* 和 Δ*NVL* 突变体为致病力减弱。本研究结果为进一步阐明甘蔗梢腐病流行机制和镰孢菌致病机理奠定了新的基础。

关键词：甘蔗梢腐病；甘蔗镰孢菌；致病性；突变体库

---

* 资助项目：广西蔗糖产业协同创新中心项目（桂教科研〔2014〕13 号）资助
** 通信作者：陈保善；E-mail：chenyaoj@ gxu. edu. cn

黄连根腐病病原鉴定及其对杀菌剂敏感性测定

程欢欢[1*]，漆梦雯[1]，李　忠[1,2]，彭丽娟[3]，丁海霞[1**]
（1. 贵州大学农学院，贵阳　550025；2. 贵州省药用植物繁育与种植重点实验室，贵阳　550025；3. 贵州大学烟草学院，贵阳　550025）

摘　要：黄连（*Coptis chinensis*）为毛茛科、黄连属多年生草本植物，其根茎常作中药药材，具清热、燥湿、泻火和解毒等良好抗菌作用，目前主产于重庆市石柱县，该地区黄连常年连作导致黄连根腐病发生严重，甚至绝收，严重影响药农收入。目前国内关于黄连根腐病及其病原菌的报道较少，对黄连根腐病的防治方法研究几乎没有。因此，本研究通过调查重庆石柱黄连根腐病病害，明确其致病菌，并于室内进行杀菌剂敏感性测定，为田间黄连根腐病的防治提供理论依据。

为明确重庆石柱黄连根腐病致病病原，采集具有典型腐烂症状的发病黄连块茎，利用组织分离法、柯赫氏法则、形态学特征观察及 rDNA-ITS 和 *tef*1 多基因序列分析对分离得到的菌株进行了鉴定。在 PDA 培养基上该菌菌落突起呈絮状，初期无色或浅紫色，后期紫红色，菌丝白色质密。通过多基因序列分析表明该菌株与 *Fusarium oxysporum* 单独聚集成一支，支持率达 100%，能与其他种明显区分开。根据柯赫氏法则，从接种发病的块茎上再次分离得到的菌株与原接种菌株相同，证实了该菌株为致病菌。综合以上结果表明引起该病害的病原菌为尖孢镰刀菌（*Fusarium oxysporum*），这与陈姗姗等报道的黄连根腐病病原菌 *Fusarium solani* 有所不同。

室内杀菌剂敏感性测定结果表明：60%嘧菌·代森联、80%甲基硫灵菌、50%多菌灵·福美双可湿性粉、1%申嗪霉素悬浮剂、30%咪鲜胺乳油、40%氟硅唑、30%唑醚·戊唑醇、75%肟菌·戊唑醇水分散粒剂等 8 种药剂对病原菌均有抑制作用，其中 30%咪鲜胺乳油和 1%申嗪霉素悬浮剂抑制效果最好，EC_{50}值分别为 0.037 6μg/mL 和 0.060 2μg/mL；40%氟硅唑、75%肟菌·戊唑醇水分散粒剂、50%多菌灵·福美双可湿性粉、30%唑醚·戊唑醇、80%甲基硫灵菌次之，其 EC_{50}值分别为 0.230 2μg/mL、0.461 0μg/mL、0.921 1μg/mL、1.834 8μg/mL、4.185 9μg/mL；60%嘧菌·代森联抑制效果较差，其 EC_{50}值为 38.688 5μg/mL。因此田间施用药剂推荐为 30%咪鲜胺乳油和 1%申嗪霉素悬浮剂。

关键词：黄连；根腐病；杀菌剂；敏感性测定

* 第一作者：程欢欢，硕士研究生；E-mail：513396976@ qq. com
** 通信作者：丁海霞，博士，讲师，主要从事植物病理学和植物病害生物防治研究；E-mail：hxding@ gzu. edu. cn

一株辣椒炭疽病生防芽孢杆菌的筛选

程欢欢[1*]，余　水[2]，彭丽娟[2]，李　忠[1]，丁海霞[1**]
（1. 贵州大学农学院，贵阳　550025；2. 贵州大学烟草学院，贵阳　550025）

摘　要：从贵州省安龙县辣椒炭疽病发病地块采集健康辣椒根际土壤，通过平板对峙法获得具有明显拮抗辣椒炭疽病菌（*Colletotrichum capsici*）的 1 株芽孢杆菌，通过形态学和生理生化特征观察，该菌在 LB 液体培养基表面可产生形态较为复杂的生物膜结构，为革兰氏阳性细菌，能够分解葡萄糖、鼠李糖和甘露醇，能够产生过氧化氢酶和硝酸还原酶，甲基红染色阳性，还具有明胶液化明显呈阳性等特征。并结合 16S rDNA、*gyrA* 基因序列分析结果构建系统发育树，该菌株能与 *Bacillus velezensis* 聚集成一支，且支持率为 100%，能与其他种明显区分开，最终确定该菌株为贝莱斯芽孢杆菌（*Bacillus velezensis*）。采用比浊法测定该菌株生长速率，其生长曲线显示该菌的延滞期较为短暂，很快进入对数生长期，在 6.5~7.5 h 达到稳定期，生长速度较快。该菌株可产生蛋白酶、纤维素酶、嗜铁素和磷酸酯酶等生防相关酶，对高粱叶斑病菌（*Alternaria alternata*）、烟草黑胫病菌（*Phytophthora nicotianae*）、石榴干腐病菌（*Phomopsis* sp.）、油菜菌核病菌（*Sclerotinia sclerotiorum*）、稻瘟病菌（*Magnaporthe oryzae*）共 5 种重要植物病原真菌也有很强的抑菌活性，可作为生防制剂进一步研究开发。

关键词：芽孢杆菌；生物防治；辣椒炭疽病

---

* 第一作者：程欢欢，硕士研究生；E-mail：513396976@ qq. com
** 通信作者：丁海霞，博士，讲师，主要从事植物病理学和植物病害生物防治研究；E-mail：hxding@ gzu. edu. cn

拟轮枝镰孢和层出镰孢竞争侵染玉米果穗的研究*

渠　清**，刘　俊，刘　宁，杨贝贝，贾　慧，曹志艳***，董金皋***
（河北省植物生理与分子病理学重点实验室，河北农业大学真菌
毒素与植物分子病理学实验室，保定　071000）

摘　要：玉米穗粒腐病是由病原菌侵染引起果穗或籽粒病变的一种真菌性病害。病原菌侵染玉米果穗后会导致果穗腐烂，影响玉米产量和品质。穗粒腐病病原菌组成复杂多变，但目前研究大多局限于单一病原菌的侵染。本试验通过 ATMT（*Agrobactirium tumfacience* mediated-transformant）技术将绿色荧光蛋白（GFP）标记到层出镰孢、红色荧光（蛋白）标记到拟轮枝镰孢，通过荧光检测初步探索了两种病原菌单独和复合侵染玉米果穗的时空差异性。通过花丝通道注射法，将两种病原菌的孢悬液单独和复合接种到吐丝期的健康玉米果穗上，荧光学和组织解剖学观察侵染情况。结果表明：单独接种 72 h 后，层出镰孢侵染进入雌花外稃和内稃细胞，子房基部未检测到荧光信号，而拟轮枝镰孢可继续侵染子房基部细胞；复合接种 48 h 后，果穗穗尖 1 cm 处出现明显褐变，荧光观察显示菌丝定殖范围及菌丝量上拟轮枝镰孢均明显大于层出镰孢；接菌 72~96 h 后，穗腐病的病症进一步蔓延，拟轮枝镰孢和层出镰孢侵染的范围和菌丝量相当，但拟轮枝镰孢出现汇聚现象；接种 120~144 h 后，穗轴出现空隙，穗基部软化且检测到大量红色荧光信号。上述结果表明，同层出镰孢相比，拟轮枝镰孢在侵染玉米果穗后表现出竞争优势。上述研究结果为明确玉米穗腐病菌侵染机制及其防控提供了参考。

关键词：玉米穗腐病；拟轮枝镰孢；层出镰孢；竞争侵染

---

* 基金项目：现代农业产业技术体系专项（CARS-02）；国家重点研发计划—粮食丰产增效科技创新专项（2016YFD0300704）

** 第一作者：渠清，硕士研究生，研究方向为穗腐病致病性的研究；E-mail：qu_qing@126.com

*** 通信作者：曹志艳；E-mail：caoyan208@126.com
董金皋；E-mail：dongjingao@126.com

玉米大斑病菌漆酶家族基因表达模式及功能解析*

刘　宁**，贾　慧，渠　清，杨贝贝，曹志艳，董金皋***
（河北省植物生理与分子病理学重点实验室，河北农业大学
真菌毒素与植物分子病理学实验室，保定　071000）

摘　要：玉米大斑病菌（*Setosphaeria turcica*）是一种半活体寄生子囊真菌，可以侵染玉米、高粱和苏丹草等植物，形成水浸状灰褐色病斑，从而影响了植物光合作用进而降低产量。漆酶作为含金属离子的多酚氧化酶，在生物体中以多个同源基因共存的方式存在，具有重要的生理功能，本研究在玉米大斑病菌基因组中鉴定得到了9个漆酶样多铜氧化酶，其蛋白氨基酸序列同源性低，仅为19.79%~48.70%，聚类在5个不同的超家族。在9个基因上游2 000 bp的启动子区域发现有*NIT*2（主要的正向氮调控基因）CRE-A（负责碳代谢产物阻遏的元件）、*ADR*1（正向调控过氧化物酶体蛋白基因）、*StuAp*（发育复杂性的调控基因）、STRE（压力响应元件）、XRE（异物质响应元件）和MRE（金属应答元件），表明漆酶基因表达受到不同的营养条件、外源物和生长发育的影响。进一步通过转录组数据分析发现，玉米大斑病菌菌丝中*StLAC*2、*StLAC*6的基因表达量最高，*StLAC*3较高，其余基因表达量均相对较低；同时漆酶家族基因转录水平受培养时间、碳源、氮源、铜离子、铁离子的影响，且在接种玉米叶片后，检测侵染的早期、中期和后期基因表达水平，发现除了*StLAC*6和*StLAC*8的表达量下降，其他基因在不同时期相对表达量均增加。通过分析基因敲除突变体发现，*StLAC*1基因与病菌形态发育相关，*StLAC*2主要在黑色素代谢中起作用，*StLAC*6基因参与毒素的次级代谢。漆酶基因转录水平上的复杂调控与病菌生长、致病及环境因素紧密相关，需要进一步确认其蛋白表达水平及活性，为研究玉米大斑病菌主要表达漆酶的功能提供依据。

关键词：玉米大斑病菌；漆酶；表达模式；功能

* 基金项目：河北省高等学校科学技术研究项目（ZD2014053）；现代农业产业技术体系专项（CARS-02）；国家重点研发计划—粮食丰产增效科技创新专项（2016YFD0300704）

** 第一作者：刘宁，博士，讲师，主要从事病原真菌致病机理研究；E-mail：lning121@126.com

*** 通信作者：董金皋，博士，教授，主要从事病害防控研究；E-mail：dongjingao@126.com

谷瘟病菌无毒基因 *PWL* 家族分布及变异分析*

李志勇[1**]，白　辉[1]，任世龙[2]，全建章[1]，董志平[1***]
(1. 河北省农林科学院谷子研究所，石家庄　050031；
2. 河北农业大学生命科学学院，保定　071001)

摘　要：谷瘟病是由灰梨孢菌 *Magnaporthe oryzae* 引起的，是谷子生产中重要流行性病害之一，严重时可造成大面积减产甚至绝收，抗性品种的选育与推广种植是防治谷瘟病最经济有效的方法。谷瘟病菌的无毒基因与谷子抗病基因互作与稻瘟病菌和水稻抗病基因互作相同，都遵循基因对基因假说。田间谷瘟菌株无毒基因的变化常常使寄主抗病性丧失，因此鉴定各地区谷瘟病菌群体包含的无毒基因类型，了解谷瘟病菌无毒基因的分布及变异情况，可为深入研究谷瘟病菌群体遗传结构及其无毒基因变异机制奠定基础。*PWL* 基因家族作为寄主特异性无毒基因，是谷瘟病菌无毒基因的重要组成部分，至今，关于我国谷瘟病菌中 *PWL* 无毒基因家族分布和变异未见报道。

为了解无毒基因 *PWL* 家族在谷瘟病菌群体中的分布和变异情况，本研究利用 5 对特异性引物对 252 株谷瘟病菌单孢菌株进行扩增及测序分析。以 252 株谷瘟病菌单孢菌株 DNA 为模板，利用设计的 5 对 *PWL* 无毒基因特异性引物进行 PCR 扩增，扩增结果表明：*PWL* 基因家族在谷瘟病菌中存在变异，变异的形式主要为基因的缺失和插入。谷瘟病菌未能扩增出 *PWL*1 基因的特异性目的片段，说明谷瘟病菌菌株均不含有 *PWL*1 基因。共 65 个谷瘟病菌菌株扩增出无毒基因 *PWL*2 的特异性目的片段，扩增率为 25.8%。而无毒基因 *PWL*4 和 *PWL*3 扩增率均为 100%，但谷瘟病菌与稻瘟病菌相比 *PWL*3 基因均存 849bp 碱基的插入。利用谷瘟病菌与稻瘟病菌 *PWL*3 基因的差异可通过 PCR 技术快速区分谷瘟病菌与稻瘟病菌。*PWL*2 基因存在单核苷酸变异，共划分为 14 个单倍型，单倍型 P1 为绝对优势单倍型。谷瘟病菌无毒基因 *PWL* 家族成员分布及变异分析对于深入研究谷瘟病菌群体遗传多样性，及谷瘟病菌与寄主谷子之间的互作机制开展研究具有重要价值。

关键词：谷子；谷瘟；无毒基因；*PWL* 家族

---

* 基金项目：国家自然科学基金资助项目（31872880）；国家现代农业产业技术体系（CARS-07-13.5-A8）；河北省农林科学院创新工程（2019-4-02-03）

** 第一作者：李志勇，研究员，主要从事谷子病害研究；E-mail：lizhiyongds@126.com

*** 通信作者：董志平，研究员，主要从事谷子病害研究；E-mail：dzping001@163.com

芸薹根肿菌初侵染期分泌蛋白鉴定和功能分析*

陈　旺**，燕瑞斌，徐　理，任　莉，刘　凡，曾令益，
杨　欢，池　鹏，王秀珍，陈坤荣，方小平***
（农业农村部油料作物生物学与遗传育种重点实验室，
中国农业科学院油料作物研究所，武汉　430062）

摘　要：根肿病是十字花科作物上最为严重的病害之一，全世界分布广泛。十字花科根肿病病原为芸薹根肿菌，其与寄主的识别发生在初侵染期，但是其内在的互作机制尚不明确。分泌蛋白是效应蛋白的候选者，在病原与寄主识别和互作中发挥关键作用。本研究通过初侵染期转录组分析、分泌蛋白预测和酵母信号肽鉴定系统获得 33 个在初侵染期表达的芸薹根肿菌分泌蛋白。进一步通过植物病毒表达系统鉴定到 2 个能够在诱导本氏烟叶片细胞坏死的分泌蛋白 PBCN_002550 和 PBCN_005499，其中通过白菜轴原生质体表达系统证实仅 PBCN_002550 能够诱导白菜细胞的坏死。除了以上两个诱导细胞坏死的分泌蛋白外，在剩余的 31 个分泌蛋白中，24 个能够抑制 BAX 诱导的细胞坏死，28 个能够抑制 PBCN_002550 诱导的细胞坏死。以上研究初步揭示了分泌蛋白在芸薹根肿菌侵染中的功能，为研究深入芸薹根肿菌的致病机制奠定了基础。

关键词：芸薹根肿菌；初侵染；分泌蛋白；植物免疫

---

* 基金项目：国家重点研发计划项目（2018YFD0200900）；湖北省青年自然科学基金（2018CFB255）

** 第一作者：陈旺，博士，助理研究员，主要从事油菜根肿病研究；E-mail：chenwang@ caas. cn

*** 通信作者：方小平，研究员，主要从事油菜病虫害研究；E-mail：xpfang2008@ 163. com

幼园无症状枝条苹果树腐烂病菌带菌率检测及树体不同月份抗病性差异分析*

郭斐然**，贺艳婷，郭　衍，李　晨，黄丽丽***，冯　浩***
（旱区作物逆境生物学国家重点实验室，西北农林科技大学植物保护学院，杨凌　712100）

摘　要：苹果树腐烂病是由黑腐皮壳属真菌 *Valsa mali* 侵染引起的枝干病害。病菌在田间可周年传播，且入侵苹果组织后具有潜伏侵染的特性。因此，早期监测预警对苹果树腐烂病的综合防控具有重要意义。本研究按照五点取样法分别选取陕西省宝鸡市扶风县幼园和苗圃以及洛川县幼园的 5 株果树，每个月每株果树按东南西北中采集一年生无症状枝条，利用巢氏 PCR 技术对枝条的带菌率，及枝条致病力进行了检测。结果表明，陕西省宝鸡市扶风县的幼园和洛川县试验示范站的幼园中的枝条均不同程度的带有苹果树腐烂病菌。扶风县的苗圃带菌率极低。表明树龄与树体带菌率有相关性。幼园带菌率分析发现，不同月份检测到的带菌率存在差异，3—4 月带菌率较低，10—12 月带菌率最高，可见，病菌潜伏侵染随时间推移逐渐加强。另外同一树体不同月份的抗性水平也存在差异，在 4 月的抗病性弱，11—12 月抗病性强，可能与寄主营养差异水平存在一定关系。本研究证明苹果树腐烂病菌可以潜伏在寄主植物组织中越冬，同时树体的抗病性会在不同月份存在差异，可见，研究结果为病害综合防控提供了重要理论依据。

关键词：苹果树腐烂病菌；幼园无症状果树；带菌率；抗病性

---

* 基金项目：国家重点研发计划（2016YFD0201108）
** 第一作者：郭斐然，西北农林科技大学植物保护学院硕士研究生；E-mail 244068071@ qq. com
*** 通信作者：黄丽丽，教授；E-mail：huanglili@ nwsuaf. edu. cn
冯浩，教授；E-mail：xiaosong04005@ 163. com

凸脐蠕孢菌玉米专化型和高粱专化型比较基因组分析*

马周杰**，朱飞宇，何世道，王禹博，高增贵***
（沈阳农业大学植物保护学院，沈阳　110866）

摘　要：玉米、高粱等禾本科作物大斑病是由凸脐蠕孢菌（*Setosphaeria turcica*）侵染引起的真菌叶部病害，经常造成严重的经济损失。玉米专化型（*Setosphearia turcica* f. sp. *zeae*）和高粱专化型（*Setosphearia turcica* f. sp. *sorghi*）表现出明显的寄主专化性，两者具有密切的进化关系。目前，对凸脐蠕孢菌专化型的致病专化机理尚不明确。本研究通过对高粱专化型进行全基因组测序，并将其与已公布的玉米专化型基因组进行比对分析。结果显示，高粱专化型中存在更多的蛋白编码基因，两个专化型都分别存在4个独特的蛋白家族。单核苷酸多态性（SNP）分析发现，两个专化型之间存在5 803个SNP位点，且有1 197个SNP发生在基因的编码区域。重点分析分泌蛋白基因后发现，高粱专化型存在704个分泌蛋白基因，其中有161个被预测为效应蛋白基因，而玉米专化型仅存在521个分泌蛋白基因和137个效应蛋白基因，进一步研究表明玉米专化型中特异存在8个功能注释的效应蛋白基因，涵盖糖苷水解酶、碳水化合物酯酶等家族。利用荧光定量PCR分析了这8个效应蛋白基因在病菌侵染过程中的表达情况，发现纤维素酶基因在侵染玉米叶片72h后比0h上调150余倍，表明纤维素酶在致病专化性过程中起到关键性作用。以上结果为凸脐蠕孢菌致病分化机制提供了重要的理论信息，为防治大斑病和抗病品种的遗传育种提供了有效的参考。

关键词：凸脐蠕孢菌；比较基因组；效应蛋白；生物信息学

---

* 基金项目：国家重点研发项目（2018YFD0300307，2017YFD0300704，2016YFD0300704）

** 第一作者：马周杰，博士研究生，主要从事玉米病害研究

*** 通信作者：高增贵，研究员，博士生导师；E-mail：gaozenggui@ sina. com

玉米大斑病菌 *StRALF* 基因 cDNA 序列分析*

朱飞宇**，马周杰，孙艳秋，何世道，姚　远，高增贵***
（沈阳农业大学植物保护学院，沈阳　110866）

摘　要：由玉米大斑病菌（*Setosphaeria turcica*）引起的玉米大斑病（Northern corn leaf blight），在全球发生普遍并造成玉米严重损失。目前，在尖孢镰刀菌（*Fusarium oxysporum*）以及多种植物中已明确快速碱化因子（Rapid Alkalization Factor，RALF）的功能，其在尖孢镰刀菌中（*F. oxysporum*）可以使真菌侵染寄主的环境发生碱化，并且增强真菌对寄主的致病力。但在玉米大斑病菌（*S. turcica*）中暂未报道此因子。通过 JGI 网站（https：//genome. jgi. doe. gov/portal/）中 tblastn，用 NCBI 中已知的尖孢镰刀菌（*F. oxysporum*）中的快速碱化因子（RALF）的蛋白序列（FOXG_21151），在玉米大斑病菌（*S. turcica*）中比对出一段高度相似的基因序列。本研究通过 SMARTer™ RACE 技术获得了 *StRALF* 基因的 cDNA 全长 279bp，其编码 92 个氨基酸，是亲水性的分泌蛋白。此蛋白不含有螺旋区域，所以不会执行代谢调节、识别分子、膜通道等功能。经过氨基酸同源性分析比对发现其与已公布的 RALF 具有一定的同源性，但不完全保守。本文的研究为此因子的蛋白原核表达以及相关功能性研究奠定基础。

关键词：*StRALF* 基因；玉米大斑病菌；RACE 技术；生物信息学

---

* 基金项目：国家重点研发项目（2018YFD0300307，2017YFD0300704，2016YFD0300704）

** 第一作者：朱飞宇，硕士研究生，主要从事玉米病害研究
*** 通信作者：高增贵，研究员，博士生导师；E-mail：gaozenggui@ sina. com

玉米大斑病菌木聚糖酶基因 *StXYL1* 生物信息学分析*

王禹博**，马周杰，何世道，朱飞宇，高增贵***
（沈阳农业大学植物保护学院，沈阳　110866）

摘　要：玉米大斑病（NCLB）是威胁世界玉米产量的重要叶部真菌病害，我国北部及高海拔的冷凉地区相对发病较为严重。该病由大斑刚毛座腔菌［*Setosphaeria turcica*（Luttrell）Leonard et Suggs］侵染玉米叶片于低温高湿的环境下发病。通过大斑刚毛座腔菌的高粱专化型与玉米专化型的效应蛋白编码基因进行对比，得出 8 个已注释基因功能的差异序列。其中的 *Xylanase* 基因序列通过同源性搜索的方法得出。命名玉米大斑病菌木聚糖酶基因 *StXYL1*。其主要的功能是形成木聚糖酶，分解细胞壁异型木聚糖主链中的 β-1，4 糖苷键。设计特异性引物扩增出该基因并克隆。经过对比得出该基因序列 DNA 全长为 985 bp，其中包含了两段内含子，分别有 46 个碱基、54 个碱基。外显子共编码 298 个氨基酸，其编码的蛋白分子质量约为 32.74 ku，理论等电点 pI 为 7.74，总共包括4 559个原子，预测分子式为 $C_{1448}H_{2253}N_{403}O_{445}S_{10}$；在组成的 20 种氨基酸中苏氨酸（Thr）所占比例最高，达到 10.1%，半胱氨酸（Cys）所占比例最低，为 1.0%；蛋白的不稳定指数为 34.49，脂肪指数为 75.60。蛋白质序列有两个结构域，第 75—290 是个高度保守的结构功能域，属于糖基水解酶 10 家族。生物信息学分析表明，该基因序列有 1 个潜在的 N-糖基化位点，未发现 O-糖基化位点。本试验经过克隆大斑病菌的 *StXYL1* 基因，并进行生物信息学的初步分析，为接下来的研究该基因的缺失对大斑病菌的影响打下基础。

关键词：玉米大斑病菌；木聚糖酶基因；生物信息学分析

* 基金项目：国家重点研发项目（2017YFD0300704，2016YFD0300704）

** 第一作者：王禹博，硕士研究生，主要从事玉米病害研究

*** 通信作者：高增贵，研究员，博士生导师；E-mail：gaozenggui@ sina. com

玉米瘤黑粉菌交配型种类鉴定及遗传多样性分析*

刘小迪**，马周杰，车广宇，王禹博，高增贵***
（沈阳农业大学植物保护学院，沈阳 110866）

摘　要：为明确玉米主要产区玉米瘤黑粉病菌的交配型类型的种类及玉蜀黍黑粉菌遗传多样性，通过分子生物学方法对从 11 个省采集的 82 份玉米瘤黑粉菌株进行交配型鉴定，并通过 ISSR 分子标记技术对所分离得到的 82 株玉蜀黍黑粉菌进行遗传多样性分析。试验结果显示：玉米黑粉菌交配型类型主要有 3 种：mfa1，mfa2 以及 mfa1、mfa2 两种交配型同时存在，82 株菌株中，32 株交配型为 mfa1，占鉴定株数的 39.02%；21 株为 mfa2，占鉴定株数的 25.61%；其中 29 株交配型为 mfa1、mfa2 同时存在，占总数的 35.37%。通过筛选出的 8 条引物扩增出清晰条带 61 条，特异性条带 55 条，多态性位点所占比例为 90.2%；供试菌株在当遗传相似系数为 0.90 时被全部分开，说明各玉蜀黍黑粉菌的种内遗传显著。聚类分析结果显示，玉蜀黍黑粉菌 ISSR 类群的划分与其来源地无密切关系，来自同一地区的玉蜀黍黑粉菌不能被完全划分到同一类群，说明同一来源地的玉蜀黍黑粉菌间存在遗传差异性。

关键词：玉蜀黍黑粉菌；交配型；遗传多样性

* 基金项目：国家重点研发项目（2017YFD0300704，2016YFD0300704）
** 第一作者：刘小迪，硕士研究生，主要从事玉米病害研究
*** 通信作者：高增贵，研究员，博士生导师；E-mail：gaozenggui@ sina. com

大豆疫霉胞外多糖及其外泌蛋白处理下大豆的代谢组学研究*

胡九龙**，屈　阳，李坤缘，刘　冬，姜庆雨，潘月敏，高智谋***

（安徽农业大学植物保护学院，合肥　230036）

摘　要：在大豆的生产过程中，大豆疫霉（*Phytophthora sojae*）引致的大豆疫病是影响大豆产量品质的最重要的病害之一。研究大豆疫霉的致病机理对于该病的有效控制十分重要。而在大豆疫霉侵染大豆过程中，大豆疫霉分泌的多糖及蛋白就如同侵染大豆的先锋部队，所以通过研究大豆疫霉的胞外粗多糖及分泌蛋白对大豆的生理生化影响可为研究大豆疫霉的致病机理提供实验依据。

本研究以大豆疫霉 GY8-3 菌株为材料，提取胞外粗多糖及分泌蛋白，用它们处理大豆愈伤组织；采用 RT-PCR，GC-MS 方法，评价粗多糖和分泌蛋白对大豆代谢的影响。结果表明，大豆疫霉胞外粗多糖及其分泌蛋白处理组的 NO 和活性氧含量较对照组均明显升高，其中胞外粗多糖处理组中的大豆 10 个抗性相关基因在粗多糖处理 6 h 后有 6 个基因明显上调；分泌蛋白处理组中有 4 个上调。GC-MS 鉴定出的代谢物共有 58 种，其中大豆疫霉胞外粗多糖处理组有 20 多种代谢物显著变化（$P \leq 0.05$，$VIP > 1$）；其中葡萄糖苷和脯氨酸的相对含量（log2（T/C））升高了 251 %，甘露糖升高了 219%，丁二酸（琥珀酸）升高了 122%；缬氨酸、天冬酰胺、苏氨酸相对含量分别下降了 216%、128%、54%，丙酸、松醇、肌醇分别下降了 91%、75%、54%。分泌蛋白处理组有 30 多种代谢物显著变化；其中葡萄糖苷相对含量升高了 477%，甘露糖升高 271%，这二者都比胞外粗多糖处理组中的升高幅度大，脯氨酸升高了 250%，这与胞外粗多糖处理组升高倍数相近；而缬氨酸、天冬酰胺相对含量并无较大变化。

上述研究结果表明，以大豆疫霉胞外粗多糖及其分泌蛋白处理大豆后，不同处理组的大豆体内代谢都发生了变化。二者相同之处是都使大豆系统抗性得到了提高，且都是诱导大豆通过减弱糖酵解途径、增加体内葡萄糖苷、脯氨酸及尸胺的含量等一系列的内部调节来实现对它们刺激的响应；不同之处在于大豆疫霉胞外粗多糖是通过减弱大豆的氮代谢途径、增强大豆的三羧酸循环来影响大豆的生长发育。本实验基于代谢组学的研究，揭示了大豆疫霉 GY8-3 菌株胞外粗多糖及其分泌蛋白处理下的大豆体内代谢规律，为研究这两类物质在大豆疫霉侵染大豆期间的作用机制提供理论依据，也为研究大豆疫霉的致病机理提供了新的研究思路和方法。同时，也对研究大豆在受到外界刺激时体内的代谢变化规律具有一定的参考意义。

关键词：大豆疫霉；胞外粗多糖；分泌蛋白；大豆；代谢组学

---

* 基金项目：公益性行业（农业）科研专项（201303018）

** 第一作者：胡九龙，硕士生，研究方向为真菌学及植物真菌病害

*** 通讯作者：高智谋，教授，主要研究方向为真菌学及植物真菌病害；E-mail：gaozhimou@ 126. com

湖北省小麦茎秆腐与穗腐病原菌种群与毒素类型分析及对多菌灵的敏感性比较研究*

陈婷婷，刘美玲，杨立军，龚双军**

（湖北省农业科学院植保土肥研究所，农业部华中作物有害生物综合治理重点实验室，农作物重大病虫草害防控湖北省重点实验室，武汉　430064）

摘　要：为明确引起湖北小麦秆腐与穗腐病原菌的组成与差别，于 2018 年 5 月在湖北省主要麦区（襄阳、枣阳、随州、荆州和武汉）采集小麦秆腐与穗腐病样品，以组织分离法进行病原物的分离培养，对分离得到的镰孢菌菌落进行纯化和单孢分离后，以形态学为基础，参照 Leisle 分类系统和分子生物学特异性引物进行鉴定。结果表明，湖北省小麦茎秆腐病病原存在亚细亚镰孢（*Fusarium asaticum*）、禾谷镰刀菌（*F. graminearum*）和黄色镰刀菌（*F. culmorum*）3 种类型，前者占绝大多数（69.73%）；毒素化学型鉴定的结果表明：76 株菌株产生 NIV，3-AcDON 和 15-AcDON 3 种类型，其中 3-AcDON 占绝大多数（59.21%）；这些菌株对多菌灵的敏感性呈连续性分布，EC_{50}为 0.51201±0.16198 mg/L。经 SAS 软件 W 法正态性检验得 $W=0.98012$，$P>\alpha$（P=0.3898，$\alpha=0.05$）。这些研究结果表明：湖北省主要麦区小麦秆腐病病原以 *F. asticum* 为主，其中 3-AcDON 是主要毒素类型，与镰刀菌造成的穗腐无差别。造成秆腐的原因主要是由于早春气温高，田间成熟子囊孢子释放同时遇到雨水，造成在扬花期前的侵染症状。

* 基金项目：食品和饲料产业链中真菌毒素控制关键技术的集成与创新（2016YFE0112900）；湖北省农业科技创新项目（2016-620-000-001-15）；国家小麦产业技术体系（CARS-03-04B）

** 通信作者：龚双军，副研究员；E-mail：gsj204@126.com

玉米大斑病菌 *StCHS6* 基因的功能研究*

王小敏**，薛江芝，毕欢欢，巩校东，刘玉卫，谷守芹***，韩建民***，董金皋***
（河北省植物生理与分子病理学重点实验室，河北农业大学
真菌毒素与植物分子病理学实验室，保定 071001）

摘 要：玉米大斑病（Corn northern leaf blight）是世界各玉米产区严重威胁玉米生产的一种真菌性病害，在我国以东华北春玉米区和南方海拔较高、气温较低的山区较易流行，直接影响玉米的产量和品质，常造成较为严重的经济损失。然而，其致病菌玉米大斑病菌（*Setosphaeria turcica*）变异频繁、生理分化明显，一旦抗病品种丧失抗性，将直接威胁到玉米生产安全。近年来，立足于对病原菌的致病性调控机制分析、探讨更加有效的防治途径已引起了植物病理学家和植物遗传育种学家的高度关注。

几丁质是构成真菌细胞壁的重要组分之一，由位于细胞膜上的几丁质合成酶 CHS（Chitin Synthase）合成。由于在植物和动物中不存在几丁质和几丁质合成酶基因，因此以几丁质合成酶为靶标的抗真菌药物对于高等真核生物来说具有较高的安全性，但是目前专门针对细胞壁几丁质成分的杀菌剂有待于进一步的开发。

本课题组在前期研究中利用生物信息学技术从玉米大斑病菌基因组中鉴定出了 8 个几丁质合成酶基因（*StCHS*1-*StCHS*8），本研究在此基础上，进一步利用 RNA-Seq 技术分析了病菌中几丁质合成酶家族基因在菌丝、分生孢子、芽管、附着胞和侵入钉等5个关键发育时期基因的表达模式。结果表明，*StCHS*1、*StCHS*2、*StCHS*5、*StCHS*6 在 5 个时期均有较高的表达量；*StCHS*3、*StCHS*4、*StCHS*7、*StCHS*8 表达量较低（FPKM<1）或者不表达；*StCHS*6 在 5 个时期的表达量均为最高，预示 *StCHS*6 在病菌的生长发育及致病过程中可能发挥重要作用。

本研究还发现，*StCHS*6 位于玉米大斑病菌基因组 scaffold_1：2066293-2069876（-）的位置，基因全长为 3 584 bp，CDS 序列大小为 2 703 bp。利用基因敲除技术获得了 2 株 *StCHS*6 基因敲除突变体 Δ*StCHS*6-1 和 Δ*StCHS*6-2，分析发现，与野生型菌株相比，*StCHS*6 基因敲除突变体菌落颜色变浅且菌丝更为致密，菌落生长速率减慢，菌丝细胞长度变短，不产生分生孢子，表明 *StCHS*6 基因参与病菌的生长发育过程。进一步研究发现，*StCHS*6 基因敲除突变体由菌丝诱导出附着胞的过程延迟且穿透玻璃纸膜的能力减弱，在完整玉米叶片及刺伤叶片上都形成较小的病斑，表明 *StCHS*6 基因参与病菌的致病过程。该研究结果明确了 *StCHS*6 基因的部分功能，为深入研究玉米大斑病菌几丁质合成酶基因家族调控的分子机制奠定基础。

关键词：玉米大斑病菌；几丁质合成酶基因家族；*StCHS*6；基因功能

---

* 基金项目：国家自然科学基金项目（31671983，31701741）；河北省引进留学归国人员资助项目（CN201705）

** 第一作者：王小敏，硕士研究生，从事植物病原真菌 MAPK 信号途径的功能研究；E-mail：2397370636@qq.com

*** 通信作者：谷守芹，教授，博士生导师；主要从事病原真菌与寄主互作研究；E-mail：gushouqin@126.com
韩建民，教授，硕士生导师；主要从事病原真菌与寄主互作研究；E-mail：hanjianminnd@163.com
董金皋，教授，博士生导师；主要从事病原真菌与寄主互作研究；E-mail：dongjingao@126.com

玉米大斑病菌 *StSLT2* 基因的功能研究*

薛江芝**，王小敏，张晓雅，巩校东，刘玉卫，韩建民***，谷守芹***，董金皋***
（河北省植物生理与分子病理学重点实验室，河北农业大学真菌毒素与植物分子病理学实验室，保定 071001）

摘　要：玉米大斑病（Northern Corn Leaf Blight，*NCLB*）是由大斑突脐蠕孢（*Setosphaeria turcica*）引起的叶部病害。近年来研究植物病原菌的发育与致病性的分子机制、探讨更有效的防治措施已成为植物病理学和植物遗传育种领域最热门的研究课题之一。

前人研究表明，许多真核生物的生长、发育和致病性都受到胞外信号转导途径的调控。信号转导途径主要有 3 类，包括 MAPK、cAMP 及 Ca^{2+} 信号转导途径等，尤其是对 MAPK 信号转导通路的研究成为了近年来研究细胞信号转导的热点。MAPK 级联途径通常由 3 个依次激活的蛋白激酶（PKs）组成，具体的说是，丝裂原活化蛋白激酶（MAPK）是由丝裂原活化蛋白激酶（MEK）磷酸化保守的苏氨酸-x-酪氨酸（TXY）基序而激活的，而 MEK 又是由丝裂原活化蛋白激酶（MEKK）激活的，而活化的 MAPK 又可以磷酸化下游的转录因子，进而调控细胞的生理生化反应。在植物病原真菌中发现了与酿酒酵母 Bck1-Mkk1/Mkk2-Slt2 同源的保守的 MAPK 级联途径，其中 *SLT2-like* 基因是 CWI-MAPK 途径中的 MAPK 基因，该类基因普遍存在于病原真菌中并参与其细胞壁发育及致病过程。但在玉米大斑病菌中有关该类基因的功能尚未见报道。

本课题组研究发现，在玉米大斑病菌数据库中鉴定得到了 *SLT2-like*，将其命名为 *StSLT2*。进一步以 pBluescript Ⅱ SK（-）载体为骨架，构建了含有草铵膦抗性的 *StSLT2* 基因敲除载体 *SLT2*-pBluescript Ⅱ SK（-）。通过 PEG 介导的遗传转化方法，将敲除载体转入到病菌的原生质体中，通过草铵膦抗性筛选、*StSLT2* 特异引物 PCR、RT-PCR 验证，获得了 2 株 *StSLT2* 基因敲除突变体，将其命名为 Δ*StSLT2*-1、Δ*StSLT2*-2。通过比较突变体与野生型发现，突变体生长速率显著降低、没有分生孢子的产生、黑色素含量显著降低；突变体中几丁质含量均显著升高，而 β-葡聚糖含量显著降低；突变体附着胞发育延迟了 24 h，且侵染丝的形成率显著降低；突变体均可以侵染并定殖 B73 玉米叶片，但形成的病斑时间延迟且病斑面积小于野生型菌株。结果表明，*StSLT2* 基因参与病菌菌丝发育、分生孢子及附着胞发育及致病过程。

关键词：玉米大斑病菌；*StSLT2* 基因；生长发育；致病性

---

* 基金项目：国家自然科学基金项目（31671983，31701741）；河北省引进留学归国人员资助项目（CN201705）
** 第一作者：薛江芝，硕士研究生，从事植物病原真菌 MAPK 信号途径的功能研究；E-mail：1337938121@qq.com
*** 通信作者：韩建民，教授，硕士生导师；主要从事病原真菌与寄主互作研究；E-mail：hanjmnd@163.com
谷守芹，教授，博士生导师；主要从事病原真菌与寄主互作研究；E-mail：gushouqin@126.com
董金皋，教授，博士生导师；主要从事病原真菌与寄主互作研究；E-mail：dongjingao@126.com

PEG 介导枸杞内生真菌 NQG8Ⅱ4 菌株原生质体转化*

闫思远**，顾沛雯***
（宁夏大学农学院，银川　750021）

摘　要：为建立 PEG 介导的枸杞内生真菌 NQG8Ⅱ4（*Fusarium nematophilum*）的遗传转化体系。以 NQG8Ⅱ4 的幼嫩菌丝为材料，通过 PEG 介导的原生质体转化法，将含有潮霉素标记的质粒 PDL2 转入 NQG8Ⅱ4 的原生质体中；对获得的转化子进行 PCR 检测。试验获得 NQG8Ⅱ4 的最优原生质体制备条件为：菌龄 16 h 的菌丝 0.05 g 于含有 3%崩溃酶+1%溶壁酶的混合酶液中反应 2.5 h；整个试验的渗透压稳定剂为 0.7 mol/L NaCl，原生质体获得量达到最大为 6.70×10^7个/mL。试验共获得 57 个转化子，转化效率为 2.85 个/μg。对转化子进行 PCR 检测，表明外源的 hph 基因已经整合到 NQG8Ⅱ4 的基因组中。本试验成功建立了稳定的 PEG 介导的 NQG8Ⅱ4 菌株遗传转化体系，可用于研究菌株 NQG8Ⅱ4 在枸杞中的侵染定殖以及对枸杞促生抗病机理。

关键词：枸杞内生真菌；NQG8Ⅱ4；原生质体；转化

---

* 基金项目：国家自然科学基金——枸杞内生真菌与宿主植物及炭疽菌互作的超微结构和细胞化学研究（31460484）
** 第一作者：闫思远，硕士，从事生物防治与菌物资源利用研究；E-mail：2047277674@ qq. com
*** 通信作者：顾沛雯，教授，主要从事植物病理学研究；E-mail：gupeiwen2019@ nxu. edu. cn

基于 ArcGIS 的贺兰山东麓酿酒葡萄霜霉病发生流行的预测预报技术初步探究*

李嘉泓**，李文学，顾沛雯***
（宁夏大学农学院，银川 750021）

摘 要：通过对贺兰山东麓葡萄种植区霜霉病发病流行规律的调查，结合温湿度、降水量和雨日等气象因素，研究了关键物候因子对贺兰山东麓葡萄霜霉病发生和流行的影响。改进并完善了贺兰山东麓酿酒葡萄霜霉病 GIS 预测模型，模型经检验符合率为 72.28%；利用 GIS 模型进行插值分析，模拟了贺兰山东麓酿酒葡萄种植区葡萄霜霉病盛发期的预测地图，预警准确率平均达 76.60%，准确地预测了葡萄霜霉病发生程度和分布范围，预测结果直观、精准地反映病害发生范围、重发区域及不同发生程度的面积和比例。模型预测结果与田间调查结果基本一致，证实了葡萄霜霉病的发生与酿酒葡萄的集中连片种植、种植年限长以及山下小气候多雨湿润度高密切相关，为相关部门进行葡萄霜霉病的早期预警、重点防治以及制定宏观防控决策提供了重要依据。

关键词：葡萄霜霉病；GIS 预测模型；气候因子

* 基金项目：宁夏回族自治区“十三五”重大科技项目——酿酒葡萄安全生产关键技术研究（2016BZ06）
** 第一作者：李嘉泓，硕士研究生，研究方向资源利用与植物保护；E-mail：912014528@qq.com
*** 通信作者：顾沛雯，教授，主要从事植物病理学与生物防治方面的研究；E-mail：gupeiwen2013@126.com

香蕉真菌性鞘腐病病原鉴定*

黄穗萍[1,2,3]**，莫贱友[1,2]，韦继光[3]，李其利[1,2]，唐利华[1,2]，郭堂勋[1,2]***
（1. 广西农业科学院植物保护研究所，南宁　530007；2. 广西作物病虫害生物学重点实验室，南宁　530007；3. 广西大学农学院，南宁　530004）

摘　要：粉蕉（Pisang Awak *Musa* ABB）具有较强的抗逆性，是我国第二大香蕉栽培品种。2016年6月，广西都安县的一个粉蕉园出现了严重的香蕉鞘腐病。2017年2月，超过90%的香蕉发病。附近的的三个粉蕉园也出现同样的病情。发病香蕉先是下部老叶与茎杆连接的鞘部出现点状黑色的病斑，病斑渐渐扩大，最后叶片枯萎、折断，整片鞘变黑、枯萎、腐烂。切开发病的鞘部，可见紫红色的坏死。病原菌先侵染老叶，随着时间推移，嫩叶也渐渐发病。受害植株茎杆变小，容易倒伏，产量减少。

采用常规组织分离法分离发病部位的病原菌。采用形态学和分子测序相结合的方法对香蕉鞘腐病病原菌进行鉴定。该菌在PDA培养基上有丰富的白色菌丝，随着培养时间的增加，菌落变成紫蓝色。该菌的分生孢子没有隔膜，棍棒型，（4.25～8.57）μm×（1.7～3.21）μm（平均5.90μm×2.34 μm）。在康乃馨培养基上，大孢子细长，具有3～4个隔膜，（32.9～57.6）μm×（2.51～4.55）μm（平均44.53μm×3.57 μm）。选择两个代表性菌株XJSF和XJSFB进行分子鉴定。将XJSF和XJSFB的ITS（internal transcribed spacer，NCBI登录号分别是MF083155和MG557985）和EF-1α（translation elongation factor 1-alpha，NCBI登录号分别是MF083156和MG557986）基因序列，在NCBI上与其他序列比对后，发现与*F. proliferatum* strains U34558 and AF160280.1的相似性分别为100%和99%。基于形态学和ITS、EF-1α序列分析结果，该菌株被鉴定为*Fusarium proliferatum*（Matsushima）Nirenberg。

将菌株XJSF和XJSFB的孢子液（1×10^6 conidia/mL）接种于刺伤的粉蕉组培苗鞘部，对照为非致病性的*Fusarium oxysporium*的孢子液（1×10^6 conidia/mL）。接种部位用薄膜覆盖保湿，2d后移除薄膜。接种的香蕉组培苗放于28℃培养箱，12h光照培养3个月。结果显示，接种XJSF和XJSFB孢子液的香蕉鞘部出现枯萎症状，而对照只有轻微的伤疤。从发病的鞘部重新分离到*Fusarium proliferatum*。将XJSF孢子液接种于巴蕉（Cavendish *Musa* AAA）的鞘上，巴蕉鞘也同样发病。据报道*Fusarium proliferatum*可以引起香蕉果实的颈腐病。然而，这是第一次发现*Fusarium proliferatum*可以引起香蕉鞘腐病。

关键词：香蕉；鞘腐病；*Fusarium proliferatum*

---

* 基金项目：广西自然科学基金（2018GXNSFBA281077）；广西创新驱动发展专项基金（桂科AA18118028）
** 第一作者：黄穗萍，硕士，助理研究员，研究方向为植物真菌病害；E-mail：361566787@qq.com
*** 通信作者：郭堂勋，硕士，副研究员，主要研究方向为植物病害防治；E-mail：415979439@qq.com

核桃炭疽病菌侵染核桃叶片的细胞学观察分析*

祝友朋**，韩长志***

（西南林业大学生物多样性保护学院，云南省森林灾害预警与控制重点实验室，昆明 650224）

摘　要：核桃作为世界四大重要坚果树种之一，广泛种植于中国、美国等地。2017 年底，中国核桃产量 192.5 万 t，产量约占世界核桃产量的一半。云南省是我国核桃种植的第一大省，全省 129 个县（市）中有 110 多个县（市）种植核桃，2017 年核桃种植面积高达 286.67 万 hm^2，总产量 115 万 t。近些年，由核桃炭疽病菌（*Colletotrichum gloeosporioides*）引起的核桃炭疽病是核桃上最主要的真菌病害之一，可致果实坏死、叶片焦枯，同时还为害嫩芽、嫩梢、叶柄、果柄等，发病严重时可使 50%以上的青果脱落，导致产量损失。然而，目前就核桃炭疽病菌侵染云南大泡核桃品种叶片的细胞学观察尚未见学术报道，因此，本研究以前期分离于大理州漾濞县核桃产区的核桃炭疽病菌 Cg1 为研究对象，采用菌丝块接种核桃叶片，并利用整体组织透明法（饱和水合氯醛水溶液中透明、1%苯胺蓝染色）处理核桃叶片以及显微观察不同时间核桃叶片组织的变化情况，从而明确核桃炭疽病菌侵染核桃叶片的细胞学变化过程，结果表明，接种 24 h 后，核桃叶片细胞出现淡蓝色，核桃炭疽病菌开始侵染核桃叶片细胞；接种 48 h 后，核桃叶片细胞出现深蓝色，开始出现坏死现象，此时核桃炭疽病菌的分生孢子在叶片表面开始萌发，并形成附着胞附着于叶片表面；接种 72 h 后，附着胞上产生侵染菌丝并伸入到核桃叶片细胞内，周围细胞中出现少量的侵染菌丝；接种 120 h 后，此时核桃炭疽病菌侵染部位叶片出现褐色病斑，细胞出现大量坏死现象，显微观察发现侵染部位表皮细胞破损，内部组织细胞大量消解。本研究为进一步明确核桃炭疽病菌侵染不同品种核桃叶片上的侵染过程，以及解析不同侵染时期核桃炭疽病菌致病基因的功能研究提供重要的理论基础。

关键词：核桃炭疽病菌；核桃；细胞学观察；云南

---

* 基金项目：西南林业大学大学生创新创业项目（201825）；国家自然科学基金项目（31560211）；云南省森林灾害预警与控制重点实验室开放基金项目（ZK150004）

** 第一作者：祝友朋，在读硕士；E-mail：3420204485@qq.com

*** 通信作者：韩长志，博士，副教授，研究方向：经济林木病害生物防治与真菌分子生物学；E-mail：hanchangzhi2010@163.com

菜心炭疽病新病原鉴定*

于　琳[1,2**]，佘小漫[1]，蓝国兵[1]，汤亚飞[1]，李正刚[1]，邓铭光[1]，何自福[1,2,***]
（1. 广东省农业科学院植物保护研究所，广州　510640；2. 广东省植物保护新技术重点实验室，广州　510640）

摘　要：菜心（*Brassica parachinensis* L. H. Bailey），又名菜薹，属十字花科芸薹属白菜亚种，是我国华南地区的特色蔬菜种类，在广东省大面积种植，也是大陆供港供澳的主要蔬菜品种之一。2018—2019 年期间，在广东省惠州市和连州市菜心黑斑病（病原菌为 *Alternaria* sp.）发病叶片的病斑上分离获得 3 个单孢分离物菌株。在 25℃下 PDA 培养基上培养，这 3 个菌株的菌落形态一致：菌落呈灰白色，圆形，产生大量蓬松的气生菌丝；培养后期菌落呈棕褐色，表面散生橘黄色黏稠的分生孢子堆。光学显微镜下观察分生孢子呈单细胞，透明，短棒状，大小为（41.4~68.8）μm×（14.0~22.8）μm。克隆上述 3 个菌株的核糖体内转录间隔区（ITS）、肌动蛋白（ACT）和 β-微管蛋白（TUB）基因序列，经 NCBI 数据库 BLASTn 分析发现，该病原菌与果生刺盘孢（*Colletotrichum fructicola*）的 ACT、TUB 和 ITS 基因序列一致性分别为 100%、100% 和 99.86%~100%。使用 10^6 个孢子/mL 的分生孢子悬浮液喷雾接种菜心（品种：碧绿粗苔菜心）植株叶片，26℃下保湿培养 7 d 后，接种无菌水的对照植株叶片上未出现病斑，而接种病原菌的叶片上可观察到“麻点”状、灰白色至褐色病斑，发病部位叶肉组织凹陷、变薄，与菜心炭疽病的病状类似（卢博彬和杨暹，2009）。通过组织分离法分离发病部位的病原真菌，获得了 18 个分离物均与上述 3 个菌株的菌落形态一致。前人研究发现，希金斯刺盘孢（*C. higginsianum*）和平头刺盘孢（*C. truncatum*）可以引起菜心炭疽病。根据本研究结果，鉴定果生刺盘孢（*C. fructicola*）是引起菜心炭疽病的新病原。

关键词：菜心；炭疽病；果生刺盘孢

---

* 基金项目：广东省自然科学基金（2018A030310194）；广州市科技计划项目（201804010268）
** 第一作者：于琳，博士，助理研究员，主要研究方向为蔬菜真菌病害；E-mail：yulin@ gdaas. cn
*** 通信作者：何自福，博士，研究员，主要研究方向为蔬菜病害；E-mail：hezf@ gdppri. com

我国主要稻作区水稻种子携带恶苗病菌检测初探*

岳鑫璐[1**]，程唤奇[1]，黄玉婷[1]，李平东[1]，李志强[1***]，胡茂林[1,2 ***]
（1. 深圳市农业科技促进中心，深圳 518055；2. 深圳市作物分子设计育种研究院，深圳 518107）

摘　要：水稻恶苗病是水稻上的重要种传病害，其病原菌主要有藤仓镰孢（*Fusarium fujikuroi*）、层出镰孢（*Fusarium proliferatum*）、拟轮枝镰孢（*Fusarium verticillioides*）等，该病害发生遍及亚洲及其他水稻产区，严重时会引起水稻减产 50%。本研究对来源于我国主要稻作区的 78 份水稻种子样品进行种子内外部带菌检测，以初步明确我国水稻种子携带恶苗病菌的优势菌群，试验结果如下：

（1）种子带菌率：分别采用平板培养、洗涤法法，挑取疑似病原物及发病种子提取 DNA，分别采用真菌 rDNA-ITS 通用引物 ITS1、ITS4 及基于 *TEF* 基因设计的 4 种特异性引物进行分子检测。结果表明，78 份水稻种子中，种子内部携带恶苗病菌检出率为 5. 45%，种子外部携带恶苗病菌的检出率为 25. 45%，内外部同时携带恶苗病菌的检出率为 2. 56%。

（2）优势种：水稻种子携带恶苗病菌的优势种主要为 *F. fujikuroi* 和 *F. proliferatum*。其中，水稻种子外部携带恶苗病菌 *F. fujikuroi*、*F. proliferatum* 的检出率为 12. 82%、11. 54%，水稻种子内部携带恶苗病菌 *F. fujikuroi*、*F. proliferatum* 的检出率为 3. 85%、7. 70%。

关键词：水稻恶苗病；病菌检测；优势菌群；筛选

---

* 基金项目：国家重点研发计划（2017YFD0201602-6）；市技术攻关项目（20170434）

** 第一作者：岳鑫璐，硕士，主要从事种子病理学与作物病害生物防治研究；E-mail：yxl20071029@126. com

*** 通信作者：李志强；E-mail：zqlee2008@qq. com

胡茂林；E-mail：maolin522612@126. com

国内橡胶树炭疽病病原菌种群多样性分析

刘先宝，郑肖兰，李博勋，江 涛，冯艳丽，黄贵修
（中国热带农业科学院环境与植物保护研究所，海口 571101）

摘 要： 炭疽病是橡胶树上重要的叶部病害之一。一直以来胶孢炭疽菌被认为是橡胶树炭疽病的病原菌，直到2008年尖孢炭疽菌在中国云南省被首次报道。近年来关于橡胶树尖孢炭疽菌的报道也越来越多。为了弄清2种炭疽菌的为害及其分布，笔者对国内橡胶树主栽区炭疽病的样品进行了收集，利用胶孢炭疽（CgInt / ITS4）和尖孢炭疽（CaInt / ITS4）复合种特异引物对不同症状的样品进行分子检测，通过多基因序列分析和形态观察对收集的菌株进行了鉴定。通过结果分析，笔者发现为害古铜期叶片并引起叶片皱缩和脱落的炭疽菌主要为尖孢炭疽复合种，部分叶片存在胶孢炭疽复合种的复合侵染；在淡绿期叶片上，尖孢炭疽复合种的侵染产生黑褐色病斑，病健交接处皱缩，病斑发展后期中间穿孔，而胶孢炭疽复合种的侵染产生浅褐色病斑，病健交接处叶片平整；在稳定期叶片上，尖孢炭疽复合种侵染产生凸起症状，而胶孢炭疽复合种侵染表现为炭疽症状或纸质状；古铜期至淡绿期尖孢炭疽复合种为优势种群，而在淡绿期至稳定期胶孢炭疽复合种为优势种群。我们对收集的菌株进行多基因序列比对，并构建ML树和贝叶斯树，发现为害橡胶树的胶孢炭疽复合种包括 *C. fructicola*、*C. siamense* 和 *C. ledongense*，尖孢炭疽复合种包括 *C. bannanense* 和 *C. australisinense* 2个种，同时还发现2个独立的遗传进化分支，推测可能为尖孢炭疽复合种的新种；从云南和海南收集的菌株中还发现了博宁炭疽复合种的2个 *C. karstii* 菌株。

关键词： 橡胶树炭疽病；胶孢炭疽；尖孢炭疽；优势种群；多样性分析

内生真菌 HND5 菌株多种生防机制解析*

杨　扬，蔡吉苗，刘　静，王　宝，黄贵修**

（中国热带农业科学院环境与植物保护研究所，农业部热带作物有害生物综合治理重点实验室，海南省热带农业有害生物监测与控制重点实验室，海南省热带作物病虫害生物防治工程技术研究中心，海口　571101）

摘　要：我们在前期从臂形草中分离得到的内生帚枝霉属真菌 *Sarocladium brachiariae* HND5 可有效抑制香蕉枯萎病菌的生长，并具有良好的盆栽和大田防效。但由于帚枝霉属真菌 *Sarolacdium* 相关研究少，HND5 菌株生防作用机理一直不清，限制了该菌株商品化开发。为解析 HND5 菌株生防作用机制，进一步促进该菌株的商品化开发和大规模推广，我们通过挥发性抑菌物质鉴定、全基因组测序分析以及抗菌基因簇鉴定对其生防机理进行了初步解析。利用顶空法固态微萃取收集 HND5 菌株培养物上方的挥发性气体，并利用 GC-MS 进行分析，发现 HND5 菌株可产生 17 种不同的挥发性物质，其中 3 种物质可有效抑制香蕉枯萎病菌的生长，2 种物质可有效抑杀根结线虫二龄幼虫。进一步的全基因组测序及比较基因组学分析结果显示，与同属的水稻鞘腐病菌 *S. oryzae* 相比，HND5 菌株含有大量的真菌细胞壁降解酶，并含有多种未知功能的次生代谢产物合成基因簇。推测该菌株可通过外泌真菌细胞壁降解酶以及具有抑菌活性的次生代谢产物防止病原菌的入侵。在全基因组数据支撑下，对 HND5 菌株中的非核糖体多肽合成酶（NRPS）表达基因进行缺失突变，发现 NRPS30 基因被缺失突变后，HND5 丧失抑制真菌活性；UPLC-Q-TOF-MS 分析野生型和突变体次生代谢产物，发现 HND5 NRPS30 缺失突变体丧失一个多肽类次生代谢产物，推测该物质是 HND5 菌株主要的抑菌活性物质。全基因组数据还显示，HND5 基因组编码一个外泌丝氨酸蛋白酶，该酶与顶孢霉属真菌激发子蛋白 AsES 具有较高的同源性，由于 AsES 蛋白可有效诱导草莓对炭疽病的抗性，推测该丝氨酸蛋白酶也具有诱导植物抗性的活性。

本文通过挥发性抑菌物质鉴定、全基因组测序分析以及抗菌基因簇鉴定，发现内生帚枝霉属真菌 *Sarocladium brachiariae* HND5 菌株具有多种生防机制，可通过产生挥发性抑菌物质、真菌细胞壁降解酶及抗菌非核糖体多肽抑制病菌真菌的生长，并可产生具有诱导抗性活性的激发子蛋白提高寄主机制抗性，为该菌株的商品化开发和大规模推广提供理论支撑。

关键词：内生真菌；HND5；生防机制；诱导抗性

* 基金项目：中国热带农业科学院基本科研业务费专项资金（1630042019006&1630042018016）

** 通信作者：黄贵修；E-mail：hgxiu@ vip. 163. com

棉花黄萎病菌 *VdSP1* 基因的功能分析

王春巧，孙　琦，何　芳，黄家风
（石河子大学农学院／新疆绿洲农业病虫害治理与植保资源利用重点实验室，石河子　832003）

摘　要：由大丽轮枝菌（*Verticillium dahliae* Kleb.）引起的棉花黄萎病是棉花生产上最具毁灭性的一种土传真菌维管束病害。由于黄萎病菌致病机制复杂、微菌核存活年限长等特点，传统的防治手段很难对棉花黄萎病进行有效防治。因此鉴定与微菌核形成及致病力相关基因的功能对防治该病害至关重要。本课题组前期研究发现 1 个受棉花根系诱导后明显上调表达的基因，分泌性预测结果表明，该基因编码的蛋白具有分泌性，是大丽轮枝菌潜在的分泌蛋白基因，命名为 *VdSP*1 基因。因此本研究针对棉花黄萎病菌 *VdSP*1 基因的功能进行了分析：以棉花黄萎病菌落叶型强致病力菌株 V592 的基因组 DNA 和 cDNA 为模板，对 *VdSP*1 基因全长进行克隆并测序，结果表明，*VdSP*1 基因的全长为 1 072 bp，包含 1 个外显子，编码 1 个具有 232 个氨基酸的蛋白；基因检索结果表明，*VdSP*1 基因与 GenBank 中的已注释的基因没有任何的序列相似性；*VdSP*1 基因编码蛋白与轮枝菌属真菌同源基因编码蛋白的氨基酸序列相似性均达 95%以上，而与其他真菌同源基因编码蛋白的序列相似性均低于 50%。构建针对 *VdSP*1 基因的敲除载体和过表达载体，通过农杆菌介导的遗传转化筛选获得 2 个 *VdSP*1 基因敲除体菌株和 2 个 *VdSP*1 基因过表达菌株。以野生型菌株 V592 为对照，对 *VdSP*1 基因敲除突变体和过表达菌株的菌落生长速率、产孢量、微菌核形成及对棉花的致病力进行测定。结果显示，野生型菌株 V592 形成黑色菌核型菌落，而 *VdSP*1 敲除体菌株形成白色菌丝型菌落，生长速度显著高于野生型菌株 V592 和过表达菌株，产孢量显著低于 V592 菌株和过表达菌株。对微菌核形成进行诱导培养及显微观察，结果表明，*VdSP*1 基因敲除导致棉花黄萎病菌不产生微菌核。致病性测定结果表明，*VdSP*1 敲除体菌株对棉花的致病力显著低于 V592 和过表达菌株，且发病时间延迟。通过实时荧光定量逆转录 PCR（Reverse transcription-quantitative real time PCR，RT-qPCR）测定其他致病相关基因在 *VdSP*1 敲除突变体中的表达量，结果显示，在 *VdSP*1 基因敲除突变体中，大丽轮枝菌微菌核形成相关基因 *VDH*1、*VMK*1、*VdPKAC*1、*VGB* 和 *VdHog*1 的表达量均显著上调。上述结果表明，*VdSP*1 基因是棉花大丽轮枝菌微菌核形成的关键基因，与大丽轮枝菌的生长速率和产孢量密切相关，并参与大丽轮枝菌致病。

关键词：棉花黄萎病菌；大丽轮枝菌；*VdSP*1 基因；基因敲除；过表达菌株

苹果树腐烂病菌 CAP 超家族蛋白 VmPR1c 降解功能域和降解途径探究*

孟 香**，尹志远，聂嘉俊，黄丽丽***
（西北农林科技大学植物病理学系，旱区作物逆境生物学国家重点实验室，杨凌 712100）

摘 要：真菌在与寄主互作过程中常分泌多种效应蛋白，植物病原菌分泌的 CAP 超家族蛋白在与寄主互作中发挥着多种功能，但关于植物病原真菌 CAP 超家族蛋白的蛋白特征仍知之甚少。课题组前期对苹果树腐烂病菌 *Valsa mali* 中的一个 CAP 蛋白 VmPR1c 进行研究发现该基因敲除突变体致病力明显下降，但 VmPR1c 蛋白在烟草瞬时表达过程中被明显降解。本研究借助 BLAST、NCBI CDD web server、SignalP 4.1、TMHMM 2.0 等对 VmPR1c 蛋白进行序列分析，利用缺失突变、点突变等方法对 VmPR1c 蛋白的降解功能域进行研究，使用不同种类蛋白酶抑制剂对该蛋白降解途径进行探索。C 端区段缺失突变后免疫印迹结果显示，突变涉及 CAP 结构域中的 α-helix 结构和 CBM 区域时，Western blot 条带明显加深；涉及 CTE 区域及 CAP 结构域中的半胱氨酸时，Western blot 条带则不同程度地减弱；替换 VmPR1c 的信号肽或突变其信号肽切割位点序列后，Western blot 条带有不同程度地加深；分别对 CTE 和 NTE 中的脯氨酸进行点突变后，蛋白降解更加严重。在蛋白表达过程中加入蛋白酶体抑制剂 MG132 和溶酶体抑制剂 chloroquine，与对照组相比，Western blot 结果无明显变化。表明 VmPR1c 蛋白序列 C 端区域中的 CTE、信号肽及其切割位点序列，以及 CAP 结构域中的 α-helix 结构介导该蛋白降解，其中 CTE 和 NTE 中的脯氨酸对该蛋白具保护作用。此外，VmPR1c 的降解不通过蛋白酶体和溶酶体途径。

关键词：*Valsa mali*；CAP 蛋白；免疫印迹；降解

---

* 基金项目：国家自然科学基金项目（No. 31871917，31671982）
** 第一作者：孟香，硕士研究生；E-mail：1520547844@ qq. com
*** 通信作者：黄丽丽，教授；E-mail：huanglili@ nwsuaf. edu. cn

金边虎皮兰软腐病的病原菌鉴定*

黄思良**，安金萍，王 路，郑雪玲，王 坦，庞发虎，陶爱丽
（南阳师范学院农业工程学院，南阳 473061）

摘 要：金边虎皮兰（*Sansevieria trifasciata* Prain）是虎皮兰属中具有较高观赏价值的品种之一。2016 年 9 月，从河南省南阳市当地花卉市场购买的金边虎皮兰的叶片上发生一种罕见的软腐病。发病初期病部组织呈水渍状，病斑边缘背光呈半透明状，多始发于叶片边缘，呈圆弧状向叶片中部扩展，湿度大时病情扩展迅速，致组织软腐，严重时叶片折倒甚至整株腐烂，低湿度下病斑扩展明显受限，病部表皮组织皱缩，病部呈灰白至灰褐色。从病斑中部常见污白色或黑褐色霉状物。剪取病健交界组织块（约 3~4mm^2），用 75%酒精对病叶组织处理 10 s，升汞表面消毒 50~80s，经无菌水冲洗 8 次后置于 PDA 平板上 26℃培养 4~5d，获得形态一致的曲霉分离物。经单孢分离后随机取一株（An-1）进行鉴定。在 PDA 培养基上 An-1 的菌落呈圆形，初期白色后转黑褐色至深黑色。分生孢子着生在分生孢子头上，单胞，球形，无色透明至浅褐色，直径约 3.5~3.7 μm（n=100）。用刺伤接种法对 3 个虎皮兰品种（金边虎皮兰、广叶金边虎皮兰、短叶金边虎皮兰）做致病性测定。结果表明，菌株 An-1 对长叶的金边虎皮兰的致病力最强，病斑扩展最快，其次为广叶金边虎皮兰。短叶金边虎皮兰对该菌的抗性最强（接种点伤口不形成病斑）。从金边虎皮兰与广叶金边虎皮兰叶片接种点产生的病斑上再分离获得了与接种菌 An-1 形态一致的曲霉，依据柯赫法则证明了菌株 An-1 为金边虎皮兰软腐病的致病菌。用引物 Beta-F（5′-cagctcgagcgtatgaacgtct-3′）和 Beta-R（5′-cggaagtcggaagcagccatc-3′）扩增菌株 An-1 的 β-微管蛋白基因序列，对该菌进行分子鉴定。扩增获得的菌株 An-1 的 β-微管蛋白基因序列长度为 935bp。BLAST 比对结果表明，菌株 An-1 的 β-微管蛋白基因序列与 GenBank 上登记的多株黑曲霉（*Aspergillus niger*）的相关序列的最大相似性达 99%。在构建的基于 β-微管蛋白基因序列的系统发育树中，菌株 An-1 与 *A. niger*（GenBank 登录号：MF150906 and MG701893）以 100%自举值相聚一群，明显区别于其他 *Aspergillus* spp.，支持菌株 An-1 属 *A. niger*。黑曲霉主要引起瓜果类作物的采后病害，该菌引起金边虎皮兰软腐病为首次报道。

关键词：金边虎皮兰；黑曲霉；软腐病；分子鉴定

---

* 基金项目：河南省高校科技创新团队支持计划项目（2010JRTSTHN012）

** 第一作者：黄思良，主要从事植物病害生物防治研究；E-mail：silianghuang@126.com

黏质沙雷氏菌（*Serratia marcescens*）TC-1 产几丁质酶条件的响应面分析*

陶爱丽**，郑雪玲，徐茹新，黄思良***，刘凤琴，李佳康，魏乙斌，王志清

（南阳师范学院农业工程学院，南阳　473061）

摘　要：许多病原真菌和昆虫以几丁质作为基本结构成分，几丁质酶在植物病虫害防治中具有重要作用。本实验室从罹病的蛴螬中分离出一株具有抑菌抗虫的红色细菌 TC-1，经鉴定为黏质沙雷氏菌（*Serratia marcescens*），且具有产几丁质酶的活性。为开发利用该菌的几丁质酶产能，本论文利用响应面分析，确定黏质沙雷氏菌 TC-1 高产几丁质酶的最佳条件。响应面分析结果为：酶活性 = 20.90+0.73 A（胶体几丁质含量）-0.12 B（培养时间）+0.25 C（培养温度）-0.26 AB+0.14AC-0.13 BC-2.55 A^2-2.82 B^2-2.36 C^2；此菌株高产几丁质酶的最优方案为：胶体几丁质含量为 9 g/L、培养时间为 64 h、培养温度为 28℃。在此种条件下，酶活力的预测值为 20.964 U/mL。经过调整验证试验，实际酶活力达到 18.637 U/mL。

关键词：几丁质酶；黏质沙雷氏菌；响应面分析；酶活力

---

* 基金项目：河南省科研服务平台专项（16105）

** 第一作者：陶爱丽，讲师，主要从事作物病虫害生物防治研究；E-mail：taltal02@ qq. com

*** 通信作者：黄思良，博士，教授，主要从事植物病害生物防治研究；E-mail：silianghuang@ 126. com

女贞褐斑病病原菌的分离与鉴定*

杨　迪[1,2**]，郑雪玲[1]，黎　鹏[1]，徐东亚[1]，王　路[1]，付　岗[2]，黄思良[1***]
（1. 南阳师范学院农业工程学院，南阳　473061；2. 广西农业科学院植物保护研究所，南宁　530007）

摘　要：女贞（*Ligustrum lucidum*）是传统的园林观赏植物，因其属常绿灌木，且具有树形整齐，适应性好等特征，近年作为城市绿化树种被广泛种植。2019 年 2 月，于贵州省麻江县宣威镇翁保村乌羊麻寨（N：26°23′8″，E：107°44′56″；海拔 617 m）女贞绿篱上发生一种褐斑病，极大地影响了其观赏效果。发病叶片上有近圆形或不规则褐色病斑，病斑中心为褐色，周边颜色较浅，叶片正面病斑前缘为黑褐色，边缘有黄色晕圈，叶片背面病斑总体为褐色，边缘为淡紫色。取发病叶片，采用组织分离法分离获得菌株 NP GY-1。通过针刺接种法依据柯赫氏法则验证其为女贞褐斑病的病原菌。在 PDA 培养基上，该病菌菌落初期为白色，有茂密的气生菌丝，后期菌落逐渐变为墨绿色，培养 10d 后，在显微镜下可观察到念珠状厚垣孢子。对该菌株进行 rDNA-ITS 序列和 β-微管蛋白基因序列鉴定，结果表明：菌株 NP GY-1 的 rDNA-ITS 序列（GenBank 登录号为 MK852167）与小新壳梭孢（*Neofusicoccum parvum*）相似度最高，达到 99.65%，其 β-微管蛋白基因序列（GenBank 登录号 MK952193）同样也与 *N. parvum* 的相似度最高，为 93.72%。在构建的基于 rDNA-ITS 和 β-微管蛋白基因序列的系统发育树中，菌株 NP GY-1与 *N. parvum* 相聚一群，与其他 *Neofusicoccum* spp. 可明显区分。结合形态学观测结果，将该菌鉴定为小新壳梭孢。用离体叶片针刺接种法，在 48 科 72 种植物上对该菌进行了寄主范围测定。结果表明：其中 44 科 66 种供试植物的叶片接种口上在试验观察期内出现不同程度的发病症状，而锦葵科的木槿（*Hibiscus syriacus* Linn.）、黄蜀葵［*Abelmoschus manihot*（L.）Medik.］、唇形科的薄荷（*Mentha canadensis* L.）、紫苏［*Perilla frutescens*（Linn.）Britt.］、堇菜科的堇菜（*Viola verecunda* A. Gray）和葫芦科的栝楼（*Trichosanthes kirilowii* Maxim.）在试验观察期（10d）内接种部位未见明显症状。*N. parvum* 可引起葡萄（*Vitis vinifera* L.）、杧果（*Mangifera indica* L.）、核桃（*Juglans regia* L.）等多种植物病害，该菌侵染引起的女贞褐斑病为首次报道。

关键词：女贞；褐斑病；小新壳梭孢；寄主范围

---

* 基金项目：河南省高校科技创新团队支持计划项目（2010JRTSTHN012）

** 第一作者：杨迪，硕士，研究方向为植物病害生物防治；E-mail：1464596962@ qq. com
*** 通信作者：黄思良；E-mail：silianghuang@ 126. com

引起番茄果实内部腐烂的链格孢菌的鉴定*

白永振**，王　璐，郑雪玲，王　坦，庞发虎，陶爱丽，焦铸锦，黄思良***
（南阳师范学院农业工程学院，南阳　473061）

摘　要：番茄（*Solanum lycopersicum* L.）富含对人体健康的营养成分。2017 年 7 月，从河南省南阳市当地超市购买的番茄果实上发现一种未知的病害。发病初期，番茄果实外部无症状，与健康果实没有明显区别，切开病果后，其内部霉变腐烂症状才得以显现。染病的维管束呈黑色或黑褐色，部分维管束变细或断裂，中果皮产生褐色的病斑。将染病的番茄果实内部组织用灭菌的解剖刀切取后置于马铃薯葡萄糖琼脂（PDA）平板上，在 28℃下培养 5d，从并组织上长出形态一致的链格孢（*Alternaria* sp.）。随机选取 2 株（At1 和 At2）经单孢分离纯化后进行致病性测定及形态和分子鉴定。两株供试菌在 PDA 培养基上产生圆形或近圆形具较茂密气生菌丝的菌落，菌落初为灰白色，培养 3~4d 后转黑褐色至墨绿色，在水琼脂（WA）平板上菌丝体可稀疏生长并产孢。WA 上产生的分生孢子多数卵形至椭圆形，黑褐色，呈长链状着生在分生孢子梗上。菌株 At1 和 At2 的分生孢子分别有 42.3%和 56.8%具长约 4~5μm 的喙（调查孢子数 109）。每个分生孢子有 0~6（平均 3.4）个横隔和 0~5（平均 0.8）个纵隔或斜隔。随机调查 110 个样本，菌株 At1 和 At2 的分生孢子大小分别为（11.4~33.3）（平均 20.3）μm×（7.3~15.1）（平均 10.6）μm 和（10.0~39.3）（平均 22.7）μm×（4.0~16.4）（平均 10.9）μm。扩增供试菌株的 rDNA-ITS（internal transcribed spacer）和 β-微管蛋白基因序列，将相关序列在 GenBank 登记后得到的菌株 At1 的 rDNA-ITS 和 β-微管蛋白基因序列登录号分别为 MG558002 和 MG558003，菌株 At2 的相应序列登录号分别为 MG925323 和 MG925327。在构建的基于 β-微管蛋白基因序列的系统发育树中，菌株 At1 和 At2 与多株细交链孢菌（*Alternaria alternata*）相聚同一群，与其他 *Alternaria* app. 明显区分别开来。BLAST 比对表明，供试菌株的 rDNA-ITS 序列与 GenBank 中多株 *A. alternata* 的相同序列的最大相似性达 100%。将供试菌株在 PDA 平板上 28℃培养 6d，用无菌水制备分生孢子悬浮液（$5×10^8$ spores/mL），用无菌注射器注射到健康的离体成熟的番茄上，以注射无菌水为对照。在 28℃培养 7d 对接种番茄和对照番茄进行解剖观察，在接种番茄上产生了内部腐烂症状，外表无症状，与原初病果症状相似；对照果实内外部均无症状。从接种发病番茄的内部病组织中再分离获得了与接种菌株形态完全一致的链格孢分离物。遵循柯赫法则验证了供试菌株对番茄的致病性。细交链孢菌引起番茄果实外部腐烂症状已有报道，该菌引起番茄内部腐烂而外部不显症的现象未见报道，笔者推测引起该症状的细交链孢菌很可能是通过维管束进行系统侵染的、在番茄内部组织厌氧环境中具有较强适应能力的新的 *A. alternata* 菌系。

关键词：番茄；果实内部腐烂；细交链格孢菌；分子鉴定

---

* 基金项目：河南省高校科技创新团队支持计划项目（2010JRTSTHN012）

** 第一作者：白永振，硕士生，主要从事植物病害生物防治研究；E-mail：1304602460@ qq. com

*** 通信作者：黄思良，博士，教授，主要从事植物病害生物防治研究；E-mail：silianghuang@ 126. com

云南普洱蔗区甘蔗白叶病的分子鉴定*

张荣跃**，李文凤，黄应昆***，王晓燕，单红丽，李　婕，仓晓燕，罗志明，尹　炯
（云南省农业科学院甘蔗研究所，云南省甘蔗遗传改良重点实验室，开远　661699）

摘　要：研究旨在明确2018年在云南普洱蔗区发现的甘蔗病害是否为植原体引起的甘蔗白叶病。使用植原体16S rRNA基因序列通用引物P1/P7和R16F2n/R16R2对13份采自云南普洱蔗区的疑似甘蔗白叶病样品进行巢氏PCR检测，并将巢氏PCR产物进行克隆测序和序列分析。巢氏PCR结果表明，13份样品中有10份样品扩增得到1 240bp左右的片段。测序结果表明扩增获得的10条序列大小均为1 247bp，序列间的核苷酸一致性为100%。BLASTN分析表明从病株扩增到的所有核苷酸序列与先前云南临沧甘蔗白叶病植原体分离物LC7和LC9的16S rRNA基因序列（Genbank登陆号：KR020691和KR020692）的核苷酸序列一致性为100%。虚拟RFLP分析表明本研究获得的16S rRNA基因序列的酶切图谱与16SrXI-B亚组植原体相同。本研究结果表明云南普洱蔗区发现的甘蔗病害为16SrXI-B亚组植原体引起的甘蔗白叶病。

关键词：甘蔗；赤腐病；发生流行特点；防控对策

* 基金项目：国家自然科学基金项目（31760504）；云南省农业基础研究联合专项［2017FG001（-054）］；国家现代农业产业技术体系（糖料）建设专项资金（CARS-170303）；云岭产业技术领军人才培养项目“甘蔗有害生物防控”（2018LJRC56）；云南省现代农业产业技术体系建设专项资金

** 第一作者：张荣跃，硕士，助理研究员，主要从事甘蔗病害研究；E-mail：rongyuezhang@hotmail.com
*** 通信作者：黄应昆，研究员，从事甘蔗病害防控研究；E-mail：huangyk64@163.com

甘蔗鞭黑粉菌 Ram1 与 Ras 调控有性配合与细胞壁完整性

孙书荃，邓懿祯，蔡恩平，李玲玉，常长青*，姜子德*
（华南农业大学农学院，华南农业大学群体微生物研究中心，广州 510642）

摘 要：由担子菌门甘蔗鞭黑粉菌（*Sporisorium scitamineum*）引起的甘蔗黑穗病对甘蔗产业影响巨大。有性配合和维持细胞壁完整性在该病原菌侵染寄主的过程中十分重要。经鉴定发现法尼酸转移酶β亚基（Ram1）对信息素前体 Mfa1 成熟加工具有关键作用，缺失突变体显著影响甘蔗鞭黑粉菌的有性配合。进一步的基因互补、表型分析和基因在不同时期的表达分析发现 Ram1 的法尼基修饰是性信息素成熟必须的，影响单倍体的有性配合和致病力。对 Ram1 潜在修饰的靶蛋白 Mfa1、Ras1 和 Ras2 蛋白编码的基因进行敲除、互补和表型分析，发现 3 个基因均影响有性配合。进一步发现 RAM1、RAS1 和 RAS2 基因突变都影响细胞壁的完整性，其机制与胞内海藻糖含量及细胞壁完整性通路关键基因的表达相关，而 Ram1 和 Ras 对细胞壁完整性存在不同的调控机制。以上结果推测甘蔗鞭黑粉菌中存在与酿酒酵母性信息素前体不同的修饰和成熟机制，此外存在与 Ras 功能不同的未知异戊二烯化底物，该底物参与有性配合和细胞壁完整性胁迫的响应。

关键词：甘蔗鞭黑粉菌；有性配合；细胞壁完整性；法尼酸转移酶；Ras

---

* 通信作者：常长青；姜子德

Peronophythora litchii RXLR effector PlAvh142 can trigger cell death in *Nicotiana benthamiana* and contribute to the virulence

Situ Junjian, Jiang Liqun, Fan Xiaoning, Xi Pinggen, Kong Guanghui *, Jiang Zide *

(*Department of Plant Pathology / Guangdong Province Key Laboratory of Microbial Signals and Disease Control*, *South China Agricultural University*, *Guangzhou* 510642, *China*)

Abstract: *Peronophythora litchii* causes downy blight on litchi fruits as well as tender leaves and panicles rot of litchi plants. RXLR effectors secreted by *Phytophthora* species play a central role in pathogen-plant interactions. Though 245 RXLR effectors have already been predicted in this litchi downy blight pathogen before, the function of them is still unknown. Here, we identified a RXLR effector PlAvh142 that could trigger strong cell death in *Nicotiana benthamiana*. In addition, heterologous expression experiments showed that PlAvh142 could induce ROS accumulation, callose deposition and genes transcription involved in plant hormone pathways in *N. benthamiana*. Next, We found that PlAvh142 localize both in cytoplasm and nucleus. The cytoplasmic localization was required for its cell death - inducing activity. Moreover, deletion of the two internal repeats in *PlAvh*142 also abolished the cell death - inducing activity. Through aligning the ORF of thirty strains, five SNPs were found in *PlAvh*142 and only three of them led to nonsynonymous substitutions, but none of the version affect cell death. The results of expression pattern revealed that *PlAvh*142 is highly transcribed during zoospores and the early stages of infection. Finally in the pathogenicity assay, *PlAvh*142 knockout mutants attenuated *P. litchii* virulence on litchi plants, whereas the overexpressed mutants are more aggressive. In conclusion, PlAvh142 is an important virulence RXLR effector of *P. litchii* and recognized by *N. benthamiana*.

Key words: *Peronophythora litchii*; PLAvh142; *Nicotiana benthamiana*

* Corresponding author: Jiang Zide; junjian. st@ hotmail. com

大循环锈菌羽茅柄锈菌鉴定及其有性阶段发育形态

马心瑶，刘　尧，陈　文，杜志敏，康振生*，赵　杰*
（西北农林科技大学植物保护学院，旱区作物逆境生物学国家重点实验室，杨凌　712100）

摘　要：禾本科杂草锈菌种类多，其中许多已明确存在转主寄生，有着完整的生命史。然而，还有其他许多锈菌，只知其无性繁殖，而不知其有性循环。羽茅柄锈菌（*Puccini achnatheri-sibirici*）是一种侵染禾本科芨芨草属杂草（*Achnatheherum* sp.）的锈菌，目前仅知道其存在夏孢子和冬孢子阶段，但是否存在有性循环尚不清楚。本研究通过该菌冬孢子萌发产生的担孢子人工接种陕西小檗（*Berberis shensiana* Ahrendt）叶片，该菌侵染小檗叶片先后分别产生性孢子和锈子器。产生的锈孢子接种在芨芨草叶片上产生明显的橙黄色的夏孢子堆，然而在小麦品种'铭贤169'上却只产生褪绿斑而没有夏孢子。羽茅柄锈菌的 ITS 序列在 NCBI 数据库比对，没有与之完全匹配的锈菌种类，仅与短柄草柄锈菌（*P. brachypodii*）同源性最高，为 94%。然而，在进化树中与短柄草柄锈菌和其他锈菌属分离，独立组成一个分支。通过光学显微镜与扫描电镜观察羽茅柄锈菌的夏孢子、冬孢子及其担孢子阶段、性孢子阶段和锈孢子阶段的病原菌形态，发现其特征与其他柄锈菌属种类不同。由此证实羽茅柄锈菌是转主寄生的全型锈菌，小檗是其转主寄主。

关键词：柄锈菌；转主寄主；小檗；有性生殖

---

* 通信作者：赵杰；E-mail：jiezhao@ nwsuaf. edu. cn
康振生；E-mail：kangzs@ nwsuaf. edu. cn

小麦条锈菌与披碱草条锈菌和冰草条锈菌的有性杂交

郑 丹*，左淑霞*，陈 文，杜志敏，康振生**，赵 杰**
（西北农林科技大学植物保护学院，旱区作物逆境生物学国家重点实验室，杨凌 712100）

摘 要：条锈病是威胁中国小麦产区的重要病害，流行时常造成巨大产量损失，严重影响小麦生产安全。病原菌毒性变异产生新小种是导致品种抗性丧失与后继病害流行的根本原因。近年研究证实小檗（*Berberis*）是小麦条锈菌（*Puccinia striiformis* f. sp. *tritici*，*Pst*）的转主寄主，揭示有性生殖是小麦条锈菌毒性变异的主要途径。然而，目前，尚无有关小麦条锈菌与其他条锈菌专化型的有性杂交与毒性变异的研究。本研究利用小麦条锈菌（CYR32）与披碱草条锈菌（*P. striiformis* f. sp. *elymi*，*Pse*）和冰草条锈菌（*P. striiformis* f. sp. *agropyri*，*Psa*）侵染陕西小檗（*Berberis shensiana* Ahrendt）后分别进行杂交，获得了小麦条锈菌×披碱草条锈菌（*Pst*×*Pse*）有性杂交单锈孢后代群体 61 个；小麦条锈菌×冰草条锈菌（*Pst*×*Psa*）有性杂交单锈孢后代群体 52 个。其中 *Pst*×*Pse* 的 61 个后代群体产生了 37 个不同的致病类型，包括 59 小种，变异率达 96%，仅有 2 个与亲本菌系 CYR32 相同；*Pst*×*Psa* 的 52 后代群体产生了 36 个不同致病类型包括 50 个小种，变异率达 96%；仅有 2 个与亲本 CYR32 相同。毒力测定结果与分析表明两个杂交后代群体在单个基因（*Yr*）位点的遗传复杂，变异率不同。研究证实，小麦条锈菌与披碱草条锈菌和冰草条锈菌可以完成有性遗传重组，产生高度毒性变异，是小麦条锈菌毒性变异的新途径。

关键词：条锈病；有性遗传重组；毒性变异；新小种；杂草

* 共同第一作者：郑丹、左淑霞同等贡献
** 通信作者：赵杰；E-mail：jiezhao@ nwsuaf. edu. cn
康振生；E-mail：kangzs@ nwsuaf. edu. cn

自然条件下秋季小麦条锈菌侵染小檗完成有性循环

杜志敏，姜舒畅，黄淑杰，陈 文，张根生，康振生*，赵 杰*
（西北农林科技大学植物保护学院，旱区作物逆境生物学国家重点实验室，杨凌 712100）

摘　要：条锈病是威胁小麦生产的毁灭性病害，由条形柄锈菌小麦专化型（*Puccinia striiformis* f. sp. *tritici*）引起。近年证实，该病原菌在小檗上完成其有性循环，是全型转主寄生锈菌，而且在自然条件下春季小麦条锈菌在野生小檗上完成其有性循环并传播锈孢子侵染小麦引起条锈病。但是，小麦条锈菌的有性循环是否在其他季节发生，目前尚无研究报道。本研究通过调查发现自然条件下在中国小檗存在秋季受侵染产生锈子器。通过接种小麦感病品种‘铭贤 169’获得了小麦条锈菌菌系 4 个，其后建立其单孢堆菌系 65 个。经中国小麦条锈菌鉴别寄主测定，4 个分离菌系均与已知条锈菌小种不匹配，为新小种且为不同致病类型。65 个单孢菌系中有 15 个为已知小种，分别为 Su11-3、Su11-4 和 Su11-14，其余的均为新小种。后代群体在单基因系鉴别寄主上的测定结果，将此单孢群体分为 28 个不同的毒性类型。9 对多态性 SSR 引物将单孢群体区分为 14 个不同基因型。本研究首次证实，自然条件下，在中国一些地区，小麦条锈菌有性生殖在野生小檗上秋季发生。

关键词：条锈病；有性生殖；毒性变异；病害流行；新小种

* 通信作者：赵杰；E-mail：jiezhao@ nwsuaf. edu. cn
康振生；E-mail：kangzs@ nwsuaf. edu. cn

Host-induced gene silencing of an important pathogenicity factor PsCPK1 in *Puccinia striiformis* f. sp. *tritici* enhances resistance of wheat to stripe rust

Qi Tuo, Guo Jun*, Kang Zhensheng**

(*State Key Laboratory of Crop Stress Biology for Arid Areas and College of Plant Protection*, *Northwest A&F University*, *Yangling* 712100, *China*)

Abstract: Rust fungi are devastating plant pathogens and cause a large economic impact on wheat production worldwide. Using resistant wheat varieties is the most economical, effective, and environmentally friendly approach for controlling the pathogen. However, most race-specific host resistance genes have transient protection, probably due to the rapid evolution of new virulent rust fungal isolates. RNA interference is a prospective tool for control of pathogenic fungi that has been experimentally validated in a large number of studies. To overcome the rapidly loss of varieties resistance of *Pst*, we generated stable transgenic wheat plants expressing siRNAs targeting potentially vital genes of *Puccinia striiformis* f. sp. *tritici* (*Pst*). Protein kinase A (PKA) has been proved to play important roles in regulating the virulence of phytopathogenic fungi. *PsCPK*1, a PKA catalytic subunit gene from Pst, is highly induced at the early infection stage of *Pst*. The instantaneous silencing of *PsCPK*1 by barley stripe mosaic virus (BSMV) -mediated host-induced gene silencing (HIGS) results in a significant reduction in the length of infection hyphae and disease phenotype. These results indicate that PsCPK1 is an important pathogenicity factor by regulating *Pst* growth and development. Two transgenic lines (L12 and L18) expressing the RNAi construct in a normally susceptible wheat cultivar displayed high levels of stable and consistent resistance to *Pst* throughout the T_3 to T_4 generations. The presence of the interfering RNAs in transgenic wheat plants was confirmed by northern blotting, and these RNAs were found to efficiently down-regulate *PsCPK*1 expression in wheat. The present study addresses important aspects for the development of fungal-derived resistance through the expression of silencing constructs in host plants as a powerful strategy to control cereal rust diseases and contribute to environmentally friendly and sustainable agriculture.

Key words: wheat; stripe rust; PKA; HIGS; transgene; RNA interference

---

* Corresponding author: 康振生; E-mail: kangzs@ nwsuaf. edu. cn

郭军; E-mail: guojunwgq@ nwsuaf. edu. cn

花椒锈病症状类型差异比较及其鞘锈菌的群体遗传结构分析*

刘　林[1,2**]，浦仕献[1]，杨　静[2]，岂桂芝[1]，王　娜[1]，郭建伟[3]，李成云[1***]

（1. 云南生物资源保护与利用国家重点实验室，昆明　650091；2. 云南农业大学烟草学院，昆明　650091；3. 红河学院生命科学与技术学院，红河　661100）

摘　要：花椒具有重要的经济价值、药用价值和生态价值，而花椒锈病是为害花椒树的主要病害之一，常可引起花椒叶片变色、坏死、脱落，严重时导致花椒全树叶片落光，反复侵染，致使树势衰弱，甚至死亡，严重影响花椒的产量和质量。本研究通过对具不同花椒锈病症状的花椒锈菌进行孢子形态观察，以及对其 ITS 序列进行分析对比的结果表明：花椒锈病症状类型多样，在花椒锈病发病的初期，花椒叶片产生水渍状大小不等的一到多个病斑，随着侵染程度的加深，花椒锈病的病斑有的分布不规则，单独一个夏孢子堆形成一个病斑（野花椒锈病）；有的以一点为中心，其余的以此为中心，围成一圈（曲靖花椒锈病和绿花椒锈病）；还有的以一点为中心，外面围成两圈，到了发病的后期，位于外面一圈的孢子堆变为红褐色（红花椒锈病）。通过病原鉴定，采集自不同花椒上具不同症状的花椒锈病的病原菌均属于担子菌亚门（Basidio－mycotina）、冬孢菌纲（Teliomycetes）、锈菌目（Uredinales）、鞘锈菌科（Coleosporaceae）、鞘锈菌属（*Coleosporium* sp.）、花椒鞘锈菌（*Coleosporium zanthoxyli*）；利用 ISSR 分子标记分析了不同花椒上的鞘锈菌的群体遗传结构，以及花椒上的锈菌与石斛上的锈菌的群体遗传结构，结果表明：野花椒上锈菌的群体遗传多态性更丰富，其观测等位基因数 Na（1.629 6），有效等位基因数 Ne（1.246 5），基因多样性指数 H（0.157 7），Shannon's 信息指数 I（0.251 9）、多态性位点数（204）和多态性位点百分比（62.96%）均最大。

关键词：花椒锈病；鞘锈菌；群体遗传结构；分子标记

* 基金项目：国家自然科学基金（31560046）；国家重点研发计划项目（2018YFD0200500）

** 第一作者：刘林，博士，副教授，主要从事植物病理研究；E-mail：liulin6032@163.com

*** 通信作者：李成云，学士，研究员，主要从事植物病理研究；E-mail：li.chengyun@163.com

Foc 侵染巴西蕉根系的比较转录组学分析*

董红红**，李华平***
（华南农业大学农学院，广州 510642）

摘　要：香蕉枯萎病是由尖孢镰刀菌古巴专化型（*Fusarium oxysporum* f. sp. *cubense*，Foc）侵染引起的一种香蕉上的毁灭性病害，严重威胁着世界范围内香蕉的生产。然而，有关其不同生理小种之间的致病性差异及其与寄主互作的研究仍然是非常有限的。笔者假设在侵染过程中，巴西蕉根系对 Foc 的识别发生在侵染的早期阶段。在本研究中，笔者通过比较转录组学分析证实了 Foc 生理小种 1 号（Foc1）和生理小种 4 号（Foc4）侵染巴西蕉根系早期阶段的基因表达和通路的变化，分别在 Foc1 和 Foc4 接种后 48 h 的巴西蕉根部鉴定到了 1 862个和 226 个差异表达的基因。GO 和 KEGG 通路分析表明，Foc 侵染早期阶段能够引起黄酮类化合物和木质素合成途径增强，硫代葡萄糖苷类、萜类和原花青素积累，许多激素和受体激酶相关基因表达。因此，巴西蕉和 Foc 之间的早期互作在整个侵染过程中起着重要作用。此外，参与和生物应激有关途径的差异表达基因在 Foc1 和 Foc4 侵染后也明显不同。进一步的分析表明，参与细胞壁代谢和苯丙烷代谢的基因也差异表达。通过测定与病害防御相关的基因的表达模式，笔者发现一个注释为可能引起过敏性细胞死亡的基因在 Foc1 侵染后被上调，推断植物免疫应答可能发生在这个侵染阶段。进一步的基因功能分析正在进行中。

关键词：香蕉；香蕉枯萎病；互作；转录组学；差异表达

* 基金项目：现代农业产业技术体系建设专项（CARS-31-09）
** 第一作者：董红红，博士研究生，植物病理学；E-mail：1245240840@ qq. com
*** 通信作者：李华平，教授；E-mail：huaping@ scau. edu. cn

Ceratocystis fimbriata 全基因组中分泌蛋白的预测

张治萍*，郝志刚，李迎宾，罗来鑫，李健强**
（中国农业大学植物保护学院，种子病害检验与防控北京市重点实验室，北京 100193）

摘 要：甘薯长喙壳菌（*Ceratocystis fimbriata*）是一种寄主范围较广的植物病原真菌，可为害 14 个属 30 余种木本及草本植物，也是国内外重要的进境检疫性有害生物，主要引起寄主植物的枯萎、溃疡及根茎腐烂等症状，严重时导致死亡。植物病原真菌分泌蛋白可与植物受体蛋白相互作用，如已报道多种效应子或激发子蛋白等。研究分泌蛋白对明确植物与病原微生物互作的分子机制具有重要意义。目前，对于 *C. fimbriata* 的研究主要集中在生物学特性、种的鉴定以及遗传多样性分析等方面，鲜有对其分泌蛋白的报道。基于此，本研究采用生物信息学方法，以菌株 *C. fimbriata* CBS 114723 的基因组（登录号：GCA_000389695. 3）为参考，对可能存在的分泌蛋白进行了预测分析，结果如下：

依据 N 端含有信号肽、不含跨膜结构域、无 GPI（Glycosyl-phosphatidyl-inositol）锚定位点、富含半胱氨酸、小分子蛋白（≤400 aa）等 4 个特征，采用 SignalP-5. 0、TMHMM Serverv. 2. 0、Phobius、PredGPI 等软件对其全基因组的 7 266个蛋白序列进行了分析，得到 93 个符合条件的候选蛋白；利用 eggNOG 5. 0 和 EffectorP 等软件进一步筛选，最终获得 9 个假定的候选分泌蛋白，主要功能有分泌金属蛋白酶、果胶裂解酶、阿拉伯糖酶、坏死诱导蛋白等。

初步研究结果为进一步探究 *C. fimbriata* 与寄主互作的分子机制提供了参考，后续将对这几个分泌蛋白的功能进行深入研究。

关键词：*Ceratocystis fimbriata*；分泌蛋白；生物信息学；预测

* 第一作者：张治萍，在读博士研究生，主要从事种子病理学和作物病害生物防治研究；E-mail：m18788425134@163. com
** 通信作者：李健强，博士，博士生导师，主要从事种子病理及杀菌剂药理学研究；E-mail：lijq231@cau. edu. cn

苹果树腐烂病菌（*Valsa mali* var. *mali*）全基因组候选效应蛋白预测与分析

郝志刚*，李迎宾，罗来鑫，李健强**
（中国农业大学植物病理学系，种子病害检验与防控北京市重点实验室，北京 100193）

摘 要：苹果树腐烂病是一种典型的枝干病害，病原菌为 *Valsa mali* var. *mali*，主要引起树皮腐烂，严重时可致整树枯死。之前关于该病害体系的报道主要集中在病原菌鉴定、病害传播规律研究及病害综合防控等方面，而对于该菌效应蛋白相关研究尚未见报道。

本研究采用生物信息手段，基于已公布的苹果树腐烂病菌全基因组信息（03-8），以 N 端含有信号肽、不含有跨膜结构域、没有 GPI 锚定位点、富含半胱氨酸的小分子蛋白等 4 个特征为依据，对该病菌基因组的候选效应蛋白进行了预测。具体方法及结果为：在 Linux 系统环境中，①利用 SignalP-5.0 对苹果腐烂病菌数据库中的蛋白序列信号肽存在与否进行分析，发现共有 955 个蛋白具有信号肽序列；②通过 TMHMM Server v.2.0、Phobius 程序选取无跨膜结构和只有一个跨膜结构域并与信号肽区域高度重合的序列，两者取交集后共获得 720 条蛋白序列；③根据其序列大小（≤400aa）和半胱氨酸含量（Cys≥4）进一步筛选得到 245 条蛋白序列；④通过 PredGPI 程序进行预测不含 GPI-anchor 位点蛋白序列 210 个；⑤利用 eggNOG 5.0 蛋白质数据库进行基因功能注释，最终得到 138 条具有功能注释的基因；⑥利用 EffectorP 进一步筛选获得 14 个候选效应蛋白，编号为 VmSEP-N（N=1，2，3，…，14），其中的功能注释包括磷脂酶、肽还原酶、SCP 结构域等。

综上，本研究采用生物信息学分析方法对苹果树腐烂病菌全基因组的 11 284个蛋白序列进行了分析，预测得到 14 个符合条件的候选效应蛋白，这为进一步研究效应蛋白的功能和苹果腐烂病菌与寄主互作研究奠定基础。

关键词：苹果树腐烂病菌；全基因组；候选效应因子；生物信息学

* 第一作者：郝志刚，博士研究生，主要从事种子病理学和植物病原物与寄主互作研究；E-mail：haozhigang@cau.edu.cn
** 通信作者：李健强，博士，博士生导师，主要从事种子病理及杀菌剂药理学研究；E-mail：lijq231@cau.edu.cn

板栗疫病菌乙酰化蛋白质组的研究*

田始根[1,2]**，林榆淞[1,2]，陈保善[2]，李　茹[1,2]***

（1. 广西大学生命科学与技术学院，南宁　530004；2. 亚热带农业生物资源保护与利用国家重点实验室，南宁　530004）

摘　要：低毒病毒（Hypovirus）侵染可导致板栗疫病菌（*Cryphonectria parasitica*）表型发生改变并显著降低真菌的毒力，因而该系统被视为解析病原真菌致病机理的一个优秀模型。蛋白质的赖氨酸乙酰化修饰是一种普遍存在的、可逆的翻译后修饰方式，它广泛参与调节基因转录、新陈代谢、蛋白合成与降解、细胞周期及应激反应等多种生命活动。本研究以板栗疫病菌野生强毒株 EP155 和低毒病毒侵染的弱毒株 EP713 为实验材料，应用基于抗体免疫富集、高分辨率质谱鉴定和 Label-Free 的定量蛋白组学，获得了板栗疫病菌乙酰化蛋白图谱。结果显示，在板栗疫病菌中共鉴定到了 329 个乙酰化蛋白和 660 个乙酰化位点。其中 41 个乙酰化蛋白（83 个乙酰化位点）仅在 EP155 中被鉴定到；26 个乙酰化蛋白（90 个乙酰化位点）仅在 EP713 中被鉴定到。与 EP155 相比有 157 个蛋白的乙酰化水平在 EP713 中显著上调，125 个下调。此外，本研究通过同源双交换敲除基因的方法对差异乙酰化蛋白柠檬酸合成酶（CS）的功能进行了研究。初步结果表明，*cs* 基因缺失突变株菌落产生少量色素，生长速度缓慢，致病性显著下降，表明 cs 参与色素合成及致病性。本研究结果将为揭示乙酰化修饰与真菌毒力的关系奠定基础。

关键词：板栗疫病菌；低毒病毒；乙酰化

---

* 基金项目：国家自然科学地区基金（31760498）

** 第一作者：田始根，在读硕士研究生；E-mail：1440213310@ qq. com

*** 通信作者：李茹，研究员；E-mail：liruonly@ 163. com

山西省蠕孢类玉米叶斑病菌种类研究*

王　迪**，吉　佩，姜晓东，郭雪梅，郝晓娟，李新凤***，王建明***
（山西农业大学农学院，晋中　030801）

摘　要：山西省是我国玉米主产省份之一，玉米种植面积居全省之首。近年来，随着玉米种植面积的扩大、耕作方式及气候条件的改变，玉米病害已成为玉米生产的制约因子之一，由蠕孢类真菌引起的玉米叶斑类病害发生也日趋严重。

为明确山西省玉米叶斑病的发生情况，本研究于2017—2018年自山西省36个县市采集玉米叶斑病罹病叶片，用组织分离法分离病原菌，经致病性试验验证，并结合形态学和分子生物学方法确定病原菌的分类地位。共鉴定出6种蠕孢类玉米叶斑病菌，分别为玉米大斑凸脐蠕孢（*Exserohilum turcica*）、玉米生平脐蠕孢（*Bipolaris zeicola*）、玉蜀黍平脐蠕孢（*Bipolaris maydis*）、玉米平脐蠕孢（*Bipolaris zeae*）、新月弯孢（*Curvularia lunata*）和嘴突凸脐蠕孢（*Exserohilum rostratum*）。其中嘴突凸脐蠕孢（*Exserohilum rostratum*）玉米叶斑病在我国并未见有报道。该病主要为害玉米叶片，初期在叶片上出现近圆形或不规则形的黄色斑点，呈水浸状，随着病情发展，病斑颜色加深，呈黄褐色，并逐渐扩大，常连成片状，严重时整株叶片下垂、枯死。这些研究结果为进一步研究山西省玉米叶斑病的发生流行及其综合治理提供了可靠的理论依据。

关键词：玉米叶斑病；分离；鉴定；嘴突凸脐蠕孢菌

---

* 基金项目：山西省应用基础研究项目（201701D121100）

** 第一作者：王迪，硕士研究生，植物病理学专业；E-mail：1484568370@ qq. com
*** 通信作者：李新凤，副教授，主要从事植物真菌病害、植物病理生理及分子植物病理学；E-mail：lxf1309@ 163. com
王建明，教授，植物病害综合防治，植物病理生理及分子植物病理学；E-mail：jm. w@ 163. com

北京地区草莓疫霉根腐病病原鉴定*

孙 倩[1**]，张 玮[2]，王 琦[1]，燕继晔[2]，李兴红[2***]

（1. 中国农业大学植物保护学院，北京 100193；
2. 北京市农林科学院植物保护环境保护研究所，北京 100097）

摘 要：草莓为蔷薇科草莓属草本植物，具有较高的经济价值和营养价值。草莓根腐病是草莓生产中十分常见的一种病害，往往由多种病原物与环境互相作用引起，造成田间大面积死苗，带来毁灭性的经济损失。2019 年 1 月笔者在北京市昌平区草莓园区发现，部分草莓萎蔫倒伏，几天后死亡，横切根茎，可见深褐色变色。通过组织分离与纯化，分离得到 4 个菌株。形态学鉴定显示，4 个菌株在 PDA 培养基上气生菌丝少，边缘明显，未见菌丝膨大体。孢子囊顶生，近球形或卵形，基部圆形。孢子囊具有明显乳突，藏卵器球形，雄器侧生。结合分子生物学鉴定手段，利用核糖体 DNA 转录内间隔区（*ITS*）、细胞色素氧化酶亚基Ⅱ（*Cox*Ⅱ）、β 微管蛋白（*β-tubulin*）三个基因，以从 NCBI 网站获得的 65 个疫霉属标准菌株为参考，钟器腐霉（*Pythium vexans*）为外群，基于最大似然法构建多基因系统发育树。结果显示，获得的 4 个菌株均与 *Pytophthora cactorum* 聚为一枝，bootstrap 支持率为 98%，因此 4 个菌株鉴定为恶疫霉（*P. cactorum*）。通过致病性测定，最终确定北京地区草莓疫霉根腐病的病原菌为恶疫霉（*P. cactorum*）。

此前，北京地区由疫霉引起的草莓根腐病的报道较少，盛茹媛等于 2011—2012 年在北京周边区县分离得到了疫霉属，但并未明确到种。本研究明确了北京地区草莓疫霉根腐病是由恶疫霉（*P. cactorum*）引起的，为草莓疫霉根腐病的防控提供参考。

关键词：草莓；根腐病；恶疫霉；多基因系统发育树

---

* 基金项目：北京市科技计划（D171100001617002）

** 第一作者：孙倩，硕士研究生，主要从事草莓病害鉴定

*** 通信作者：李兴红，研究员，主要从事植物真菌病害鉴定与防控技术研究；E-mail：lixinghong1962@163.com

稻瘟菌钒氯代过氧化物酶的基因功能分析*

冷梅钦[1,2]，聂燕芳[1,3]，王振中[1,2]，李云锋[1,2]**

（1. 华南农业大学广东省微生物信号与作物病害重点实验室，广州 510642；2. 华南农业大学农学院，广州 510642；3. 华南农业大学材料与能源学院，广州 510642）

摘 要：由稻瘟菌（*Magnaporthe oryzae*）引起的稻瘟病是水稻生产上最重要的病害之一。前期研究中，笔者发现在稻瘟菌分生孢子附着孢形成期，钒氯代过氧化物酶（Vanadium chloroperoxidase，MoVAN）的表达量显著上调。以此为基础，笔者对 *MoVAN* 基因进行了克隆及测序，获得了大小为 2 080 bp 的 DNA 序列和 1 790 bp 的 cDNA 序列；该基因编码蛋白为 595 个氨基酸，属于 PAP2 家族中的 PAP2_haloperoxidase 型蛋白。采用同源重组策略，利用 *MoVAN* 两端侧翼序列设计特异性引物，构建了稻瘟菌 *MoVAN* 基因敲除载体；采用原生质体转化法，对 *MoVAN* 基因进行敲除；经过潮霉素抗性筛选、PCR 分析、Southern blot 鉴定，成功获得了 6 个 *MoVAN* 基因敲除突变株。对 Δ*Movan* 敲除突变体进行了表型分析，发现其在 PDA 培养基中生长速度减慢，气生菌丝稀疏，黑色素明显减少；此外，其产孢量减少，分生孢子萌发率降低、萌发形成的附着胞减少；但分生孢子和菌丝形态与野生型菌株没有明显区别。胁迫试验结果表明，Δ*Movan* 敲除突变体对 NaCl、山梨醇、SDS 和 H_2O_2 均表现敏感，菌落生长速率明显降低，黑色素合成明显减少。细胞壁完整性试验结果表明，Δ*MoVAN* 对刚果红表现敏感，菌落生长速率明显降低。采用离体和活体接种法分别进行致病性分析，结果表明 Δ*Movan* 敲除突变体的致病力显著降低。

关键词：稻瘟菌；钒氯代过氧化物酶；基因敲除；致病性分析

* 基金项目：国家自然科学基金（31671968）、广州市科技计划项目（201804010119）和广东省科技计划项目（2016A020210099）

** 通信作者：李云锋；E-mail：yunfengli@ scau. edu. cn

香蕉枯萎病菌分泌蛋白质的差异表达分析*

聂燕芳[1,2**]，周　淦[1,3]，李华平[1,3]，王振中[1,3]，李云锋[1,3]
（1. 华南农业大学广东省微生物信号与作物病害重点实验室，广州　510642；2. 华南农业大学材料与能源学院，广州　510642；3. 华南农业大学农学院，广州　510642）

摘　要：由尖孢镰刀菌古巴专化型（*Fusarium oxysporum* f. sp. *cubense*，Foc）引起的香蕉枯萎病是香蕉生产上最重要的病害之一。Foc 有 4 个生理小种，其中为害我国植蕉区的主要是 4 号生理小种（Foc4）。分泌蛋白作为一类重要的致病因子，在 Foc4 侵染香蕉过程中起着重要作用。

采用体外模拟植物与病原菌的互作条件，以 Foc4 为研究对象，用巴西蕉组织提取物进行诱导，采用非标记蛋白定量技术（Label-free）分析了其菌丝的分泌蛋白质表达变化，并结合生物信息学等方法对差异表达的分泌蛋白进行了功能预测分析。结果表明，采用 Label-free 的蛋白质组学技术共鉴定了 1 685个分泌蛋白。与对照相比，诱导条件下的 Foc4 有 66 个分泌蛋白质差异表达，其中 27 个上调表达，39 个下调表达。利用 SignalP4. 1、WOLF PSORT、TargetP 1. 1 Server 和 big-PIPredictor 等软件进行分析，发现经典分泌蛋白有 43 个，非经典分泌蛋白有 23 个。生物信息学分析结果表明，差异表达的分泌蛋白功能主要涉及细胞壁降解、蛋白修饰、氧化胁迫反应过程、氧化还原过程和能量代谢等过程。

关键词：香蕉枯萎病菌；尖孢镰刀菌古巴专化型；分泌蛋白质；非标记蛋白定量技术

* 基金项目：国家自然科学基金（31600663）；广东省科技计划项目（2016A020210098）

** 通信作者：聂燕芳；E-mail：yanfangnie@ scau. edu. cn

生姜茎基腐病原菌环介导等温扩增技术快速检测体系的建立*

王静琰[1,2]**，张　博[1]，祁　凯[1]，张悦丽[1]，马立国[1]，齐军山[1]，李长松[1]***

（1. 山东省农业科学院植物保护研究所，济南　250000；
2. 烟台大学生命科学院，烟台　264000）

摘　要：生姜茎基腐病是生姜主产区发生严重的病害之一，为典型的土传病害和种子传播病害，为害造成根部和茎基部腐烂，一般损失20%~30%，严重可达50%以上。经分离鉴定发现是由多种腐霉菌（*Pyhtium*）侵染引起，以群结腐霉（*P. myriotylum*）、林栖腐霉（ *P. sylvaticum*）为主。为了进一步研究病害的发生规律与预警，建立病菌快速检测技术十分必要，目前针对生姜茎基腐病菌常用的检测技术为普通PCR等分子生物学方法，存在对仪器设备及操作人员要求高且耗时长等缺点，不利于向基层技术部门推广。本研究所采用的环介导等温扩增（loop-mediated isothermalamplification，LAMP）技术，通过比对林栖腐霉与其他腐霉菌的基因序列，筛选有序列特异性的基因作为靶标来设计LAMP特异性引物，从而建立一种可通过颜色判定的检测体系，能在短时间内实现DNA的大量扩增，具有快速、准确、简便的特点。

该体系为25μL，扩增前在反应管管盖内侧加入SYBR Green I作为指示剂，在等温条件65 ℃下进行核酸扩增反应60 min，80℃ 10min灭活DNA Polymerase，将反应管离心1min后即可通过肉眼直接观察结果，SYBR Green I显示绿色即为阳性反应；同时，扩增产物经琼脂糖凝胶电泳验证为梯形条带，也可判定为阳性反应。该体系对DNA的最低检测限为100pg/μL。本研究建立了一种针对林栖腐霉的快速检测技术，为其他病原腐霉菌快速检测体系的建立和以腐霉菌侵染引起生姜茎基腐病的快速诊断提供了技术支撑和理论依据。

关键词：生姜茎基腐病；林栖腐霉；环介导等温扩增；快速检测技术

---

* 基金项目：国家重点研发计划（2017YFD02016005）；山东省重大科技创新工程（017CXGC0207-1）；公益性行业（农业）科研专项（201503112）；山东省自然科学基金（ZR2017YL013）

** 第一作者：王静琰，硕士研究生，主要从事腐霉快速检测技术研究；E-mail：1018817094@ qq. com

*** 通信作者：李长松，研究员，从事植物病害研究；E-mail：lics1011@ sina. com

谷瘟病菌生理小种的组成和分布*

王　璐[1,2]**，白　辉[1]，王永芳[1]，全建章[1]，董志平[1]，李志勇***，董　立[1]***
（1. 河北省农林科学院谷子研究所，石家庄　050031；2. 河北工程大学
园林与生态工程学院，邯郸　056006）

摘　要：谷瘟病是谷子生产中的重要病害之一，其中叶瘟和穗瘟发生最为普遍，造成的损失也最为严重。谷瘟病菌生理小种的研究始于 20 世纪 70 年代，由曹功懋等第一次报道了谷瘟病菌存在致病性分化，之后闫万元等于 80 年代建立了一套谷瘟病菌生理小种鉴别品系，将采集分离于 10 个省份共 711 个谷瘟病菌单孢菌株分为 7 个类群 32 个生理小种。近几年，由于谷子种植品种中抗病品种缺乏和谷瘟病菌生理小种的改变，谷瘟病发病有逐渐加重的趋势，而目前我国不同地区谷瘟优势生理小种和分布尚不清楚。

为了解我国谷瘟病菌不同生理小种的分布和组成，本研究将采自河北、河南、山东、山西等 10 个省份的 204 株谷瘟病菌菌株进行分离、接种鉴定和调查，通过闫万元等筛选的 6 个鉴别寄主，将我国 204 株谷瘟病菌的生理小种分成 7 个群（A、B、C、D、E、F、G）28 个小种。通过分析对比闫万元等 1985 年的生理小种鉴定结果，较闫万元等测定的小种组成增加了一些新的小种，如 A52（1.0）、B25（2.5）、C15（1.5）、C13（10.8）、C10（1.0）、D6（2.0），但同时 A62、A47、A41、B27、B24、B22、B21、E2 等小种未能在此次鉴定中检出。7 个种群出现频率最高为 C 群，出现频率为 40.20%，是本次研究的优势菌群；其次为 D 种群，出现频率为 20.60%；之后依次是 E 群、A 群、F 群、B 群和 G 群，出现频率分别为 10.29%、9.80%、8.82%、7.35%和 2.94%。在 28 个生理小种中出现频率最高的为 C17，出现频率为 22.05%，为优势小种；其次为 D7，出现频率为 13.72%；生理小种 C13 和 E3 的出现频率分别为 10.78%和 10.29%，其余生理小种的出现频率均小于 10.00%。谷瘟病菌生理小种地区分布结果显示不同地区谷瘟病菌生理小种组成具有差异性，但不同地区也存在共同的生理小种，其中有些生理小种分布范围十分广泛，如生理小种 C17 存在于所有 10 个谷子主要产区，生理小种 C13 存在于除陕西地区以外的其他 9 个谷子产区。此外，不同谷子产区的谷瘟病菌生理小种有着地区的特异性，但并不遵循严格的地理限制。

关键词：谷子；谷瘟；生理小种

---

* 基金项目：国家自然科学基金资助项目（31872880）；国家现代农业产业技术体系（CARS-07-13.5-A8）；河北省农林科学院创新工程（2019-4-02-03）

** 第一作者：王璐，在读研究生，主要从事谷子病害研究；E-mail：wanglu921009@163.com

*** 通信作者：李志勇，研究员，主要从事谷子病害研究；E-mail：lizhiyongds@126.com

董立，研究员，主要从事谷子病害研究；E-mail：guzisuodong@126.com

ITS-RFLP 分析花生白绢病原菌遗传多样性*

宋万朵，晏立英，康彦平，雷　永，淮东欣，王志慧，廖伯寿
（中国农业科学院油料作物研究所，油料作物生物学与遗传育种重点实验室，武汉　430062）

摘　要：花生白绢病（Southern stem rot）是花生上重要的枯萎性真菌病害，由齐整小核菌（*Sclerotium rolfsii* Sacc.）引起。该病害在我国各大花生产区都有发生，对花生生产造成了严重的危害和经济损失。RFLP（Restriction Fragment Length Polymorphism，限制性内切酶片段长度多态性）技术可用于在不同环境中微生物多样性的研究，本研究应用 ITS-RFLP 技术分析花生白绢病原菌遗传多样性，以期了解我国不同地区花生白绢病菌地理居群的遗传变异情况，为该病害的防治提供理论依据。

提取 39 个菌株的菌丝体总 DNA，使用 ITS1 和 ITS4 引物对 rDNA-ITS 区进行 PCR 扩增，用 4 种限制性内切酶对扩增产物进行酶切。结果显示，所有菌株均扩增出了一条约 750bp 大小的片段，用 *Alu* Ⅰ、*Rsa* Ⅰ酶切后，均产生了 2 种酶切表型，*Hpa* Ⅱ酶切产生了 3 种酶切表型，*Mbo* Ⅰ则酶切产生了 5 种酶切表型。对酶切结果进行聚类分析，这些菌株分为了 4 个类群，多数地理来源相同的菌株都聚在同一类群，但也有同一地理来源的菌株分布在不同的 2 个或 3 个类群。该研究结果说明我国不同地区间的白绢病原菌之间存在着关联与差异，为花生白绢病的科学防治提供了指导意义。

关键词：花生白绢病；*Sclerotium rolfsii*；ITS-RFLP；遗传多样性

* 基金项目：国家花生产业技术体系（CARS-14）；中国农业科学院创新工程（CAAS-ASTIP-2013-OCRI）

小麦叶锈菌分泌蛋白的预测*

张 悦**，齐 悦[1]，韦 杰[1]，李建嫄[1,2]，杨文香[1***]，刘大群[3 ***]
（1. 河北农业大学植物保护学院，河北省农作物病虫害生物防治工程技术研究中心，
国家北方山区农业工程技术研究中心，保定 071001；2. 邢台学院，邢台 054001；
3. 中国农业科学院研究生院，北京 100081）

摘 要：小麦叶锈病是由小麦叶锈菌（*Puccinia triticina* ，*Pt*）引起的一类破坏性强、循环侵染、专性寄生的重大真菌病害，严重威胁人类粮食安全并且造成重大产量损失。植物病原物在与寄主互作过程种会产生效应蛋白，它们通过“伪装”躲避宿主的防御反应，或者直接杀死宿主细胞，或者通过干扰、抑制寄主的防卫反应，引发植物感病。因此理解和鉴定效应蛋白对于发展作物抗病机制至关重要。本课题利用RNA-Seq技术对3个小麦叶锈菌单胞菌系08-5-9-2（KHTT）、13-5-28-1（JHKT）和13-5-72-1（THSN）进行了测序，并利用多个计算机软件对获得的CDS序列做了初步分析。

根据3个单胞菌系的测序结果，共得到155 873个CDS，将其进行比对发现39 741个与小麦叶锈菌数据库BBBD同源性较高的序列。利用信号肽预测软件SignalP v4.1对这39 741个CDS序列进行分析，发现其中含有编码信号肽的基因有5 279个，进一步利用TargetP v1.1分析基因的亚细胞定位，发现有4 317个含有分泌途径的信号肽，189个定位在线粒体“M”且RC值范围为1~3。将这4 506个序列通过TMHMM v2.0分析其是否含有跨膜结构域，发现3 116个序列不含有跨膜结构域。最后通过EffectorP v1.0和v2.0两个版本进一步进行排除，发现636个基因被预测为“effector”。通过上述4个软件进行预测，最终预测到636个候选效应蛋白。

随机选取其中的60个基因在本氏烟草上进行瞬时表达分析，发现其中53个基因能够有效抑制BAX诱导产生的坏死反应，抑制率达到88%。对其中的10个基因利用酵母转化酶缺陷型菌株YTK12中的SUC2系统进行了信号肽分泌功能验证，发现其中8个基因的信号肽能够引导蛋白分泌，为分泌蛋白。虽然用该方法从测序结果中仅获得636个候选效应蛋白，但准确率高，对于今后的研究更具针对性。当然严苛的筛选条件可能会使某些特殊的效应蛋白被忽略，但是随着越来越多的病原菌基因组被测序完成，获得的基因数量越来越大，如何在庞大的数据库中获得效应蛋白是非常重要的环节，利用这些计算机软件对效应蛋白进行预测，则可以在很短的时间内发现大部分的效应蛋白，同时利用异源表达系统对专性寄生菌效应蛋白进行筛选，使得筛选效率高，不失为高效的效应蛋白筛选工具。研究结果对揭示植物病原体相互作用的生物多样性的机制具有重要意义，并为发现利用效应生物学进行病害控制的新方法奠定基础。

关键词：小麦叶锈菌；效应蛋白；生物信息学；预测软件；异源系统；BAX

---

* 基金项目：国家自然科学基金项目（301571956；301871915）

** 第一作者：张悦，在读硕士研究生，主要从事分子植物病理学研究；E-mail：YueZhangND2018@163.com

*** 通信作者：杨文香，教授，主要从事小麦叶锈菌致病机制研究；E-mail：wenxiangyang2003@163.com
刘大群，教授，主要从事生物防治与分子植物病理学研究；E-mail：ldq@hebau.edu.cn

禾谷炭疽菌 Pth11 GPCR 预测及生物信息学分析*

刘宏莉**，祝友朋，韩长志***

（西南林业大学，云南省森林灾害预警与控制重点实验室，昆明　650224）

摘　要：目前，炭疽菌属真菌约包括 600 个种，可以侵染 3 200多种单了叶植物和双子叶植物。作为该属中重要的病菌——禾谷炭疽菌 *Colletotrichum graminicola*（Cesati）Wilson，其可以侵染玉米、小麦、高粱等禾本科农作物引起炭疽病，给各国农业生产造成巨大的经济损失。自 20 世纪 70 年代以来，由该病菌引起的玉米炭疽病在美国、印度等国家非常普遍。2012 年，随着禾谷炭疽菌全基因组序列的公布，人们对该炭疽菌开展了诸多方面的研究工作，主要涉及该菌的 MAPK 途径蛋白预测及其生物信息学分析，以及致病基因鉴定、功能解析、分泌蛋白预测及 RGS、14-3-3 蛋白、磷酸二酯酶、AC、PITP 等 G 蛋白信号途径相关蛋白的生物信息学分析等。G 蛋白偶联受体蛋白（GPCR）一直是学术界基础研究的热点和重要的药物靶点，以 GPCR 作为靶点的处方药物占据全世界药物市场的 50%。前人对诸如酿酒酵母、裂殖酵母、粗糙脉孢霉、构巢曲霉、新型隐球酵母等模式真菌以及大豆疫霉、橡树疫霉、致病疫霉等卵菌 GPCR 的研究开展了较多工作，然而，尚未见有关禾谷炭疽菌中 Pth11 GPCR 预测及理化性质分析的研究报道。因此，本研究通过关键词搜索及 Blastp 比对获得该菌中具有的 20 条 Pth11 GPCR 蛋白，并以此序列为基础，采用 SignalP v4. 0、ProtCompB、TMHMM、Phobius、SMART 等在线分析程序对氨基酸序列进行预测分析，对该蛋白的理化性质、二级结构、疏水性、转运肽及保守结构域等性质进行分析，以期为深入解析该病菌侵染机制的研究打下坚实的理论基础。

关键词：禾谷炭疽菌；Pth11 GPCR；跨膜区；信号肽；生物信息学

---

* 基金项目：云南省大学生创新创业训练计划项目（项目编号：S201710677013）

** 第一作者：刘宏莉，本科生；E-mail：2685989360@ qq. com

*** 通信作者：韩长志，博士，副教授，研究方向：经济林木病害生物防治与真菌分子生物学；E-mail：hanchangzhi2010@ 163. com

禾谷镰刀菌 Cdc2A 有性生殖阶段特异性磷酸化底物的鉴定与功能验证*

栾巧巧[1**]，宋真真[1]，江　聪[1]，许金荣[2]，刘慧泉[1***]
（1. 西农-普度大学联合研究中心，旱区作物逆境生物学国家重点实验室，西北农林科技大学植物保护学院，杨凌　712100；2. 美国普渡大学植物及植物病理系，印第安纳州　IN47907）

摘　要：由禾谷镰刀菌（*Fusarium graminearum*）引起的小麦赤霉病是一种影响小麦产量和品质的穗部真菌病害。禾谷镰刀菌通过有性生殖产生的子囊孢子是病害发生的初侵染源。有性发育过程受细胞周期调控，细胞周期蛋白依赖性激酶 Cdc2 是细胞周期调控中最重要的因子。大多数真菌只编码一个 *CDC*2 基因，实验室前期研究发现禾谷镰刀菌编码两个 *CDC*2 基因（*CDC2A* 和 *CDC2B*），只有 *CDC2A* 特异性调控有性生殖。为了明确 *CDC2A* 特异性调控有性生殖的机制，本研究通过磷酸化蛋白质组学分析筛选到 40 个在 *cdc2A* 突变体有性生殖阶段相比 *cdc2B* 突变体磷酸化水平下降的蛋白，包括前期研究发现的子囊和子囊孢子发育的重要基因 *PUK*1 和 *AMD*1。通过基因敲除对部分基因进行了功能验证，研究发现 *FgATG*20 缺失后突变体有性生殖产生的子囊壳数量减少，部分子囊中子囊孢子数目不足 8 个，且子囊孢子发育后期隔膜处会出现缢裂现象。目前正在通过磷酸化抗体杂交实验验证 Amd1、FgAtg20 和 Puk1 是否为 Cdc2A 磷酸化底物，并对鉴定到的磷酸化位点进行突变以验证其功能。

关键词：小麦赤霉病；禾谷镰孢菌；有性生殖；CDC2；磷酸化底物

---

* 基金项目：国家自然科学基金面上项目（3167110907）

** 第一作者：栾巧巧，硕士研究生，主要从事禾谷镰刀菌分子生物学研究
*** 通信作者：刘慧泉，研究员

稻瘟病菌两个 MAX 类效应蛋白的重组表达和纯化

赵　鹤，刘　洋，张　鑫，马梦琪，彭友良，刘俊峰
（中国农业大学植物保护学院植物病理学系，北京　100193）

摘　要：稻瘟病是由稻瘟病菌（*Magnaporthe oryzae*）侵染引起的在栽培水稻中最严重的病害之一，对全球粮食安全构成严重威胁。在稻瘟菌侵染过程中，水稻中的抗病蛋白 R 蛋白直接或间接识别稻瘟菌分泌的效应蛋白，从而激发寄主防卫反应。目前研究发现水稻中存在可以被抗病蛋白识别的一类序列一致性较低但结构上由六股方向平行的 β-片层排列成三明治型的效应蛋白，它们统称为 MAX 类效应蛋白。解析稻瘟菌中两个新的 MAX 类效应蛋白的结构，将为研究 MAX 类效应蛋白的结构及作用机制，设计识别更广谱持久抗性的免疫受体提供基础。

本研究利用大肠杆菌原核表达系统，将两个编码 MAX 类效应蛋白的全长基因分别构建到不同的表达载体，并将构建好的载体转化到大肠杆菌 Bl21（DE3）、Rosetta-gami（DE3）等不同的表达菌株中，经 IPTG 诱导目的蛋白的表达来筛选到合适的表达体系。通过亲和层析、离子交换层析和凝胶过滤层析等技术纯化符合晶体生长要求的蛋白样品。并利用气相扩散坐滴法对纯化好的蛋白样品进行大量晶体生长条件的筛选，以期获得衍射数据较好的蛋白晶体，进而为解析其结构和分析其与抗病蛋白的互作机制提供结构学数据和基础。

关键词：MAX 类效应蛋白；原核表达；蛋白纯化

云南东部地区玉米锈病类型的调查及其锈菌的群体遗传结构分析*

刘　林[1,2**]，浦仕献[1]，王　娜[2]，李玉林[1]，杨　静[1]，郭建伟[3]，赵　婧[1]，李成云[1***]

（1. 云南生物资源保护与利用国家重点实验室，昆明　650091；2. 云南农业大学烟草学院，昆明　650091；3. 红河学院生命科学与技术学院，红河　661100）

摘　要：玉米是世界三大粮食作物之一，也是我国重要的粮食、经济和饲料作物。玉米锈病是玉米上常见的病害之一，根据引发玉米锈病的症状类型和病原菌不同可以把玉米锈病分为：玉米普通型锈病（病原为：*Puccinia sorghi* Schw）、玉米南方型锈病（病原为 *Puccinia polysora* Uned-rw）、玉米热带型锈病［病原为 *Physoplla zeae*（mains） Cummins and Pamaxhar］和玉米秆锈病（病原为 *Puccinia giaminis* Pers）4 种。本文通过症状观察、病原形态鉴定、ITS 序列、玉米多堆柄锈菌与玉米柄锈菌两种不同锈菌的特异分子标记分析采集自云南东部地区的 270 份玉米锈病样品，结果表明采集自云南曲靖、昭通、昆明、玉溪和红河弥勒、开远、建水等地区采集的 180 份玉米锈病均为由玉米柄锈菌引起的玉米普通锈病，红河屏边和文山两个地区采集的 90 份玉米锈病均为由玉米多堆柄锈菌侵染引起的南方型锈病；而且通过分析和比较两种锈菌的夏孢子形态和长宽比发现，玉米多堆柄锈菌的夏孢子多为椭圆形，长宽比>1.2 以上的夏孢子占总孢子数 81.6%，玉米柄锈菌的夏孢子多近圆形，其长宽比在 1.0~1.3。另外，本文利用 ISSR 分子标记分析了锈菌的群体遗传结构。结果表明：采集自文山地区的玉米多堆锈菌遗传多样性较红河丰富；采集自昭通的玉米柄锈菌的群体遗传多样性最丰富，采集自昆明的玉米柄锈菌群体遗传多样性最低。

关键词：玉米多堆柄锈菌；玉米柄锈菌；病原鉴定；群体遗传结构

---

* 基金项目：国家自然科学基金（31560046）；国家重点研发计划项目（2018YFD0200500）

** 第一作者：刘林，副教授，主要从事植物病理研究；E-mail：liulin6032@ 163. com

*** 通信作者：李成云，研究员，主要从事植物病理研究；E-mail：li. chengyun@ 163. com

海南菠萝可可毛色二孢（*Lasiodiplodia theobromae*）叶斑病病原菌的分离与鉴定以及多基因序列比较分析*

唐中发**，秦春秀，林春花，郑服丛，缪卫国***，刘文波***
（海南大学植物保护学院，热带农林生物灾害绿色防控教育部重点实验室，海口　570228）

摘　要：为明确菠萝可可毛色二孢叶斑病病原菌在海南省16个市县的亲缘关系及遗传差异。从海南省的海口、澄迈、儋州、三亚、保亭等18个市县进行病样采集和病样分离，根据科赫氏法则鉴定，获得18株致病菌株，观察形态特征，并基于多基因联合序列分析其遗传多样性。结果表明，通过形态学鉴定和ITS-TUB2基因序列联合进化树分析，其中来自海口、澄迈、儋州、三亚、保亭等16个市县的16株病原菌均鉴定为可可毛色二孢（*Lasiodiplodia theobromae*），分生孢子平均大小为：（22.06~31.07）μm×（11.77~16.48）μm；来自乐东的1株为假可可毛色二孢属（*L. pseudotheobromae*），其分生孢子平均大小为：（21.52~29.77）μm×（12.04~15.42）μm；来自临高的1株为画眉草弯孢（*Curvularia eragrostidis*），其分生孢子平均大小为：（17.16~26.77）μm×（8.44~16.32）μm；对16株可可毛色二孢菌株（*L. theobromae*）进行ITS-TUB2-EF-1α-GAPDH-CHS-1-ACT基因拼接序列聚类分析，结果分为3个类群，海南岛的中部（儋州、昌江、白沙、五指山、万宁、琼海等地）聚为一个类群、中北部（屯昌、临高）聚为一个类群、西南部（东方、乐东、三亚等地）聚为一个类群。该结果说明来源于不同产地不同菠萝品种上的可可毛色二孢遗传多样性丰富。

关键词：菠萝；可可毛色二孢；叶斑病；多基因序列分析法

* 基金项目：海南省重大科技项目（ZDKJ201817-23）；国家自然科学基金（31560495，31760499）；国家重点研发计划（2018YFD0201105）；海南大学青年基金项目（hdkyxj201708）

** 第一作者：唐中发，在读硕士，研究方向：热带植物病理研究；E-mail：852375085@ qq. com

*** 通信作者：刘文波，硕士，副教授，研究方向：植物病理学；E-mail：saucher@ hainanu. edu. cn
缪卫国，博士，教授，研究方向：分子植物病理学；E-mail：weiguomiao1105@ 126. com

橡胶树白粉菌启动子 WY172 的克隆及功能鉴定*

殷金瑶**，王　义，朱　利，王　晨，刘文波，林春花，缪卫国***

（海南大学植物保护学院，海南大学热带农林生物灾害绿色
防控教育部重点实验室，海口　570228）

摘　要：启动子驱使外源基因在受体中启动转录是外源基因能够表达的必要条件，其通过参与基因的转录环节来调控基因表达的水平、部位及方式。其中，诱导型启动子可以实现对目的基因的精细、可控表达，避免了对植物体过量的营养消耗，在实际生产应用中显示出了极大地开发潜力。通常，大多数丝状真菌在表达外源蛋白的量比较低，为实现外源基因的高效表达，就需要强大的启动子来驱动基因的转录，但丝状真菌启动子目前研究、应用都比较少，主要是木霉属和黑曲霉属的一些启动子。本研究基于实验室研究基础，通过对专性寄生真菌橡胶树白粉菌全基因组进行预测及与经典启动子 35S、Actin1 比对，得到了一个诱导型启动子 WY172，它在单子叶植物和双子叶植物中均可以高度表达。本研究中通过将 WY172 替换 pBI121 中的 35S 启动子，构建植物表达载体 pBI121-WY172，通过三亲杂交法转化农杆菌菌株，通过 ATMT 实现 WY172 驱动 *GUS* 基因在三生烟中的稳定表达，获得转基因烟草植株。对 T1 代转基因烟草植株进行 IAA 处理，后对整株进行 GUS 染色，结果显示，IAA 处理后的转基因烟草植株显示出蓝色的阳性结果，而未经 IAA 处理的转基因烟草植株无蓝色，因此预测其为 IAA 诱导型启动子。本研究不仅可以促进橡胶树白粉菌分子致病机理的研究，也为转基因工程提供了新的工具和方法。

关键词：橡胶树白粉菌；诱导型启动子；ATMT；PCR-Southern；GUS 染色

* 基金项目：国家自然科学基金（31660033，31560495，31760499）；国家重点研发计划（2018YFD0201105）；海南省重点研发计划项目（ZDYF2018240）

** 第一作者：殷金瑶，在读硕士，研究方向：热带植物病理研究；E-mail：1050376472 @ qq. com

*** 通信作者：缪卫国，博士，教授，研究方向：分子植物病理学；E-mail：weiguomiao1105@ 126. com

橡胶白粉菌内源启动子 WY193 的克隆、功能鉴定及应用初探*

朱　利**，王　义，殷金瑶，王　晨，林春花，刘文波，缪卫国***
（海南大学植物保护学院，海南大学热带农林生物灾害绿色
防控教育部重点实验室，海口　570228）

摘　要：启动子是调控基因表达重要的顺式作用元件，是分子生物学研究的热点。本研究中从橡胶白粉菌基因组当中预测得到一个疑似启动子，并命名为 WY193，通过农杆菌介导法，将 WY193 启动子整合到模式生物三生烟（*Nicotianatabacum* var. *samsun* NN），获得转基因烟草，并进行转基因烟草植株瞬时表达与稳定表达。结果表明，瞬时表达证明了 WY193 的启动子功能，RT-PCR 结果显示 WY193 的瞬时表达量比经典启动子 CaMV35S 高得多，表明 WY193 是强启动子。通过分子手段进行验证，PCR 结果，Southern blot 检测结果均证明 WY193 启动子已经整合到受体基因组中。取 T1 代的转 WY193 基因烟草进行 GUS 染色，没有任何组织或器官染色呈阳性，说明 WY193 为诱导型启动子。经 PlantCare 分析，WY193 除了启动子基序外，还含有参与 ME-JA 反应、光反应、热应激反应、低温反应、水杨酸反应，以及胚乳表达所需顺式调节元件，后续实验将进行验证具体是哪种诱导类型启动子。通过农杆菌介导法，将 WY193 启动子基因整合到单子叶植物水稻叶片、椰子叶片和双子叶植物中火龙果基因组中，进行瞬时表达，染色结果表明，WY193 不仅能在单子叶植物水稻叶片、椰子叶片中表达，也能在双子叶植物火龙果中表达，其表达范围广，效率之高，不失为一个优秀的启动子，其应用前景广阔。

关键词：启动子；WY193 启动子；CaMV35S 启动子；诱导型启动子；农杆菌介导法；瞬时表达；转基因植株

* 基金项目：国家自然科学基金（31660033，31560495，31760499）；国家重点研发计划（2018YFD0201105）；海南省重点研发计划项目（ZDYF2018240）

** 第一作者：朱利，在读硕士，研究方向：热带植物病理研究；E-mail：1050376472 @ qq. com

*** 通信作者：缪卫国，博士，教授，研究方向：分子植物病理学；E-mail：weiguomiao1105@ 126. com

橡胶树炭疽菌 *Colletotrichum siamense* 疏水蛋白 Hydr 基因的克隆与原核表达[*]

王记圆[**]，方思齐，廖小淼，何其光，李　潇，
刘文波，张　宇，林春花[***]，缪卫国[***]
（海南大学植物保护学院，海南省热带生物资源可持续利用国家重点
实验室培育基地，海口　570228）

摘　要：疏水蛋白属于丝状真菌所特有的一类小分子分泌型蛋白，对丝状真菌的分生孢子生成、分布和附着以及病原菌的致病性有一定的影响。项目组前期通过酵母双杂技术，以脂滴包被蛋白 CsCap20 为诱饵调取获得一疏水蛋白 Hydr，为了进一步研究 Hydr 在橡胶树炭疽菌 *Colletotrichum siamense* 中的功能，本研究利用同源克隆法克隆获得橡胶树炭疽菌的疏水蛋白 Hydr 基因，并利用原核表达系统对其进行了表达。结果表明：*Hydr* 基因大小为 417nt，编码 139 个氨基酸，在蛋白的 N 端含有由 19 个氨基酸组成的一段信号肽；分别构建了含信号肽的 Hydr-His 和缺失信号肽的 Hydr（-19aa）-His 的融合表达载体，利用 SDS-PAGE 和 Western Blot 检测蛋白表达情况，表明该蛋白在去除信号肽的情况下在 16℃、0.4mmol/L、0.6mmol/L、0.8mmol/L、1.0 mmol/L IPTG 诱导可成功表达，但未去除信号肽时，在菌体上清和包涵体中均未获得融合蛋白。该研究结果为进一步验证疏水蛋白与脂滴包被蛋白的互作，开展该疏水蛋白在橡胶树炭疽菌的功能分析奠定了基础。

关键词：橡胶树炭疽菌；Hydr 蛋白；原核表达；信号肽

---

* 基金项目：国家自然科学基金（31560495，31760499）；国家重点研发计划（2018YFD0201105）；海南省重点研发计划项目（ZDYF2018240）；现代农业产业技术体系建设专项资金项目（CARS-33-GW-BC1）

** 第一作者：王记圆，在读硕士，研究方向：热带植物病理研究；E-mail：865683591@qq.com

*** 通信作者：林春花，博士，副教授，研究方向：分子植物病理学；E-mail：lin3286320@126.com
缪卫国，博士，教授，研究方向：分子植物病理学；E-mail：weiguomiao1105@126.com

美国进境高羊茅种子携带真菌的分离与鉴定*

高慧鸽[1,2**]，张博瑞[1]，高文娜[2]，刘西莉[1***]

（1. 中国农业大学植物病理学系，北京　1001093；2. 北京检验检疫局，北京　101300）

摘　要：牧草草坪草种子是我国口岸进口量较大的几类重要草种之一，其中高羊茅（*Festuca arundinacea*）是我国目前使用量最大的冷季型草坪草之一，其适宜于在我国华北、华东、华中、西北中南部以及西南高海拔较凉爽地区等多地区种植，使用范围广，可用于家庭花园、公共绿地、公园、足球场以及高尔夫球场等，因此其具有极高的生态价值和商业价值。

进境高羊茅种子携带病原物不仅会造成严重的经济损失，还直接影响我国的生态环境安全，是我国口岸检疫的重要对象，目前关于进境高羊茅种子携带真菌的研究鲜有报道。因此，本研究使用美国进境高羊茅种子为材料，分别通过PDA培养法和洗涤法分离了其种子内部和外部携带的真菌，并利用ITS1和ITS4引物对其转录间隔区进行了扩增、测序和BLAST比对，结合形态学观察，对分离获得的真菌进行了初步鉴定。

结果表明，从2 400粒高羊茅种子中共检测到7个属112株真菌，其中高羊茅种子内部带菌率为8.25%，主要携带 *Alternaria* sp.（83.33%），*Arthrinium* sp.（7.58%），*Clonostachy rosea*（4.55%），*Stemphylium* sp.（1.52%），*Bionectria* sp.（1.52%），*Gibellulopsis nigrescens*（1.52%）；高羊茅种子外部带菌率为2.88%，主要携带 *Alternaria* sp.（91.30%），*Periconia* sp.（6.52%），*Gibellulopsis nigrescens*（2.17%）。

上述研究结果为明确美国进境高羊茅种子的带菌情况及带菌种类提供了参考，为高羊茅种子的检疫提供了依据。

关键词：高羊茅；种子带菌；检疫

* 基金项目：国家重点研发计划（2017YFD0201602）

** 第一作者：高慧鸽，在读硕士研究生；E-mail：jiayoulalalahei@163.com

*** 通信作者：刘西莉，教授，主要从事植物病原卵菌与杀菌剂互作；E-mail：seedling@cau.edu.cn

小麦散黑粉菌冬孢子储藏条件筛选研究*

钟　珊**，王治文，高翔，刘西莉***
（中国农业大学植物保护学院植物病理学系，北京　100193）

摘　要：小麦是全球三大粮食作物之一，由小麦散黑粉菌（*Ustilago tritici*）引起的小麦散黑穗病发生普遍，严重影响小麦的产量和品质。该菌主要为害小麦穗部，在病穗内部产生大量黑色粉末即病原菌的冬孢子堆，冬孢子对于小麦散黑穗病的病害循环具有重要作用，因此抑制冬孢子的萌发是防治小麦散黑穗病的可行途径。目前对于该病的防治主要以戊唑醇和苯醚甲环唑等进行种子处理，缺乏对冬孢子萌发具有良好抑制活性的药剂，而且由于小麦散黑粉菌冬孢子在离体条件下活力易于丧失，因此如何保持小麦散黑粉菌冬孢子活力是药剂筛选中的重要前提。

本实验室利用采集自河北保定的小麦散黑粉菌的冬孢子粉作为研究材料，通过比较不同储藏温度和时间周期对冬孢子萌发的影响，以期获得良好的冬孢子储藏条件。其中，储存温度分别设置为-20℃、4℃、12℃和25℃，并于7d、14d、21d、30d、45d、60d、75d和90d后测定冬孢子的萌发率。结果表明4℃处理7d后冬孢子的萌发率为69.2%，为冬孢子短期储藏的最适条件；-20℃处理90d后冬孢子萌发率仍维持50.0%左右，为冬孢子长期储藏的最适温度；而12℃和25℃处理7d后，冬孢子几乎不萌发，表明高温不利于冬孢子的萌发。本研究结果可为小麦散黑穗病的有效防治和杀菌剂筛选提供理论参考。

关键词：小麦散黑穗病；冬孢子；萌发条件

* 基金项目：国家重点研发计划（2017YFD0201602）

** 第一作者：钟珊，博士研究生，植物病理学专业；E-mail：zhongshan@cau.edu.com
*** 通信作者：刘西莉，教授，主要从事杀菌剂药理学及病原物抗药性研究；E-mail：seedling@cau.edu.cn

杧果炭疽病菌漆酶基因 *Cglac*3 序列特征和敲除突变体获得*

钟昌开**，肖春丽[1]，张 贺[2]，蒲金基[2]，刘晓妹[1]***

（1. 海南大学植物保护学院，热带农林生物灾害绿色防控教育部重点实验室，海口 570228；2. 中国热带农业科学院环境与植物保护研究所，农业部热带作物有害生物综合治理重点实验室，海口 571101）

摘 要：杧果产业已成为中国热区农业支柱性产业之一。由胶孢炭疽菌（*Colletotrichum gloeosporioides* Penz. & Sacc.）引起的杧果炭疽病是影响该产业健康发展的主要因素之一。该菌主要靠芽管顶端产生的黑色素化的附着胞直接穿透寄主表皮侵入为害，DHN 黑色素是四大类黑色素之一，其合成途径的最后一步需要被漆酶氧化才能形成。漆酶是一类含铜的多酚氧化酶，与真菌的生长发育、分生孢子形成、黑色素合成、致病性等相关，常以多基因家族形式存在，漆酶蛋白序列并不具有较强的保守性，但都含铜离子结合保守结构域，即便来源同一病原菌漆酶家族中的不同漆酶基因，功能差异也较大。

本课题组通过在橡胶树胶孢炭疽菌 HBCg01 全基因组参考数据库搜索发现了 12 个漆酶基因。在杧果炭疽病菌中同源克隆后，研究表明漆酶基因 *LAC*1 主要调控菌丝的生长、发育、分化、黑色素的沉着，分生孢子、漆酶和纤维素酶的产生，对环境的适应性和对寄主的致病力；漆酶基因 *LAC*2 在接种后 6 h 表达量最低，9 h 最高，可能在杧果炭疽病菌侵染寄主的过程中发挥着重要作用。

本文同源克隆获得了漆酶基因 *Cglac*3，DNA 全长 1 842bp，含有 3 个内含子，开放读码框大小为 1 677 bp，编码 558 个氨基酸，分子量 61. 38 ku，等电点 4. 50，是一种碱性蛋白质，富含极性氨基酸，含铜离子结合保守结构域，不存在跨膜结构，存在 1 个信号肽，属于分泌型蛋白；二级结构中 α-螺旋占 12. 72%，延伸链占 27. 78%，β-转角占 7. 53%，无规则卷曲占 51. 97%，与草莓炭疽病菌（*C. gloeosporioides*）登录号为 XP _ 007273628. 1 的漆酶序列的相似性为 98%；qRT-PCR 分析显示在接种叶片后 12h 时表达量最高，24h 时最低，72h 内的平均相对表达量为 9. 80，可见与侵染过程密切相关；通过用绿色荧光蛋白-潮霉素 B 抗性基因（*gfp*：：*hygB*）替换 *Cglac*3 编码区序列，成功获得了 *Cglac*3 基因的敲除载体 pCglac3GH8；将其用 PEG 介导法转化的杧果炭疽病菌原生质体，经过含潮霉素平板筛选、PCR 鉴定及实时荧光定量分析，初步确定获得了 1 个敲除突变体 *ΔCglac3GH*，为后续测定表型、确定 *Cglac*3 的功能打下了重要的材料基础。

关键词：杧果炭疽病菌；漆酶基因 *Cglac*3；克隆；敲除

---

* 基金项目：国家自然科学基金（31860479）

** 第一作者：钟昌开，硕士研究生，研究方向：热带果树病理学；E-mail：913633622@ qq. com

*** 通信作者：刘晓妹

2018—2019 年江苏省稻瘟病菌种群结构特征分析*

齐中强**，杜　艳，于俊杰，张荣胜，俞咪娜，曹慧娟，刘永锋***

（江苏省农业科学院植物保护研究所，南京　210014）

摘　要：稻瘟病菌（*Magnaporthe oryzae*）引起的稻瘟病是水稻生产上一种最具破坏性的真菌病害，严重威胁我国的水稻生产安全；种植抗病品种是稻瘟病防控中最经济有效的措施，但是田间稻瘟病菌群体较为复杂，严重制约着抗病品种的科学布局。为了明确江苏稻区稻瘟病菌群体结构特征，分别从江苏省 9 个地市的监测点采集稻瘟病标样，实验室单孢分离菌株 160 株（2018 年）、224 株（2019 年），利用 Pot2-rep-PCR 对上述菌株进行了 PCR 检测，分析产生的指纹图谱显示群体遗传的结构类型及其多样性。结果发现所有菌株共扩增处 11 条分子量不等的 DNA 片段，单菌株可以产生 0~11 条扩增片段，大小在 250~ 5 000 bp。根据 Pot2-rep-PCR 扩增结果，结果完全相同的菌株归为 1 个单元型，上述菌株获得不同的单元型，利用 Ntsys 软件对指纹图谱进行聚类分析，以彼此间遗传距离小于 75%为界，将 2018 年和 2019 年菌株分别比较，2018 年菌株共划分为 15 个遗传谱系，其中优势谱系所占比例为 31. 67%；2019 年菌株共分为 14 个遗传谱系，其中优势谱系所占比例为 29. 82%。进一步分析发现江苏北部地区（连云港、盐城、徐州、淮安）稻瘟病菌划分的遗传谱系最多，平均为 5 个（2018 年）和 5. 25 个（2019 年），江苏中部地区（南通、扬州）稻瘟病菌划分的遗传谱系居中，平均为 3. 5 个（2018 年）和 4. 5 个（2019 年），江苏南部地区（南京、常州、苏州）稻瘟病菌划分的遗传谱系最少，平均为 2. 7 个（2018 年）和 3. 7 个（2019 年），表明稻瘟病菌遗传谱系受到地理环境和气候影响较大。综上所述，2018 年和 2019 年江苏省稻瘟病菌种群遗传谱系变化不大，但是遗传谱系与地域差异密切相关。

关键词：稻瘟病菌；遗传谱系；Pot2

---

* 基金项目：国家自然科学基金面上项目（31871921）

** 第一作者：齐中强，副研究员，主要从事植物病理学研究；E-mail：qizhongqiang2006@ 126. com

*** 通信作者：刘永锋，研究员，主要从事植物病理学研究；E-mail：Liuyf@ jaas. ac. cn

稻曲病菌 RasGTP 酶激活蛋白 UvGap1 的功能研究*

曹慧娟**，张瑾瑾，俞咪娜，于俊杰，雍明丽，刘永锋***
（江苏省农业科学院植物保护研究所，南京　210014）

摘　要：由稻曲病菌（*Ustilagornoidea virens*）侵染引起的水稻稻曲病（Rice false smut）是我国水稻生产的重要穗部病害，严重威胁水稻质量和安全，阐明稻曲病菌不同于其他病原真菌的致病机制具有重要意义。前期在稻曲病菌 T-DNA 插入突变体库中筛选到菌株 B2510（T-DNA 以双拷贝形式插入）表现致病能力减弱，其中一个 T-DNA 插入基因 Uv8b_1386（编码 RasGTP 酶激活蛋白 UvGap1）的启动子区。UvGap1 在稻曲病菌侵染致病过程中可能扮演重要角色。

为了解析 UvGap1 及其介导的 Ras 信号途径在稻曲病菌生长发育及致病过程的功能，利用 CRISPR-Cas9 系统和同源重组相结合的方法对稻曲病菌 *UvGAP*1 基因进行敲除工作，得到基因敲除突变体 Δ*UvGap*1。Δ*Uvgap*1 在 PSA 和 TB3 固体平板的生长速率与野生型菌株 P1 没有显著变化，在固体产孢平板培养或者液体 PS 培养基摇培后，Δ*Uvgap*1 的分生孢子产量与野生型相当，说明 *UvGAP*1 基因不参与调控稻曲病菌菌丝生长和分生孢子产生过程。接种实验结果显示，野生型菌株 P1 接种水稻穗部后每穗平均形成 30 个稻曲球，而 Δ*Uvgap*1-23 完全丧失形成稻曲球的能力。进一步在接种水稻穗部 15d 后剖开水稻颖壳观察，发现 Δ*Uvgap*1 接种的水稻颖壳中无白色菌丝，野生型和互补菌株侵染后在水稻颖壳内形成白色菌丝，互补菌株可以恢复突变体致病能力的缺陷，说明 *UvGAP*1 参与稻曲病菌致病过程。UvGap1 及其介导的 Ras 信号途径对稻曲病菌致病过程的具体调控网络有待进一步的深入研究。

关键词：稻曲病菌；RasGTP 酶激活蛋白 UvGap1；致病机制

---

* 基金项目：江苏省自然科学基金（BK20180296）
** 第一作者：曹慧娟，副研究员，主要从事水稻稻曲病菌的致病机制研究；E-mail：huijuancao@ yeah. net
*** 通信作者：刘永锋，研究员，主要从事水稻病害病理学及其生物防治技术研究；E-mail：liuyf@ jaas. ac. cn

稻曲病菌 velvet 蛋白 UvVelB 参与病原菌营养生长和分生孢子产生*

俞咪娜[1,2**]，于俊杰[1]，曹慧娟[1]，雍明丽[1]，黄世文[2]，刘永锋[1***]
（1. 江苏省农业科学院植物保护研究所，南京 210014；2. 中国水稻研究所，杭州 310006）

摘 要：由稻曲病菌［*Ustilaginoidea virens*（Cooke）Tak.］侵染水稻引起的水稻稻曲病（rice false smut）是一种世界性的水稻穗部病害，该病害的发生可严重影响水稻产量和品质。

Velvet 家族蛋白具有典型的 DNA 结合结构域 VELVET，在镰刀菌中作为 global regulators 参与调节营养生长、细胞发育和次生物质代谢多种真菌的生物学功能中参与调节菌的。在 *Aspergillus nidulans* 中，velvet 蛋白 VelB-VeA-LaeA 形成聚合复合体参与调节有性生殖、次生物质代谢等。

为研究 velvet 蛋白家族在稻曲病菌中的生物学功能，我们通过同源序列比对得到 VelB 蛋白在稻曲病菌中的同源蛋白 UvVelB，利用同源双交换获得 *UvVelB* 基因敲除突变体。生物学表型结果分析表明，Uv*VeLB* 突变体与野生型相比表现为气生菌丝减少，产孢量显著增加，对水稻萌发的胚根生长具有抑制作用等，这与 *Fusarium graminearum* 的 *FgVeLB* 突变体表型相似。然而稻曲病菌 *UvVelB* 基因敲除突变体的色素积累、菌丝疏水性等表型没有发生变化，与 *FgVelB* 突变体表型不同；在 *Aspergillus nidulans* 中，VelB 突变体表现为产孢量减少、色素积累等表型，与稻曲病菌 *UvVelB* 基因功能存在差异。由此可以推测，VelB 作为 global regulators 参与菌的多个生物学功能，但是在不同菌中其参与的途径存在差异。因此，本研究后续将深入解析 *UvVeLB* 及其他 velvet 蛋白在稻曲病菌中的生物学功能。

关键词：稻曲病菌；velvet 蛋白家族；营养生长；分生孢子

* 基金项目：江苏省自然科学基金（BK20151386）

** 第一作者：俞咪娜，副研究员，主要从事水稻稻曲病菌的致病机制研究；E-mail：zjpsyu@163.com
*** 通信作者：刘永锋，研究员，主要从事水稻病害病理学及其生物防治技术研究；E-mail：liuyf@jaas.ac.cn

稻曲病菌突变体 B-766 中厚垣孢子形成调控基因的克隆和功能研究*

于俊杰**，俞咪娜，曹慧娟，潘夏艳，雍明丽，
宋天巧，齐中强，杜　艳，张荣胜，尹小乐，刘永锋***
（江苏省农业科学院植物保护研究所，南京　210014）

摘　要：本研究克隆了稻曲病菌（*Ustilaginoidea virens*）突变菌株 B-766 中与厚垣孢子形成相关的基因，并通过基因敲除验证该基因在稻曲病菌分生孢子形成中的功能，为进一步揭示稻曲病菌的厚垣孢子形成机制提供理论基础。研究通过利用 hiTAIL-PCR、Southern 杂交和半定量 PCR 等方法克隆到 B-766 中与厚垣孢子形成相关基因编码转录因子 UvHox2，该基因为 homeobox 家族基因。进一步通过基于农杆菌介导转化和 CRISPR 技术的基因敲除方法确证了该基因在厚垣孢子形成过程中具有重要的调控功能，同时也参与调控分生孢子形成和致病性。观察发现，UvHOX2 通过调控厚垣孢子和分生孢子的小梗形成，从而影响厚垣孢子和分生孢子的形成。通过 RNA-seq 技术比较厚垣孢子形成初期的野生型菌株 P-1 和 UvHox2 基因敲除突变体发现，UvHOX2 可能影响多种功能基因的表达，涉及信号传导途径、细胞壁合成、泛素化和细胞自噬、渗透压和细胞膜完整性。其中，BrlA-AbaA-WetA 信号传导途径被发现处于 UvHOX2 调控途径下游，该信号传导途经通常被认为参与真菌分生孢子小梗形成的调控，本研究发现在稻曲病菌中转录因子 UvHOX2 可能通过 BrlA-AbaA-WetA 信号传导途径同时调控厚垣孢子和分生孢子小梗的形成。

关键词：稻曲病菌；厚垣孢子；调控基因

* 基金项目：国家自然科学基金（31301624，31571961）

** 第一作者：于俊杰；E-mail：jjyu@ jaas. ac. cn
*** 通信作者：刘永锋；E-mail：liuyf@ jaas. ac. cn

江苏省稻瘟病菌对稻瘟灵的敏感性研究*

尹小乐**，于俊杰，张荣胜，刘永锋***
（江苏省农业科学院植物保护研究所，南京　210014）

摘　要：为了明确江苏省稻瘟病菌对稻瘟灵的敏感性，利用菌丝生长速率法测定了 2018 年从该省 8 个地市分离的 92 个稻瘟病菌单孢菌株的毒力。试验结果表明：稻瘟灵对江苏省稻瘟病菌抑制中浓度 EC_{50} 为 1. 3523～11. 1256μg/mL，敏感性差异达 8. 23 倍；各地市平均 EC_{50} 值为 3. 70～8. 28μg/mL，敏感性差异达 2. 24 倍。敏感性频率分布图显示，稻瘟病菌群体中存在着对稻瘟灵敏感性较低的亚群体，但 94. 57%供试菌株敏感性频率呈正态分布，将此部分菌株 EC_{50} 平均值 4. 14 μg/mL 作为稻瘟病菌对稻瘟灵的敏感性基线，全省敏感菌株出现频率为 51. 09%，低抗菌株出现频率为 48. 91%。结果显示，尽管江苏省稻瘟病菌群体中存在着对稻瘟灵敏感性较低的亚群体，但稻瘟灵仍可作为江苏省防治水稻稻瘟病的有效药剂。

关键词：江苏省；稻瘟病菌；稻瘟灵；敏感性

---

* 基金项目：国家重点研发计划（2016YFD0300706）；江苏省自主创新资金项目 CX16（1001）

** 第一作者：尹小乐，助理研究员，主要从事主要从事水稻病害病理学及其生物防治技术研究；E-mail：yinjsha@ 163. com
*** 通信作者：刘永锋，研究员，主要从事水稻病害病理学及其生物防治技术研究；E-mail：liuyf@ jaas. ac. cn

交配型基因 *MAT*1-1-3 对稻曲病菌的菌丝生长和有性生殖具有重要作用*

雍明丽**，于俊杰，俞咪娜，曹慧娟，刘永锋***
（江苏省农业科学院植物保护研究所，南京　210014）

摘　要：稻曲病菌引起的稻曲病，在全世界水稻种植区成为重要水稻病害之一，严重影响水稻的产量和食用安全，研究稻曲病菌的生活史有助于制定有效的防治策略。有性生殖过程中产生的菌核在生活史中具有重要作用，交配型基因对有性生殖起着关键性作用。稻曲病菌交配型基因 *MAT*1-1-3 在有性生殖和无性生殖中表达具有差异，*MAT*1-1-3 具有 HMG-box 功能域，而且 *MAT*1-1-3 仅在少数真菌中存在。本文利用 CRISPR/Cas9 技术对交配型基因 *MAT*1-1-3 进行靶向敲除，通过 qRT-PCR 和测序确定转化子。突变体 Δ*mat*1-1-3 生长速度下降，形态变化较大，与 *MAT*1-2 型菌株混合接种后无法形成菌核，对水稻致病性影响不明显。敲除突变体对渗透、氧化应激适应和细胞壁完整性都具有一定的影响。*MAT*1-1-3 既能够定位到细胞质中也能够定位到细胞核中，而且还能和交配型基因 *MAT*1-1-1 进行较强的互作，敲除 *MAT*1-1-3 基因导致交配型基因 *MAT*1-1-1、*MAT*1-1-2 及信息素基因 *UvPpg*1 的表达下调，但是信息素受体基因 *UvPre*1 和 *UvPre*2 却上调表达。上述结果说明交配型基因 *MAT*1-1-3 对稻曲病菌的菌丝生长和有性生殖具有重要的作用。

关键词：稻曲病；有性生殖；*MAT*1-1-3；菌丝生长

---

* 基金项目：国家自然科学基金面上项目（31571961）；中国博士后科学基金面上资助（2018M632257）
** 第一作者：雍明丽，助理研究员，主要从事水稻稻曲病菌的研究；E-mail：m17366282150@163.com
*** 通信作者：刘永锋，研究员，主要从事水稻病害病理学及其生物防治技术研究；E-mail：liuyf@jaas.ac.cn

葡萄灰霉病抗性基因 VpNBS1 表达载体的构建与亚细胞定位分析*

王泽琼[1]**，蔡 珺[1]，刘 勇[2]***

（1. 武汉生物工程学院，武汉 430415；2. 湖北省农业科学院果树茶叶研究所，武汉 430064）

摘 要：灰霉病是葡萄上的一种重要病害，在我国南方设施葡萄生产中发生尤为严重，每年因灰霉病造成的损失在 20%左右，在发病严重的葡萄园中，病穗率甚至达到 40%以上。目前，对该病害的防治以化学方法为主，不仅防效有限，且易产生抗药性及农药残留的问题。随着生物技术的飞速发展，基因工程与常规育种方法相结合，大大推动了农作物品种的遗传改良。从现有的葡萄种质资源中筛选抗性资源，挖掘葡萄抗灰霉病基因对于葡萄生产和育种具有重要的指导意义。本团队在前期的研究中以对灰霉病表现高抗的‘摩尔多瓦’葡萄作为研究材料，通过 RNA-seq 测序技术从中筛选到一个高丰度差异表达的 NBS-LRR 类抗性基因 VpNBS1，该基因在 *Botrytis cinerea* 侵染的‘摩尔多瓦’叶片中显著性上调表达，表明 VpNBS1 在葡萄与 *Botrytis cinerea* 互作过程中可能扮演着重要作用。

为进一步从分子水平上明确其在葡萄与 *Botrytis cinerea* 互作过程中的作用机制，本研究以‘摩尔多瓦’第一链 cDNA 为模板，采用 PCR 法克隆出 *VpNBS*1 基因，进一步通过常规载体的构建方法将其构建到 pBI121-EGFP 载体上，获得 pBI121-VpNBS1-EGFP 植物表达融合载体，经农杆菌介导转染烟草，激光共聚焦显微镜结果显示目标基因 VpNBS1-GFP 标签荧光分布在细胞膜上，表明该基因编码的蛋白定位在细胞膜上，初步明确了该基因的基本信息。本研究克隆 *VpNBS*1 基因并进行了亚细胞定位分析，为后期进行该基因的遗传转化、基因表达和生物学功能分析奠定了基础。

关键词：葡萄；灰霉病；抗性基因；载体构建；亚细胞定位

---

* 基金项目：国家重点研发计划（2018YFD0201300）；农业部华中作物有害生物综合治理重点实验室开放课题（2018ZTSJJ9）；湖北省自然科学基金（2017CFB163）

** 第一作者：王泽琼，博士研究生，研究方向为植物病理学

*** 通信作者：刘勇，博士研究生，研究方向为植物病理学

茶树炭疽病菌鉴定及其对不同药剂的敏感性研究

陆铮铮*，孙雅楠，吴鉴艳**
（浙江农林大学农业与食品科学学院，临安　311300）

摘　要：炭疽菌引起的茶树炭疽病在中国各产茶区均有发生，各产茶区炭疽病的炭疽菌种类复杂不一，且不同的炭疽菌对杀菌剂的敏感性不同。本文从浙江省3个茶园分离得到10个炭疽菌，通过菌落形态、分生孢子形态、分生孢子附着胞形态等观察，同时基于β-Tub2、ITS、GPDH和ACT基因序列的多基因系统进化分析，确定分离到3种炭疽菌，分别为*Colletotrichum fruticola*、*Glomerella cigulata* f. sp. *camelliae*、*C. siamense*；致病性测定表明，3类*Colletotrichum*的致病性较弱，在龙井“43号”寄主上，需要通过伤口才能引起侵染。菌丝生长法测定3类*Colletotrichum*对苯醚甲环唑、吡唑醚菌酯和百菌清的敏感性表明，3类菌对苯醚甲环唑最为敏感，EC_{50}值为0.070~1.061 mg/L，平均EC_{50}值为0.334 ± 0.312 mg/L，其次是吡唑醚菌酯，EC_{50}值为0.174~5.090 mg/L，平均EC_{50}值为1.275 ± 1.424 mg/L，而大部分对百菌清的敏感性较差，平均EC_{50}值为14.512 ± 4.038 mg/L。

关键词：*Colletotrichum*；形态学特征；多基因序列分析；致病力；敏感性测定

* 第一作者：陆铮铮，本科生，植物保护学；E-mail：m15988800048@163.com
** 通信作者：吴鉴艳，博士，讲师，研究方向为植物病理学；E-mail：wujianyan@zafu.edu.cn

稻瘟病菌抗稻瘟灵的分子机理研究*

孟凡珠[1**]，王佐乾[1]，阴伟晓[1,2]，罗朝喜[1,2***]
（1. 华中农业大学植物科学技术学院，武汉 430070；
2. 湖北省作物病害监测和安全控制重点实验室，武汉 430070）

摘 要：稻瘟病是水稻三大病害之一，稻瘟病流行将会对粮食产量和安全造成重大影响。稻瘟灵于 20 世纪 70 年代开发并成功用于稻瘟病的控制，具有高效、低毒的特点，由于与有机磷杀菌剂异稻瘟净具有交互抗性，又被认为是磷脂酰胆碱合成抑制剂。

笔者实验室前期通过药剂筛选获得 3 个抗药性突变体，并通过全基因组测序和遗传转化分析，明确了一个编码 Zn2Cys6 锌指蛋白转录因子的基因 *MoIRR* 负调控稻瘟病菌对稻瘟灵的抗性。本研究通过对野生型菌株 H08-1a、抗药性突变体 1a_mut、*MoIRR* 敲除突变体 ΔMoIRR-1 进行了转录组分析。与野生型菌株 H08-1a 相比，在 1a_mut 和 ΔMoIRR-1 之中出现共差异的基因数目 434 个。其中共同上调表达的基因为 220 个，富集 37 个代谢通路；共同下调表达的基因为 214 个，富集 9 个代谢通路。本研究对富集在磷脂酰胆碱合成与代谢过程中的两个基因 *MGG*_01523、*MGG*_05804 进行了过表达实验，但过表达转化子并没有对稻瘟灵抗性产生差异表现，所以上述两个基因不是稻瘟灵的作用靶标。

关键词：稻瘟病菌；稻瘟灵；杀菌剂抗性；转录组

---

* 基金项目：国家重点研发计划项目（2016YFD0300700）；华中农业大学基础研究支持项目（2662015PY195）

** 第一作者：孟凡珠，硕士研究生；E-mail：mengfanzhu123@163.com

*** 通信作者：罗朝喜，教授，主要从事病原真菌抗药性机理、果树病害防控及稻曲病与水稻互作研究；E-mail：cxluo@mail.hzau.edu.com

中国桃黑星菌生物学特性研究

周 扬[1,2]，赵 阳[1]，朱宜庭[1]，李西城[1]，罗梦珂[1]，阴伟晓[1,2]，罗朝喜[1,2]
（1. 华中农业大学植物科技学院，武汉 430070；2. 园艺植物生物学教育部重点实验室，湖北省作物病害检测及安全控制重点实验室，武汉 430070）

摘 要：由桃黑星菌（*Venturia carpophila*，又名嗜果黑星菌，属于子囊菌门球壳目黑星菌属真菌）引起的疮痂病是桃产区重要的病害之一，桃疮痂病的发生严重影响了桃的品质和产量。由于桃黑星菌纯培养时难分离、生长慢、易污染，在寄主上潜伏期长，因此优化桃黑星菌的分离方法、筛选培养条件对开展后续研究至关重要。采用组织分离法、孢子稀释法及改进的显微镜挑单胞分离法对病样进行分离，结果表明只有改进的显微镜挑单胞分离法能够分离出病原菌，组织分离法和孢子稀释法均有大量污染。采用改进的显微镜挑单胞分离法得到全国 16 个省的 750 株桃黑星菌单胞菌株。筛选了 10 种培养基，结果表明桃黑星菌在 MEA 培养基上生长最快，在 OMA 培养基上产孢量最大，在 PDA 培养基中加入桃叶片能够显著加快桃黑星菌的生长。在 OMA 培养基中桃黑星菌的最佳生长及产孢温度为 21℃，最佳生长 pH 值为 4.5，最佳产孢 pH 值为 5.5。优化分离方法、筛选合适培养条件为桃黑星病的进一步研究提供基础。

关键词：桃黑星菌；*Venturia carpophila*；分离方法；生物学特性

海南槟榔炭疽病病原菌的分离鉴定*

曹学仁**，车海彦，罗大全***
（中国热带农业科学院环境与植物保护研究所，农业部热带作物
有害生物综合治理重点实验室，海口 571101）

摘 要：槟榔是海南省第一大特色经济作物。炭疽病是槟榔主要病害之一，在海南各地均有发生。一直以来，胶孢炭疽菌（*Colletotrichum gloeosporioides*）被认为是海南槟榔炭疽菌的病原菌，但研究表明胶孢炭疽菌是一个复合种。为了明确海南槟榔炭疽菌的病原菌种类，从海南万宁、琼海、文昌、海口、儋州、琼中、三亚和保亭等 8 个市（县）的 11 个槟榔园采集病害标样，室内经组织分离和单孢分离纯化共获得 43 个炭疽菌菌株。利用多基因位点序列（ITS、*ACT*、*CHS*-1、*GAPDH*、*TUB*2 和 *ApMat*）分析，并结合形态学和致病性测定对所分离菌株进行了鉴定。结果表明，引起海南槟榔炭疽病的病原菌共有 8 个种，包括 6 个已报道种：柯氏炭疽菌 *C. cordylinicola*、果生炭疽菌 *C. fructicola*、胶孢炭疽菌 *C. gloeosporioides*、喀斯特炭疽菌 *C. karsti*、暹罗炭疽菌 *C. siamense*、热带炭疽菌 *C. tropicale*，1 个疑似新种和 1 个待定种，菌株个数分别为 2 个、3 个、1 个、2 个、16 个、1 个、15 个和 3 个。致病性结果表明所有鉴定出来的炭疽菌种都导致槟榔叶片出现症状，但不同菌株之间的致病力存在差异。本研究明确了海南槟榔炭疽病病原菌的种类，从而为病害防治提供了理论依据。

关键词：槟榔炭疽病；病原；鉴定

* 基金项目：海南省槟榔病虫害重大科技计划项目（ZDKJ201817）；中国热带农业科学院基本科研业务费（1630042017023）

** 第一作者：曹学仁，博士，主要从事植物真菌病害研究

*** 通信作者：罗大全，研究员，主要从事热带作物病理学研究

京郊市售西甜瓜种子携带的主要真菌的分离与鉴定*

王丽云，蒋 娜，李健强，罗来鑫**
（中国农业大学植物保护学院，种子病害检验与防控北京市重点实验室，北京 100193）

摘 要：中国是西瓜和甜瓜的生产与消费大国，西瓜种植面积占世界总面积的60%以上，产量占70%以上；甜瓜面积占世界总面积的45%以上，产量占55%以上，且近年来西甜瓜的种植面积和产量都呈逐年增加的趋势（中国国家统计局数据）。西瓜和甜瓜在生产中受到许多病害为害，已报道西瓜和甜瓜种子携带的真菌有10个属，包括枯萎病菌、蔓枯病菌、根腐病菌等。镰刀菌属下的不同种会导致瓜类根腐病、枯萎病、果实腐烂等病害，在生产上造成严重损失。

为掌握北京及周边地区的西瓜和甜瓜种子携带的主要真菌病原物种类，本研究收集了52批次西瓜和甜瓜种子，进行了种子携带真菌的检测。检测结果表明，供试的40批次西瓜种子中，外部带菌36批次，占比90%；外部孢子负荷量范围在0.1~275 CFU/粒；内部带菌27批次，占比67.5%；西瓜种子内部带菌率百分比为1%~26%。供试的12批次甜瓜种子中，外部带菌3批次，占比25%，外部孢子负荷量范围在0.2~275 CFU/粒；内部带菌2批次，占比16.7%，甜瓜种子内部带菌率百分比为2%~3%。共分离纯化了123株真菌，通过形态学和ITS序列测定分析结果，将这123株真菌初步鉴定为10个属，包括毛壳菌属（*Chaetomium* spp.）、青霉属（*Penicillium* spp.）、曲霉属（*Aspergillus* spp.）、链格孢属（*Alternaria* spp.）、篮状菌属（*Talaromyces* spp.）、枝孢属（*Cladosporium* spp.）、镰刀菌属（*Fusarium* spp.）、木霉属（*Trichoderma* spp.），共头霉属（*Syncephalastrum* spp.）和 *Stagonosporopsis* 属。对分离得到的镰刀菌菌株M2、M30、V32进行了进一步鉴定，使用EF1/EF2、H3-1a/H3-1b、mtSSUF/mtSSUR的扩增序列拼接后建树分析，结果表明：M2为胶孢镰刀菌（*Fusatium subglutinans*），M30为变红镰刀菌（*Fusarium incarnatum*）、V32为轮枝镰刀菌（*Fusarium verticillioides*）。检测结果为后续种子处理和病害防治提供了理论依据。

关键词：西瓜甜瓜；种子带菌；鉴定；镰刀菌

* 第一作者：王丽云，硕士研究生；E-mail：liyunwang@cau.edu.cn
** 通信作者：罗来鑫，副教授，博士生导师，主要从事种子病理学研究；E-mail：luolaixin@cau.edu.cn

环介导等温扩增（LAMP）技术检测根串珠霉菌研究

王甲军*，李珊珊，窦彦霞，马冠华**
（西南大学植物保护学院，重庆 400725）

摘 要：为建立一种快速、灵敏的检测根串珠霉菌（*Thielaviopsis basicola*）的环介导等温扩增（loop-mediated isothermal amplification，LAMP）方法，以根串珠霉菌核糖体大亚基 RNA 基因的部分基因序列为靶基因，用在线引物设计软件 Primer Explorer V5.0 设计并合成 4 条特异性引物 F3、B3、FIP 和 BIP 进行 LAMP 扩增试验，分别设置 Mg^{2+} 终浓度为 2mmol/L、4mmol/L、6mmol/L、8mmol/L、10mmol/L、12mmol/L，d NPTs 终浓度为 0.2mmol/L、0.4mmol/L、0.6mmol/L、0.8mmol/L、1.0mmol/L、1.2mmol/L，内引物 FIB/BIF 终浓度为 0.25mmol/L、0.5mmol/L、0.75mmol/L、1mmol/L、1.25mmol/L、1.5mmol/L，甜菜碱终浓度为 0、0.2mol/L、0.4mol/L、0.6mol/L、0.8mol/L、1.0mol/L，*Bst* DNA 聚合酶终浓度为 0.08U/μL、0.16U/μL、0.24U/μL、0.32U/μL、0.40U/μL、0.48U/μL，反应温度为 50℃、53℃、56℃、59℃、62℃、65℃、68℃、71℃，反应时间为 10min、20min、30min、40min、50min、60min、70min、80min。扩增产物用 SYBR Green I 和琼脂糖凝胶电泳检测 LAMP 反应的结果。以烟草茎点霉（*Phoma tabaci* Em. Sousa da Camara）、烟草拟盘多毛孢（*pestalotiopsis nicotiae*）、烟草白星病菌（*phyllosticta nicotianae* Ell. etev）、烟草碎叶病菌（*Mycosphaerella nicotiana*）、烟草赤星病菌（*Alternaria alternata*（Fries）Keissler）、腐霉（*Pythium*）、根霉（*Rhizopus*）、毛霉（*Mucor*）、柑橘青霉（*P. italicum* Wehmer）、黄曲霉（*Aspergillus flavus*）、烟草炭疽病菌（*Colletotrichum micotianae* Averna）和烟草灰霉病菌（*Botrytiscinerea* Per. etFries）基因组 DNA 为模板进行特异性验证。通过将根串珠霉菌基因组 DNA 稀释 10 倍梯度，分别获得 DNA 浓度 1.62×10^2ng/μL、1.62×10ng/μL、1.62ng/μL、1.62×10^2 pg/μL、1.62×10pg/μL、1.62pg/μL、1.62×10^2 fg/μL、1.62×10fg/μL、1.62fg/μL 作为模板进行 LAMP 灵敏度测定，并与常规 PCR 进行比较。结果表明：25μL 反应体系最佳终浓度分别为外引物浓度 0.2μmol/L、内引物浓度 1.6μmol/L，Mg^{2+}浓度 6mmol/L，dNTP 浓度 0.6mmol/L，甜菜碱浓度 0.6mol/L，*Bst* DNA 聚合酶 0.08U/μL，反应温度 62℃，反应时间 60 min。特异性结果表明，只有根串珠霉菌基因组 DNA 有黄绿色和特异性梯状条带，其他供试菌株 DNA 无黄绿色且不能检测出任何条带。LAMP 检测根串珠霉菌基因组 DNA 的最低检出浓度为 1.62×10^2fg/μL，比常规的 PCR 检测高出两个数量级，并且方法简单、节省时间。成功建立了准确、快速、低成本的 LAMP 检测根串珠霉菌体系，为相关研究和应用提供了技术支持。

关键词：环介导等温扩增；根串珠霉菌；快速检测；LAMP

---

* 第一作者：王甲军，硕士研究生，主要从事植物病理学研究；E-mail：571934151@ qq. com
** 通信作者：马冠华，副教授，主要从事植物病理学研究；E-mail：nikemgh@ swu. edu. cn

Toxicity and mechanism of UDP-KDG, a transient metabolite of rhamnose synthesis in *Botrytis cinerea*

马 良*

(浙江农林大学农业与食品科学学院，临安 311300)

Abstract: Can accumulation of a normally transient metabolite affect fungal biology? UDP-4-keto-6-deoxyglucose (UDP-KDG) represents an intermediate stage in conversion of UDP-glucose to UDP-rhamnose. Normally, UDP-KDG is not detected in living cells, because it is quickly converted to UDP-rhamnose by the enzyme UDP-4-keto-6-deoxyglucose-3, 5-epimerase/-4-reductase (ER). We previously found that deletion of the *er* gene in *Botrytis cinerea* resulted in accumulation of UDP-KDG to levels that were toxic to the fungus due to destabilization of the cell wall. Here we show that these negative effects are at least partly due to inhibition by UDP-KDG of the enzyme UDP-galactopyranose mutase (UGM), which reversibly converts UDP-galactopyranose (UDP-Gal *p*) to UDP-galactofuranose (UDP-Gal *f*). An enzymatic activity assay showed that UDP-KDG inhibits the *B. cinerea* UGM enzyme with a Ki of 221.9 μM. Deletion of the *ugm* gene resulted in strains with weakened cell walls and phenotypes that were similar to those of the *er* deletion strain, which accumulates UDP-KDG. Gal *f* residue levels were completely abolished in the Δ*ugm* strain and reduced in the Δ*er* strain, while overexpression of the *ugm* gene in the background of a Δ*er* strain restored Gal *f* levels and alleviated the phenotypes. Collectively, our results show that the antifungal activity of UDP-KDG is due to inhibition of UGM and possibly other nucleotide sugar-modifying enzymes and that the rhamnose metabolic pathway serves as a shunt that prevents accumulation of UDP-KDG to toxic levels. These findings, together with the fact that there is no Gal *f* in mammals, support the possibility of developing UDP-KDG or its derivatives as antifungal drugs.

Key words: Botrytiscinerea; UDP-KDG; UDP-galactopyranose mutase; Cell wall

* 第一作者/通信作者：马良，博士，副教授，主要从事植物病原真菌研究；E-mail：liangm2008@ yahoo. com

基于通径分析的宁夏小麦条锈病流行预测模型*

杨璐嘉**，张克瑜[1]，邓　杰[1]，刘　媛[2]，马占鸿[1]***

（1. 中国农业大学植物病理学系，北京　100193；2. 宁夏回族自治区农业技术推广总站，银川　75001）

摘　要：宁夏南部山区属黄土高原边缘，是宁夏小麦的主要种植区，也是我国小麦条锈病的常发区。由条形柄锈菌小麦专化型（*Puccinia striiformis* f. sp. *tritici*）引起的小麦条锈病是宁夏小麦生产中发生面积最广，危害最严重的典型远程气传性病害之一。本研究以宁夏发病最重的西吉县为例，通过整理 20 年（1998—2018 年）小麦条锈病病情数据，利用 SPSS 相关分析、逐步回归分析、通径分析研究方法，探讨西吉县气象因子与小麦条锈病发病率、病情指数及 AUDPC 间的关系，并建立了小麦条锈病流行预测模型。研究中，首先利用 1998—2016 年的气象资料和各年度的病情指数组建了多元线性回归模型，再用预留的 2017 年、2018 年的调查结果用于检测该模型的预测准确度。分析结果表明：平均气温、日平均最高气温和日平均最低气温与田间病情发病率、病情指数及 AUDPC 相互之间均为极显著相关。采用旬平均气温（X）与病情指数（Y）建立了多元线性回归方程为 $Y=4.018X-54.302$，直接和间接通径系数 R^2 分别为 0.414 和 0.178。通过逐步回归和通径分析表明，温度对当年田间病情的发展具有决定性作用，这对宁夏小麦条锈病大区流行预测预报具有重要参考意义。

关键词：宁夏西吉；小麦条锈病；逐步回归分析；通径分析；预测预报

---

* 基金项目：宁夏回族自治区重点研发计划（东西部合作项目 2017BY080）

** 第一作者：杨璐嘉，博士研究生，主要从事植物病害流行学研究；E-mail：ylj0818@ 126. com

*** 通信作者：马占鸿，教授，博士生导师，主要从事植物病害流行和宏观植物病理学研究；E-mail：mazh@ cau. edu. cn

云南省不同小麦品种条锈菌群体遗传结构分析*

江冰冰[1**]，努尔阿丽耶·麦麦提江[1]，苟学莉[2]，王翠翠[1]，初炳瑶[1]，马占鸿[1***]
（1. 中国农业大学植物病理学系，北京　100193；2. 中国农业大学农学院，北京　100193）

摘　要：小麦条锈病由条形柄锈菌小麦专化型［*Puccinia striiformis* f. sp. *tritici*（*Pst*）］引起，危害小麦生产，影响粮食安全。云南省地处云贵高原，地形、地势复杂，小麦条锈菌能在此地越冬、越夏，完成周年循环，加之各地区小麦品种布局的多样性和播期的不一致，使得各地区间小麦条锈菌群体遗传结构较为复杂，因此明确云南省的小麦条锈菌群体遗传结构与品种和地理环境的相关性尤为重要。本试验利用 11 对 SSR 引物对 2018 年春季来自 10 个县 13 个小麦品种的 258 个条锈菌样品进行群体遗传结构分析，表明不同小麦品种上的条锈菌群体遗传多样性存在差异，遗传分化较小且主要来源于不同小麦品种亚群体个体内部；分别来源于施甸县和砚山县的两个小麦品种的条锈菌亚群体间遗传多样性差异不大，遗传分化较大；分别来源于师宗县和宣威市同一小麦品种不同采样点条锈菌亚群体之间遗传多样性差异较大，不同品种的遗传分化程度不同。总的结果显示不同小麦品种、不同地区间条锈菌群体遗传结构均有差异，可能是由于小麦品种的抗性差异和地理环境差异导致，此研究结果对明确云南省小麦条锈菌群体结构和小麦抗病品种选育及地区间联防联治具有指导意义。

关键词：小麦条锈菌；小麦品种；群体遗传结构；云南省

---

* 基金项目：国家重点研发计划（2017YFD0200400，2016YFD0300702）
** 第一作者：江冰冰，在读博士生，主要从事小麦条锈病分子流行学研究
*** 通信作者：马占鸿，教授，博士生导师，主要从事植物病害流行与宏观植物病理学研究工作

链霉菌的筛选鉴定、发酵条件优化及防病效果的研究

冯艳娟*，秦　旭，沈鹏飞，羊国根，潘月敏**
（安徽农业大学植物保护学院，合肥　230036）

摘　要：小麦全蚀病又称黑脚病，是由子囊菌亚门禾顶囊壳属小麦变种引起的病害。该病害在全国各地均有发生，幼苗期开始发病，随后根部腐烂，严重时可造成大量减产甚至绝收。为了有效防治该病害，本研究从安徽北部小麦病害区的土壤中筛出对小麦全蚀病具有抑制作用的生防菌，经过平板对峙实验筛出一株颉颃效果较好的生防菌，经过生理生化和分子生物学鉴定，这株生防菌为链霉菌（*Streptomvces*）。在对峙培养中，链霉菌青 9 可以使小麦全蚀病菌的菌丝扭曲分化和原生质体渗透。链霉菌青 9 的发酵液对小麦全蚀病菌的抑制率高达 80%，进一步实验表明，链霉菌青 9 的无菌发酵液对小麦全蚀病病菌抑制率也达到 70%。盆栽实验表明，链霉菌青 9 的发酵液对于小麦的生长有一定的促生作用，对侵染小麦全蚀病的植株有一定治愈效果。因此链霉菌青 9 有潜在的生防功能，其生防物质主要存在于发酵液中。

关键词：小麦全蚀病；生物防治；链霉菌；抑菌率

---

* 第一作者：冯艳娟，在读硕士研究生，研究方向为资源利用与植物保护；E-mail：1793873624@ qq. com
** 通信作者：潘月敏，教授；E-mail：panyuemin2008@ 163. com

咖啡叶枯病病原菌鉴定及生物学特性研究*

巩佳莉[1,2**]，陆　英[2]，贺春萍[2]，吴伟怀[2]，梁艳琼[2]，
黄　兴[2]，郑金龙[2]，习金根[2]，易克贤[2***]
（1. 南京农业大学植物保护学院，南京　210000；
2. 中国热带农业科学院环境与植物保护研究所，海口　571101）

摘　要：从海南省澄迈咖啡园采集典型的咖啡叶枯病叶，进行病原菌分离鉴定并对其致病性和生物学特性进行研究。通过对其培养特性、形态特征的观察及 rDNA－ITS 序列测定，在 Genbank 进行同源性比对，确定引起咖啡叶枯病的病原菌为拟茎点霉属（*Phomopsis*）。该病原菌通过分生孢子器产生两种类型的分生孢子（甲型和乙型分生孢子），甲型分生孢子单细胞，长 4～6μm，宽 2～3μm，呈卵形至椭圆形，内含一到两个油滴，乙型分生孢子呈线性一端呈钩状。采用菌丝生长速率法比较不同温度、pH 值、碳源、氮源、光照条件对菌丝生长的影响，并评价不同培养条件下菌株的产孢情况。生物学特性研究结果表明，菌株生长最适温度为 25～30℃，致死温度为 55℃，光照对其影响不大，最适 pH 值为 6，最佳碳源为麦芽糖，最佳氮源为硝酸钾；温度及光照对其产孢无显著影响，pH 值为 11 和碳源为丙三醇最有利于产孢。

关键词：咖啡叶枯病；拟茎点霉；形态特征；生物学特性

---

* 基金项目：国家重点研发计划项目“特色经济作物化肥农药减施技术集成研究与示范”（2018YFD0201100）资助
** 第一作者：巩佳莉，在读研究生，研究方向为资源利用与植物保护；E-mail：2018802222@ njau. edu. cn
*** 通信作者：易克贤，研究员，博士生导师，研究方向为植物病理学；E-mail：yikexian@ 126. com

北京地区番茄灰霉病菌对啶酰菌胺抗药性检测*

乔广行**，黄金宝，周　莹，李兴红
（北京市农林科学院植物保护环境保护研究所，北京　100097）

摘　要：啶酰菌胺（boscalid）是新型烟酰胺类杀菌剂，具有活性高、作用机理独特、杀菌谱广、不易产生交互抗性、对作物安全，几乎对所有类型的真菌病害都有活性，主要防治灰霉病与菌核病，并且对其他药剂的产生抗药性病原菌亦有效，而且啶酰菌胺具有耐雨冲刷性和渗透传导作用，持效期长，可减少施药次数，是值得重视的新型高效低毒杀菌剂。啶酰菌胺的作用靶标为线粒体呼吸链酶复合体 II 琥珀酸脱氢酶（Succinate DeHydrogenase，SDH）），巴斯夫欧洲公司 2005 年登记该农药进入中国市场，我国 2007 年有番茄灰霉病菌对啶酰菌胺敏感性研究报道，国外近几年有研究报道灰霉病菌对该药剂产生抗药性，随该药剂在北京地区推广应用有必要开展该药剂的抗药性检测，本研究采用菌丝生长抑制法测定啶酰菌胺对北京地区番茄灰霉病菌生物活性，测定结果表明：供试的 158 株菌株中有 68 株为相对敏感菌株所占比例 43.04%，78 株表现为低水平抗药性所占比例 49.37%，低水平抗药性菌株所占的比例较大，并发现 6 株高抗药性与 5 株非常高抗药性菌株，说明北京地区存在番茄灰霉病菌对啶酰菌胺的抗药性风险，针对高抗药性菌株利用分子生物学进行靶标基因克隆测序，得到靶标基因全长，通过序列比对明确琥珀酸脱氢酶亚基 B（SDHB）基因非常高抗药性菌株中 SHYyh1507 外显子的 815 位脱氧核苷酸由 A（腺嘌呤）突变为 G（鸟嘌呤），该菌株对应 SDHB 蛋白氨基酸的 272 位氨基酸由 H（组氨酸）突变为 R（精氨酸），该突变与国外已报道高抗药性突变类型一致，进一步证明啶酰菌胺在北京地区存在抗药性风险，生产中应注意轮换用药，每个生长季节限制使用次数。为农业生产中番茄灰霉病的有效防治提供理论依据，对于延长该药剂的使用寿命应该加以重视，并且为啶酰菌胺的有效使用提出了预警，达到病害的有效防治以及杀菌剂减量增效使用的目的。

关键词：番茄灰霉病菌；啶酰菌胺；抗药性

---

* 基金项目：北京市农林科学院创新能力建设专项（区域协同创新：KJCX20170709）

** 第一作者：乔广行，博士，高级农艺师，主要从事园艺病害诊断与综合治理研究；E-mail：qghang98@126.com

宁夏马铃薯镰刀菌根腐病 ISSR 遗传多样性分析*

王喜刚[1**]，杨　波[2]，郭成瑾[1]，沈瑞清[1***]
（1. 宁夏农林科学院植物保护研究所，银川　750002；2. 宁夏大学农学院，银川　750021）

摘　要：镰刀菌属是一类具有重要经济意义的真菌，由于镰刀菌的多型性与易变异的特点，使得镰刀菌一直都是真菌类群中最难鉴定的属之一。随着分子生物学的迅速发展，DNA 分子标记技术已被广泛地应用于生物群体遗传多样性研究，目前 DNA 多态性分子标记在镰刀菌上的应用主要有 RFLP、RAPD、ISSR、SSR、SRAP 等，其中 ISSR 分子标记技术是由 Zietkiewicz 等创建的一种简单序列重复区间扩增多态性分子标记方法，其操作简单、成本低，又有反应稳定、重复性好的优点，ISSR 标记技术被应用到各种病原菌中，以揭示病原菌群体间的遗传多样性。为明确宁夏马铃薯镰刀菌根腐病种内各菌间的遗传差异与亲缘关系，通过 ISSR 分子标记技术聚类分析 30 株地理来源不同的镰刀菌的遗传多样性。结果表明：对 30 株镰刀菌 ISSR 的遗传多样性分析，选用的 13 条 ISSR 引物共扩增出 84 条条带，多分布在 250~ 2 000bp 之间，其中多态性条带为 82 条，多态性平均比例为 98.3%。遗传相似性与聚类分析结果表明，供试的 30 株镰刀菌的遗传相似系数在 0.476~1.000。在 0.476 水平上，分为 2 个类群（IG），其中 IG1 包括 1~16 号和 23~30 号镰刀菌，为尖孢镰刀菌、木贼镰刀菌、茄病镰刀菌和锐顶镰刀菌；IG2 包括 17~22 号菌株，全部为接骨木镰刀菌。在遗传相似系数为 0.933 时，供试的 30 个菌株可被全部区分开。供试的镰刀菌基因组在 SSR 区域具有丰富的多态性，菌种之间存在一定相关性，同一类群中，不同菌株之间的遗传相似性与菌株的地理来源存在一定的相关性，分离自同一地区同一菌种，其菌株间也存在一定的遗传差异性。

关键词：马铃薯根腐病；镰刀菌；ISSR；遗传多样性

---

* 基金项目：宁夏农林科学院全产业链创新示范项目（NKYZ16-0104）；公益性行业（农业）科研专项（201503112-7）；宁夏农林科学院科技创新引导项目（NKYG-18-07）

** 第一作者：王喜刚，助理研究员，主要从事植物病理学及植物病虫害防治方面的研究；E-mail：wxg198712@163.com
*** 通信作者：沈瑞清，研究员，研究方向为主要从事植物病理学和真菌学研究；E-mail：srqzh@sina.com

灰葡萄孢转录因子 BcCLR-1 的功能研究*

马 良**，时浩杰***
（浙江农林大学农业与食品科学学院，杭州 311300）

摘 要：近年来，灰霉病在生产和储运销过程中产生的危害日益严重，其病原物灰葡萄孢（*Botrytis cinerea*）是一种典型的死体营养型真菌，在侵染初期能够分泌大量的细胞壁降解酶（Cell wall degrading emzymes，CWDEs）来降解寄主的细胞壁。CWDEs 是灰葡萄孢侵染初期的主要致病因子，然而目前对灰葡萄孢 CWDEs 的分子调控机制还知之甚少。纤维素为植物细胞壁的主要成分，本项目组研究发现灰葡萄孢一个锌指转录因子 BcCLR-1 的敲除突变体 Δ*bcclr*-1 在纤维素作为唯一碳源的平板上不能生长，且在寄主上的致病力显著下降，这暗示 BcCLR-1 在调控细胞壁纤维素降解的过程中发挥关键的作用，但其分子机制有待深入研究。本研究将应用 RNA-seq 技术系统分析灰葡萄孢野生型与 Δ*bcclr*-1 侵染过程中的差异表达基因，并通过 ChIP-seq 和 ChIP-qPCR 进一步验证分析 BcCLR-1 所调控的靶基因，然后通过 BcCLR-1 的亚细胞定位、互作蛋白的筛选等方法对其进行功能分析，解析 BcCLR-1 对纤维素降解酶基因调控的分子机制。本研究将有助于在分子层面上更清晰地认识在侵染寄主植物的过程中灰葡萄孢 CWDEs 的分子调控机理，并为深入地了解灰葡萄孢致病关键因子奠定基础。

关键词：灰葡萄孢；转录因子；细胞壁降解酶

* 基金项目：浙江省自然科学基金（LY19C140004）
** 第一作者：马良，副教授，研究方向：植物病原真菌及真菌病害；E-mail：liangm2008@ yahoo. com
*** 通信作者：时浩杰，副教授，研究方向：植物病原真菌及真菌病害；E-mail：mvp8851@ 126. com

粉红螺旋聚孢霉 67-1 菌株 Crmapk 互作蛋白的筛选*

吕斌娜**，Rakibul Hasan，江 娜，李世东，孙漫红***
（中国农业科学院植物保护研究所，北京 100193）

摘 要：粉红螺旋聚孢霉（*Clonostachys rosea*，异名：粉红粘帚霉 *Gliocladium roseum*）是一类重要的菌寄生菌，能够侵染丝核菌、核盘菌、镰刀菌和灰葡萄孢等多种植物病原真菌，具有极大的生防潜力。笔者课题组通过前期对高效菌株 67-1 寄生核盘菌转录组测序分析，获得了一个显著上调表达的编码丝裂原活化蛋白激酶（MAPK）的基因 *crmapk*，敲除该基因后对核盘菌寄生能力下降，推测 *crmapk* 可能参与调控粉红螺旋聚孢霉菌寄生过程。生物信息学分析表明，*crmapk* 编码一个含有 355 个氨基酸的蛋白质，大小 41 ku，等电点 6.64，为亲水性蛋白，无信号肽和跨膜区。功能结构域分析表明，该蛋白含有一个保守的蛋白激酶结构域，属于 Ser/Thr 类蛋白激酶，并与酵母中的 Fus3/Kss1 同源，主要参与 Fus3/Kss1-MAPK 途径。

为进一步解析 Crmapk 调控粉红螺旋聚孢霉菌寄生的分子机制，笔者提取了核盘菌诱导条件下 67-1 总 RNA，纯化得到 mRNA，利用 Gateway 技术构建了粉红螺旋聚孢霉 cDNA 文库。经检测，初级文库库容量为 1.6×10^7CFU/mL，重组率 96%，扩增文库平均插入片段长度>1 kb，表明该 cDNA 文库质量较高。采用双酶切法构建了 *crmapk*-BD 诱饵载体，对其自激活和毒性检测显示，*crmapk*-BD 可以在 SD/-Leu/-Trp/X-α-Gal 培养基中正常生长且不变蓝，说明诱饵蛋白无自激活能力、无毒性。将 *crmapk*-BD 载体与 cDNA 文库共转化 Y2HGold，在 SD/-Ade/-His/-Leu/-Trp/X-α-Gal/AbA 培养基平板上筛选培养，共得到 149 个阳性克隆。从中随机挑取 50 个菌落摇瓶培养，提取酵母质粒，转化大肠杆菌 DH5α 并测序。对 AD 质粒与 BD 质粒回转 Y2H Gold 酵母菌一对一互作验证，得到 29 个阳性基因。经序列比对，在 67-1 中获得 17 个 Crmapk 候选互作蛋白，包括转录因子、转运蛋白、信号转导体等，主要参与了基因的表达调控、营养物质代谢和信号转导等重要途径，下一步将采用 Co-IP 和 Pull-down 等技术对这些候选互作蛋白进一步验证。本研究初步表明 Crmapk 与互作蛋白通过多个生物学途径共同介导了粉红螺旋聚孢霉菌寄生的调控，为揭示粉红螺旋聚孢霉生防作用机制奠定了理论基础。

关键词：粉红螺旋聚孢霉；菌寄生；酵母双杂交；互作蛋白

* 基金项目：国家重点研发计划（2017YFD0201102）；现代农业产业技术体系项目（CARS-25-D-03）；院重大选题项目（CAAS-ZDXT2018005）

** 第一作者：吕斌娜，博士研究生，从事生防真菌功能基因研究

*** 通信作者：孙漫红，副研究员，从事生防真菌防治植物病害机制研究

Identification of three *Berberis* species as potential alternate hosts for *Puccinia striiformis* f. sp. *tritici* in wheat-growing regions of Xinjiang, China

Zhuang Hua*, Zhao Jing, Huang Lili, Kang Zhensheng**, Zhao Jie**

(*State Key Laboratory of Crop Stress Biology for Arid Areas, College of Plant Protection, Northwest A&F University, Yangling* 712100, *China*)

Abstract: Since the recent discovery of barberry (*Berberis* spp.) as an alternate host for the stripe rust pathogen *Puccinia striiformis*, many Chinese *Berberis* species have been identified as alternate hosts for *P. striiformis* f. sp. *tritici*. However, little is known about *Berberis* species and their distribution in wheat-growing regions in Xinjiang, China, where stripe rust is endemic. As the largest province, Xinjiang represents a relatively independent epidemic region for wheat stripe rust in China. In this study, we conducted a survey of barberry plants in the main wheat-growing areas of Xinjiang. We identified three *Berberis* species, *B. heteropoda*, *B. nummularia* and *B. kaschgarica*, and confirmed their roles as potential alternate hosts for *P. striiformis* f. sp. *tritici* in the laboratory.

Key words: barberry; alternate host; sexual reproduction; stripe rust; yellow rust

* First author: Zhuang Hua; E-mail: zhuanghuaok@nwsuaf.edu.cn

** Corresponding authors: Kang Zhensheng; E-mail: kangzs@nwafu.edu.cn

Zhao Jie; E-mail: jiezhao@nwafu.edu.cn

Simplicillium obclavatum, a hyperparasitism affects the infection dynamics of *Puccinia striiformis* f. sp. *tritici* on wheat

Fan Xin*, He Mengying*, Zhang Shan*,
Tang Chunlei, Kang Zhensheng, Wang Xiaojie**
(*State Key Laboratory of Crop Stress Biology for Arid Areas and College of Plant Protection*, *Northwest A&F University*, *Yangling* 712100, *China*)

Abstract: Wheat stripe rust caused by *Puccinia striiformis* f. sp. *tritici* (*Pst*), is one of the most serious economic losses in wheat production worldwide. To assess the effects of plant pathogens' enemies on their host, which is potential for use the parasites in biological control strategies of plant pathogenic fungus. Here, we focus on a novel hyperparasitic fungus strain of this obligate bitrophic pathogen *Pst. This* hyperparasitic was identified as *Simplicillium obclavatum* by molecular data and morphological features observation. Moreover, this hyperparasite just affected the growth and development of the rust urediniospore, which caused the rust production reduced. Consequently, S. obclavatum has the potential to be used as a biological control to fend off of *Pst*.

Key words: *Puccinia striiformis*; *Simplicillium obclavatum*; hyperparasite; biological control; wheat stripe rust

* First authors: Fan Xin, Postgraduate Student, Focus on phytoimmunology; E-mail: 15029926806@163.com
He Mengying, Postgraduate Student, Focus on phytoimmunology; E-mail: 18392053924@163.com
Zhang Shan, Postgraduate Student, Focus on phytoimmunology; E-mail: dorothyzl@163.com
** Corresponding author: Wang Xiaojie, professor, focus on interaction between wheat and *Pst*; E-mail: wangxiaojie@nwsuaf.edu.cn

Transcriptomic reprogramming of wheat stripe rusturediniospores and basidiospores during infect wheat and barberry

Zhao Jing*, Duan Wanlu, Wang Long, Tian Song, Zhuang Hua, Kang Zhensheng**

(*State Key Laboratory of Crop Stress Biology for Arid Areas and College of Plant Protection*, *Northwest A&F University*, *Yangling* 712100, *China*)

Abstract: *Puccinia striiformis* f. sp. *tritici* (*Pst*) is the causal agent of wheat stripe rust, which cause severe yield losses to wheat production all over the world. As a macrocyclic heteroecious rust, *Pst* can infect two exclusive and unrelated plant hosts. Urediniospores infect wheat and cause disease epidemic, while basidiospores infect barbery to complete sexual stage. This complex life cycle poses interesting questions in the different mechanisms of pathogenesis underlying the infection of two hosts. In the present study, transcriptomes of *Pst* from leaves from wheat and barberry were compared. As a result, 198 genes specifically expressed in wheat (WEG) were identified, which were far less than 2575 genes specifically expressed in barberry (BEG). BEGs are mainly enriched in "gene expression", "cellular protein metabolic process", "nucleic acid metabolic process" and "biosynthesis of antibiotics", while WEGs are related to "carbohydrate catabolic process". Additionally, the percentage of new genes in WEGs was significantly greater than that in BEGs, implying that the original host of *Pst* is barberry other than wheat or grass. Expression patterns of genes encoding cell wall degrading enzymes were compared, and the results demonstrated that urediniospores and basidiospores use different strategies to destroy the host cell walls according to their constituent. These results represent the first analysis of the transcriptional profile of the *Pst* basidiospores in barberry, and will contribute to a better understanding of evolutionary and pathogenic strategies for different type of rust spores during infecting wheat and barberry.

Key words: wheat stripe rust; barberry; transcriptome; urediniospore; basidiospore

* First author: Zhao Jing, Focus on Plant Pathology; E-mail: zhaojing@ nwsuaf. edu. cn

** Corresponding author: Kang Zhensheng, Focus on Plant Pathology; E-mail: kangzs@ nwsuaf. edu. cn

自然条件下中国小麦条锈菌冬孢子的产生和萌发力

陈　文[1,2*]，孟　岩[1]，刘　尧[1]，赵　杰[1**]，康振生[**]
（1. 西北农林科技大学植物保护学院 旱区作物逆境生物学国家重点实验室，杨凌　712100；
2. 贵州省农业科学院植物保护研究所，贵阳　550006）

摘　要：小麦条锈菌（*Puccinia striiformis* Westend. f. sp. *tritici* Eriks. & Henn.）引起的条锈病是严重威胁我国小麦安全生产的重要病害之一，一般流行年份致使小麦减产10%~20%，重发年份可达40%~60%，甚至绝收。小麦条锈菌是一种转主寄生、全型锈菌，其夏孢子、冬孢子发生在小麦上，萌发的冬孢子产生担孢子，侵染感病转主寄主小檗（*Berberis*）进行有性生殖过程，完成其性孢子和锈孢子阶段。因此，有活力的冬孢子对有性生殖的发生至关重要。目前，已报道在中国自然条件下在感病小檗和小麦共生区域存在小麦条锈菌有性生殖过程，表明小麦收获后条锈菌冬孢子能存活到翌年春季为感病小檗提供初始菌源，然而缺乏冬孢子存活的直接证据。此外，条锈菌冬孢子除在小麦生长后期产生外，是否在小麦其他生长阶段产生并保持活力也不明确。因此，本研究于2018年和2019年，在中国小麦条锈病的主要流行区四川、甘肃、陕西、云南、贵州等区域开展了条锈菌冬孢子的产生和萌发力调查与研究。两年的研究表明：在这些区域在秋季和冬季小麦生长期条锈菌均能产生冬孢子，在0.25 m^2范围冬孢子平均产生率为14.33%（各地介于0.93%~44.44%），冬孢子平均萌发率为6.80%（各地介于4.39%~12.06%）。在小檗抽嫩芽和新梢的2—7月条锈菌冬孢子平均萌发率为5.70%（介于2.07%~9.70%）。对室外条件下小麦秸秆草堆中叶片或叶鞘中残留的条锈菌冬孢子萌发力研究，两年研究表明：在我国高海拔省份青海从1—5月条锈菌冬孢子均具有不同程度的萌发率，1月的平均萌发率为22.24%（14.19%~19.73%），3月的为8.5%（5.78%~10.73%），4月的为3.73%（3.12%~4.33%），5月的为3.72%（1.41%~7.65%），6月的为15.01%（1.03%~15.13%）。本研究结果表明，自然条件下在小麦生长期各阶段均可产生条锈菌冬孢子且具有萌发力；麦秸垛为条锈菌冬孢子的越冬存活提供了良好的场所。具有活力的条锈菌冬孢子与小檗生长的物候期重叠，通过风传播到感病小檗，在适宜的气候条件下，萌发产生担孢子侵染小檗，完成有性循环。

关键词：条锈病；条锈菌；冬孢子；产生与萌发；有性循环

* 第一作者：陈文；E-mail：cw0708@163.com
** 通信作者：康振生；E-mail：kangzs@nwsuaf.edu.cn
赵杰；E-mail：jiezhao@nwsuaf.edu.cn

稻曲病菌关键候选效应蛋白 SCRE6 的毒性功能研究*

郑馨航**，方安菲，邱姗姗，张　楠，李月娇，高　涵，孙文献***
（中国农业大学植物保护学院，农业部作物有害生物监测与绿色
防控重点实验室，北京　100193）

摘　要：稻曲病是由 *Ustilaginoidea virens*（Cooke）Takah 所引起的，发生在水稻开花期至乳熟期的一种穗部病害，近年来该病害的发生日益严重，不仅降低水稻产量，而且由稻曲菌产生的稻曲菌素严重影响稻米品质。然而，对于该病害致病机制的理解十分有限。病原菌分泌的效应蛋白能够抑制寄主植物的免疫反应，因而在致病过程中扮演着重要角色。笔者实验室前期借助水稻病原细菌谷枯菌与本氏烟的互作系统，在谷枯菌中表达并分泌稻曲菌效应蛋白，筛选、获得大量可以抑制谷枯菌诱导的本生烟过敏反应的效应蛋白。本研究进一步探究了其中一个关键候选效应蛋白 SCRE6 的毒性功能及其在寄主中的潜在靶标。在本生烟中 SCRE6 能够抑制由 BAX、INF1 诱导的细胞死亡。在水稻里异源表达 SCRE6 能够抑制 PAMPs 诱导的 PR 基因表达及 MAPK 的激活。通过酵母双杂交、荧光素酶互补实验和免疫共沉淀实验验证了 SCRE6 与水稻中的 MAPK 发生互作，并且可以特异的增强其稳定性，目前推测 SCRE6 很可能通过靶标 PUB 家族的 E3 泛素连接酶的途径来行使这一功能，但具体的机制尚不可知。这些研究结果为后续揭示 SCRE6 的毒性功能及其致病分子机理奠定了重要基础。

关键词：稻曲菌；效应蛋白；靶标

---

* 基金项目：稻曲病菌致病关键效应蛋白的鉴定及其毒性功能的分子机理（21026050）

** 第一作者：郑馨航，博士研究生，研究方向：植物与病原真菌互作分子机理研究；E-mail：1761393448@qq.com
*** 通信作者：孙文献，教授，主要从事水稻与病原细菌、真菌的互作分子机理研究；E-mail：wxs@cau.edu.cn

稻曲病菌效应蛋白 SCRE10 的功能研究*

邱姗姗**，方安菲，张　楠，郑馨航，李月娇，孙文献***
（中国农业大学植物保护学院，农业部作物有害生物监测与
绿色防控重点实验室，北京　100193）

摘　要：稻曲病菌在侵染水稻过程中，能够分泌大量的效应蛋白来参与水稻的分子互作与协同进化。本课题前期基于稻曲病菌基因组与预测的效应蛋白组，利用非寄主本氏烟瞬时表达体系，大规模筛选出能抑制植物免疫的候选效应蛋白。本研究选取了其中的候选效应蛋白作为研究对象，其中 SCRE10 能够抑制 INF1 和 BAX 预表达 6h 后诱导的细胞死亡，在稻曲病菌侵染过程中的表达模式分析发现，SCRE10 在侵染前期能够大量表达。利用酵母转化酶（invertase）缺陷菌株 YTK12 中 SUC2 分泌系统，验证了该候选效应蛋白信号肽均能够引导蛋白的分泌。其次通过酵母双杂交筛选体系，从含 300 个抗病相关候选基因的小型文库筛选出互作蛋白 OsSLR1，通过荧光素互补技术（Luciferase Complementation Imaging）和水稻 Co-IP 实验进一步验证。通过截断体分析发现，SCRE10 与 OsSLR1 通过 GRAS Domain 发生互作。通过筛选水稻中 15 个 JAZs 蛋白发现，SCRE10 能够与 OsJAZ3、OsJAZ5、OsJAZ8、OsJAZ9 和 OsJAZ11 等多个 JAZs 蛋白互作，用 MeJA 处理转基因株系后，SCRE10 能够增强 *OsLox*2 的表达。但是 SCRE10 与 OsSLR1、JAZs 蛋白互作的生物学意义需要进一步研究。

关键词：稻曲病菌；效应蛋白；JAZs 蛋白

* 基金项目：国家自然科学基金项目 31471728

** 第一作者：邱姗姗，博士研究生，研究方向：植物与病原真菌互作；E-mail：13121954743@163.com

*** 通信作者：孙文献，教授，主要从事植物与病原真菌互作；E-mail：wxs@cau.edu.cn

稻曲病菌中稻绿核菌素生物合成基因簇的鉴定与功能分析*

李月娇**，王　明，刘朝辉，张　亢，王黎锦，崔福浩，孙文献***
（中国农业大学植物保护学院，农业部作物有害生物监测与绿色防控重点实验室，北京　100193）

摘　要：稻曲病目前已成为我国水稻生产上的三大病害之一，该病原菌在田间稻穗上形成体积数倍于稻粒的稻曲球，严重危害我国水稻的产量。此外，稻曲球中含有一类对植物、动物以及人类有害的真菌毒素——稻绿核菌素。目前，对稻曲病菌中稻绿核菌素的生物合成途径尚不明确。本研究结合生物信息学的研究基础，通过基因敲除与生化功能研究确定在稻曲菌中编码聚酮合酶的基因 *UvPKS*1 所在的基因簇负责稻绿核菌素生物合成。该基因簇全长 49.7 kb，包含 14 个基因。UvPKS1 负责稻绿核菌素生物合成的第一步聚酮反应，在 *UvPKS*1 的敲除突变体中，未检测到任何稻绿核菌素与中间产物。在 *ugsO* 的敲除突变体中，稻绿核菌素的产量下降且稻绿核菌素 N/E 与 M/D 的产量比值相对于野生型明显增大，推测 ugsO 参与生物合成过程中氧化还原反应。*ugsT* 编码转运蛋白，敲除该基因后丧失了产生稻绿核菌素的能力。在 Δ*ugsJ* 中，检测到缺少 C3 甲基化修饰的稻绿核菌素 F、G 和 A，证明 *ugsJ* 编码甲基转移酶并负责 C3 甲基化修饰。*ugsL* 编码漆酶，该基因敲除后未检测稻绿核菌素，检测到稻绿核菌素的单体化合物 3-methyl-dihydro-nor-rubrofusarin，证明 *ugsL* 负责生物合成途径中的二聚化过程。此外，*ugsR*2 预测为转录因子，该基因敲除后对稻绿核菌素的产量以及基因簇中相关基因的表达没有明显影响。综上，结合生物信息学预测与生化功能研究鉴定了稻曲菌中稻绿核菌素的生物合成途径，为进一步研究稻绿核菌素在稻曲菌中的生理功能奠定基础。

关键词：稻绿核菌素；基因敲除；生物合成；基因簇

---

* 基金项目：国家自然科学基金项目（31471728）

** 第一作者：李月娇，博士研究生，研究方向：植物与病原真菌互作；E-mail：molly_lyj@126.com

*** 通信作者：孙文献，教授，主要从事植物与病原细菌、真菌互作；E-mail：08042@cau.edu.cn

The rice false smut pathogen *Villosiclava virens* secretes a class v chitinase to suppress plant immunity*

Li Guobang**, Fan Jing, Li Yan, Wang Wenming ***

(*Rice Research Institute*, *Sichuan Agricultural University*, *Chengdu*, 611130, *China*)

Abstract: Rice false smut (RFS), caused by *Villosiclava virens* (*Vv*) (anamorph: *Ustilaginoidea virens* (Cooke) Takahashi), is an important panicle disease in rice causing huge yield loss. The fungal pathogen *Vv* infects the developing spikelets of rice panicle at booting stage and converts grains into false smut balls, contaminating grains with its mycotoxins, which threaten food safety of rice. Therefore, investigation on *Vv* fungus-rice interactions becomes one active research area in plant pathology and rice biology.

We identified a candidate effector secreted by *Vv*, designated Uv5918, specifically up-regulated upon infection of rice spikelets. Overexpression of Uv5918 resulted in significantly enhanced pathogenicity of *Vv*. Ectopic expression of Uv5918 in *Arabidopsis* and rice led to compromised defensive responses and enhanced disease-susceptibility. In addition, Uv5918 encodes a chitinase that binds to chitin and results in decrease of chitin *in vitro*. Taken together, our data indicate that Uv_5918 may play dual roles in pathogenesis by suppression of host defense.

Key words: Rice false smut (RFS); *Villosiclava virens*; v chitinase; plant immunity

---

* Funding: This work was supported by the National Natural Science Foundation of China (grants 31501598 to J. Fan)

** First author: Li Guobang, Graduate student; E-mail: guobangli_power@163.com

*** Corresponding author: Wang Wenming; E-mail: j316wenmingwang@sicau.edu.cn

四川盆地稻瘟病菌无毒基因分布频率与变异分析

胡孜进*，李 燕，樊 晶，黄衍焱，王文明**
（四川农业大学水稻研究所，成都 611130）

摘 要：稻瘟病是水稻生产上的重大病害之一，每年都造成水稻产量的严重损失。四川盆地寡日照、小温差及高湿度的气候特点，使得稻瘟病常年多发，成为制约水稻安全生产的瓶颈问题。

为明确四川盆地无毒基因的分布及变异情况，以采自四川盆地东北部营山、通江和大竹地区的共 238 个稻瘟菌菌株的基因组 DNA 为模板，用 5 个无毒基因（AVR-Pik，AVR-Pi9，AVR-Pii，AVR-Pib，AVR-Pi54）的特异引物进行 PCR 检测及测序分析。结果表明，这 5 个基因的检出率分别为 100%，98.32%，34.03%，100%和 100%。其中通江和大竹菌株的 AVR-Pik 基因型均为含有 AVR-Pik-A 和 AVR-Pik-D 两个拷贝的混合型，而营山的菌株主要由 AVR-Pik-E 组成。对 AVR-Pib 基因 PCR 产物测序分析显示，有 135 个菌株在启动子区域存在转座子 Pot3 的插入；而检测出的 AVR-Pii，AVR-Pi9 和 AVR-Pi54 的 PCR 产物测序分析暂未发现插入、缺失及单碱基变异等。后续将对四川盆地其他地区的无毒基因分布及变异情况进行分析，从而对水稻抗病品种的合理布局提供参考。

关键词：稻瘟菌；无毒基因；分布频率；基因变异；品种布局

* 第一作者：胡孜进，博士研究生；E-mail：592517252@ qq. com
** 通信作者：王文明，教授；E-mail：j316wenmingwang@ sicau. edu. cn

一种由可可球二孢菌引起的槟榔新病害*

唐庆华[1**]，王慧卿[2]，许才得[2]，余凤玉[1]，于少帅[1]，宋薇薇[1]，牛晓庆[1]，覃伟权[1***]
（1. 中国热带农业科学院椰子研究所，文昌 571339；
2. 海南热带海洋学院生态环境学院，三亚 572022）

摘　要： 槟榔是海南省最具特色的重要经济作物之一，全省200多万农民以槟榔为主要经济来源。近年来，以槟榔黄化病为代表的“黄化”灾害日趋严重，现已制约着海南槟榔产业的健康发展。随着研究的深入，发现除黄化病外，炭疽病、病毒病等多种病害均可以在槟榔叶片产生黄色病斑，且多种叶部病害症状类似、难以区分。因此，对病原进行准确鉴定、准确区分不同病害对“黄化”病害的有效防控尤为重要。2018年12月，笔者从采集于定安县富文镇与保亭县南林乡的病样上分别获得4个和10个菌株，用引物ITS1/ITS4对2个代表性强致病力菌株DAFW-8与BTNL-1的核糖体DNA内转录间隔区（internal transcribed spacer）进行了PCR扩增。序列比对结果表明，2条序列与可可球二孢菌 *Botryodiplodia theohromae* 菌株HY-23的ITS序列（MK370861）同源性高达99.42%。同时，该菌初始菌落呈白色或米白色，绒毛状，生长2 d后开始产生绿色色素，3 d后菌落变成灰绿色，5 d后菌落呈灰色至黑褐色。30 ℃恒温光照培养12 d后产生碳黑色的子座，分生孢子初始为单细胞、无色，成熟后为双细胞、褐色。结合ITS序列和形态学特征，笔者将DAFW-8与BTNL-1鉴定为可可球二孢菌（亦有文献称为可可毛色二孢菌）*B. theohromae*。选取菌株DAFW-8做了进一步的生物学特性研究。结果表明，菌株DAFW-8菌丝最适生长温度为30 ℃，最适生长碳源和氮源分别为蔗糖和蛋白胨，pH值为5~8时较适于菌丝生长。迄今，尚无可可球二孢菌侵染槟榔引起病害的报道。

关键词： 槟榔；黄化灾害；可可球二孢菌；分子鉴定

---

* 基金项目：海南省农业厅重点项目：“槟榔黄化灾害防控及生态高效栽培关键技术研究与示范”（ZDKJ201817）；中国热带农业科学院基本科研业务费专项：槟榔产业技术创新团队-槟榔病虫害综合防控技术研究（1630152017015）

** 第一作者：唐庆华，博士，助理研究员，研究方向为植原体病害综合防治及病原细菌-植物互作功能基因组学；E-mail：tchuna129@163.com

*** 通信作者：覃伟权，研究员；E-mail：QWQ268@163.com

自育型致病疫霉菌株有性生殖发生及后代分离特点*

梁静思[1,2,3**]，张荣英[1]，陶　宇[1]，汤淑丽[1]，唐　唯[1,2,3***]

（1. 云南师范大学生命科学学院，昆明　650500；2. 云南师范大学马铃薯科学研究院，昆明　650500；3. 云南省教育厅马铃薯生物学重点实验室，昆明　650500）

摘　要：由致病疫霉引起的晚疫病是云南省马铃薯生产中最严重的病害之一。为了探讨致病疫霉有性生殖后代群体遗传特点，本研究通过从昆明市嵩明县采集致病疫霉菌株 C88SM 和昆明市寻甸县采集 127P，经单孢分离及纯化后，通过对峙培养确定其交配型为自育型并通过流式细胞仪确定其染色体倍性为二倍体。随后，利用前期探索的卵孢子培养和诱导萌发条件，以 C88SM 为亲本利用卵孢子萌发得到 F_1，在 F_1代中随机选取 1 株交配型为自育型的菌株，诱导得到 F_2，以此类推得到 F_3-F_5后代群体共 93 株，包括 F_1代（12 株），F_2代（12 株），F_3代（24 株），F_4代（19 株）和 F_5代（26 株）；以 127P 菌株为亲本做相同处理，共得到自育后代共 83 株，包括 F_1代（11 株），F_2代（12 株），F_3代（16 株），F_4代（20 株）和 F_5代（24 株）。倍性检测结果表明，176 株后代染色体倍性和亲本一致，均为二倍体，没有发生分离。交配型检测结果表明，C88SM 的自育后代 F_1-F_4代交配型均为自育型，F_5代交配型发生分离，发现了 3 株 A0 交配型；127P 的自育后代中，F_1、F_2、F_4代后代群体交配型均为自育型，没有发生分离，在 F_3代群体及 F_5代群体共发现 2 株 A0 交配型。本研究可为解析致病疫霉有性生殖分离特点提供理论支持，也可为预测云南省致病疫霉群体结构动态变化和病害流行趋势提供指导。

关键词：马铃薯；致病疫霉；有性生殖；自育型；染色体倍性

---

* 基金项目：国家自然科学基金项目（31660503）；云南师范大学大学生科研训练基金项目（ky2018-136）

** 第一作者：梁静思，硕士研究生，研究方向：植物病理；E-mail：569020457@ qq. com

*** 通信作者：唐唯，讲师，研究方向：植物病理；E-mail：4311@ ynnu. edu. cn

禾谷镰刀菌 MAPK 信号通路间互作关系的初探*

任静毅[1**]，李程亮[1]，高承宇[1]，许金荣[2]，江　聪[1,2]，王光辉[1,2***]
（1. 西农-普度大学联合研究中心，旱区作物逆境生物学国家重点实验室，西北农林科技大学植物保护学院，杨陵　712100；2. 普度大学植物及植物病理系，印第安纳州　IN47907）

摘　要：小麦赤霉病（Fusarium head blight）是小麦的一项毁灭性真菌病害，禾谷镰刀菌（*Fusarium graminearum*）为该病的主要病原菌。丝裂原活化蛋白激酶 MAPK（Mitogen-activated protein kinase）级联途径在禾谷镰刀菌营养生长，有性生殖及致病过程中扮演重要作用。前人的研究工作分别揭示了禾谷镰刀菌中三条 MAPK 通路的主要功能：其中 *MGV*1 通路主要参与调控细胞壁完整性，*FgHOG*1 通路主要调控细胞的渗透压调节，*GPMK*1 通路参与一些细胞壁降解酶的合成。为揭示禾谷镰刀菌中三条 MAPK 途径之间的关系，在本研究中构建了 *gpmk*1 *mgv*1，*gpmk*1 *Fghog*1，*mgv*1 *Fghog*1 双敲突变体，并进行一系列表型观察。其中 *GPMK*1 和 *FgHOG*1 的双重缺失表现为生长减慢和对渗透压敏感；*GPMK*1 的缺失加剧了 *mgv*1 突变体的生长速率和细胞壁完整性缺陷；值得一提的是，*FgHOG*1 的缺失部分恢复了 *mgv*1 突变体菌丝生长和细胞壁完整性的缺陷。另外，Mgv1 的缺失也减轻了 *FgHog*1 突变体在菌丝生长时期对高渗透压的敏感性。TEY 实验进一步表明在 *mgv*1 *Fghog*1 双敲突变体中，特别是在细胞壁压力胁迫的条件下，Gpmk1 的蛋白磷酸化活性升高，这些结果表明 3 条 MAPK 通路间存在遗传互作。

关键词：细胞壁完整性；渗透压调节；MAPK 信号通路；禾谷镰刀菌

*　基金项目：国家自然科学基金（No. 31801684）；国家大学生创新创业计划（No. 201810712103）
**　第一作者：任静毅，硕士研究生
***　通信作者：王光辉，主要从事小麦赤霉病研究；E-mail：wgh2891458@163. com

梨轮纹病和干腐病病原菌 *Botryosphaeria dothidea* 的基因型鉴定*

肖 峰，洪 霓，王国平**

（华中农业大学植物科学技术学院，农业微生物国家重点实验室，武汉 430070）

摘 要：目前已研究证实在我国主栽梨树发生的轮纹病和干腐病，均由 *Botryosphaeria dothidea* 所致，为同一种病原菌引起的两种症状表现。本研究从我国 16 个省份的梨主产区采集梨树轮纹病和干腐病样品，通过组织分离法获得 461 个 *B. dothidea* 菌株。从所获得的菌株的菌丝分别提取 DNA，根据其小亚基核糖体 rDNA 中 group Ⅰ内含子插入类型的不同，通过 PCR 鉴定分析，这些菌株分为 2 类基因型，即 genotype Ⅰ和 genotype Ⅱ，其中属于 genotype Ⅰ型的菌株 69 个，属于 genotype Ⅱ型的菌株 392 个。依据菌株地域来源的不同，挑选其中的 5 个 genotype Ⅰ型菌株和 15 个 genotype Ⅱ型菌株进行 ITS 序列结合 *EF*1-α 基因系统进化发育树分析，结果显示，两种基因型的代表菌株分别聚于两个不同的进化分支，证实 genotype Ⅰ型菌株与 genotype Ⅱ型菌株存在一定的遗传距离。生物学特性比较分析发现，两种类型的菌株在菌落形态、菌丝生长速率、分生孢子形成及其形态上均存在明显差异，genotype Ⅰ型菌株在 PDA 培养基上较 genotype Ⅱ型菌株更易诱导产生分生孢子，但 genotype Ⅱ型菌株在离体梨叶片上形成的分生孢子器则比 genotype Ⅰ型菌株明显多。两种类型的菌株接种于离体和活体梨枝条的结果看出，genotype Ⅰ和 genotype Ⅱ菌株都能引起梨干腐病，但 genotype Ⅱ型菌株的致病力明显强于 genotype Ⅰ型菌株，而 genotype Ⅱ型菌株还可产生梨轮纹病。本研究结果为进一步认识梨轮纹病和干腐病的病原及 *B. dothidea* 的致病多样性提供了新的有用信息。

关键词：梨；轮纹病；干腐病；*Botryosphaeria dothidea*；基因型鉴定

* 基金项目：农业产业技术体系（CARS-28-15）；国家重点研发项目（2018YFD0201406）

** 通信作者：王国平；E-mail：gpwang@ mail. hzau. edu. cn

中国梨胴枯病病原间座壳菌的种类多样性研究*

郭雅双，白　晴，傅　敏，洪　霓，王国平**
（华中农业大学植物科学技术学院，武汉　430070）

摘　要：梨胴枯病是近年来在中国栽培梨上新发生的一种枝干病害，严重威胁梨产业的发展。经本实验室的前期鉴定研究，梨胴枯病系由间座壳菌（*Diaporthe* spp.）所致。为明确我国不同梨产区及不同的梨栽培种的胴枯病病原菌种类，及其种类多样性与菌株的地域和梨种来源之间的关系，本研究从中国12个省份的不同地区栽培的砂梨（*Pyrus pyrifolia*）、白梨（*P. bretschneideri*）、秋子梨（*P. ussuriensis*）和西洋梨（*P. communis*）上，采集表现胴枯病典型症状的病枝样品，通过组织分离法获得453个间座壳菌分离株。选取其中的113个分离株作为代表菌株，进行形态学（菌落形态、无性态及有性态等）观察比较和多基因位点（包括ITS、TEF1、CAL、HIS和TUB）序列系统发育进化树分析，结果显示，这些代表菌株分属于间座壳属的19个种，其中 *Diaporthe eres*、*D. hongkongensis*、*D. sojae*、*D. unshiuensis*、*D. fusicola*、*D. velutina*、*D. ganjae*、*D. caryae*、*D. padina*、*D. citrichinensis*、*D. taoicola*、*D. cercidis* 和 *D. pescicola* 等13个为已报道的种，*D. acuta*、*D. chongqingensis*、*D. fulvicolor*、*D. parvae*、*D. spinosa* 和 *D. zaobaisu* 等6个为本研究新描述的种。以上已鉴定明确的19种间座壳菌中，除 *D. eres* 外，其余均为侵染梨树的首次报道。研究结果表明，引起梨胴枯病的间座壳菌具有丰富的种类多样性，不同种类的流行率与地域和梨种来源密切相关。19种间座壳菌接种于离体和活体梨枝条的结果显示，它们均可使梨枝致病，但其致病力之间存在明显差异。本研究为深入认识间座壳菌的形态变化和分子变异提供了新的科学证据，并为我国梨胴枯病的防治提供了重要的理论依据。

关键词：梨胴枯病；间座壳菌；种类多样性

---

* 基金项目：农业产业技术体系（CARS-28-15）；国家重点研发项目（2018YFD0201406）

** 通信作者：王国平；E-mail：gpwang@ mail. hzau. edu. cn

山西省红芸豆根腐病病原菌鉴定*

翟雅鑫[1**]，薛丽芳[1]，任美凤[2]，王燕[3]，李新凤[1]，郝晓娟[1***]，王建明[1***]
（1. 山西农业大学农学院，太谷 030801；2. 山西省农业科学院植物保护研究所，太原 030032；3. 山西省农业科学院农作物品种资源研究所，太原 030031）

摘 要：红芸豆是芸豆（*Phaseolus vulgaris*）的一种，籽粒硕大，营养丰富，具有较高的食用价值和经济价值。山西省是我国红芸豆主产地之一，目前全省种植面积已达 3.3 万 hm^2左右。随着种植面积和年限的增加，红芸豆根腐病的发生日益加重，给当地生产造成了严重损失。本研究采用组织分离法对采自山西省岢岚县的红芸豆根腐病病株进行病原菌分离，通过形态学与分子生物学方法进行病原菌种类鉴定，利用柯赫氏法则验证其致病性。结果表明，病原菌在 PDA 培养基上气生菌丝卷毛状，菌落白色至粉色，背面初为橘黄色，后期变为酒红色；单瓶梗产孢；小型分生孢子长椭圆形，0~1 分隔，（7.89~17.14）μm×（2.58~4.63）μm；大型分生孢子镰刀形，多为 3 分隔，（20.45~32.19）μm×（2.58~6.32）μm。结合形态特征和 ITS 序列分析结果，将其鉴定为三线镰刀菌（*Fusarium tricinctum*）。以往报道中，山西省红芸豆根腐病病原菌被认为是单一的腐皮镰刀菌（*Fusarium solani*）。但根据课题组前期研究结果可知，红芸豆根腐病病原菌种类比以往认知的要复杂。本结果为进一步研究山西省红芸豆根腐病病原菌种群结构及该病害的防治奠定了基础。

关键词：红芸豆；根腐病；三线镰刀菌（*Fusarium tricinctum*）；分离；鉴定

---

* 基金项目：山西省重点研发计划项目（201803D221012-1）；山西省农业科学院特色农业技术攻关项目（YGG17108）；山西省面上青年基金项目（201701D221207）

** 第一作者：翟雅鑫，硕士研究生，研究方向为植物病理学；E-mail：yaxinzhai@126.com
*** 通信作者：郝晓娟，副教授，研究方向为植物病原与分子病理；E-mail：xiaojuanhao@126.com
王建明，教授，植物病原与分子病理；E-mail：jm.w@163.com

玉米大斑病菌 *Stflo8* 基因的功能研究*

王 擎[1**]，龙 凤[1]，郝志敏[1]，董金皋[1,2]，申珅[1***]
（1. 河北农业大学生命科学学院，河北省植物生理与分子病理学实验室，
保定 071001；2. 河北农业大学植物保护学院，保定 071001）

摘 要：玉米是我国重要的粮食作物，玉米大斑病频发于我国的冷凉地区，暴发严重时会导致玉米减产 50%，给农业生产带来巨大的经济损失。由于玉米大斑病菌小种较多且变异频繁，抗病品种的研发种植和药物的防治效率较低，给防治带来巨大挑战，因此，对病菌致病机制的研究成为当前研究的重点。cAMP 途径是病原真菌中较为普遍的信号通路，*Stflo*8 是 cAMP 信号通路下游的重要转录因子，在玉米大斑病菌的致病过程中发挥重要的调控作用。为明确 *Stflo*8 在病菌侵染过程中发挥作用的机制，本研究通过构建 *Stflo*8 敲除载体、PEG 介导丝状真菌转化获得 67 株抗性转化子，通过 DNA 水平的验证获得 2 株阳性 *Stflo*8 缺失突变体，分析突变体表型以明确该基因的功能。结果显示，突变体菌丝形态出现不规则膨大，生长速率降低了 25%，分生孢子产量显著下降，仅为野生型菌株的 3. 2%，将菌丝孵育在玻璃纸上诱导，12h 时，野生型菌株附着胞生成，24h 时，侵入钉产生，*Stflo*8 缺失突变体则一直呈现菌丝状态，不具备生成侵入钉的能力，同时穿透能力也丧失，进一步测定胞内黑色素含量，发现突变体黑色素较野生型显著下降，说明其合成能力有所弱化。结果表明 *Stflo*8 在玉米大斑病菌的产孢、萌发、侵入等侵染过程中发挥一定的调控作用；本研究为明确病菌侵染过程中主要功能基因的作用机制、有效防治玉米大斑病奠定了一定的理论基础。

关键词：玉米大斑病菌；*Stflo*8 基因；病菌侵染；缺失突变体

---

* 基金项目：国家自然科学基金项目（No. 31601598）；河北省自然科学基金项目（No. 2019204120）
** 第一作者：王擎，硕士研究生，研究方向为微生物与生化药学；E-mail：690586631@ qq. com
*** 通信作者：申珅，博士，讲师，研究方向为植物病原真菌防治；E-mail：shenshen0428@ 163. com

基于转录组学分析茶叶与茶白星病菌的互作研究

周凌云[1,2]，刘红艳[1]，李 维[1]，向 芬[1]，银 霞[1]，曾泽萱[1]，王振中[2]
（1. 湖南省农业科学院 茶叶研究所，长沙 410125；2. 华南农业大学农学院，广州 510642）

摘 要：茶白星病是一种高海拔茶区普遍发生的真菌病害，其发生规律及防控已被广泛研究，但茶白星病菌与茶树的分子机理仍不清楚。本研究选取茶白星病菌侵染茶树 0，7 d，14 d 的对照组和试验组 5 个样本为研究对象，进行了 RNA-seq 分析。结果表明，茶白星病菌侵染茶叶的转录组测序共产生了 33. 78Gb Clean Data，其中 81. 88%～82. 83%的测序数据定位到茶树参考基因组上。通过对 5 个样品基因差异表达水平进行分析，共得到 2 180条差异表达基因，1 042条上调，1 138条下调。其中有 2 个基因是病变后一直存在的差异性表达基因，一个为细胞色素 P450，一个为未知新基因。通过 GO 功能显著性富集分析发现接种叶片在 3 个功能区均有大量基因分布，其中接种 7 d 的差异 DEGs 富集区主要为细胞成分区，以上调基因为主，尤其是接种 14 d差异 DEGs 细胞成分全部为富集的上调基因。KEGG 代谢通路分析，接种 7 d（P-Value<0. 05）的硫代谢通路中，3 个差异表达基因富集。而接种 14 d（P-Value<0. 05）主要通路中有 28 个在次级代谢通路中，21 个基因富集于生物合成通路中，19 个在次生代谢产物的生物合成，9 个基因富集于类黄酮生物代谢通路中，8 个基因富集于光合作用-触角蛋白质通路中。茶叶与茶白星病菌互作是一个成本高的过程，第 7 d 应对病菌胁迫的应激方法较为强烈，而且在细胞成分上也有相应的富集，并明确其中关键上调基因分别为编码 Hsp90. 1、Hsp70 及 TT4 等基因。为进一步发掘与茶白星病抗性相关的重要基因及生物学通路提供依据，对茶叶抗病育种工程具有重要意义。

关键词：茶；茶白星病；转录组；互作

莲藕腐败病病原菌的分离鉴定及其致病力测定*

邓 晟**，王 锦，魏利辉***
（江苏省农业科学院植物保护研究所，南京 210014）

摘 要：在2017—2018年，本研究团队从江苏扬州、浙江金华、湖北武汉、潜江和江西广昌等地收集了6个批次的发生莲藕腐败病的莲藕样品，从中分离出21株疑似莲藕腐败病病原菌，其中扬州1株、金华6株、武汉10株、潜江2株、广昌3株。笔者对部分菌株进行了病原菌回接、致病力测定以及保守序列的测序分析。

通过对来自上述5个地点的菌株进行藕节菌碟接种以及实生苗接种，并以西瓜枯萎病菌、番茄枯萎病菌和水处理作为对照，发现：①分离到的菌株均可以导致藕节接种部位出现扩展的、褐化腐烂的病斑，但非莲藕分离的对照菌株则只能形成较小、颜色较浅的病斑；②莲藕上分离的菌株接种实生苗后，均不同程度导致实生苗枯死，但对照菌株则几乎不影响实生苗的存活率；③在上述测试条件下，‘广昌2-2’菌株的致病力最强，‘扬州藕1st’菌株的致病力次之，非莲藕寄主的尖孢镰刀菌致病力最弱。

在进行致病力测定以后，笔者扩增了部分菌株的ITS、mtSSU、EF1α这3个基因组保守位点的序列并测序，经NCBI序列比对后发现，全部ITS序列都比对为尖孢镰刀菌（*Fusarium oxysporum*），但mtSSU序列和EF1α序列的比对结果则把莲藕腐败病菌分为2个种，其中‘广昌2-2’为*F. commune*，而‘扬州藕1st’则为*F. oxysporum*。有文献表明，*F. commune*和*F. oxysporum*是两个亲缘关系非常近的种，而且也已经有国内同行报道了*F. commune*可以引起莲藕腐败病（植物病理学报，2017，曾莉莎等）。

本研究中接种方法的确立以及相关菌株的鉴定，将为后期莲藕腐败病病原菌的快速鉴定奠定稳定可靠的方法，也为莲藕抗性品种的筛选奠定基础。

关键词：莲藕腐败病；分离鉴定；致病力测定；基因序列比对

---

* 基金项目：国家特色蔬菜产业技术体系CARS-24-C-01

** 第一作者：邓晟，博士，副研究员，主要从事土传病原真菌致病机理的研究；E-mail：xunikongjian@163.com

*** 通信作者：魏利辉，研究员，主要从事植物病原线虫、植物土传病害及其防控技术研究；E-mail：weilihui@jaas.ac.cn

水稻纹枯病菌（AG1-IA）LAMP 检测体系的建立*

张照茹**，王海宁，李昕洋，刘　伟，王应玲，李　帅，魏松红***
（沈阳农业大学植物保护学院，沈阳　110866）

摘　要：水稻纹枯病是一种普遍性病害，在我国南北方水稻种植区普遍发生，近年来在北方，病害发病趋势加重，已成为限制水稻高产和稳产的主要障碍之一。由于水稻纹枯病致病菌主要为茄丝核菌的 AG1-IA 融合群，因此建立水稻纹枯病菌（AG1-IA）田间快速检测技术体系就显得尤为重要。

本研究基于环介导等温扩增技术（Loop-mediated isothermal amplification，LAMP），以 rDNA ITS 为靶标基因序列，设计了 4 条特异性的 LAMP 引物和 1 条环引物，建立了水稻纹枯病菌（AG1-IA）的 LAMP 检测体系，并对反应体系的反应温度、反应时间、Mg^{2+} 浓度、内/外引物浓度、环引物浓度、dNTPs 浓度以及甜菜碱浓度等因素进行优化，确定最佳反应体系。利用最佳反应体系进行灵敏度和特异性试验。结果表明：特异性试验中，水稻纹枯病菌菌株 DNA 扩增后均呈阳性反应，而其他水稻常见病害以及一些土壤习居菌的供试菌株 DNA 扩增后均呈阴性反应，引物特异性良好。在灵敏度试验中，ITS-LAMP 技术最低检测限为 1 pg/μL，比常规 PCR 技术高出两个数量级，具有较高灵敏度。应用结果显示，田间发病植株与人工接种发病植株病部 LAMP 检测结果均呈阳性。该方法的建立为水稻纹枯病菌的检测提供了新的技术支持，实现了对田间水稻纹枯病菌的快速检测。

关键词：水稻纹枯病菌；rDNA ITS；环介导等温扩增技术（LAMP）；分子检测

* 基金项目：辽宁省“百千万人才工程”资助项目；现代农业产业技术体系建设专项资金资助（CARS-01）
** 第一作者：张照茹，硕士研究生，研究方向为植物病理学；E-mail：1483645128@ qq. com
*** 通信作者：魏松红，博士，教授，主要从事植物病原真菌学及水稻病害研究；E-mail：songhongw125@ 163. com

玫瑰天竺葵丽赤壳叶斑病的病原鉴定以及防治药剂筛选*

邢玉姣[1]**，张桂军[1]**，温　浩[1]**，毕　扬[1]***
（1. 北京农学院生物与资源环境学院，农业农村部华北都市农业重点实验室，北京　102206）

摘　要：2018年从北京延庆玫瑰天竺葵种植基地采集分离到了 *Calonectria canadensis*，经柯赫氏验证，该病原引起玫瑰天竺葵叶斑病，在叶片上初期侵染症状为形成浅褐色病斑，病斑面积较小，随侵染加重，病斑面积逐渐扩大，颜色加深，严重时可造成全部叶片脱落。在PDA培养基上菌丝茂密，颜色为红褐色，后期产生大量分生孢子，分生孢子无色，直圆柱型，两端钝圆，无隔膜或具1个隔膜。

C. canadensis 在玫瑰天竺葵上引起的叶斑病还未有过报道，已报道的寄主主要有蓝莓（*Vaccinium* spp.）、兰花（*Boat orchids*）等，主要侵染寄主的根、茎、叶。现有的研究主要是关于其生物学特性以及形态特征，有关防治方面的研究报道较少。本文用从北京延庆采集到的 *C. canadensis* 菌株开展实验，用菌丝生长速率的方法对腈菌唑（*Myclobutanil*）、百菌清（*Chlorothalonil*）、戊唑醇（*Tebuconazole*）、咪鲜胺（*Prochloraz*）、苯醚甲环唑（*Difenoconazole*）、氟环唑（*Epoxiconazole*）、双苯菌胺（SYP-14288）这7种杀菌剂进行药效测定和筛选。

结果显示双苯菌胺药剂对供试菌株的防治效果最好，平均 EC_{50} 值为 0.48±0.06μg/mL；咪鲜胺药剂的防治效果仅次于双苯菌胺药剂，平均 EC_{50} 值为 0.91±0.21μg/mL；戊唑醇、苯醚甲环唑在供试药剂中的防治效果较好，平均 EC_{50} 值分别为 9.59±2.25μg/mL、4.84±0.24μg/mL；腈菌唑、百菌清和氟环唑防治效果较差，平均 EC_{50} 值分别为 60.46±13.63μg/mL、38.67±0.39μg/mL、30.37±3.60μg/mL。

关键词：玫瑰天竺葵；丽赤壳叶斑病；病原鉴定；药剂筛选

---

* 基金项目：国家重点研发计划（2017YFD0201601）；北京市教委科研计划项目（KM201610020007）

** 第一作者：邢玉姣，硕士研究生；E-mail：1531457922@qq.com
张桂军，硕士研究生；E-mail：1769483624@qq.com
温浩，硕士研究生；E-mail：971091867@qq.com
*** 通信作者：毕扬，讲师，植物病害化学防治与病原菌抗药性研究；E-mail：biyang0620@126.com

河南省不同土壤类型耕作区小麦根、茎部内生真菌的多样性*

杨　岚**，李华奇，何　姗，李美霖，秦玉佳，徐建强***
（河南科技大学林学院，洛阳　471003）

摘　要：为系统地研究小麦根、茎部内生真菌的多样性，筛选到对小麦常见土传病害——纹枯病、茎基腐病有抑制作用的生防菌株，达到菌药联合作用控制小麦土传病害的目的，笔者于 2017 年 3 月与 5 月、2019 年 3 月从河南省不同土壤类型耕作区，共 14 个地市共 22 个县区采集健康小麦植株，采用常规组织分离法进行分离，共得到 426 株内生真菌；通过对不同年份、不同土壤类型小麦茎部、根部内生真菌定殖率、分离率进行比较，分析其多样性。结果发现：五种土壤类型耕作区中，黄褐土区小麦内生真菌的定殖率、分离率均最低，水稻土区最高。不同土壤类型耕作区小麦茎部内生真菌的定殖率、分离率最低为黄褐土区，最高为褐土区；根部内生真菌的定殖率、分离率最低也为黄褐土区，最高则为水稻土区。相同土壤类型耕作区小麦根部内生真菌的定殖率、分离率与茎部相比均较高；其中褐土区根部与茎部的内生真菌分离结果最为接近，水稻土区分离结果差异最大。2019 年采集的小麦植株内生真菌定殖率、分离率与 2017 年相比较低，这可能是由于越冬期受外界温度影响，真菌的活跃度与繁殖速度降低，后随着温度升高真菌渐渐活跃，2017 年包含 5 月采集的植株，因此较 2019 年 3 月植株样本分离率高。由此可以推测，小麦根、茎部内生真菌的定殖、数量受土壤类型、分离部位、生长时期等的影响。本文初步明确了小麦内生真菌的分布多样性，为进一步开展对有益内生真菌的开发利用和潜在病原菌的防治研究提供了理论基础。

关键词：小麦；内生真菌；不同土壤类型；多样性

---

* 基金项目：河南科技大学专项科研基金；国家级大学生创新创业训练计划（201710464027）；河南科技大学大学生研究训练计划（2017320）

** 第一作者：杨岚，硕士研究生，从事小麦内生真菌及生物防治研究；1435927301@ qq. com

*** 通信作者：徐建强，副教授，主要从事植物病害综合防治研究；E-mail：xujqhust@ 126. com

禾谷镰孢菌 *FgStuA* 与 cAMP-PKA 信号通路在无性孢子发育中的协同调控*

黄俊锜[1**]，曹心雨[1]，王晨芳[1***]，许金荣[1,2***]
（1. 西农-普度大学联合研究中心，旱区作物逆境生物学国家重点实验室，西北农林科技大学植物保护学院，杨凌 712100；2. 美国普渡大学植物及植物病理系，印第安纳州 IN47907）

摘 要：禾谷镰孢菌（*Fusarium graminearum*）是造成小麦赤霉病的主要病原真菌。cAMP-PKA 信号通路参与着多项发育调控，禾谷镰孢菌 PKA 激酶的两个催化亚基 *cpk*1、*cpk*2 双敲除突变体 DKO 在生长产孢致病力有性生殖等方面有着严重的缺陷。前期研究发现，cAMP-PKA 下游转录因 FgSFL1 的 C 末端 501 位氨基酸后区段缺失即可以完全恢复 DKO 的菌丝生长但却不能恢复孢子产量。*Fgsfl*1 对菌丝生长和无性孢子形成都不是必须的。有趣的是，在收集角变子的过程中，笔者收集到了 2 个菌落颜色发生变化但生长无恢复的角变菌株 JQ112 及 JQ72，经全基因组测序，笔者检测到 *FgstuA* 出现移码突变。*cpk*1 *cpk*2 双敲除突变体 DKO 和 *FgstuA* 突变体均表现为分生孢子萌发延迟，但 JQ112 完全失去了萌发能力，分生孢子中存在大量多核细胞。综合实验结果可知，PKA 和 FgSTUA 在分生孢子萌发过程中均起重要作用，并且存在功能冗余。

此外，JQ112 仍易发生角变，得到的二次角变子 JQ112s5 在 *Fgsfl*1 基因上出现了 C 端 425 位移码突变，并发现 *Fgsfl*1 的 C 端缺失并不能完全恢复 *FgstuA* 功能丧失造成的菌丝生长缺陷。通过同源重组的方法，同时敲除 *FgstuA* 与 *Fgsfl*1 基因，*FgstuA Fgsfl*1 双敲菌株与 JQ112s5 一致，表现出了比 *FgstuA* 和 *Fgsfl*1 单独缺失更严重的孢子产量下降。

以上表明 *FgstuA* 不仅与 PKA 协同调控着分生孢子的萌发，还与 cAMP-PKA 的下游转录因子 SFL1 同源共同调控着分生孢子的产生。该研究从功能重叠角度阐述 cAMP-PKA 通路存在的协同调控关系。

关键词：cAMP-PKA 信号通路；产孢；分生孢子萌发；协同作用

---

* 基金项目：国家自然科学基金面上项目（31571953）；国家级大学生创新创业训练计划项目（201810712103）
** 第一作者：黄俊锜，硕士研究生，主要从事禾谷镰孢菌分子生物学研究
*** 通信作者：王晨芳，副研究员
许金荣，教授

禾谷镰孢菌 *FSY1* 的功能及其对 *YNG2* 介导的组蛋白乙酰化的分子调控*

颜　明[1**]，江　航[1]，黑若楠[1]，金巧军[1]，刘慧泉[1]，江　聪[1***]，许金荣[1,2 ***]
（1. 西农-普度大学联合研究中心，旱区作物逆境生物学国家重点实验室，西北农林科技大学植物保护学院，杨凌　712100；2. 美国普渡大学植物及植物病理系，印第安纳州　IN47907）

摘　要：组蛋白乙酰化是真核生物中分布最广、研究最多的一类组蛋白修饰，介导组蛋白乙酰化的复合体有 SAGA、NuA3、NuA4 等，其中 NuA4 复合体作为进化中非常保守和必需的多亚基复合体，具有乙酰化组蛋白 H4/H2A 和激活转录的能力，广泛参与了基因转录激活、DNA 损伤修复、细胞周期等重要的细胞生理过程。Yng2 是 NuA4 复合体主要组分，对 NuA4 复合体的功能发挥具有至关重要的作用。笔者实验室前期研究发现敲除禾谷镰孢菌（*Fusarium graminearum*）*FgYNG2* 后，突变体生长极慢，培养在 PDA 上数日后易产生角突变。全基因组重测序发现其中两个角变子在丝状真菌特有基因 *FSY1*（Filamentous fungi-specific Suppressor of *Fgyng2* mutant 1）上分别发生了移码突变和终止突变，推测 *FSY1* 很可能参与到 *FgYNG2* 介导的丝状真菌组蛋白乙酰化调控中。为了明确 *FSY1* 和 *FgYNG2* 之间的关系以及 *FSY1* 的功能，本研究首先在 *yng2* 突变体中敲除了 *FSY1*，研究结果发现 *Fgyng2/fsy1* 双突变体与角变子生长一致，均部分回复了 *Fgyng2* 的营养生长缺陷。但是，*Fgyng2* 突变体在侵染和有性发育阶段的缺陷则未得到回复。然后在野生型背景下敲除了 *FSY1*，并对 *fsy1* 突变体进行表型的观察，发现 *fsy1* 突变体菌丝生长滞缓，几乎不产色素，无法形成子囊壳，致病力丧失，表明 *FSY1* 在生长发育和致病中均十分重要。目前笔者正在用 H4 的特异性乙酰化抗体分析 Fsy1 对 *Fgyng2* 突变体组蛋白乙酰化水平的回复情况，并通过对 Fsy1 互作蛋白的鉴定，进一步揭示该蛋白参与的调控网络。此外，研究还发现 Fsy1 是一个富含谷氨酰胺的蛋白，氨基酸组成中有 18.18%为谷氨酰胺，而且在序列中有大量连续的谷氨酰胺，最长的有连续的 8 个氨基酸为谷氨酰胺，这种连续的谷氨酰胺与 Fsy1 功能的关系有待进一步的探究。

关键词：组蛋白乙酰化；禾谷镰孢菌；角突变；Q-rich 蛋白

---

* 基金项目：陕西省创新人才推进计划-青年科技新星项目（2018KJXX-068）
** 第一作者：颜明，硕士研究生，主要从事禾谷镰孢菌分子生物学研究
*** 通信作者：江聪，副研究员
许金荣，教授

禾谷镰孢菌中 *TUB5* 负调控 *TUB2* 在无性生长时期功能

郝超峰[1*]，王　欢[1]，张　菊[1]，王晨芳[1]，刘慧泉[1]，许金荣[1,2**]

（1. 西农-普度大学联合研究中心，旱区作物逆境生物学国家重点实验室，西北农林科技大学植物保护学院，杨凌　712100；2. 美国普渡大学植物及植物病理系，印第安纳州　IN47907）

摘　要：禾谷镰孢菌引起的小麦赤霉病是重要的小麦病害，病害流行年份不仅造成粮食大量减产，侵染小麦过程中产生的 DON 毒素严重威胁人畜健康。禾谷镰孢菌中有 2 个 β-tubulin（Tub1，Tub2）和 2 个 α-tubulin（Tub4，Tub5）构成微管骨架，在细胞分裂，物质运输，菌丝极性生长和延伸等过程中具有重要作用。笔者实验室前期发现 *TUB*1 缺失引起菌落生长速率减慢，有性生殖时期产生较小子囊壳且无子囊孢子产生，*TUB*2 缺失引起菌落生长和致病力严重降低，有性生殖与野生型无异，*tub*4 缺失突变体的菌落生长速率和致病力严重降低，有性生殖产生较小子囊壳并且没有子囊孢子产生，而 *tub*5 缺失突变体在菌落生长，致病力有性生殖过程均与野生型无异。*tub*2 自发恢复突变菌株的全基因组测序发现，菌株 9-4s、11-1 在 *TUB*5 上的移码突变会部分恢复 *tub*2 的菌落生长速率。*tub*2*tub*5 双敲突变体与自发回复菌株 9-4s，11-1 在菌落生长速率上相近，说明 *TUB*5 缺失能部分恢复 *tub*2 缺失突变体在生长时期的缺陷，但小麦穗部接种实验发现，菌株 9-4s、11-1 和 *tub*2*tub*5 双敲突变体没有恢复 *tub*2 缺失突变体在致病力上的缺陷。*TUB*5 缺失并不影响 Tub1 的定位。*tub*1*tub*5 双敲突变体可以部分恢复 *tub*1 菌落生长速率，但不能恢复 *tub*1 在有性时期的缺陷。*tub*1*tub*4 双敲突变体菌落生长速率和有性生殖与 *tub*4 突变体无异。实验结果暗示 *TUB*5 负调控 β-tubulin 在无性生长时期的功能但对致病力和有性生殖无影响。

关键词：禾谷镰孢菌；微管蛋白；β-tubulin；α-tubulin

* 第一作者：郝超峰，博士研究生

** 通信作者：许金荣，国家千人计划特聘教授，主要从事病原真菌功能基因组学研究；E-mail：jinrong@ purdue. edu

禾谷镰刀菌剪接体蛋白 FgPrp6 与蛋白激酶 FgPrp4 的关系研究[*]

范芝丽[1**]，李朝晖[1]，江　聪[1]，刘慧泉[1]，金巧军[1***]，许金荣[1,2***]

（1. 西农-普度大学联合研究中心，旱区作物逆境生物学国家重点实验室，西北农林科技大学植物保护学院，杨凌　712100；2. 美国普渡大学植物及植物病理系，印第安纳州　IN47907）

摘　要：剪接体蛋白 Prp6 是三聚体 U4/U6. U5 的重要组分。在人类细胞和裂殖酵母中，Prp6 为蛋白激酶 Prp4 磷酸化，且这个过程对剪接复合体 B 的形成很重要。笔者课题组在研究禾谷镰刀菌蛋白激酶功能时发现，FgPrp6 上的突变部分恢复 *Fgprp*4 突变体的表型。为了研究禾谷镰刀菌中 Prp6 与 Prp4 激酶的关系，笔者进一步鉴定了 FgPrp6 上的 10 个可以恢复 *Fgprp*4 突变体的表型的突变位点，这些位点都位于与裂殖酵母或人类细胞 Prp6 上保守的 Prp4 磷酸化位点附近，表明突变对 FgPrp6 功能的影响与 FgPrp4 磷酸化相似。为了研究 FgPrp6 上保守磷酸化位点的功能，我们构建了阻止磷酸化突变菌株 *Fgprp*6/*FgPRP*6$^{T199A\&T200A}$、*Fgprp*6/*FgPRP*6$^{T219A\&T221A}$、*Fgprp*6/*FgPRP*6^{T261A}、*Fgprp*6/*FgPRP*6^{T252A}、*Fgprp*6/*FgPRP*6$^{\Delta 199-221}$ 和 *Fgprp*6/*FgPRP*6$^{\Delta 250-262}$。对这些菌株进行表型分析发现它们的菌落形态、生长速率以及有性生长过程中的子囊壳、子囊孢子数量和形态均与野生型相似。但是 *Fgprp*6/*FgPRP*6$^{T199A\&T200A}$、*Fgprp*6/*FgPRP*6$^{T219A\&T221A}$、*Fgprp*6/*FgPRP*6^{T261A}、*Fgprp*6/*FgPRP*6$^{\Delta 199-221}$ 和 *Fgprp*6/*FgPRP*6$^{\Delta 250-262}$ 菌株在小麦胚芽鞘和玉米须上的致病力下降；而 *Fgprp*6/*FgPRP*6^{T252A} 菌株致病力下降不明显，表明 T199、T200、T219、T221、T261 五个位点对 FgPrp6 的功能重要。为了明确 T199、T200、T219、T221、T261 位点也为 FgPrp4 磷酸化，笔者构建了模拟磷酸化菌株 *Fgprp*4/*FgPRP*6$^{T199D\&T200D}$、*Fgprp*4/*FgPRP*6$^{T219D\&T221D}$ 和 *Fgprp*4/*FgPRP*6^{T261D}，笔者将通过观察这些菌株的表型来判断 T199、T200、T219、T221、T261 位点是否为 FgPrp4 磷酸化。

关键词：Prp6；Prp4；前体 RNA 剪接；磷酸化

---

* 基金项目：国家自然科学基金-青年科学基金（31600117）

** 第一作者：范芝丽，硕士研究生，主要从事禾谷镰刀菌分子生物学研究

*** 通信作者：金巧军，副教授

许金荣，教授

禾谷镰刀菌转录因子*ACE*1与微管蛋白TUA1的功能相关性研究

王　欢[1*]，王晨芳[1]，许金荣[1,2**]
（1. 西农-普度大学联合研究中心，旱区作物逆境生物学国家重点实验室，西北农林科技大学植物保护学院，杨凌　712100；2. 美国普渡大学植物及植物病理系，印第安纳州　IN47907）

摘　要：禾谷镰刀菌是引起小麦赤霉病的主要病原菌，其产生DON等毒素能严重危害人畜健康。微管是目前我国生产上广泛使用的苯并咪唑类杀菌剂的作用靶标，同时也是重要细胞骨架，在真菌极性生长，物质运输、细胞器定位、有丝分裂和维持细胞形态等众多生物学过程中具有重要功能。禾谷镰刀菌具有2个α-微管蛋白基因（*FgTUA*1，*FgTUA*2），课题组前期研究过程中发现，*FgTUA*1与营养生长和有性生殖密切相关，而*FgTUA*2无明显表型缺陷。微管发挥作用可能依赖于其微管相关蛋白（microtubule associated proteins），这种功能分化可能与两个微管蛋白各自的微管相关蛋白调控有关。*FgTUA*1突变体生长不稳定，容易形成角变子。角变子的产生可能与*FgTUA*1相关基因上发生突变有关。通过对*FgTUA*1角变子进行重测序分析，发现有8个角变子在转录因子*ACE*1上发生了终止突变。

笔者课题组研究发现，*ACE*1和*TUA*1双敲突变体和其角变子的菌落形态类似，证明*ACE*1基因的缺失可以回复*tua*1突变体的生长缺陷。与野生型相比，*ACE*1基因缺失突变体，在生长，产毒，压力筛选和致病方面没有明显缺陷，主要影响有性生殖，其子囊壳较小，没有成型的子囊和子囊孢子。*ACE*1基因定位在细胞质，用10mmol/L氯化钙处理之后，可以进入细胞核。同时，*ACE*1基因缺失不影响4个微管蛋白的定位。另外，课题组研究发现在*ace*1突变体和其角变子中，*TUB*1和*TUA*2基因的表达量明显上升，这说明*ACE*1转录因子主要调控有性生殖，并且通过调控*TUB*1和*TUA*2两个基因的表达量来调控微管基因的表达。

关键词：禾谷镰刀菌；微管相关蛋白；微管辅因子；TUA1；ACE1；有性生殖

* 第一作者：王欢，博士研究生
** 通信作者：许金荣，国家千人计划特聘教授，主要从事病原真菌功能基因组学研究；E-mail：jinrong@ purdue. edu

ClVf19 对玉米弯孢叶斑病菌生长发育及致病力的影响*

徐靖茹**，王　芬，肖淑芹，薛春生***
（沈阳农业大学植物保护学院，沈阳　110161）

摘　要：植物病原真菌分泌多种细胞壁降解酶解聚细胞壁而达到侵染寄主植物的目的，因此细胞壁降解酶是植物病原真菌产生的一类重要致病因子。据 Akhil Srivastava 等报道，引起十字花科黑斑病的芸薹生链格孢（*Alternaria brassicicola*）的 *AbVf*19（*A. brassicicola* virulence factor 19）调控细胞壁降解酶的生物合成，*AbVf*19 缺失后 *A. brassicicola* 产生的 26 种水解酶如角质酶、糖苷水解酶和果胶裂解酶等的表达量显著下降，同时对果胶的利用率降低，致病力显著下降。

玉米弯孢叶斑病是中国玉米生产中重要的叶部病害之一，其主要致病菌为新月弯孢菌［*Curvularia lunata*（Wakker）Boed.］。*C. lunata* 产生多种细胞壁降解酶，是该病菌的致病因子之一，但调控机制不清楚。本研究根据玉米弯孢叶斑病菌全基因组序列信息（JFHG01001253.1），以野生型菌株 CX-3 为模板，PCR 扩增获得玉米弯孢叶斑病菌 *ClVf*19，核苷酸序列全长为 1 504bp，编码 334 个氨基酸，含有串联的 C2H2-锌指结构域。利用多片段克隆的方法，将该基因的两个片段及标记基因 NPTⅡ与线性化双元载体 pPZP100 连接，获得重组质粒 pPZP100ClVf19。采用电击法将 pPZP100ClVf19 敲除载体质粒导入农杆菌 AGL-1 中，通过农杆菌介导遗传转化（ATMT）方法获得玉米弯孢叶斑病菌 *ClVf*19 缺失突变体。与野生型 CX-3 比较，Δ*Clvf*19 菌丝稀疏、生长速度下降 41%，分生孢子萌发时间滞后 2 h，分生孢子小，接种感病自交系黄早四后致病力下降。

关键词：玉米弯孢叶斑病菌（*Curvularia lunata*）；*ClVf*19；生长发育；致病力

---

* 基金项目：国家重点研发计划（2018YFD0300300，2017YFD0201805）

** 第一作者，徐靖茹，硕士研究生，从事玉米病害研究

*** 通信作者：薛春生，教授，博士生导师；E-mail：cshxue@ sina. com

玉米弯孢叶斑病菌 *TFIIEβ* 基因的载体构建与敲除突变体获得*

高维达**，路媛媛，肖淑芹，薛春生***
（沈阳农业大学植物保护学院，沈阳 110866）

摘 要：转录因子 IIE（TFIIE）是异四聚体，由 α 和 β 亚基组成。TFIIE 直接与 TFIIF、TFIIB 和 PolⅡ相互作用后结合靶基因的启动子区，同时 TFIIE 调控 TFIIH。TFIIE β 翼状螺旋结构域位于 TFIIE β 的核心区域，能够结合双链 DNA *TFIIE β* 缺失后 TFIIE 复合物的表达显著下降，影响转录的起始和延伸。

前期研究发现玉米弯孢叶斑病菌两条铁离子吸收途径中关键酶基因 *Ftr*1 和 *NPS*6 缺失后，*TFIIEβ* 基因表达量均明显下调。到目前为止，在玉米弯孢叶斑病菌中并未有该基因的相关研究。本研究从 NCBI 和 JGI 数据库中的新月弯孢菌株 m118 和 CX-3 全基因组中获得该基因的同源序列，该基因全长 855bp，包含一个开放阅读框，编码 264 个氨基酸。设计带有 *EcoR* Ⅰ、*Kpn* Ⅰ和 *Xba* Ⅰ、*Pst* Ⅰ酶切位点的特异性引物，利用 PCR 方法克隆得到 717bp、509bp 的侧翼片段，连接至含有潮霉素抗性基因的骨架载体 pXEH，最终获得重组质粒 pXEHTFIIEβ，采用冻融法将 pXEHTFIIEβ 转入农杆菌菌株 AGL-1 中，利用 ATMT 技术获得了玉米弯孢叶斑病菌 *TFIIEβ* 基因缺失突变体，菌落颜色较浅。

本研究拟对突变体与野生型菌株的表型及致病力进行比较，以明确铁离子对 *TFIIEβ* 基因的影响及其调控机制，为玉米弯孢叶斑病菌致病分子机制的研究提供理论基础。

关键词：玉米弯孢叶斑病菌；TFIIE；载体构建；ATMT

---

* 基金项目：国家重点研发计划（2017YFD0201805，2018YFD0300300）
** 第一作者，高维达，硕士研究生，从事玉米病害研究
*** 通信作者：薛春生，教授，博士生导师；E-mail：cshxue@ sina. com

我国南方地区花生烂果病病原的初步鉴定*

康彦平**，雷　永，淮东欣，王志慧，宋万朵，晏立英***，廖伯寿***
（中国农业科学院油料作物研究所，油料作物生物学与遗传改良重点实验室，武汉　430062）

摘　要：花生烂果病是我国花生产业近年来发生较为普遍的一种真菌病害，主要为害花生地下荚果，使荚果严重腐烂，而受害植株的地上部分与健康植株无异，很难通过地上部分来判断花生荚果是否已感病，给防治烂果病增加了一定的难度，而且该病在我国南方花生产区呈逐年加重之势。明确花生烂果病病原，对开展针对不同病原培育相对应的抗病品种工作具有重要的指导作用。笔者对采自海南三亚花生种植区不同程度的烂果病样进行病原菌分离鉴定，目前已获得 91 株真菌菌株。通过对其进行形态学特征观察和 rDNA-ITS 序列分析，共鉴定到镰刀菌属（*Fusarium*）、壳针孢属（*Septoria*）、葡萄座腔菌属（*Botryosphaeria*）和丝核菌属（*Rhizoctonia*）4 个属的真菌。其中镰刀菌属共鉴定到 5 个种，包括 *Fusarium solani*，*Fusarium oxysporum*，*Fusarium chlamydosporum*，*Fusarium equiseti* 和 *Fusarium incarnatum*，以 *F. solani* 分离得到的菌株最多，*F. chlamydosporum* 次之。壳针孢属分离得到 4 株球壳孢菌株（*Macrophomina phaseolina*），葡萄座腔菌属分离得到 3 株可可毛色二孢菌株（*Lasiodiplodia theobromae*），丝核菌属分离得到 2 株丝核菌株（*Rhizoctonia bataticola*）。采用菌丝块在离体花生荚果上进行接种的结果初步显示，*F. solani*、*F. chlamydosporum*、*F. oxysporum*、*F. incarnatum*、*F. equiseti* 和 *L. theobromae* 均可使花生荚果发病，其中 *F. chlamydosporum* 和 *L. theobromae* 的致病力最强。通过研究分析，初步明确了我国南方地区花生烂果病的病原，为进一步开展花生抗烂果病种质筛选鉴定提供了依据。

关键词：花生；烂果病；病原菌；分离鉴定

---

* 基金项目：国家花生产业技术体系（CARS-14）；中国农业科学院创新工程（CAAS-ASTIP-2013-OCRI）

** 第一作者：康彦平，硕士，从事油料作物病害研究；E-mail：kangyanping@caas.cn
*** 通信作者：晏立英，副研究员，从事油料作物病害研究；E-mail：yanliying2002@126.com
廖伯寿，研究员，从事花生病害研究；E-mail：lboshou@hotmail.com

北京地区甜樱桃叶斑病病原菌的分离鉴定

周悦妍[1,2*]，Chethana K W T[2]，张 玮[2]，李兴红[2]，吴学宏[1]，燕继晔[2**]
（1. 中国农业大学植物保护学院植物病理学系，北京 100193；
2. 北京市农林科学院植物保护环境保护研究所，北京 100097）

摘 要：樱桃是蔷薇科李属多年生木本植物，在世界温带地区广泛种植。近年来，我国甜樱桃产业发展迅速，据中国园艺学会樱桃分会统计，2016 年我国甜樱桃栽培面积为 270 万亩，其中北京地区为 5.8 万亩。然而随着甜樱桃在我国种植面积的不断扩大，病害问题也日益凸显。通过田间调查发现，北京地区樱桃叶斑病发病较重，发病叶片表现紫色或褐色斑点，后期病斑干枯脱落，造成穿孔，病叶变黄，提早脱落。叶斑病会影响叶片的光合作用、削弱树势，降低果实的产量和品质。本研究通过对北京地区甜樱桃叶斑病样品进行病原的分离鉴定，明确引起北京地区樱桃叶斑病的主要病原菌，从而为其田间诊断和绿色防控提供理论依据。采用组织分离法从病样中分离出 102 个真菌菌株，通过形态学以及多基因整合系统发育分析对病原菌进行初步鉴定，结果表明，分离的真菌来自 9 个属，病原菌种类分别为 *Alternaria* sp.、*Colletotrichum* sp.、*Stagonosporopsis* sp.、*Epicoccum* sp.、*Nothophoma* sp.、*Botryosphaeria* sp.、*Nigrospora* sp.、*Cladosporium* sp. 和 *Mycosphaerella* sp.，其中 *Alternaria* sp.、*Colletotrichum* sp. 和 *Stagonosporopsis* sp. 为主要病原菌种类；目前正在进行柯赫氏法则验证和菌株的群体结构分析，相关结果为分析引起我国北京地区樱桃叶斑病的优势病原菌及其致病性差异提供基础。

关键词：樱桃；叶斑病；鉴定；链格孢属；刺盘孢属

* 第一作者：周悦妍，硕士研究生；E-mail：yueyan_zhou@163.com
** 通信作者：燕继晔，副研究员；E-mail：jiyeyan@vip.163.com

可可毛色二孢菌外泌蛋白 LtGhp1 的功能分析*

彭军波，Chethana K W T，李兴红，张　玮，邢启凯，燕继晔**
（北京市农林科学院植物保护环境保护研究所，北京　100097）

摘　要：由可可毛色二孢菌侵染引起的葡萄溃疡病在我国葡萄产区广泛发生且造成严重的损失，但关于该病原真菌侵染致病的分子机制并不清楚。本研究室在探究可可毛色二孢菌的致病机理时，克隆到一个假定的外泌蛋白编码基因 *LtGHP*1。预测 LtGhp1 蛋白 N 端具有一个信号肽，酵母互补实验表明该信号肽能够引起酵母蔗糖酶的外泌。*LtGHP*1 基因的超表达、RNAi 及病原菌接种实验表明 LtGhp1 是一个重要的致病因子，不同侵染阶段的 RNA 积累量分析显示 *LtGHP*1 基因在病原菌侵染寄主初期表达量显著升高，并能引起寄主中抗性反应相关基因的上调表达。此外，瞬时表达及活体注射 LtGhp1 蛋白均能引起烟草细胞的坏死。通过筛选寄主植物的 cDNA 文库，获得了 LtGhp1 蛋白在寄主中的互作靶标，关于 LtGhp1 与互作靶标的调控机理正在分析中。

关键词：可可毛色二孢菌；外泌蛋白；细胞坏死

---

* 基金项目：北京市青年拔尖个人项目（2016000021223ZK29）

** 通信作者：燕继晔，副研究员；E-mail：jiyeyan@ vip. 163. com

可可毛色二孢菌效应子 LtCrel 寄主靶标筛选与功能分析*

曹　阳，邢启凯，李铃仙，张　玮，彭军波，李兴红，燕继晔**
（北京市农林科学院植物保护环境保护研究所，北京　100097）

摘　要：在我国，可可毛色二孢菌（*Lasiodiplodia theobromae*）是枝干病害葡萄溃疡病的主要致病菌之一。前期研究中，笔者对一个可可毛色二孢菌侵染响应富半胱氨酸效应子 *LtCRE*1 的进行了克隆和生物学功能初步分析，证明其是可可毛色二孢菌的致病力的正调控因子。为了深入研究效应子 LtCrel 的作用机制，采用酵母双杂交技术，以可可毛色二孢菌效应因子 LtCrel 为诱饵蛋白，在葡萄的 cDNA 文库中进行寄主靶标的筛选，并对所有的阳性候选克隆进行提质粒和测序，在葡萄基因组数据库中比对出靶标基因的全长序列与功能信息，最终比对后得到 36 个基因，其中包括 WRKY、TCP、bZIP 以及 NAC 型转录因子，ABA 受体 PYL4，1 个囊泡外泌蛋白等。对抗病信号转导途径中的重要候选基因进行了克隆及自激活活性分析，利用酵母双杂交技术与诱饵载体进行回复验证正在进行中。该研究工作将为解析可可毛色二孢菌的致病机理提供理论依据。

关键词：葡萄溃疡病；可可毛色二孢菌；效应因子；寄主靶标；酵母双杂交

---

* 基金项目：北京市自然科学基金（6184041）；北京市农林科学院科技创新能力建设专项（KJCX20190406）

** 通信作者：燕继晔；E-mail：jiyeyan@ vip. 163. com

利用酵母双杂交系统筛选可可毛色二孢菌效应因子 LtALL1 的互作蛋白*

李铃仙，曹　阳，邢启凯，张　玮，彭军波，李兴红，燕继晔**
（北京市农林科学院植物保护环境保护研究所，北京　100097）

摘　要：葡萄溃疡病（*Botryosphaeria dieback*）是重要的葡萄枝干病害，在世界各国葡萄产区均普遍发生，且每年都会造成较大的经济损失。可可毛色二孢菌（*Lasiodiplodia theobromae*）是引发我国葡萄溃疡病的优势种群之一。目前，对于可可毛色二孢菌的致病机理还不明晰。效应子是病原菌分泌到寄主细胞间隙或寄主细胞内，在病原菌的侵入、定殖和扩展过程中发挥着重要作用的蛋白因子。本研究鉴定到一个可可毛色二孢菌效应因子 LtALL1，其编码一个 146 个氨基酸的多肽，分子量为 15 ku。LtALL1 信号肽可使缺陷型酵母 YTK12 恢复蔗糖酶活性，具有分泌信号肽的功能。*LtALL1* 基因在可可毛色二孢菌中的超表达降低了病原菌的致病力。进一步利用可可毛色二孢菌侵染的酵母双杂交 cDNA 文库，通过酵母双杂交技术对效应子 LtALL1 在寄主葡萄里的互作靶标进行了筛选。经测序比对分析，最终筛选获得 66 个阳性克隆，其中包括 3 个钙调节蛋白，3 个转录因子（WRKY 型、Myb 型、bZIP 型以及 NAC 型），2 个泛素蛋白酶等。该项研究将为解析可可毛色二孢菌的致病机理奠定基础。

关键词：葡萄溃疡病；可可毛色二孢菌；效应因子；酵母双杂交

---

* 基金项目：国家葡萄产业技术体系（CARS-29）；北京市农林科学院科技创新能力建设专项（KJCX20190406）

** 通信作者：燕继晔；E-mail：jiyeyan@ vip. 163. com

山西省葡萄枝干病害病原菌的分离和鉴定

叶清桐[1,2*]，韩昌坪[1,2]，张　玮[2]，李兴红[2]，吴学宏[1]，燕继晔[2**]
（1. 中国农业大学植物保护学院植物病理学系，北京　100193；
2. 北京市农林科学院植物保护环境保护研究所，北京　100097）

摘　要：葡萄是世界上最重要的果树之一，其面积和产量均居于世界前列。2016 年世界葡萄栽培面积为 709 万公顷，位居世界果树首位（FAO）；葡萄是中国的五大果树之一，在我国具有重要的地位。我国葡萄栽培面积为 84 万公顷，是世界上第二大种植葡萄的国家（FAO）。目前，山西省葡萄栽培面积为 39. 3 万余亩，居全国第 8 位（2014 年度葡萄产业技术发展报告）。近年来，葡萄枝干病害在山西省某些葡萄产区发生严重，葡萄叶片表现为干枯、黄化、变红；花穗干枯、掉落；顶芽枯死；根部表现为腐烂，侧根表面白色菌丝；葡萄枝干内部变褐，干枯，因此，为了明确引起山西省葡萄枝干病害的病原种类，为葡萄枝干病害的精准防治提供相应的理论基础，本研究主要于 2018 年 4—9 月对山西省的太谷县、清徐县和乡宁县等地区的 16 个葡萄枝干病样进行病原鉴定。本实验采用组织分离法以及稀释分离法进行病原菌的分离，进一步利用形态学以及 ITS 序列分析对病原菌进行鉴定，鉴定结果表明，山西省部分地区的葡萄枝干病样分离出 29 个真菌菌株，其中主要是 *Diaporthe* sp. *Diplodia* sp. 、*Botryosphaeria* sp. 和 *Neopestalotiopsis* sp. ，还需要进行进一步的种的鉴定。其中太谷县的枝干病样的分离结果显示，主要分离出如链隔孢（分离比例约 17%）、木霉（分离比例约 10%）和丝核菌（分离比例约 10%）等真菌；清徐县的枝干病样的分离结果显示，主要分离出 *Diaporthe* sp. （分离比例约 60%）和 *Diplodia* sp. （分离比例约 20%）；乡宁县的枝干病样的分离结果显示，主要分离出 *Diaporthe* sp. （分离比例约 80%）、*Botryosphaeria* sp. （分离比例 10%~30%）和 *Neopestalotiopsis* sp. （分离比例约 20%）。后续作者将进一步将分离物精准鉴定到种的水平，并完成柯赫氏法则验证；与其他省份的病原鉴定情况整合，分析我国枝干病害的发生为害特点与优势种群。

关键词：葡萄；形态学；ITS；Botryosphaeriaceae；*Neopestalotiopsis* sp.

---

* 第一作者：叶清桐，硕士研究生；E-mail：qingtong_ye@ 163. com
** 通信作者：燕继晔，副研究员；E-mail：jiyeyan@ vip. 163. com

利用本生烟筛选小麦叶锈菌候选效应蛋白*

齐　悦**，张　悦[1]，李建嫄[1,2]，杨文香[1]***，刘大群[2]***

（1. 河北农业大学植物病理学系，河北省农作物病虫害生物防治工程技术研究中心，
国家北方山区农业工程技术研究中心，保定　071001；2. 河北省邢台学院，
邢台　054000；3. 中国农业科学院研究生院，北京　100081）

摘　要：小麦叶锈病是由小麦叶锈菌引起的一种气传性真菌病害，在全世界范围均有发生，严重情况下可造成小麦 15%~40%甚至更高的产量损失。最为安全、经济、有效的防治方法是建立抗病品种，但是小麦叶锈菌致病性呈现多样化，且毒性频繁发生变异，因此导致小麦抗叶锈性不断丧失。探讨小麦叶锈菌致病的分子机制对于有效控制小麦叶锈病尤为重要。本研究在前期已构建好的转录组文库基础上进行试验，该转录组文库在获得高质量 RNA 的基础上，使用 Illumina Hiseq 平台进行测，包括 3 个小麦叶锈菌单胞菌系 08-5-9-2（KHTT）、13-5-28-1（JHKT）和 13-5-72-1（THSN）分别在休眠孢子、萌发夏孢子和侵染 Thatcher6 d 时期的转录组数据，将文库数据通过 SingnalP4. 1、TargetP、TMHMM、EffectorP 进行效应蛋白的初步筛选，获得了 427 个候选效应因子，从中选取了 16 个与 *Lr*20 候选无毒基因具有同源性的基因，之后对利用本生烟和能够引起过敏性坏死反应的 Bax 和 Inf 对 16 个候选效应因子进行进一步的筛选，最终获得了 16 个候选效应因子。本研究结果表明异源表达系统是进行专性寄生菌效应蛋白筛选的有力工具，为深入开展效应因子的致病作用研究提供了参考依据。

关键词：小麦叶锈菌；效应蛋白；异源表达；Bax；Inf；致病机制

---

* 基金项目：国家自然科学基金项目（301571956；301871915）

** 第一作者：齐悦，硕士研究生，主要开展小麦叶锈菌的致病机制研究；E-mail：908952718@ qq. com

*** 通信作者：杨文香，教授，主要研究方向：小麦叶锈菌致病机制；E-mail：wenxiangyang2003@ 163. com
刘大群，教授，主要研究方向：生物防治与分子植物病理学；E-mail：ldq@ hebau. edu. cn

小麦叶锈菌 ABC 转运蛋白的功能验证*

韦　杰**，张　悦[1]，齐　悦[1]，杨文香[1***]，刘大群[2 ***]
（1. 河北农业大学植物保护学院，河北省农作物病虫害生物防治工程技术研究中心，国家北方山区农业工程技术研究中心，保定　071001；2. 中国农业科学院研究生院，北京　100081）

摘　要：小麦叶锈病是我国小麦上主要病害之一，防治小麦叶锈病的主要药剂是咪唑类、三唑类、苯吡咯类。对于这些药剂，病原菌则进化出相应的抵御措施：一方面将有毒化合物排出体外，使菌体内抵抗病原物的物质浓度降低；另一方面屏障寄主的抗病性，增强病原物的致病能力。ABC 转运蛋白被认为是病原真菌对杀菌剂和多种外源毒性物质排泄的重要通道，也是转运营养，保存菌体活力的能量中枢。本课题组在 11 个小麦叶锈菌的转录组文库中筛选到 243 个 ABC 转运子，其中 113 个转运蛋白注释到了 7 个亚家族 ABCA、ABCB、ABCC、ABCD、ABCE、ABCF 和 ABCG 中。基于小麦抗叶锈菌 TcLr19 及其突变体的转录组文库，得到 2 个在 TcLr19 突变体上表达量显著上调的真菌基因 *Pt-ABCF*13949 和 *Pt-ABCB*21878。利用大麦条纹花叶病毒（BSMV）介导的寄主诱导的基因沉默（HIGS）技术，对 *Pt-ABCF*13949 和 *Pt-ABCB*21878 进行基因沉默。接种小麦叶锈菌 THTT 后，实时定量结果显示，*Pt-ABCF*13949 和 *Pt-ABCB*21878 被有效的沉默，并且沉默 *Pt-ABCF*13949 后使得小麦叶锈菌菌株 $M_4$33-3-11-2 在 TcLr17 和 TcLr28 上侵染型由“4”变为“1”，沉默 *Pt-ABCB*21878*a*、*Pt-ABCH*21878*b* 后使得小麦叶锈菌菌株 $M_4$33-3-11-2 在 TcLr2b 和 TcLr14b、在 TcLr1 和 TcLr3ka 表型均由“4”变为“；1”，结合组织学观察结果，发现 *Pt-ABCF*13949 和 *Pt-ABCB*21878 基因沉默后抑制了病原菌的发育，在小麦叶锈菌侵染时发挥了毒性作用。明确来自小麦叶锈菌中的 ABC 转运子在小麦叶锈菌中发挥毒性作用，对于提高杀菌剂的药效，寻找新的药物靶标具有重要的理论与实践意义。

关键词：小麦叶锈菌；ABC 转运蛋白；毒性作用；基因沉默

* 基金项目：国家自然科学基金项目（301571956；301871915）
** 第一作者：韦杰，在读博士研究生，专业方向：分子植物病理学；E-mail：444256956@ qq. com
*** 通信作者：杨文香，教授，主要研究方向：小麦叶锈菌致病机制；E-mail：wenxiangyang2003@ 163. com
刘大群，教授，主要研究方向：生物防治与分子植物病理学；E-mail：ldq@ hebau. edu. cn

小麦叶锈菌与小麦互作中的效应蛋白筛选及功能验证

张瑞丰[1]，崔立平[1]，范学锋[1,2]，张　悦[1]，杨文香[1]，刘大群[2]
（1. 河北农业大学植物病理学系，河北省农作物病虫害生物防治工程技术研究中心，国家北方山区农业工程技术研究中心，保定　071001；2. 中国农业科学院，北京　100081）

摘　要：由小麦叶锈菌引起的小麦叶锈病，是世界重要的一种真菌病害，造成了我国严重的经济损失。大量的实践证明，控制小麦叶锈菌最经济环保的方法是选育和利用抗病品种，然而叶锈菌毒性变异频繁，新的毒性小种的不断出现，往往使得抗叶锈病丧失，因此研究小麦叶锈病的致病机理对防控小麦叶锈病显得尤为重要。本试验使用小麦叶锈菌致病类型为 THTT 的生理小种与 TcLr19 非亲和互作（6h、12h、和 24h、记作 RIQ）与 TcLr19 感病突变体 M433-3-11-2 亲和互作早期（6 h、12 h 和 24 h，记作 MIQ）及各自互作 6 d（记作 RI6d 与 MI6d）转录组的测序数据分析及锈菌侵染小麦期间效应蛋白的功能验证。为揭示叶锈菌的萌发过程及叶锈菌侵染小麦的机理奠定基础。利用生物信息学方法在整个 RNA 测序数据库中筛选效应蛋白，共筛选出 2816 个候选的效应蛋白。其中 126 个属于孢子萌发阶段与休眠夏孢子时期的显著差异基因；431 个属于亲和互作时期与萌发时期的差异表达基因；279 个属于互作 6d 与互作早期的显著差异基因。根据表达时期、表达量及基因功能注释选出 13 个候选效应蛋白，利用异源表达系统分析候选效应蛋白毒性。将 13 个候选效应蛋白与大鼠促凋亡蛋白（BAX）基因共同注射烟草发现有 12 个基因能在烟草上抑制 BAX 诱导的程序性细胞坏死（PCD），证明 12 个候选效应蛋白具有毒性。利用寄主诱导的基因沉默技术（HIGS）在抗叶锈近等基因系 TcLr14b、TcLr16、TcLr2b 和 TcLr3ka 上沉默基因 Cluster-19789. 84114、Cluster-19789. 21637 和 Cluster-19789. 98252，结果发现沉默 Cluster-19789. 84114 后 TcLr14b 和 TcLr3ka 上锈菌的表现型均由“4”转变为“1”；沉默Cluster-19789. 21637 后 TcLr2b 和 TcLr14b 上锈菌的表现型由“4”转变为“1”；沉默 Cluster-19789. 98252 后 TcLr16 和 TcLr14b 上锈菌的表现型由“4”转变为“1”。HIGS 实验表明三个效应蛋白在锈菌的侵染寄主过程中起到毒性作用。选择沉默效果较好的两个效应蛋白 Cluster-19789. 84114 和 Cluster-19789. 98252（去信号肽）进行亚细胞定位观察。发现两个基因均定位在细胞膜和细胞核，推测两个效应蛋白被叶锈菌分泌至寄主细胞，在寄主细胞膜和细胞核发挥作用。

关键词：小麦叶锈病，效应子；致病机制；转录组测序

由甲基磺酸乙酯诱变的小麦叶锈菌突变菌株的筛选及鉴定*

韦 杰**，齐 悦[1]，张 悦[1]，周宗悦[1]，杨文香[1]***，刘大群[2]***
（1. 河北农业大学植物保护学院，河北省农作物病虫害生物防治工程技术研究中心，
国家北方山区农业工程技术研究中心，保定 071001；
2. 中国农业科学院研究生院，北京 100081）

摘 要：小麦叶锈病是我国小麦上主要病害之一，选育抗病品种一直以来是防治该病害的最有效方式，随着小麦叶锈菌毒性变异频繁，小麦叶锈抗病品种抗性逐渐丧失。通过筛选小麦叶锈菌中的无毒基因，为寻找新的抗源和培育抗病品种开辟新途径，本研究利用甲基磺酸乙酯（EMS）以 0.04mol/L，处理 8min 作为最适诱变剂量和时间，对小麦叶锈菌 08－5－9－2（KHHT），13-5-72-1（THSN），13-2-28-1（JHKT），17-8-1465（BBBB）进行化学诱变。本试验获得了 20 株遗传稳定发生变异的小麦叶锈菌毒性突变菌株，各突变菌株的毒性均较野生菌系发生明显改变。其中，6 个 Mu08-5-9-2 菌株在 7 个近等基因系上的表型发生变化，9 个 Mu13-2-28-1 菌株在 31 个近等基因系上的表型发生变化，4 个 Mu13-2-72-1 菌株在 12 个近等基因系上的表型发生变化，2 个 Mu17-8-1465 菌株在 8 个近等基因系上的表型发生变化。值得注意的是，Mu08-5-9-2②在 TcLr28 的毒性由“4”降低为“0”；Mu13-5-28⑤在 TcLr21 上的表型由“4”变为“0”；Mu13-5-28-1④⑦⑨⑩侵染型在 TcLr28 上由“4”分别突变为“;”、“0”、“0”和“;”；Mu13-5-72-1⑦侵染型在 TcLr1 上由“4”突变为“0”。试验表明，EMS 诱导的方法可以使我国小麦叶锈菌发生毒性突变和非毒性突变，为今后无毒基因，特别是 *Lr*1、*Lr*21、*Lr*28 对应的无毒基因的筛选奠定了基础。

关键词：小麦叶锈菌；甲基磺酸乙酯（EMS）；化学诱变；毒性突变菌株

---

* 基金项目：国家自然科学基金项目（301571956；301871915）

** 第一作者：韦杰，在读博士研究生，专业方向：分子植物病理学；E-mail：444256956@qq.com

*** 通信作者：杨文香，教授，主要研究方向：小麦叶锈菌致病机制；E-mail：wenxiangyang2003@163.com
刘大群，教授，主要研究方向：生物防治与分子植物病理学；E-mail：ldq@hebau.edu.cn

4 个小麦叶锈菌效应蛋白的功能分析*

齐　悦**，韦　杰[1]，张　悦[1]，杨文香[1***]，刘大群[2***]
（1. 河北农业大学植物病理学系，河北省农作物病虫害生物防治工程技术研究中心，国家北方山区农业工程技术研究中心，保定　071001，2. 中国农业科学院研究生院，北京　100081）

摘　要：小麦叶锈病是由小麦叶锈菌侵染的一种气传性真菌病害，在全世界范围均有发生，造成小麦严重的产量损失。防治该病害最为安全、经济、有效的方法是抗病品种的利用，但由于小麦叶锈菌致病性多样化及毒性的频繁发生变异，导致小麦抗叶锈性不断丧失，因此，探讨小麦叶锈菌致病的分子机制对于有效控制小麦叶锈病尤为重要。前期在三个小麦叶锈菌单胞菌系 08-5-9-2（KHTT）、13-5-28-1（JHKT）和 13-5-72-1（THSN）休眠孢子、萌发夏孢子和侵染 6 d时期的小麦叶锈菌转录组文库中筛选到 20 个与 *Lr*20 候选无毒基因具有同源性的基因候选效应因子。通过对筛选到的 20 个候选效应因子利用全套的近等基因系进行检测，结果发现基因 *Pt*13024 在近等基因系 TcLr30、TcLr42、TcLrB、TcLr38、TcLr19 和 TcLr11 上出现了坏死反应。其中，基因 *Pt*18906 在近等基因系 TcLr10+27+31 和 TcLr42 上出现了坏死反应，基因 *Pt*20911 在近等基因系 TcLr25 和 TcLr38，*Pt*3863 在近等基因系 TcLr18、TcLr1、TcLr10 和 TcLr2a 上出现了坏死反应。研究表明这 4 个效应因子触发了寄主相应的防御反应，小麦中存在与效应蛋白互作的分子靶标。之后我们利用 qRT-PCR 检测效应蛋白基因在小麦叶锈菌生理小种 13-5-28-1 和 13-5-72-1 侵染 Thatcher 叶片 0 h、6 h、12 h、18 h、24 h、36 h、48 h、72 h、96 h、6 d、9 d 和 12 d后的表达量，发现 4 个基因均在 24h 表达量达到了最高峰，而接种后 24 小时是小麦叶锈菌产生吸器母细胞和形成吸器的阶段，也是病原与寄主作用激烈的阶段，因此猜测这 4 个基因可能与叶锈菌吸器的形成有关。为了知道基因抑制坏死以及在寄主上产生坏死的作用部位，我们利用烟草进行亚细胞定位试验得知 4 个基因定位在细胞核以及细胞质，效应蛋白在细胞内起作用。之后为了明确效应蛋白的真正作用位点，我们对 4 个效应蛋白进行了缺失突变，发现功能区域在去掉信号肽后 N 末端的前 20 个氨基酸。然后我们进行了信号肽分泌功能的验证，发现基因 *Pt*13024、*Pt*20911 和 *Pt*3863 的信号肽是有分泌功能的，基因 *Pt*18906 的信号肽没有分泌功能。最后我们对 4 个基因进行了原核表达载体的构建，发现 4 个基因都能够表达成蛋白。后续我们将利用 HIGS 沉默技术验证效应蛋白在寄主与病原菌互作中的作用，利用酵母双杂技术筛选效应蛋白的互作靶标，这对我们揭示病菌的致病机理具有重要意义。

关键词：小麦叶锈菌；过敏性坏死反应；效应蛋白

* 基金项目：国家自然科学基金项目（301571956；301871915）

** 第一作者：齐悦，硕士研究生，主要开展小麦叶锈菌的致病机制研究；E-mail：908952718@qq.com

*** 通信作者：杨文香，教授，主要研究方向：小麦叶锈菌致病机制；E-mail：wenxiangyang2003@163.com
刘大群，教授，主要研究方向：生物防治与分子植物病理学；E-mail：ldq@hebau.edu.cn

多基因序列法分析中国咖啡炭疽病菌遗传种群

陆　英[1]，巩佳莉[1,2]，贺春萍[1]，吴伟怀[1]，梁艳琼[1]，黄　兴[1]，
郑金龙[1]，习金根[1]，谭思北[1]，易克贤[1]
（1. 中国热带农业科学院环境与植物保护研究所，海口　571101；
2. 南京农业大学植物保护学院，南京　210000）

摘　要：咖啡是重要的农业贸易商品之一，其产量、产值及消费量均位居世界三大饮料作物（咖啡，茶叶，可可）之首，每年零售额达 700 亿美元。近年我国咖啡产业迅速发展，2016—2017 年我国咖啡种植面积约为 12 万 hm^2，咖啡豆产量 16.03 万 t，位列全球第 13 位，亚洲 4 位。由炭疽菌属（*Colletotrichum* spp.）引起的炭疽病是咖啡的一个重要病害。炭疽菌种类繁多，遗传多态性丰富，关于炭疽菌种属的分类一直较为混乱，仅根据菌落的形态特征、培养性状、生理学和病理学等特征很难进行精准的分类鉴定，随着分子生物学技术被广泛应用于炭疽菌属真菌的遗传种群研究，使得多个炭疽菌种在属内的分类地位发生了很大变化，为了弄清我国咖啡炭疽病菌的种类，笔者从我国云南、海南咖啡产区采集炭疽病标本，采用组织分离法分离得到 28 个咖啡炭疽菌菌株。利用 ITS（Internal transcribed spacer）、GAPDH（Glyceraldehyde-3-phosphate dehydrogenase）、CHS-1（Chitin synthase）和 ACT（Actin）基因的部分序列，对这些菌株进行基因谱系比较分析。对各菌株分别进行 ITS、ACT、CHS-1 和 GAPDH 单基因聚类分析，同时将各菌株的基因按 ITS-ACT-CHS-1-GAPDH 顺序进行序列拼接后，构建系统进化树，分析单基因与多基因序列拼接聚类结果的异同及咖啡树炭疽菌种群差异。鉴定结果表明：中国咖啡炭疽菌有 *C. tropicale*、*C. fructicola*、*C. gloeosporioides*、*C. boninense*、*C. siamense*、*C. brevisporum*、*C. theobromicola*、*C. acutatum*。其中 *C. gloeosporioides* 占所鉴定菌株的 50%，该种为优势种。

关键词：多基因序列；咖啡；炭疽菌；遗传种群

球孢白僵菌诱导的柑橘木虱免疫相关基因鉴定*

宋晓兵**，崔一平，程保平，凌金锋，彭埃天***
（广东省农业科学院植物保护研究所，广东省植物保护新技术重点实验室，广州 510640）

摘　要：采用 Illumina 高通量测序平台对感染球孢白僵菌 24h、48h、72h 与健康柑橘木虱的转录组进行了测序，基于目前已知的昆虫免疫应答相关基因，通过基因序列比对、结构域分析、系统进化树构建等生物信息学方法，筛选出柑橘木虱体内免疫相关基因。本研究共鉴定了 80 个免疫相关基因，其中包括 21 个免疫识别基因、12 个信号转导基因、39 个调制器基因和 8 个效应分子基因。分析发现上述 80 个免疫相关基因在进化上相对保守，但在抵御球孢白僵菌入侵过程中表现有多种转录谱，相关基因的功能还需要进一步研究。利用 qRT-PCR 测定了 15 个免疫相关基因随时间变化的相对表达量，研究结果显示免疫相关基因在柑橘木虱抵御外来真菌侵染的过程中发挥着重要作用，在球孢白僵菌侵染柑橘木虱的不同时间节点，免疫相关基因的上调或者下调的波动，一方面反映了基因表达量的变化趋势，另一方面体现了基因在不同时期参与免疫反应的程度。

关键词：柑橘木虱；球孢白僵菌；免疫基因；鉴定

---

* 基金项目：国家重点研发计划项目（2018YFD0201500、2017YFD0202000）；广东省柑橘柑果产业技术体系创新团队项目
** 第一作者：宋晓兵，副研究员，研究方向：柑橘木虱综合防控；E-mail：xbsong@126.com
*** 通信作者：彭埃天，研究员；E-mail：pengait@163.com

橡胶树红根病菌 LAMP 检测方法的建立*

贺春萍[1]**，董文敏[2]**，吴伟怀，梁艳琼，谢　立[3]，李　锐，易克贤***

（1. 中国热带农业科学院环境与植物保护研究所，农业农村部热带农林有害生物入侵检测与控制重点实验室，海南省热带农业有害生物检测监控重点实验室，海口　571101；2. 南京农业大学植物保护学院，南京　210095；3. 海南大学植物保护学院，海口　570228）

摘　要：为了建立一种快速、高效、准确检测橡胶树红根病菌的技术。本研究以 SYBR Green I 为指示剂，建立了橡胶树红根病菌即橡胶灵芝菌（*Ganoderma pesudoferreum*）的环介导等温扩增（LAMP）可视化快速检测方法。根据橡胶树红根病菌特异的线粒体转录间隔区序列设计了 4 条引物（两条内引物、两条外引物）进行 LAMP 扩增。对 LAMP 反应条件和体系进行优化，并进行特异性和灵敏度验证。研究确定最佳反应体系（25μL 中模板 100ng/μL，外引物 F3/B3 各为 0.2μmol/L，内引物 FIP /BIP 各为 1.2μmol/L，dNTPs 为 1.4mmol/L，Mg^{2+}为 4 mmol/L，甜菜碱为 0.8mol/L，Bst DNA polymerase 为 8U/μL,）和反应条件（最佳反应温度 64℃，反应 1h）。特异性检测结果显示，橡胶树红根病病菌 LAMP 产物均呈阳性（绿色），而其他病原菌组织 LAMP 产物均为阴性（橙色）。灵敏度验证结果显示，该体系的 DNA 的最低检测限为 1×10^{-5}ng/μL，是普通 PCR 的 10^3倍。

关键词：橡胶树红根病；灵芝菌；环介导等温扩增；检测

---

* 基金项目：国家重点研发计划项目（No. 2018YFD0201100）；国家天然橡胶产业技术体系建设专项资金资助项目（No. CARS-33-GW-BC1）；海南省科协青年科技英才创新计划项目（No. QCXM201714）

** 第一作者：贺春萍，硕士，研究员，研究方向：植物病理学；E-mail：hechunppp@ 163. com

董文敏，硕士研究生；专业方向：植物病理学；E-mail：18260063153@ 163. com

*** 通信作者：易克贤，博士，研究员，研究方向：植物病理学；E-mail：yikexian@ 126. com

核盘菌木聚糖酶 SsXyl2 激发植物免疫反应研究*

王娅波**，黄志强，蔡俊松，方安菲，杨宇衡，毕朝位，余　洋***
（西南大学植物保护学院，重庆　400715）

摘　要：核盘菌（*Sclerotinia sclerotiorum*）是一种分布广泛，可侵染多种植物的重要植物病原真菌，由其引起的作物菌核病每年给我国农业生产带来了巨大经济损失。木聚糖酶是由病原菌在侵染植物时产生的一种细胞壁降解酶（Cell wall-degrading enzymes，CWEDs），近年来的研究表明其一方面由于能分解植物细胞壁而促进病原菌侵染，另一方面还可能激发植物的免疫反应。本研究前期利用农杆菌介导的瞬时表达系统发现核盘菌木聚糖酶 SsXyl2 靶向质外体，可激发植物细胞坏死。SsXyl2 蛋白共有 281 个氨基酸，N 端具有一个典型的 GH11 家族糖基水解酶结构，C 端具有纤维素结合域。进一步通过注射纯化 SsXyl2 蛋白表明其可诱导本氏烟、大豆和玉米等植物细胞坏死，且导致大量活性氧积累。通过定点突变产生酶活缺失突变体，发现其仍可诱导植物细胞坏死，表明 SsXyl2 激发植物免疫反应不依赖其木聚糖酶酶活。此外，GH11 结构中的 96 个氨基酸肽段足以诱导植物细胞死亡。*SsXyl2* 基因在核盘菌侵染寄主 36~48h 时表达量显著性升高，其敲除转化子的致病力显著降低，而互补转化子的致病力恢复正常，表明该基因与核盘菌的致病性密切相关。本研究正进一步通过免疫共沉淀和酵母双杂交等技术鉴定 SsXyl2 在本氏烟中的互作蛋白，为进一步明确 SsXyl2 蛋白作用机理提供线索。

关键词：核盘菌；木聚糖酶；植物免疫

---

* 致谢：该研究受中央高校基本科研业务费重点项目（XDJK2019B034）资助

** 第一作者：王娅波，硕士研究生，主要从事分子植物病理学研究；E-mail：2919875695@qq.com

*** 通信作者：余洋；E-mail：zbyuyang@swu.edu.cn

基于 VIGS 技术的小麦白粉菌丝氨酸/苏氨酸蛋白激酶基因 *STPK2* 的功能研究*

曾凡松[1]，蔺瑞明[2]，朱红艳[3]，袁 斌[1]，杨立军[1]，龚双军[1]，史文琦[1]，喻大昭[1**]

（1. 农业部华中作物有害生物综合治理重点实验室，农作物重大病虫草害防控湖北省重点实验室，湖北省农业科学院植保土肥研究所，武汉 430064；2. 中国农业科学院植物保护研究所，植物病虫害生物学国家重点实验室，北京 100193；3. 湖北省农业科学院果树茶叶研究所，武汉 430064）

摘 要：小麦白粉病是由禾布氏白粉菌小麦专化型（*Blumeria graminis* f. sp. *tritici*）引起的小麦重要病害之一。产孢阶段转录组分析筛选出 1 个在分生孢子梗形成阶段（5dpi）表达量显著上调的基因。序列分析结果表明，该基因全长 1 923bp，编码 640 个氨基酸残基的丝氨酸/苏氨酸蛋白激酶（serine/threonine protein kinase），暂命名为 *STPK2*。该激酶具有推定的 ATP 结合位点和丝氨酸/苏氨酸蛋白激酶催化结构域，且在 N 段含有 1 段核定位信号序列。与其他 19 种植物病原真菌的 29 个同源蛋白比较分析结果表明，STPK2 属于 KSP1 类丝氨酸/苏氨酸蛋白激酶。为进一步明确该基因在小麦白粉菌产孢过程中的作用，本研究采用大麦条纹花叶病毒（BSMV）诱导的基因沉默（virus-inducing gene silencing，VIGS）技术，对小麦白粉菌内源的 *STPK2* 基因进行了沉默。定量 PCR 结果证明 *STPK2* 在 5dpi 时显著下降（$P<0.01$）。组织学观察结果表明，该基因的沉默导致白粉菌足细胞形成率，分生孢子梗形成率以及产孢量均显著下降（$P<0.01$），病害症状明显减轻。这些结果说明，*STPK2* 在小麦白粉菌产孢过程中起重要作用，且可以作为用于小麦持久抗白粉病育种的候选靶标基因。

关键词：小麦白粉菌；产孢；蛋白激酶；基因沉默

---

* 基金项目：国家小麦产业技术体系专项（CARS0304B）

** 通信作者：喻大昭；E-mail：Dazhaoyu@ china. com

小新壳梭孢菌引起的枇杷果腐病初步研究*

廖　辉，张娇花，赵　行，冯　柳，张美鑫，翟立峰**
（长江师范学院生命科学与技术学院，重庆　408000）

摘　要： 枇杷［*Eriobotrya japonica*（Thunb.）Lindl.］又名卢橘，是蔷薇科（Rosaceae）枇杷属（*Eriobotrya*）常绿乔木。中国是枇杷生产大国，长江流域以南地区均有栽培。近年来，重庆市涪陵区果园枇杷田间烂果发生逐渐加重，为了明确枇杷果腐病的病原菌种类，对采自涪陵区不同果园的枇杷果腐病样品进行病原菌的分离与纯化，根据病原菌的形态学特征及致病性，并结合菌株的 rDNA-ITS、β-tubulin 和 EF1-α 序列分析，鉴定其病原菌的种类。结果表明：在分离纯化的 21 份菌株中，存在 4 种类型的真菌，其中有 4 株表现为气生菌丝旺盛，菌株的菌落初期为白色，随后菌落逐渐变成黑色。分生孢子为椭圆形、无隔膜、透明的单细胞，分生孢子长度为 12.29～23.11 μm，宽度为 6.11～9.30 μm，平均长宽比为 2.31。通过对这 4 株菌株的 ITS、β-tubulin 和 EF1-α 序列构建系统进化树分析表明这 4 株菌株以 96% 的支持率和小新壳梭孢菌（*Neofusicoccum parvum*）聚在一个分枝上，表明供试的 4 株菌均属于小新壳梭孢菌。将这 4 株菌株无伤和有伤接种枇杷果实（大五星）测定致病性，结果显示接种位点果皮形成棕色病斑逐渐变成黑色，果肉内部呈褐色水渍状腐烂。接种无菌 PDA 培养基的果实未发病。对发病果实进行再分离，所得病原菌与原接种病原菌形态特征和 rDNA-ITS 序列一致，符合科赫氏法则，说明分离得到的小新壳梭孢菌菌株均为引起枇杷果实腐烂的病原菌。

关键词： 枇杷；小新壳梭孢菌；鉴定

---

* 基金项目：国家自然科学基金（31701837）

** 通信作者：翟立峰，副教授，研究方向：果树病害及其防治；E-mail：zhailf@yeah.net

多肉植物彩虹黑腐病病原菌的分离鉴定

刘　浩*，杨　爽，田佩玉，刘诗婷，杨丽娜，詹家绥**
（福建农林大学植物病毒研究所，福州　350012）

摘　要：彩虹（*Echeveria* Rainbow）是景天科拟石莲属的多肉植物，是紫珍珠的锦化品种，近几年在全国范围内广泛种植。黑腐病具有较强的传染性，是彩虹种植过程中为害最严重的病害，感染后全株迅速变黑腐烂。为明确彩虹黑腐病病原菌的种类，在观察病原菌的培养性状、分生孢子的基础上，以核糖体内转录间隔区、转录延伸因子的测序分析对病原菌进行鉴定，并利用回接法检测病原菌的致病性。形态学结果观察表明，该菌在 PDA 上菌落为白色圆形。分生孢子主要是小型分生孢子，无色透明，椭圆形；大型分生孢子镰刀形，多数具 3 隔，大小为（21.8~32.0）μm ×（3.3~4.0）μm（平均值 26.8μm×3.7μm）。通过对 ITS 和 EF-1α 序列分析表明该菌为尖孢镰刀菌（*Fusarium oxysporum*），并具有致病性。结合形态学观察和分子鉴定结果，将该菌鉴定为尖孢镰刀菌。

关键词：多肉植物；黑腐病；尖孢镰刀菌

* 第一作者：刘浩，硕士研究生，植物病理学；E-mail：lhao0520@126.com
** 通信作者：詹家绥，教授，主要从事群体遗传学研究；E-mail：Jiasui.zhan@fafu.edu.cn

中国马铃薯四大耕作区的致病疫霉 *PI*02860 的群体遗传结构分析

周世豪*，沈林林，杨　爽，刘　浩，黄艳媚，刘诗婷，詹家绥**
（福建农林大学植物病毒研究所，福州　350012）

摘　要：由致病疫霉引起的马铃薯晚疫病是马铃薯上毁灭性病害，而致病疫霉效应子在侵染过程中发挥着至关重要的作用。本研究对来自全国 4 个耕作区的 490 个菌株的 *PI*02860 效应子的基因进行扩增，结合生物信息学，群体遗传学等方法探究不同耕作区的地理因素对其遗传结构的影响。结果共检测出 6 种单倍型，且其变异类型为点突变；除了南方冬作区的优势单倍型为 Hap_5，其他 3 个耕作区的优势单倍型均为 Hap_2；同时发现随着地区年均温的升高，*PI*02860 的单倍型个数，核苷酸多样性也越来越高。说明年均温越高的地区，其遗传结构更复杂。

关键词：致病疫霉；效应子；群体遗传

* 第一作者：周世豪，硕士研究生，主要从事马铃薯晚疫病的研究；E-mail：zsh11234567@ 163. com
** 通信作者：詹家绥，教授，主要从事群体遗传学研究；E-mail：Jiasui. zhan@ fafu. edu. cn

胶孢炭疽菌 *CgCDC2* 的生物学功能分析*

夏 杨**，苏初连，叶 子，刘晓妹，蒲金基，张 贺***
（中国热带农业科学院环境与植物保护研究所，农业农村部热带作物有害生物综合治理重点实验室，海南省热带农业有害生物监测与控制重点实验室，海口 571101）

摘 要：胶孢炭疽菌（*Colletotrichum gloeosporioides*）是引起炭疽病的一种重要病原真菌，其具有寄主范围广泛、侵染策略多样等特点，可侵染多种作物，对农业生产造成严重的经济损失。细胞周期蛋白依赖性激酶 CDC2 是细胞周期调控的主要因子之一，它广泛存在于丝状真菌和酵母中，对植物病原菌的生长发育有着重要作用，目前还未有对胶孢炭疽菌 CDC2 蛋白生物学功能方面的研究。本研究通过 PCR 扩增获得了 *CgCDC2* 基因序列，其编码 326 个氨基酸。利用 RNAi 技术和 PEG 介导的原生质体转化技术获得 *CgCDC2* 的 RNAi 突变体菌株，通过表型分析、TMT 蛋白组学分析、酶活性测定发现，RNAi 突变体菌株与野生型相比，生长速率减缓、分生孢子产量显著下降、对 H_2O_2 敏感性增强，PG、PL 蛋白质含量、基因相对表达量和酶活性均显著降低，致病力减弱。以上结果表明 *CgCDC2* 参与调控胶孢炭疽菌的生长，分生孢子产量，氧化应激反应以及 PG、PL 的酶活性，从而降低病原菌的致病力。

关键词：胶孢炭疽菌；CDC2；RNAi；TMT 蛋白组学；致病力

---

* 基金项目：国家重点研发专项（2017YFD0202100）；海南省重大科技计划项目（ZDKJ2017003）；中国热带农业科学院基本科研业务费专项资金（1630042017019）；农业农村部农业技术试验示范与服务支持项目“滇桂黔石漠化地区特色作物产业发展关键技术集成示范”

** 第一作者：夏杨，硕士研究生，研究方向：病原微生物；E-mail：xiayeung@163.com

*** 通信作者：张贺，硕士，助理研究员，研究方向：热带果树病理学；E-mail：atzzhef@163.com

基于核酸片段的苜蓿黄萎病菌遗传进化分析

高瑞芳[1,2]，王　颖[1,2]，章桂明[1,2]
（1. 深圳海关动植物检验检疫技术中心，深圳　518045；
2. 深圳市外来有害生物检测技术研发重点实验室，深圳　518045）

摘　要：苜蓿黄萎病菌 *Verticillium albo-atrum* Reinke et Berthold 是威胁苜蓿、蚕豆、马铃薯、草莓等种植的毁灭性病害，为种传病害。中国每年进口大量苜蓿、饲草及其种子，如果传入内地，极有可能在几年之内袭击国内主要产区，造成重大损失，被明确列入检疫性有害生物名录。本研究以该病原菌为研究对象，通过对其核酸片段的进化和聚类分析探讨不同来源的病原菌的群体遗传分群情况。候选基因片段有 ITS、GPD、18S、28S、tubulin、cytb、EF-1α、NADH、tryptophan 共 9 个，从 NCBI 和 BOLD 数据库中下载序列 10891 条序列，使用 MEGA 软件对序列进行校对。通过有效数据量、基于 K2P-distance 计算最大种内遗传距离、最小种内遗传距离和种内遗传距离平均值，比较得出 ITS 片段适合作为进化分析的基因片段，其有效数据量为 48，种内遗传距离最大值、最小值和平均值分别为 0.041、0.000 和 0.016。对挑选的分别来自于比利时、挪威、伊朗、加拿大的 13 个样品，使用 RAxML 软件 maximum likehood 算法构建了不同来源地物种的系统进化关系，使用 PLINK 软件进行主成分分析得出聚为 3 个类群，使用 ADMIXTURE 软件进行群体遗传结构分析得出来自于 3 个原始祖先。通过聚类和进化关系的分析，探讨了不同来源地菌株的群体遗传关系，为精准明确的评估其跨境传播风险，区别性的制定监管处理措施，有利于将病原菌的防控向源头延伸，提高生物安全风险预警和监测能力。

关键词：苜蓿黄萎病菌；ITS 片段；群体遗传

高温胁迫对不同地区致病疫霉生长及产孢的影响*

谷楠林**，常婧一，白家琪，赵冬梅***，杨志辉***
（河北农业大学植物保护学院，保定 071000）

摘 要：由致病疫霉引起的马铃薯晚疫病一直以来都是马铃薯生产上的重要病害，而温度是决定马铃薯晚疫病是否发生及发生程度的重要因素，温度对病原菌的生长和繁殖都有重要影响。本文以18℃恒温处理为对照，研究了在30℃、35℃和40℃分别处理3 h，6 h，9 h条件下对云南和内蒙古两个地区菌株2130和1423生长及产孢的影响。研究结果表明，3 h短时间的高温胁迫不会影响致病疫霉的生长。但随着高温胁迫时间的延长则会抑制致病疫霉的生长，当在30℃、35℃和40℃高温胁迫处理9 h时，内蒙古菌株1423的菌落直径较对照温度减少了22.60%、36.90%和47.62%，云南菌株2130的菌落直径较对照温度减少了16.47%、29.41%和44.71%。云南菌株在变温处理后，菌落直径略大于内蒙古地区菌株的菌落直径，但两个菌株间没有显著的差异。研究还发现在不同温度的高温胁迫下，对致病疫霉的产孢具有不同的作用，30℃和35℃的高温胁迫对致病疫霉的产孢具有刺激作用，但当温度达到40℃时，则对致病疫霉的产孢具有抑制作用。菌株1423在18℃时单位面积产孢量为1个/mm^2，在30℃ 9 h的温度处理下，单位面积产孢量增加到了125个/ mm^2，增加了125倍。菌株2130在18℃时单位面积产孢量为59个/ mm^2，在35℃ 3 h的温度处理下，单位面积产孢量增加到175个/ mm^2。菌株1423在40℃ 3 h处理下的单位面积产孢量为0.8个/ mm^2，菌株2130在40℃ 3 h处理下的单位面积产孢量为46个/ mm^2，都低于对照温度下菌株的单位面积产孢量。

关键词：马铃薯；致病疫霉；温度处理

* 基金项目：河北省薯类产业技术体系创新团队专项基金资助（HBCT2018080205）
** 第一作者：谷楠林，本科生，动植物检疫专业；E-mail：gunanlin@ qq. com
*** 通信作者：赵冬梅，讲师，主要从事马铃薯病害和分子植物病理学研究；E-mail：zhaodongm03@ 126. com
杨志辉，教授，主要从事马铃薯病害和分子植物病理学研究；E-mail：13933291416@ 163. com

茄链格孢致病基因 *AsSlt2* 的功能研究*

范莎莎**，赵冬梅***，杨志辉，张　岱，朱杰华***
（河北农业大学植物保护学院，保定　071000）

摘　要：中国是世界上最大的马铃薯生产国，但是早疫病逐渐成为限制我国马铃薯产业发展的重要因素，该病的发生严重降低了马铃薯品质及产量。近年来，基因组学的研究发展使得基因功能的研究成为热点。茄链格孢（*Alternaria solani*）是引起马铃薯早疫病的主要病原菌，2015 年本研究室完成了茄链格孢的全基因组测序。本研究在茄链格孢全基因组测序基础上，克隆得到 MAPK 通路中 *Slt*2 基因的同源基因，命名为 *AsSlt*2。通过同源重组方法对该基因进行敲除，获得了 4 株缺失 *AsSlt*2 基因的突变菌株，并在此基础之上获得恢复株 Δ*Slt*2-*C*。以野生型菌株 HWC-168 和恢复株 Δ*Slt*2-*C* 作为对照，对 *AsSlt*2 基因功能进行解析，主要研究结果如下：① *AsSlt*2 基因全长为 1675 bp，包含 6 个内含子和 7 个外显子，CDS 序列为 1251 bp，共编码 416 个氨基酸，其中酸性氨基酸 57 个，碱性氨基酸 46 个。该基因编码的氨基酸序列不具有信号肽结构，*AsSlt*2 基因与 *A. alternata* 中同源基因亲缘关系最近，氨基酸序列一致性高达 100%。②*AsSlt*2 基因参与调控茄链格孢菌丝的营养生长，突变株 Δ*Slt*2 生长速率明显下降，菌丝形态较弯曲、浓密，分枝减少。③与野生型和恢复菌株相比，突变株 Δ*Slt*2 丧失了产生分生孢子和黄色素的能力。④使用未创伤和创伤处理叶片进行致病性验证，结果显示突变株 Δ*Slt*2 致病力严重减弱。野生型菌株 HWC-168 侵染未创伤叶片产生病斑直径为 0.99cm，接种突变株 Δ*Slt*2 的叶片未出现病斑。创伤处理的叶片接种突变株 Δ*Slt*2 后能够产生病斑，但病斑大小与野生型和互补株具有极显著差异。由此表明，*AsSlt*2 基因参与调控茄链格孢的营养生长、产孢能力、黄色素产生能力和致病性等多方面的功能。

关键词：马铃薯；茄链格孢；致病性；*Slt*2 基因

---

* 基金项目：现代农业产业技术体系建设专项资金资助（CARS-09-P18）；河北省薯类产业技术体系创新团队专项基金资助（HBCT2018080205）

** 第一作者：范莎莎，硕士研究生，植物病理学专业；E-mail：1531017055@ qq. com

*** 通信作者：赵冬梅，讲师，博士，主要从事马铃薯病害和分子植物病理学研究；E-mail：zhaodongm03@ 126. com
朱杰华，教授，博士，主要从事马铃薯病害和分子植物病理学研究；E-mail：zhujiehua356@ 126. com

立枯丝核菌 Rs-1 代谢物对苦荞萌发生长的影响*

赵江林**，唐晓慧，吴志伟，钟灵允，赵　钢
（成都大学药学与生物工程学院，农业农村部杂粮加工重点实验室）

摘　要：荞麦（Buckwheat）是一种著名的食药同源小宗杂粮作物，其营养保健功能强、经济价值高、开发利用前景广阔。近年来，随着我国荞麦种植面积的不断扩大及连作障碍增多，荞麦病害的发生也呈逐年加重趋势，其中由立枯丝核菌（*Rhizoctonia solani*）引起的荞麦立枯病害较为突出，可导致荞麦的年产量损失高达15%~20%，严重影响了我国荞麦的产量和品质。植物病原真菌在生长过程会产生对寄主有明显致病或致毒作用的活性物质。为探明荞麦立枯病菌的致病机理，本研究采用胚根生长抑制法检测了立枯丝核菌 Rs-1 菌丝菌液有机溶剂提取物、菌丝菌液多糖和菌液蛋白对苦荞种子萌发生长的影响。

研究结果表明，立枯丝核菌 Rs-1 菌丝有机溶剂提取物对苦荞种子的萌发和生长具有一定的抑制性作用，其中菌丝正丁醇层提取物对苦荞种子萌发和生长的抑制作用较为明显。当菌丝正丁醇层提取物浓度为4.0 mg/mL 时，苦荞芽根长仅为1.66 cm，与空白对照6.88 cm 相比，其抑制率达到75.9%。菌液多糖（EPS）对苦荞种子的萌发和生长均具有一定促进作用。菌丝水提多糖（WPS）对苦荞种子萌发和生长无显著影响。菌丝酸提多糖（APS）对苦荞种子萌发和生长具有一定的抑制作用。菌丝碱提多糖（SPS）对苦荞种子萌发和生长具有明显的抑制活性，且存在一定的量效关系。当 SPS 浓度为2.0 g/L 时，苦荞芽根长仅为1.49 cm，较空白对照相比，其抑制率为79.8%。立枯丝核菌 Rs-1 菌液蛋白对苦荞种子的萌发生长具有明显的抑制作用，随着蛋白质溶液浓度的升高，其抑制作用逐渐加强。当立枯丝核菌 Rs-1 菌液蛋白浓度为72.0 mg/L 时，苦荞芽根长仅为0.53 cm，与对照相比而言，其抑制率高达93.4%。该研究结果为荞麦立枯丝核菌致病机制的深入研究提供了重要依据。

关键词：荞麦；立枯丝核菌；代谢物；致病机制

---

* 基金项目：四川省科技计划项目（2019YJ0662）；国家自然科学基金项目（31701358）

** 通信作者：赵江林，教授，主要从事植物与微生物相互作用的次生代谢生物学研究；E-mail：jlzhao@ cdu. edu. cn

苜蓿根腐病病原菌分离鉴定及苜蓿品种的抗性评价

杨剑锋*，王　娜，刘　欢，李昊宇，张　键，赵　君**
（内蒙古农业大学园艺与植物保护学院，呼和浩特　010018）

摘　要：苜蓿根腐病是苜蓿种植过程中最常见的一种病害。由于引起该病害的病原较为复杂，因此，明确不同生态环境下苜蓿根腐病的病原种类、确定优势种，对于根腐病的防治非常重要。本研究利用柯赫氏法则对采自内蒙古 3 个不同地点苜蓿田中疑似根腐病的病样进行了分离和鉴定，结果表明兴安盟科右中旗的病样上分离鉴定的根腐病菌为 XAM-1（*Fusarium equiseti*）、XAM-2（*F. acuminatum*）和 XAM-4（*F. thapsinum*）；中国农业科学院草原研究所试验基地的根腐病的致病菌为 CYS1-1（*Fusarium equiseti*）、CYS1-2（*F. incarnatum*）、CYS1-3（*F. oxysporum*）、CYS-2（*Boeremia exigua*）和 CYS-4（*Alternaria alternate*）；达拉特旗地块中病样上分离到的致病菌有 MX-1（*F. tricinctum*）和 MX-2（*F. equiseti*）两种。室内对 32 个苜蓿品种的抗根腐病水平进行了评价，结果表明人工接种 MX-1（*F. tricinctum*）和 MX-2（*F. equiseti*）后，品种康赛的的病情指数分别为 26.07 和 26.81，表现出最高的抗性水平；其次为大银河品种，接种后病情指数分别为 26.67 和 29.56；而品种肇东高感苜蓿根腐病，接种后病情指数分别为 56.89 和 66.44。部分苜蓿品种抗性水平差异显著，本研究中未发现免疫品种。

关键词：苜蓿根腐病；病原菌分离和鉴定；品种抗性鉴定

---

* 第一作者：杨剑锋，硕士研究生；E-mail：jianfengyang89@126.com
** 通信作者：赵君，教授；E-mail：zhaojun02@hotmail.com

河南信阳地区水稻穗腐病病原多样性*

陈利军[1,2]**，王春生[1,2]，田雪亮[3]，智亚楠[1,2]

（1. 信阳农林学院农学院，信阳 464000；2. 豫南农作物有害生物绿色防控院士工作站，信阳 464000；3. 河南科技学院资源与环境学院，新乡 453003）

摘　要：水稻穗腐病（rice spikelet rot disease，RSRD）是我国各水稻种植区发生为害严重的水稻穗部病害，近年来在河南信阳稻区的发生为害也呈现出逐年蔓延和加重的趋势。水稻穗腐病为害稻穗，造成谷粒腐坏、变色、结实率降低或不实、稻米畸形等，不仅影响产量、降低稻米品质，威胁水稻生产，同时病原菌还会产生毒素，对食用者的安全和健康构成危害。本研究在2016—2018年从河南省信阳市平桥区、浉河区、罗山县和光山县采集水稻穗腐病样本，通过病原菌分离纯化、形态学和*TEF*1基因序列分析鉴定，明确了信阳地区水稻穗腐病的病原菌种类及优势种群。从信阳地区水稻穗腐病样本中共分离到279株病原真菌，以镰孢菌*Fusarium* spp. 为主，分离频率达50.18%，其中包括藤仓镰孢菌复合群*F. fujikuroi* species complex、厚垣镰孢菌复合群*F. chlamydosporum* species complex、禾谷镰孢菌复合群*F. graminearum* species complex、*F. incarnatum-equiseti* species complex、尖孢镰孢复合群*F. oxysporum* species complex，分离频率分别为19.36%、12.90%、11.11%、3.58%和3.23%。细极链格孢*Alternaria tenuissima*的分离频率相对较高，达10.75%；黑孢属和弯孢属病原菌各2种，即球黑孢菌*Nigrospora sphaerica*（9.68%）和稻黑孢*N. oryzae*（4.66%）、棒弯孢*Curvularia clavata*（6.45%）和新月弯孢*C. lunata*（6.09%）；水稻胡麻斑病菌稻平脐蠕孢*Bipolaris oryzae*在穗腐病害材料中的分离频率为6.81%；在谷粒上形成蓝绿色霉层的草酸青霉*Penicillium oxalicum*分离频率为5.38%。镰孢菌属分类系统复杂，不同复合群中具体种的鉴定仍需要进一步的研究。

关键词：水稻穗腐病；病原多样性；信阳地区；镰孢菌属

---

* 基金项目：河南省高等学校重点科研项目计划（16A210053）

** 第一作者：陈利军，教授，主要从事植物真菌病害与植病生防资源的研究；E-mail：chlijun1980@163.com

希金斯炭疽菌自噬基因 *ChAtg8* 序列分析及敲除载体的构建[*]

祝一鸣[**]，曾森林，刘艳潇，舒灿伟，周而勋[***]
（华南农业大学植物病理学系，广东省微生物信号与作物病害重点实验室，广州 510642）

摘　要：炭疽菌（*Colletotrichum* spp.）是世界十大植物病原真菌之一，其为害范围广泛，造成的经济损失非常严重。由希金斯炭疽菌（*C. higginsianum*）引起的十字花科蔬菜炭疽病是蔬菜生产上的重要病害之一，对十字花科蔬菜，如广东菜心、白菜、芥蓝等，都造成了非常严重的经济损失，特别是广东高温高湿的气候条件非常适于菜心炭疽病的发生和为害，该病害已经成为制约广东菜心产业发展的重要因素之一。细胞自噬是从酵母到哺乳动物都非常保守的一个过程，在营养不良等环境下，自噬过程被诱导。细胞自噬在不同的生命活动中起着不同的作用，此前的研究表明细胞自噬与病原菌的致病力息息相关，而自噬在希金斯炭疽菌中的作用尚不清楚。

本研究在希金斯炭疽菌中克隆得到了自噬基因 *ChAtg*8 的全长序列，并进行了基因敲除载体的构建，旨在明确其生物学功能，为今后揭示自噬相关过程在希金斯炭疽菌中的作用打下基础。首先从希金斯炭疽菌基因组中扩增得到 *ChAtg*8 基因 DNA 和 cDNA 的全长序列，然后采用 NCBI 在线结构域预测工具 CDD（https：//www. ncbi. nlm. nih. gov/Structure/cdd/）分析 *ChAtg*8 基因的功能；以希金斯炭疽菌的基因组为模板，分别扩增 *ChAtg*8 基因上下游大约 1～2 kb 的序列作为上下臂，同时分别使用相应的限制性内切酶对 pFGL821 质粒进行酶切，最后，使用无缝克隆技术将上下臂依次连入经对应的限制性内切酶处理的 pFGL821 载体中，后通过 PCR 验证、测序，最终获得基因敲除载体 pFGL821-Atg8。*ChAtg*8 基因全长 623 bp，含有两个内含子区域，编码的 cds 全长 489 bp，编码 162 个氨基酸，含有一个微管结合位点、一个 Atg7 结合位点以及一个维持 Atg8 泛素样结构极为重要的泛素样蛋白 GABARAP 核心。敲除载体 pFGL821-Atg8 已构建成功，为下一步 *ChAtg*8 基因功能的深入研究奠定了基础。

关键词：希金斯炭疽菌；自噬基因 *ChAtg*8；基因克隆；序列特征分析；基因敲除载体构建

---

* 基金项目：国家重点研发计划项目（2017YFD0200900）
** 第一作者：祝一鸣，博士生，研究方向：植物病原真菌及真菌病害；E-mail：zhu_yiming1992@163. com
*** 通信作者：周而勋，教授，博导，研究方向：植物病原真菌及真菌病害；E-mail：exzhou@scau. edu. cn

RxLR 基因 *PITG*-14788 和 *PITG*-19831 促进致病疫霉对本氏烟的侵染[*]

马 英[**]，赵冬梅[***]，杨志辉，朱杰华[***]
（河北农业大学植物保护学院，保定 071000）

摘 要：由致病疫霉（*Phytophthora infestans*）引起的马铃薯晚疫病是马铃薯生产中最具毁灭性的病害。致病疫霉能够分泌效应子来促进侵染寄主植物并在寄主植物内定殖。通过效应子研究病原菌的致病及与植物的互作机制有助于理解病原菌的致病机理。本研究通过对致病疫霉中RxLR 基因 *PITG*-14788 和 *PITG*-19831 功能进行研究，以期进一步明确 RxLR 类效应子在致病过程中的作用。

经生物信息学预测发现 *PITG*-14788 和 *PITG*-19831 都含有 RxLR 结构域且 N 端都含有信号肽，*PITG*-14788 预测有 2 个功能域，而 *PITG*-19831 在 C 端有一段跨膜区，但没有预测得到结构域和活性位点。*PITG*-14788 和 *PITG*-19831 在致病疫霉侵染马铃薯过程中均上调表达，其中 *PITG*-14788 在接种 6 h 后表达量显著上调表达，12 h 后表达量达到最大值，而 *PITG*-19831 在接种 24 h 后显著上调表达并达到了最大值，说明这两个基因在致病疫霉侵染马铃薯的过程中均发挥作用，但 *PITG*-14788 在侵染前期发挥主要作用，而 *PITG*-19831 在侵染中期发挥主要作用。分别在烟草和不同马铃薯品种中瞬时表达 *PITG*-14788 和 *PITG*-19831 发现，*PITG*-14788 可诱导本氏烟和马铃薯品种“冀张薯 8 号”产生过敏性坏死反应，而表达 *PITG*-19831 仅可诱导马铃薯品种“陇薯 6 号”产生过敏性坏死反应，表明这两个效应子可以激发不同寄主植物的免疫防卫反应。在本氏烟中表达 2 个基因后均能够促进致病疫霉对本氏烟的侵染。同时，qRT-PCR 结果显示在本氏烟中表达 *PITG*-14788 后，SA、JA 和 ET 信号途径中的指示基因 *PR2b*、*LOX* 和 *ERF*1 均显著上调表达；表达 *PITG*-19831 后 SA 和 ET 信号途径中的指示基因 *PR2b* 和 *ERF*1 显著下调表达，而 JA 途径的指示基因 *LOX* 则上调表达，表明这两个效应子可以通过影响不同的信号途径干扰植物的免疫防卫反应，从而促进致病疫霉的侵染。研究结果可为进一步探索效应蛋白的转运机制、毒性功能，以及开展对植物病原卵菌致病机理的深入研究提供重要的理论依据。

关键词：马铃薯晚疫病；RxLR 基因；基因功能；免疫防卫反应

---

* 基金项目：现代农业产业技术体系建设专项资金资助（CARS-09-P18）
** 第一作者：马英，硕士研究生，植物病理学专业；E-mail：852616127@qq.com
*** 通信作者：赵冬梅，讲师，主要从事马铃薯病害和分子植物病理学研究；E-mail：zhaodongm03@126.com
朱杰华，教授，主要从事马铃薯病害和分子植物病理学研究；E-mail：zhujiehua356@126.com

茄链格孢无性产孢相关基因 *flbA* 功能研究*

石永蓉**，赵冬梅，杨志辉***，潘　阳，朱杰华***
（河北农业大学植物保护学院，保定　071000）

摘　要：马铃薯作为重要粮食作物，其种植面积呈现逐年上升趋势，在我国国民生产中占据重要地位，由茄链格孢引起的早疫病是马铃薯生产上的重要病害，能够造成巨大的经济损失。无性孢子在茄链格孢侵染、传播和病害发生中具有重要的作用，但是目前并没有对马铃薯早疫病菌无性产孢通路及产孢基因功能的相关研究，这大大限制了人们从分子水平上认识其无性繁殖过程的本质。明确产孢基因的功能，为阐明早疫病菌无性产孢机制以及寻找病害防治新靶标提供理论依据。

本研究成功克隆了产孢基因 *flbA*，该基因全长为 1 762 bp，含有 3 个内含子，编码 483 个氨基酸。通过融合 PCR 技术，构建得到了含有潮霉素抗性基因的 *flbA* 同源重组片段，通过 PEG 介导的原生质体转化法获得了 3 株缺失 *flbA* 基因的突变菌株。同时，成功构建了恢复载体，并获得了 2 株恢复菌株。在产孢量、分生孢子形态等方面比较分析了野生菌株、Δ*flbA* 突变体和恢复菌株的差异。结果发现，与野生菌株和恢复菌株相比，Δ*flbA* 突变菌株产孢量明显下降，较野生菌株减少了 98.68%，但分生孢子形态未发生改变。Δ*flbA* 突变菌株菌落颜色发白，菌丝呈绒毛状且分枝减少，菌丝穿透力丧失，但菌丝的生长速度与致病力没有发生变化。研究结果表明，*flbA* 基因在调控茄链格孢的产孢过程中起重要的作用，同时也参与调控茄链格孢的菌落颜色、菌丝形态及菌丝的体外穿透力。

关键词：马铃薯；茄链格孢；无性产孢；*flbA* 基因

---

* 基金项目：现代农业产业技术体系建设专项资金资助（CARS-09-P18）；河北省薯类产业技术体系创新团队专项基金资助（HBCT2018080205）

** 第一作者：石永蓉，硕士研究生，植物病理学专业；E-mail：2542911105@qq.com

*** 通信作者：杨志辉，教授，博士，主要从事马铃薯病害和分子植物病理学研究；E-mail：13933291416@163.com
朱杰华，教授，博士，主要从事马铃薯病害和分子植物病理学研究；E-mail：zhujiehua356@126.com

效应基因 *BdLM* 参与调控葡萄座腔菌的生长发育及致病过程*

李培航，温胜慧，陆柳伊，国立耘，朱小琼**
（中国农业大学植物保护学院，北京　100193）

摘　要：葡萄座腔菌（*Botryosphaeria dothidea*）可以危害 20 多个属的植物。该菌侵染苹果、梨引起的轮纹病是我国苹果、梨生产上发生范围广、为害大、防治困难的一种重要病害。含有溶解素基序（lysin motif，LysM）的蛋白广泛存在于细菌、真菌、植物、动物中。引起植物叶片或根部病害的真菌中含 LysM 结构域的效应子干扰了寄主植物中几丁质诱导的免疫反应并且在病菌的毒力方面具有重要作用。LysM 类效应子在葡萄座腔菌中的生物学功能尚不清楚。本研究获得了一个能够抑制 Bax 引起的烟草细胞程序性死亡的候选效应子 BdLM。BdLM 在烟草中瞬时表达后能够促进烟草疫霉的侵染，过氧化物积累增加。通过基因敲除及互补试验获得了 BdLM 突变体和回补菌株，并以野生型为对照，对突变体的生物学表型进行了分析。结果发现，敲除体的生长速率比野生型和回补体显著增加，菌落形态不变；分生孢子器数量、大小、产孢量、分生孢子大小均无显著改变。敲除体产生黑色素的时间延迟，培养 10 d 后黑色素的量与野生型无显著差异；对 125g/L Sorbitol 的敏感性与野生型相比有显著性差异；在苹果一年生离体枝条上的致病性显著降低，但在苹果果实上的致病性没有显著差异。上述研究结果表明，*BdLM* 基因参与调控葡萄座腔菌的营养生长、对外界胁迫的应答及致病过程，不参与病菌的无性繁殖；可能参与调控植物的免疫反应。

* 基金项目：国家重点研发计划—京津苹果农药减施增效技术集成研究与示范（2016YFD0201129）

** 通信作者：朱小琼，副教授，主要研究方向为植物病原真菌及致病机制；E-mail：mycolozhu@ cau. edu. cn

匍柄霉中发现引起寄主致病力衰退的新真菌病毒

刘 洪*，王 慧，陆 训，潘显婷，周 倩**
（湖南农业大学植物保护学院，植物病虫害生物学与防控湖南省重点实验室，长沙 410128）

摘 要：匍柄霉（*Stemphylium* spp.）是一类弱寄生植物病原真菌，可引起多种作物叶斑病。实验室在鉴定莴苣叶斑病病原时发现一株携带 dsRNA 病毒的番茄匍柄霉（*S. lycopersici*）菌株 WS-01，与野生型菌株相比致病力显著降低且产毒素 Altersolanol A 的能力显著下降。利用高通量测序的方法获得了该病毒的基因组全序列，含有 4 条 dsRNA 条带，病毒粒子直径约 34 nm，暂时命名为 Stemphylium lycopersici mycovirus 1。该病毒可以水平传播至囊状匍柄霉菌株 SvHN-02 中，同样导致寄主致病力衰退和毒素产量显著降低。

关键词：匍柄霉；毒素；真菌病毒

* 第一作者：刘洪，硕士研究生，研究方向为植物病原真菌病毒；E-mail：1125293383@ qq. com
** 通信作者：周倩，教授，博士，研究方向为分子植物病理学；E-mail：zhouqian2617@ hunau. edu. cn

希金斯炭疽菌效应子基因的表达模式分析及效应子 ChEP85 功能的初步分析*

刘艳潇**，皮　磊，祝一鸣，舒灿伟，周而勋***
（华南农业大学植物病理学系，广东省微生物信号与作物病害重点实验室，广州　510642）

摘　要：希金斯炭疽菌（*Colletotrichum higginsianum*）是一种重要的半活体型营养植物病原真菌，引起的炭疽病是十字花科蔬菜生产上最常见和最严重的病害之一，为害区域分布广泛，华南地区有利的气候条件使病害发生更加严重。效应子是由病原菌分泌的、可以改变寄主植物细胞的结构和代谢途径，从而促进对寄主植物成功侵染或者触发寄主防卫反应的外泌型蛋白分子。作为研究病原菌与寄主互作的一种模式病原菌，希金斯炭疽菌的致病机制尚不清楚。本研究基于本实验室前期研究的基础，首先对筛选出的 10 个候选效应子基因进行表达模式分析，然后选出其中一个引起烟草坏死的关键效应子进行功能分析，以期更好地了解效应子在病原菌致病过程中发挥的具体作用。一方面，制备希金斯炭疽菌分生孢子悬浮液，对拟南芥叶片喷雾接种后 0h、6h、12h、24h、36h、48h、60h 定点采样，提取拟南芥叶片总 RNA，反转录为 cDNA，利用实时荧光定量 RCR 技术分析炭疽菌侵染拟南芥后不同阶段效应子基因的表达情况；另一方面，构建含有效应子基因的植物表达高效载体并进行烟草瞬时表达，筛选引起烟草坏死的关键效应子。研究结果表明，在 10 个候选效应子中，效应子 ChEP85 能稳定引起烟草叶片明显的过敏性坏死反应，且在接种后 24h 的病原菌侵染的活体营养阶段高度表达。为了研究效应子 ChEP85 的功能，去除其信号肽序列，构建表达载体以研究信号肽的作用，探究去除信号肽的效应子是否还能引起烟草叶片坏死的现象；同时构建荧光表达载体研究信号肽对效应子亚细胞定位的影响；进一步构建效应子 ChEP85 基因的敲除载体，通过 ATMT 转化获得基因敲除突变体，从而深入研究其功能。本研究通过对希金斯炭疽菌效应子功能的研究，探索病原菌效应子发挥作用的机制及其关键途径，并为病原菌与寄主互作的分子机制提供重要依据。

关键词：希金斯炭疽菌；分泌蛋白；基因表达模式；效应子；功能分析

---

* 基金项目：国家重点研发计划项目（2017YFD0200900）

** 第一作者：刘艳潇，硕士生，研究方向：植物病原真菌及真菌病害；E-mail：2524340857@ qq. com
*** 通信作者：周而勋，教授，博导，研究方向：植物病原真菌及真菌病害；E-mail：exzhou@ scau. edu. cn

Identifying and characterizing the circular RNAs of tea leaves inoculated with *Phoma segeticola* var. *camelliae**

Li Dongxue**, Wang Xue, Yin Qiaoxiu, Ren Yafeng, Bao Xingtao, Dharmasena Dissanayake-saman-pradeep, Wu Xian, Song Baoan, Chen Zhuo***

(*State Key Laboratory Breeding Base of Green Pesticide and Agricultural Bioengineering*, *Guizhou University*, *Guiyang* 550025, *China*)

Abstract: As a novel class of non-coding RNAs, circular RNAs (circRNAs) have been reported to play a role in various biological processes. Nevertheless, a little attention has been focused on plant circRNAs in the study of plant-pathogen interactions. This study aims to explore the expression profile and functional role of circRNAs in tea leaf infected by *Phoma segeticola* var. *camelliae*. We used comparative transcriptome analysis of the leaves of tea plant (*Camellia sinensis*) to identify significant differentially expressed circRNAs (DECs) from leaves infected or un-infected by the pathogen. CircRNA candidates contained more than 98% circRNA arising from the exons of genes (mismatch $\leqslant 2$, back-spliced junctions reads $\geqslant 1$ and genomic distances to back-splicing sites were less than 100 kb). We identified 349 DECs including 231 up-regulated DECs and 118 down-regulated DECs ($P<0.05$, $|\log_2^{foldchange}| \geqslant 1$). Gene Ontology (GO) analysis indicated that DECs were most enriched in catalytic activity of molecular functions including 11 circRNAs, intracellular membrane-bounded organelle of cellular component including 6 circRNAs and response to stress of biological process including 4 circRNAs. KEGG enrichment analysis indicated that the most enriched DECs are the pathway of purine metabolism including 12 circRNAs, followed by fatty acid metabolism including 10 circRNAs ($P<0.05$). Based on the data of circRNAs in our study, we would provide novel insights to reveal the molecular mechanisms of plant-pathogen interactions.

Key words: circRNAs; *Phoma segeticola* var. *camelliae*; tea leaves; plant-pathogen interactions

* Funding: This work was supported by National Key Research Development Program of China (2017YFD0200308) and its Post-subsidy project (2018-5262), and the Major Science and Technology Projects inGuizhou Province (No. 2012-6012)

** First author: Li Dongxue; E-mail: gydxli@ aliyun. com

*** Corresponding author: Chen Zhuo; E-mail: gychenzhuo@ aliyun. com

Metabolome changes of tea leaf spot pathogen *Phoma segeticola* var. *camelliae* cultured on PDA with tea liquor*

Ren Yafeng[1**], Li Dongxue[1], Wang Xue[1], Yin Qiaoxiu[1],
Dharmasena Dissanayake-saman-pradeep[1], Jiang Shilong[1,2],
Wu Xian[1], Song Baoan[1], Chen Zhuo[1***]

(1. *Key Laboratory of Green Pesticide and Agricultural Bioengineering*, *Ministry of Education*, *Guizhou University*, *Guiyang*, *Guizhou* 550025, *China*; 2. *College of Agriculture*, *Guizhou University*, *Guiyang* 550025, *China*)

Abstract: The fungal pathogen *Phoma segeticola* var. *camelliae* causes leaf spot on tea (*Camellia sinensis*) in Guizhou Province, China. To study the mechanism of metabolomics during the stage of the pathogen infection, we design the experiment scheme of the pathogen culturing on PDA or PDA with tea liquor (Tea variety: Fudingdabaicha). The samples were detected using Nano - LC - MS - MS (TripleTOF 5600) under the mode of positive ion and negative ion, and then the data of the metabolites was further identified and quantified using the software XCMS and metaX. The MS and MS-MS of the metabolites were annotated using the software metaX and in-house fragment spectrum library. There were 12 624 features in the positive mode, and 524 could be identified by LC-MS-MS. Meanwhile, 1 1008 features were detected in the negative mode, and 323 could be identified by LC-MS-MS. In addition, 10850 high quality features were used to quantity following condition: ratio ≥ 2 or ratio ≤ 1/2; q value≤0. 05; VIP≥1. 122 2 and 1 048 were up-or down-regulated in the positive mode, respectively. In the negative mode, 9917 high quality features were used to quantity, and 1 192 and 1 078 features were up-or down-regulated, respectively. According to the partial least-squares discrimination analysis, R^2 value was 1. 00, and Q2 value was 0. 98 in the positive mode. The R^2 value was 1. 00, and Q2 value was 0. 99 in the negative mode. The different metabolites were identified in 117 pathways, such as Biosynthesis of secondary metabolites (204), Biosynthesis of antibiotics (128), Microbial metabolism in diverse environments (95) and Arachidonic acid metabolism (45) *et al*. These data would provide important reference for the pathogenesis of tea leaf spot.

Key words: *Phoma segeticola* var. *camelliae*; *Camellia sinensis*; metabolome

* Funding: This work was supported by National Key Research Development Program of China (2017YFD0200308) and its Post-subsidy project (2018-5262), and the Major Science and Technology Projects inGuizhou Province (No. 2012-6012)

** First author: Ren Yafeng; E-mail: renyafeng@ aliyun. com

*** Corresponding author: Chen Zhuo; E-mail: gychenzhuo@ aliyun. com

Uncovering responsive miRNAs in tea plant (*Camellia sinensis*) against leaf spot infected by *Phoma segeticola* var. *camelliae* using high-throughput sequencing and prediction of their targets through degradome*

Li Dongxue**, Wang Xue, Yin Qiaoxiu, Ren Yafeng, Bao Xingtao,
Dharmasena Dissanayake-saman-pradeep, Wu Xian, Song Baoan, Chen Zhuo***
(*State Key Laboratory Breeding Base of Green Pesticide and Agricultural Bioengineering, Guizhou University, Guiyang* 550025, *China*)

Abstract: Tea plant (*Camellia sinensis*) was widely grown in Guizhou Province, in China, with the grown area reached at the top list. Tea leaf spot caused by *Phoma segeticola* var. *camelliae* was firstly reported in tea plantation in Guizhou Province, which lead to a huge loss of tea production. The response mechanism of tea host under the stress of the pathogen would contribute for the future research on host-pathogen interactions, determination of trait-specific genes and plant-host adaptation mechanisms. 970 known miRNAs or conservative miRNAs and 4730 novel miRNAs were identified from two treatment groups inoculated by *Phoma segeticola* var. *camelliae* and health tea leaves. Comparing with the treatment group of inoculation pathogen and the healthy treatment group, the number of miRNAs with up-regulated trends or down-regulated trends was 17 and 15 at $P<0.01$. nta-MIR6149a-p3_2ss13GA18CG, pvu-MIR399a-p5_1ss9AG and fifteen PC miRNAs containing with PC-3p-3524808_3, et al were significantly up-regulated for pathogen infection. mtr-miR166a_1ss21CT, ma-MIR167h-13p3_1ss19CT and thirteen PC microRNAs containing with PC-5p-2004217_4 were found to be significantly down-regulated for pathogen infection ($P<0.01$). The Targetfinder software detected a total of 18506 potential targets from the degradome sequence data sets for both conserved and novel (total 4 072) tea miRNAs identified in this study. The number of target sites detected by target gene prediction and sequencing method was 9845. For these target sites, 3882 or 5963 were up-regulated or down-regulated, respectivly. The target gene predicted using miRNA and mRNA from the degradation group density file were combined to operate, then the number of target gene were obtained. cas-MIR157b-p3_2ss13GA18CG and cas-MIR157b-p5_2ss13GA18CG had 348 the target genes. Transcripts was predicted by TargetFinder software, then combined and calculated with mRNA from the degradation group density file. The number of transcripts corresponding with interacted miRNA was 4725 with the value of Degradome_Valid_miRNA being greater than 0. TEA006960. 1 represent 59 miRNAs can interacted the transcripts. Meanwhile, we found a transcript was recognized and cleaved by more miRNAs, such as 23 miRNAs can cleave TEA014723. 1. The target genes of miRNA were classified at the level of BP, CC and MF. The number of the target genes of miRNA at oxidation-reduction process was 75, the number of the target genes of miR-

* Funding: This work was supported by National Key Research Development Program of China (2017YFD0200308) and its Post-subsidy project (2018-5262), and the Major Science and Technology Projects in Guizhou Province (No. 2012-6012)

** First author: Li Dongxue; E-mail: gydxli@aliyun.com

*** Corresponding author: Chen Zhuo; E-mail: gychenzhuo@aliyun.com

NA at integral component of membrane was 169, and the number of the target genes of miRNA at ATP binding was 119. The target genes of miRNAs were classified into 25 kinds by KOG. Beside of the kind of [S], the number of target genes under the kind of [O] toped list, with the value being 303, followed by [K] and [T], with the value being 298 and 254. The target genes of miRNAs were enriched by KEGG. For instance, 104 transcripts were enriched in pathway of Plant-pathogen interaction. ath-MIR858b-p3_2ss13GA18CG targeted and down-regulated MYBPAR (TEA002308.1), gma-miR828a_1ss22AT targeted and down-regulated transcription factor MYB82, gma-miR828a_1ss22AT targeted and down-regulated transcription factor MYB82. Our study. Our study provided a significance data for the study of molecular mechanism of tea plant for leaf spot infected by *Phoma segeticola* var. *Camelliae*.

Key words: Biotic stress; Plant-pathogen interaction; miRNA; Target genes; molecular mechanism

Control of maize stalk rot by bacterial strain GS2 from Ginseng rhizosphere soil*

Liu Bing, Liu Jinliang, Zhang Xianghui, Pan Hongyu **, Zhang Yanhua ***

(*College of Plant Sciences, Jilin University, Changchun* 130012, *China*)

Abstract: Maize stalk rot is a worldwide soil-borne disease, which has a serious impact on the yield of maize in China, with the safety problem of chemical pesticide derivatives, the screening of bio-control bacterium of maize stalk rot and the development of bio-control agents are of great importance. In our ongoing work, a bacterium (GS2) was isolated from the suspension of Ginseng rhizosphere soil. It was found that GS2 had remarkable bacteriostasis effect by confrontation incubated with *Fusarium graminearum*, the main pathogen of maize stalk rot. Meanwhile, it was confronted with *Rhizoctonia solani* (maize), *Botrytis cinerea*, *Pyricularia grisea*, *Exserohilum turcicum* and *Sclerotinia sclerotiorum*. 16S rRNA gene sequencing, phylogenetic tree construction and BLAST of NCBI were used to identify the bacterium, and the control effect of the strain GS2 was tested in the field. The results showed that GS2 had obvious inhibitory effect on several tested pathogenic bacteria, the inhibitory rate were 35.4%, 43.8%, 48.1%, 35.3%, 57.6%, 30.4% respectively; GS2 was identified as *Asaia* sp.. The control effect of GS2 was significant in the field test, and there was moderate impact with the control agent 2% tebuconazole suspension seed coating.

Key words: Maize stalk rot; *Fusarium graminearum*; bio-control bacterium; control; effect; screening

* Funding: National Natural Science Foundation of China (31101394, 31772108, 31471730), National Key Research and Development Program of China (2017YFD0300606)

** Corresponding authors: Pan Hongyu; E-mail: panhongyu@jlu.edu.cn

Zhang Yanhua; E-mail: yh_zhang@jlu.edu.cn

第二部分　卵　菌

Effect of carbon dioxide on the production of *Phytophthora infestans* sporangia in China

Liu Hao*, Shen Linlin, Wang Yanping, Xiao Yuexuan, Yang Lina, Zhan Jiasui**

(*Institute of Plant Virology, Fujian Agriculture and Forestry University/Key Laboratory of Plant Virology of Fujian Province, Fuzhou* 350002)

Abstract: The concentration of carbon dioxide in the atmosphere has risen from around 280 mg/kg more than 100 years ago to about 405.5 mg/kg today. And the concentration is expected to reach 730~1 020 mg/kg by the end of this century. Potato late blight caused by *Phytophthora infestans* is a devastating disease in potato production, which is mainly spread by sporangium. After infection, sporangiophore grew from the stomata of the leaves, and sporangium was further produced for spreading, becoming the primary source of infection in the next season of the same year. Therefore, it is of great significance to study the effect of carbon dioxide on the sporangia yield of *Phytophthora infestans* under the elevating carbon dioxide climate condition. In this study, *Phytophthora infestans* was collected from 6 provinces in China, and 5 strains were selected from each region according to the different SSR genotypes for the test. Isolates were cultured at different carbon dioxide concentrations and then sporangium production was measured. The results showed that the yield of sporangium increased first and then decreased with the increase of carbon dioxide. Therefore, as the concentration of carbon dioxide in the atmosphere gradually increases, the incidence of potato late blight may be more serious.

Key words: *Phytophthora infestans*; Sporangia; Carbon Dioxide

* First author: Liu Hao, master student, major in plant pathology; E-mail: lhao0520@126.com

** Corresponding author: Zhan Jiasui, professor, research interests for population genetics; E-mail: Jiasui.zhan@fafu.edu.cn

恶疫霉质外体小蛋白 SCR96 的抗体制备及检测应用*

黄沈鑫**，张子辉，陈孝仁***
（扬州大学园艺与植物保护学院，扬州 225009）

摘 要：PcF/SCR 型效应蛋白是卵菌泌出到植物质外体空间的一类疏水小蛋白，但目前对其功能知之甚少。研究的困难之一就是这些蛋白分子量小，融合的蛋白检测标签不仅会影响蛋白的活性，而且也容易在质外体空间被加工降解掉，影响蛋白检测。为解决这两个难题，便于分析该类效应蛋白在植物病原卵菌侵染致病中的作用，本研究以恶疫霉（*Phytophthora cactorum*）中的 SCR96 为研究对象，把化学合成的 SCR96 成熟短肽（73 aa，MW 8058.94）作为抗原，通过免疫新西兰白兔获得了高效价的免疫多抗血清，抗体经亲和纯化后 ELISA 效价达到 1 : 128 000；该抗体可特异性地检测出酵母、本氏烟表达出的 SCR96 及其缺失突变体蛋白。本研究制备出的多克隆抗体在 SCR96 蛋白结构、功能及其与寄主的互作研究中具有重要意义。

关键词：恶疫霉；质外体效应蛋白；化学多肽；多克隆抗体

---

* 基金项目：国家自然科学基金（31671971，31871907）

** 第一作者：黄沈鑫，硕士研究生，从事植物真菌病害研究

*** 通信作者：陈孝仁，副教授，主要从事植物真菌病害研究；E-mail：xrchen@ yzu. edu. cn

大豆疫霉的菌丝和休止孢阶段差异蛋白组比较研究*

崔僮珊**，张　灿，张　凡，刘西莉***
（中国农业大学植物病理学系，北京　100193）

摘　要：大豆疫霉（*Phytophthora sojae*）是引起大豆疫病的重要病原菌，其在2015年被Molecular Plant Pathology期刊评选为最重要的10种植物病原卵菌之一。大豆疫霉生活史包括无性阶段和有性阶段，无性阶段形成的孢子囊可以直接萌发形成菌丝，也可以分化后释放游动孢子，游动孢子在几个小时后即可形成休止孢，进而萌发并侵染植物。因此，菌丝阶段和休止孢阶段在大豆疫霉的侵染过程中均发挥着重要作用。

随着卵菌基因组数据的公开，蛋白组学方法已成功应用于卵菌研究中。本研究采用TMT标记的方法进行了大豆疫霉菌丝和休止孢阶段的蛋白组学测定，共鉴定到4 682个蛋白质，其中4 050个蛋白质可定量分析。比较大豆疫霉休止孢阶段与菌丝阶段，发现共有445个蛋白发生差异性表达，其中上调倍数大于2倍的差异性蛋白有172个，下调倍数大于2倍的差异性蛋白有273个。通过GO富集分析表明，上述差异表达蛋白主要参与氧化还原、水解和核苷酸磷酸化等过程；通过KEGG通路富集分析表明，休止孢阶段与菌丝阶段相比，其差异蛋白在次级代谢产物合成、糖酵解和糖代谢等途径发生显著差异表达。

本研究分析了大豆疫霉菌丝阶段和休止孢阶段的差异蛋白组，对所获得蛋白进行功能注释和通路富集，丰富了卵菌蛋白数据库，并为新型卵菌抑制剂的创制提供了潜在的分子靶标。

关键词：大豆疫霉；菌丝阶段；休止孢阶段；差异蛋白组

---

* 基金项目：重点研发计划（2017017101307l5）

** 第一作者：崔僮珊，在读硕士研究生；E-mail：cuitongshan0619@163.com

*** 通信作者：刘西莉，教授，主要从事植物病原卵菌与杀菌剂互作；E-mail：seedling@cau.edu.cn

辣椒疫霉 *Sec4* 同源基因敲除及其功能初探*

方 媛**，彭 钦，高 翔，钟 珊，刘西莉***
（中国农业大学植物病理学系，北京 100193）

摘 要：辣椒疫霉（*Phytophthora capsici* Leonian）是一种重要的土传病原卵菌，侵染植物后会引发猝倒、茎腐、根腐以及其他营养器官腐烂等病症，发病严重时可导致植物整株死亡。辣椒疫霉寄主范围广泛，可侵染包括茄科、葫芦科以及豆科等在内的 70 多种植物，是十大重要植物病原卵菌之一，对农业生产危害严重。

胞外囊泡是由细胞分泌到胞外空间行使功能的一类结构。目前根据这些胞外囊泡的来源以及粒径分布主要分为三类，分别为外泌体、微囊泡体和凋亡小体。胞外囊泡可以对多种生物活性物质包括蛋白质、糖类、脂质以及 mRNA、非编码 RNA 进行运输。对胞外囊泡的生物学功能方面的研究发现其在细胞间信息交流、物质交换、抗原呈递以及病原菌与宿主细胞互作过程具有重要作用。

Sec4 蛋白是 Rab 家族蛋白中的一员，该家族蛋白参与了细胞内囊泡形成、运输、融合和释放等多个重要的过程，sec4 蛋白在胞外囊泡的形成过程中发挥了重要作用。本研究前期通过提取辣椒疫霉胞外囊泡并进行蛋白组鉴定分析，发现其中含有多种酶类、细胞壁形成相关蛋白以及致病相关蛋白。将分离提取的胞外囊泡和辣椒疫霉的游动孢子混合接种辣椒叶片后，可以显著促进辣椒疫霉游动孢子的致病力。为了进一步探究胞外囊泡的形成对病原菌致病力的影响，作者在 JGI 辣椒疫霉数据库中进行检索，发现了 2 个与酿酒酵母（*Saccharomyces cerevisiae*）中报道的 *Sec4* 同源的基因，通过 CRISPR/Cas9 敲除体系对其中一个 *Sec4* 同源基因进行敲除，并获得 2 株纯合敲除转化子。进一步研究了纯合敲除转化子、空载体菌株和野生型菌株的菌丝生长速率、游动孢子产量、休止孢萌发和致病力等生物学性状，结果表明，敲除 *Sec4* 同源基因后辣椒疫霉菌丝生长受到一定程度的抑制，但游动孢子产量和休止孢萌发无明显影响，离体致病力显著下降，初步推测辣椒疫霉中 *Sec4* 同源基因可能与辣椒疫霉的致病力相关。

关键词：辣椒疫霉；胞外囊泡；*Sec4* 基因；致病力；生物学功能

---

* 基金项目：国家重点研发计划（编号：2017YFD0200501）

** 第一作者：方媛，博士研究生；E-mail：fangyuan7852@163.com

*** 通信作者：刘西莉，教授，主要从事杀菌剂药理学及病原物抗药性研究；E-mail：seedling@cau.edu.cn

辣椒疫霉纤维素合酶基因功能研究*

李腾蛟**，王为镇，刘西莉***
（中国农业大学植物病理学系，北京　100193）

摘　要：辣椒疫霉是一种寄主范围广泛的植物病原卵菌，在世界范围内造成多种经济作物的严重损失，被列为十大病原卵菌之一。纤维素作为辣椒疫霉细胞壁的主要组成成分，对于辣椒疫霉的生长发育和致病过程至关重要，辣椒疫霉中存在 4 个纤维素合酶基因（*Cellulose synthetase*，*CesAs*）。已有研究表明，烯酰吗啉等羧酸酰胺类卵菌抑制剂可以通过结合卵菌 CesA3 蛋白，影响卵菌纤维素合成过程从而抑制病原菌生长。本团队在前期研究中，运用 qRT-PCR 技术检测了辣椒疫霉 4 个纤维素合酶基因在不同发育阶段的表达量，发现 4 个 *PcCesAs* 在不同发育阶段均有表达。本研究拟进一步探究辣椒疫霉纤维素合酶基因的功能，旨在为探究辣椒疫霉纤维素合成机制，以及开展以分子靶标为导向的新药剂开发提供理论基础。

基于上述研究背景，本研究通过 CRISPR-HDR 基因敲除技术，实现了 *PcCesA*1、*PcCesA*2 和 *PcCesA*4 的单基因敲除。敲除转化子的生物学性状测定发现，*PcCesA*1 和 *PcCesA*2 单敲除转化子菌丝生长速率减缓，休止孢直径增大且萌发率降低，致病力下降，推测 *PcCesA*1 和 *PcCesA*2 基因可能在辣椒疫霉菌丝生长、休止孢的形成和萌发过程，及其侵染阶段发挥着重要功能。*PcCesA*4 单敲除转化子则表现为菌丝生长速率减缓，游动孢子产量下降，致病力降低，推测 *PcCesA*4 基因可能在辣椒疫霉菌丝生长、游动孢子形成及侵染阶段发挥重要功能。通过菌丝生长速率法测定单敲除转化子对羧酸酰胺类卵菌抑制剂烯酰吗啉的敏感性，与亲本菌株相比，发现 *PcCesA*1 和 *PcCesA*2 单敲除转化子对烯酰吗啉更为敏感，推测 *PcCesA*1 和 *PcCesA*2 基因的敲除可能影响了辣椒疫霉纤维素的合成或者组装过程。本研究仅获得 *PcCesA*3 基因的杂合敲除转化子，进一步对其产生的卵孢子开展自交试验，在自交后代中也未获得 *PcCesA*3 纯合敲除转化子，推测 *PcCesA*3 基因敲除致死，其在辣椒疫霉生长发育过程中发挥至关重要的作用。

综上，CesAs 对于辣椒疫霉的生长发育以及致病都具有重要的作用，是潜在的良好杀菌剂靶标。同时，进一步探究不同 CesAs 亚基之间结合和组装形式，解析纤维素合酶合成纤维素的分子机制，是后续研究的重点。

关键词：辣椒疫霉；纤维素合酶；分子靶标；新药剂开发

---

* 基金项目：国家自然科学基金项目（31672052）
** 第一作者：李腾蛟，博士研究生；E-mail：litengjiao@ cau. edu. cn
*** 通信作者：刘西莉，教授，主要从事杀菌剂药理学及病原物抗药性研究；E-mail：seedling@ cau. edu. cn

大豆疫霉 RNA 结合蛋白 PsLARP 的生物学功能研究*

张　凡**，张　灿，崔僮珊，刘西莉***
（中国农业大学植物病理学系，北京　100193）

摘　要：植物病原卵菌侵染作物种类繁多，可给农业生产造成巨大的经济损失。其中，大豆疫霉（*Phytophthora sojae*）可导致大豆疫病的发生，危害严重，为全球十大病原卵菌之一。与辣椒疫霉、致病疫霉等其他病原卵菌一样，大豆疫霉的生活史也分为无性阶段和有性阶段，其中无性阶段产生的孢子囊和游动孢子在其病害侵染循环中发挥着重要作用。

La 相关蛋白（La－related proteins，LARPs）包括 LARP1、LARP1b、LARP3、LARP4a、LARP4b、LARP6、LARP7 等。LARPs 蛋白结构中包含了保守的 RNA 识别模体（RNA recognition motif，RRM）、运输元件和多个其他结构。该蛋白可与 RNA 相互作用，在细胞转录和翻译等方面发挥重要的作用。本研究前期比对分析发现，在大豆疫霉基因组中检索到一个 La 相关蛋白，命名为 PsLARP。采用 CRISPR/Cas9 介导的原生质体转化技术，成功获得了 *PsLARP* 基因的纯合敲除突变体，并开展了其生物学性状研究。结果表明，与大豆疫霉亲本野生型菌株相比，敲除突变体菌丝生长速率无显著性差异，孢子囊产量显著下降，较亲本减少了 57.6%；另外，敲除突变体的游动孢子释放量也明显低于亲本。推测该蛋白在大豆疫霉的无性生殖阶段发挥着主要作用。

LA 相关蛋白作为一种 RNA 结合蛋白，在其他物种中报道与 mRNA 的稳定性紧密相关，推测 PsLARP 可能影响了大豆疫霉孢子囊产生和游动孢子释放等过程中相关基因的转录和翻译，但其具体的调控机制还需要进一步研究。

关键词：大豆疫霉；无性生殖；RNA 结合蛋白

* 基金项目：国家重点研发计划（2017YFD0200501）

** 第一作者：张凡，在读硕士研究生；E-mail：843360141@qq.com

*** 通信作者：刘西莉，教授，主要从事植物病原卵菌与杀菌剂互作研究；E-mail：seedling@cau.edu.cn

致病疫霉侵染马铃薯叶片前后期的蛋白质组学研究*

肖春芳**，高剑华，张远学，王　甄，张等宏，闫　雷，沈艳芬***
（湖北恩施中国南方马铃薯研究中心，恩施土家族苗族自治州农业科学院，恩施　445000）

摘　要：由致病疫霉侵染引起的马铃薯晚疫病是马铃薯生产上危害最为严重的毁灭性病害。本文选择马铃薯品种‘米拉’，利用致病疫霉孢子悬浮液接种马铃薯叶片后，分别于4h、48h和120h三个关键时间点进行前后期的蛋白质组学分析，初步探讨了马铃薯抗晚疫病的分子反馈机制。iTRAQ比较蛋白质组学研究表明，在病菌侵染不同阶段共有1 229种差异表达蛋白（DEPS），包括侵染前期和后期的特异性DEPS以及相同DEPS。80%以上的蛋白丰度变化在侵染前期上调，而在侵染后期约61%的DEP下调。对表达模式、功能分类和富集试验的进一步研究显示，在病害侵染前期表现为细胞壁相关防御反应蛋白和R基因介导的超敏反应启动的显著协调和富集，在病程后期表现为细胞蛋白修饰过程和膜蛋白复合物形成。此外，在马铃薯晚疫病侵染后期，发现了可能与病害发生机制有关的强蛋白相互作用。

关键词：马铃薯晚疫病；米拉；致病疫霉；蛋白质组学

---

* 基金项目：湖北省技术创新专项（鄂西民族专项2016AKB052）；现代农业产业技术体系专项资金资助（CARS-09）；中央引导地方科技发展专项；农业部华中薯类科学观测实验站

** 第一作者：肖春芳，硕士，农艺师，主要从事马铃薯病害防治与遗传育种研究

*** 通信作者：沈艳芬，研究员，从事马铃薯遗传育种及病害防治研究；E-mail：13872728746@163.com

二氧化碳对我国三大耕作区马铃薯晚疫病菌生长速率的影响

刘 浩*，周世豪，黄艳媚，刘诗婷，杨丽娜，詹家绥**
（福建农林大学植物病毒研究所，福州 350012）

摘 要：大气中二氧化碳浓度已由 100 多年前 280 mg/m^3 百万分之一左右上升到目前的 405.5 mg/m^3 左右，并预计到本世纪末达到 730～1 020 mg/m^3。致病疫霉采自我国三大耕作区：北方一作区（甘肃、宁夏），西南混作区（云南、贵州）、南方冬作区（福建、广西），每个地点选取 20 株 SSR 基因型不同的菌株。分别在不同二氧化碳浓度下培养，观察记录菌落生长状况。结果发现高浓度二氧化碳致病疫霉的生长有一定的促进作用。北方一作区与南方冬作区的致病疫霉对二氧化碳的响应无差异，西南混作区在低浓度二氧化碳下生长速率显著低于其他两个耕作区。因此，伴随着大气中二氧化碳浓度逐渐升高，致病疫霉的生长速率可能会加快，马铃薯晚疫病病情将进一步加重。

关键词：致病疫霉；二氧化碳；生长速率

* 第一作者：刘浩，硕士研究生，植物病理学；E-mail：lhao0520@126.com
** 通信作者：詹家绥，教授，主要从事群体遗传学研究；E-mail：Jiasui.zhan@fafu.edu.cn

中国马铃薯四大耕作区的致病疫霉 *Pi*04089 的群体遗传结构分析

杨　爽*，沈林林，刘　浩，周世豪，丁继鹏，胡海平，詹家绥**
（福建农林大学植物病毒研究所，福州　350012）

摘　要：由致病疫霉引起的马铃薯晚疫病是一种毁灭性的卵菌病害，该病原菌通过向寄主细胞分泌效应子蛋白抑制寄主免疫系统，促使植株感病。本研究通过对采集自中国马铃薯四大耕作区的 209 株菌株的 *Pi*04089 进行克隆测序，并分析该效应子的群体遗传结构。结果共检测出 25 种不同的单倍型，四大耕作区的优势单倍型均为 Hap_1，单倍型多样性分布在 0.034 48～0.707 45，核苷酸多样性在 0.000 22～0.015 00，此外，研究表明西南混作区与北方一作区、中原二作区和南方冬作区种群的遗传分化较大。

关键词：致病疫霉；效应子；群体遗传结构

* 第一作者：杨爽，硕士研究生，主要从事马铃薯晚疫病的研究；E-mail：YS1174309622@163.com
** 通信作者：詹家绥，教授，主要从事群体遗传学研究；E-mail：Jiasui.zhan@fafu.edu.cn

云南曲靖不同品种的致病疫霉的群体遗传结构分析

蔡铭铭*，沈林林，段国华，周世豪，杨　爽，刘玉婵，黄艳媚，詹家绥**

（福建农林大学植物病毒研究所，福州　350012）

摘　要：致病疫霉所引起的马铃薯晚疫病是限制马铃薯产业发展的重要因素。明确致病疫霉的遗传多样性能为马铃薯晚疫病的防治提供理论基础。本研究利用 8 对 SSR 引物对 2017 年采自云南省曲靖市 3 个马铃薯品种上的 152 株致病疫霉菌株进行遗传多样性分析。结果共检测出 56 个SSR 基因型和 30 个等位基因。每个群体中均存在着丰富的 SSR 基因型类型，其中 2 个群体各含有 23 个 SSR 基因型，另一个群体含有 27 种 SSR 基因型。3 个群体间的平均遗传分化系数 F_{ST}为0.053 6，平均期望杂合度和平均香农指数分别为0.246 2和0.438 4。总体来看该致病疫霉群体的遗传多样性较高，但 3 个不同品种的致病疫霉群体之间的遗传多样性的差异不显著。

关键词：致病疫霉；SSR；品种；群体遗传

* 第一作者：蔡铭铭，本科生，主要从事马铃薯晚疫病的研究；E-mail：2367211983@ qq. com

** 通信作者：詹家绥，教授，主要从事群体遗传学研究；E-mail：Jiasui. zhan@ fafu. edu. cn

前作植物对烟草黑胫病防治效果的研究*

盖晓彤**，卢灿华，夏振远，方敦煌***
（云南省烟草农业科学研究院，昆明　650021）

摘　要：由烟草寄生疫霉（*Phytophthora parasitica* var. *nicotianae*）侵染引起的烟草黑胫病是烟草生产上最具毁灭性的土传病害之一。目前，前作植物对烟草黑胫病的防治效果以及前作植物根系分泌物对土壤微生物多样性的影响研究暂未见报道，相关研究对烟草黑胫病的绿色生态防控具重要意义。本研究在烤烟黑胫病发生较重的田块分别种植油菜、小麦、大麦、蚕豆、豌豆、大蒜、黑麦草等10种前作作物，利用荧光定量PCR法、叶片诱饵法和土壤病菌分离3种方法，测定不同前作种植后土壤中黑胫病菌的数量，同时调查种植感病烟草品种后的黑胫病发病情况，进行统计学分析；并通过T-RFLP方法分析不同前作植物的根际细菌群落结构及根际土壤细菌的多样性情况。综合3种测定方法的结果，本研究明确了10种不同前作植物对烟草黑胫病的防治效果：前作种植黑麦草、大蒜、大麦可有效减少土壤中烟草黑胫病菌的含量，降低病株率和病情指数；同时，利用T-RFLP方法解析了不同前作植物根系土壤中细菌群落的种群结构与群落丰富度。结合农业种植现状，最终明确在实际生产中大麦可作为较好的前作植物抑制和减少病菌，其次是黑麦草；此外，不同前作会影响其土壤根际微生物结构和群落丰富度。本研究通过筛选对烟草黑胫病具有较好防效的前作植物，并分析各前作土壤中微生物的种群结构，筛选出适合烟区种植的前作种类，将有助于实现对烟草黑胫病的绿色生态控制。

关键词：烟草黑胫病；前作；防治效果；大麦

* 基金项目：中国烟草总公司云南省公司科技计划项目（2017YN08）
** 第一作者：盖晓彤，博士，主要从事烟草植保相关研究；E-mail：gaixiaotong0617@163.com
*** 通信作者：方敦煌，研究员；E-mail：1151276925@qq.com

第三部分 病 毒

First report of *Wisteria vein mosaic virus* in Chinese wisteria in Jiangsu Province, China*

Zhu Pengxiang, Zhang Qinqin, Che Yanping, Ma Yiming, Zhu Feng**

(*College of Horticulture and Plant Protection, Yangzhou University, Yangzhou, Jiangsu* 225009, *China*)

Abstract: The ornamental Chinese wisteria (*Wisteria sinensis*) is a large deciduous climber with twining stems that is prized by gardeners for its vigorous habit, beauty and sweet fragrance. Wisteria mosaic caused by *Wisteria vein mosaic virus* (WVMV) a member of the *Potyvirus* genus has become a worldwide disease. Although the disease did not significantly reduce the vitality of infected plants, the leaves were chlorotic and mottled, making them impossible to sell. In May 2018, virus-like symptoms such as chlorosis, mottling, mosaic and narrowing were observed in Chinese wisteria in Yangzhou City in Jiangsu Province in China. In order to identify this virus, we first used double-antibody sandwich (DAS)-Enzyme-linked immunosorbent assay (ELISA) to confirm the presence of virus in wisteria plants. The results revealed that the samples from symptomatic wisteria plants were positive for WVMV. In order to further confirm the presence of WVMV in wisteria plants, we used molecular biology method by sequencing the coat protein of WVMV. The expected 834 bp amplicons were amplified from the symptomatic wisteria plants that were positive for WVMV in DAS-ELISA. Nucleotide BLAST analysis revealed that the sequence showed 100% coverage and 98% nucleotide identity to 2 isolate WVMV sequences available in GenBank (AY656816 and AY519365). To our knowledge, this is the first report of WVMV in Chinese wisteria in Jiangsu Province in China.

* Funding: This work was supported by the Qing Lan Project of Yangzhou University and the fund of Yangzhou Institute of Landscape Science and Engineering (YZYL20180015)

** Corresponding author: Zhu Feng; E-mail: zhufeng@ yzu. edu. cn

Pumpkin: A New Natural Host of *Papaya leaf distortion mosaic virus**

Peng Bin, Zhang Zhenwei, Wu Huijie, Gu Qinsheng**

(*Henan Key Laboratory of Fruit and Cucurbit Biology*, *Zhengzhou Fruit Research Institute*, *Chinese Academy of Agricultural Sciences*, *Zhengzhou* 450009, *China*)

Abstract: *Papaya leaf distortion mosaic virus* (PLDMV) of the genus *potyvirus* (family *Potyviridae*) was first found on Papaya (*Carica papaya*) grown in Okinawa Island in Japan in 1954. It has been reported on papaya in Hainan and Taiwan. In 2017, we observed in Qiliying town of Xinxiang County, Henan Province, that the pumpkin leaf showed virus disease symptoms with severe mosaic, distortion and yellowing. A total of 6 suspected plant samples was collected and the total RNA was extracted from each sample. Two mixture comprises RNA from 3 samples respectively were used to construct small RNA (sRNA) library and sequence. The 18−30 nt reads accounting for 99% of the raw data after removing of the adapter and low−quality reads were used to detect the virus by VirusDetect software. The contigs form clean data matched to *Zucchini yellow mosaic* (ZYMV), *Watermelon mosaic virus* (WMV), *Cucumber mosaic virus* (CMV), *Cucurbit aphid−borne yellows virus* (CABYV) and PLDMV by Blstan analyses. Among them, 26 out of 339 contigs aligned with PLDMV genome (Accession No. JX974555). Other 16 contigs were aligned with PLDMV polyprotein (Accession No. AGC54443) by Blstax analyses. Four pairs of primers were designed based on sRNA sequencing. Amplification products of 2 163 nt, 3 021 nt, 4 676 nt and 1 812 nt were obtained by RT−PCR from one sample (named XX7) out of 6 Suspected virus samples. The 5′ and 3′ end of PLDMV genome were completed by RACE (rapid−amplification of cDNA ends). A complete nucleotide sequence was 10 152nt in length after assembled by DNAMAN V9.0, shows that identity is 76.3%−76.6% and 82.1%−82.5% by multiple alignment with other five PLDMV isolates at the nucleotide and amino acid level, respectively. It was shown that less than 55% and 50% identity at nucleotide and amino acid level compared to other *Potyviruses* respectively. Twenty−eight pumpkin seedlings were inoculated with sap of XX7 sample plants showed mosaic symptoms after 14 dpi. PLDMV were confirmed from 3 out of 28 inoculated plants using RT−PCR, respectively. To our knowledge, pumpkin is a new natural host of PLDMV and the virus is also first reported in inland areas of China. PLDMV can transmitted through aphid, which may be an underlying threat to the cultivation of Cucurbitaceae crops. The findings reported here will assist further investigations on the epidemiology and biological characteristics of the virus in China.

Key words: New host; Pumpkin; Papaya leaf distortion mosaic virus

---

* Funding: This work was supported by Modern Agro−industry Technology Research System (Number: CARS−26−13), The Agricultural Science and Technology Innovation Program (CAAS−ASTIP−2015−ZFRI), Central Public−interest Scientific Institution Basal Research Fund (1610192019503) and The Foundation and Frontier Technology Program of Henan province (152300410138)

** Corresponding author: Gu Qinsheng; E-mail: guqinsheng@caas.cn

A *Cucumber green mottle mosaic virus*-based vectors for virus inducing gene silencing in cucurbits

Liu Mei, Gu Qinsheng
(*Zhengzhou Fruit Research Institute*, *Chinese Academy of Agricultural Sciences Zhengzhou* 450009, *China*)

Abstract: An array of virus-induced gene silencing (VIGS) vectors have been developed for gene function validation in diverse plants. However, few ideal VIGS vectors are available for cucurbit plants. Here, we describe a new VIGS system derived from *Cucumber green mottle mosaic virus*, a monopartite virus that can infect cucurbit plants in natural condition. We show that the modified CGMMV vector that can produce systemic infection and is able to induce efficient silencing of endogenous genes in the model plant *Nicotiana benthamiana* and cucurbit plants including watermelon (*Citrullus lanatus*), melon (*Cucumis melo*), cucumber (*Cucumis sativus* L.) and bottle gourd (*Lagenaria siceraria*). This is evidenced that the modified CGMMV vector, harbouring the sense-orientated phytoene desaturase (PDS) gene sequence of 100-300bp in length. To improve the efficiency of silencing endogenous genes, we further insert hairpin dsRNA structure into CGMMV vector. Silencing of *PDS* gene resulted in photobleached phenotypes in *N. benthamiana* and cucurbit plants. In addition, we evaluate the stability and propagation performance of this vector. Results show that silencing of the *PDS* gene could persist for over 2 month and the silencing effect of CGMMV-based vectors could be passaged, indicating that the CGMMV vector may serve as a powerful toolbox for identifying gene function of the cucurbit plants. Furthermore, CGMMV-based gene silencing is the first VIGS system, constructed with a monopartite and cucurbitaceae virus reported for cucurbit plants.

Key words: *Cucumber green mottle mosaic virus*; viral vector; virus-induced gene silencing; cucurbit plants

Rapid screened hypovirulence-associated mycoviruses and analyzed its possible mechanism on *Colletotrichum* spp. in mango*

Li Chunxia[1**], Li Min[1], Gao Zhaoyin[1], Chang Shengxin[2], Gong Deqiang[1], Wang Yi[3], Zhang Shaogang[3], Hu Meijiao[1***]

(1. *Environment and Plant Protection Research Institute*, *CATAS*, *Haikou*, *Hainan* 571101, *China*; 2. *Tropical Crop Germplasm Research Institute*, *CATAS*, *Haikou*, *Hainan* 571101, *China*; 3. *Department of microbiology*, *College of Tropical Agriculture and Forestry*, *Hainan University*, *Haikou* 570228, *China*)

Abstract: Anthracnose was a common disease on mango, which made a great lose on mango during production, storage and preservation. Hypovirulence - associated mycovirus could inhibit fungus pathogenicity quickly and environment friendly, and will be expected to use in controling mango anthracnose. In order to quickly screen out hypovirulence-associated mycovirus and analyze its possible mechanism on *Colletotrichum* spp. in mango, total RNA of 16 of hypovirulence strains (randomly divided into two groups, R1 and R2) and 8 of pathogenic strains were extracted for meta-transcriptome and RNA-seq. Meta-transcriptome results showed that viral genome sequences in the hypovirulence strains accounted for 7% (R1) and 0.004% (R2) of total sequences, respectively, which were much higher than those in pathogenic strains (0.0002%). We further analyzed the viral genome sequence composition of each group, results showed that 100% of the viral genome sequence in R1 was the Partitiviridae, and about 47% of R2 was the Partitiviridae, while no Partitiviridae was found in the pathogenic strains. Therefore, we speculated there will be possible a mycovirus caused the attenuated of pathogenicity in hypovirulence strains, and this mycovirus is closely related to the Partitiviridae. Furthermore, we performed a Venn analysis on all of DEGs from RNA-seq of each group and results showed that the genes related to the pathogenicity of fungi, such as NRPS, NRPS-type-I PKS fusion protein, copper amine oxidase, Rieske Fe/S protein and subtilin-like protease, their expression levels were significantly reduced in the two groups of hypovirulence strains compared with the pathogenic strains. Similarly, the expression of carbonic anhydrase gene, which maintained cell homeostasis, and the polyketonase gene, which is involved in cytochrome/toxin synthesis also had a notably decrease in hypovirulence strains. Therefore, it is reasonable to consider that the hypovirulence-associated mycovirus may reduce the pathogenicity and growth of *Colletotrichum* spp. by down-regulating the expression of genes mentioned above.

Key words: *Colletotrichum* spp.; Hypovirulence - associated mycovirus; Virulence attenuated; Mango; Meta-transcriptome and RNA-seq

---

* Funding: This research was supported by Hainan Provincial Natural Science Foundation of China (319MS095) and Central Public-interest Scientific Institution Basal Research Fund for Chinese Academy of Tropical Agricultural Sciences (1630042019041)

** First author: Li Chunxia, Ph. D., Assistant research fellow, Research on postharvest disease control of the tropical fruits and vegetables; E-mail: lichunxia2012@foxmail.com

*** Corresponding author: Hu Meijiao, Ph. D., Research fellow, Research on postharvest preservation and disease control of the tropical fruits and vegetables; E-mail: humeijiao320@163.com

Complete genomic sequenceand organization of a novel mycovirus from *Phoma matteuciicola* strain LG915

Zheng Fan[1], Xu Gang[1], Zhou Jia[1], Xie Changping[1],
Cui Hongguang[1], Miao Weiguo[1], Kang Zhensheng[2], Zheng Li[1]
(1. *College of Plant Protection, Hainan University/Key Laboratory of Green Prevention and Control of Tropical Plant Diseases and Pests, Ministry of Education, Haikou* 570228, *China*; 2. *State Key Laboratory of Crop Stress Biology for Arid Areas and College of Plant Protection, Northwest A&F University, Yangling*, 712100, *China*)

Abstract: The complete genome of a double-stranded RNA (dsRNA) mycovirus, Phoma matteuciicola partitivirus 1 (PmPV1) was sequenced. It consists of two dsRNA segments with 1 664 bp (dsRNA-1) and 1 383 bp (dsRNA-2) in length and each containing a single open reading frame (ORF) and potentially encoded 46.78 ku and 40.92 ku proteins, respectively. DsRNA-1 encodes a putative polypeptide with a conserved RNA-dependent RNA polymerase (RdRp) domain and showed similarity to that of partitiviruses. The protein encoded by dsRNA-2 has no significant similarity to the typical coat protein (CP) of the partitiviruses, while the structure analysis further suggested that might have the hypothetical function of coat protein. Purified viral particles of PmPV1 were isometric and approximately 29 nm in diameter. Phylogenetic analysis showed that PmPV1 is closely related to the *Gammapartitivirus* within the family *Partitiviridae*, but forming a separate branch with Colletotrichum acutatum RNA virus 1 and Ustilaginoidea virens partitivirus 2. This is the first report of the full-length nucleotide sequence of a novel virus classified in the genus *Gammapartitivirus*, infecting *P. matteuciicola* strain LG915, the causal agent of leaf blight of *Curcuma wenyujin*.

Key words: *Phoma matteuciicola*; Mycovirus; *Gammapartitivirus*; *Partitiviridae*

Evolutionary rates and phylogeographical analysis of *Odontoglossum ringspot virus* based on the CP gene sequence

He Zhen*, Dong Tingting, Wu Weiwen, Chen Wen, Liu Xian, Li Liangjun

(*School of Horticulture and Plant Protection, Yangzhou University, Yangzhou* 225009, *China*)

Abstract: *Odontoglossum ringspot virus* (ORSV) is a member of the genus *Tobamovirus*. It is one of the most prevalent viruses infecting orchids worldwide. Earlier studies reported the genetic variability of ORSV isolates from Korea and China. However, the evolutionary rate, timescale and phylogeographical analyses of ORSV were unclear. Twenty-one CP gene sequences of ORSV were determined in this study, and used them together with 145 CP sequences obtained from GenBank to infer the genetic diversities, evolutionary rate, timescale and migration of ORSV populations. Evolutionary rate of ORSV populations was 1.25×10^{-3} nt/site/year. The most recent common ancestors (TMRCAs) came from 30 (95% CIs, 26-40) year ago. Based on CP gene, ORSV migrated from mainland China and South Korea to Taiwan island, Germany, Australia, Singapore, and Indonesia, and it also circulated within east Asia. Our study is the first attempt to evaluate the evolutionary rates, timescales and migration dynamics of ORSV.

Key words: *Odontoglossum ringspot virus*; evolutionary rates; gene flow; migration

* Corresponding author: He Zhen; E-mail: hezhen@yzu.edu.cn

The interchangeability between satellite RNAs of tobacco bushy top virus isolates from China and Zimbabwe

Zhao Xingneng[1,2], Zhang Wei[1,3], Zhang Lifang[1,2,4], Xu Ping [1,2], Li Yanqiong[1,2], Yu Qing[1], Yu Min[3], Chen Hairu[2], Mo Xiaohan[1 *]

(1. *Yunnan Academy of Tobacco Agricultural Science*, *Kunming* 650021, *China*; 2. *College of Plant Protection*, *Yunnan Agricultural University*, *Kunming* 650201, *China*; 3. *College of Life Sciences*, *Yunnan University*, *Kunming* 650504, *China*; 4. *College of Bioresources and Food Engineering*, *Qujing Normal University*, *Qujing* 655011, *China*)

Abstract: Tobacco bushy top disease was first reported in Zimbabwe, causing significant economic losses. The complete genome sequences of tobacco bushy top virus and its satellite RNA of a Zimbabwean isolate were determined. Phylogenetic analyses showed that the Zimbabwean isolate of tobacco bushy top virus was different from the Chinese isolate of tobacco bushy top virus at the species level and the satellite RNAs from the Zimbabwean and Chinese isolates were phylogenetically unrelated. Agrobacterium - mediated infectious clones of the Zimbabwean isolate of tobacco bushy top virus and its satellite RNA were constructed to investigate the interactions between the virus and its satellite RNA. The Zimbabwean isolate of tobacco bushy top virus caused mild symptoms on *Nicotiana benthamiana* and *N. tabacum* alone, and the symptoms were intensified when the plants were coinfected with the virus and its satellite RNA. In combination with agroclones of a Chinese isolate of tobacco bushy top virus and its satellite RNA, the interchangeability between the satellites of tobacco bushy top virus isolates from China and Zimbabwe was tested. The results showed that the Zimbabwean isolate of tobacco bushy top virus could support the replification and systemic movement of the satellite RNA of the Chinese isolate of tobacco bushy top virus, and the Chinese isolate of tobacco bushy top virus could support the replification and systemic movement of the satellite RNA of the Zimbabwean isolate too. The interchangeability of the phylogenetically unrelated satellite RNAs from different umbraviruses facilitates the understanding of interactions between plant viruses and their satellite RNAs.

* Corresponding author: Mo Xiaohan; E-mail: xiaohanmo@ foxmail. com

Increased pathogenicity of the pathogen of corn southern leaf blight caused by a novel chrysovirus

Gong Mingyue [1*], Wang Haoran[1*], Jia Dongsheng[2], Yan Fei[3], Zhang Songbai[1**]
(1. *Collge of Agriculture*, *Yangtze University*, *Jingzhou* 434025, *China*; 2. *Institute of Plant Virus & Key Lab of Biopesticide and Chemical Biology*, *Ministry of Education*, *Fujian Agriculture and Forestry University*, *Fuzhou* 350002, *China*; 3. *Zhejiang Academy of Agricultural Sciences*, *Hangzhou* 310021, *China*)

Abstract: A novel double-stranded RNA (dsRNA) mycovirus, named Bipolaris maydis chrysovirus 1 (BmCV1), was isolated from the plant pathogenic fungus *Bipolaris maydis*. The genome of the BmCV1 consists of four dsRNA segments comprising 3 617 (dsRNA 1), 3 123 (dsRNA 2), 2 984 (dsRNA 3) and 2 753 (dsRNA 4) base pairs, respectively. Each dsRNA fragment contained one open reading frame which encoded the virus RdRp, CP, chryso-p3, chryso-p4 respectively. GenBank accession numbers of the genome were KY489954. 1, KY489955. 1, KY489956. 1, KY489957. 1. The molecular characteristics and evolutionary status of BmCV1 were analyzed by software Clustal2. 1 and MEGA6. 0. The results were indicated that BmCV1 grouped with known or putative chrysoviruses which forming a cluster, and the cluster could be subdivided into two clusters (Ⅰ and Ⅱ). In cluster I, BmCV1 felled in a well-supported monophyletic clade with HvV145S. Therefore, BmCV1 is considered a new member of *Chrysoviride*. In the study, it was found that BmCV1 has two-sided effects, which weakens host fungal growth and enhances fungal pathogenicity to plants (Figure).

Key words: *Bipolaris maydis*; DsRNA virus; Chrysovirus; Attenuation; Enhancement

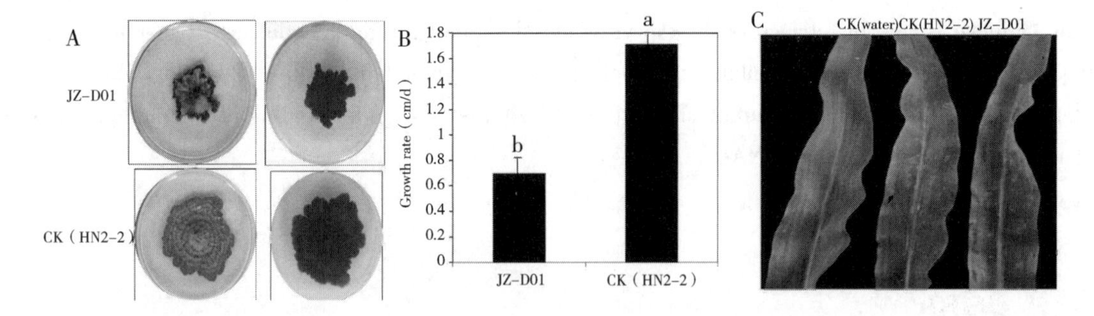

Figure Colony morphology (A), growth rate on PDA (B), and lesion extent of different strains of *B. maydis* on corn leaves (C). JZ-D01, BmCV1-infected strain; HN2-2, BmCV1-free strain

* These authors contributed equally to this work
** Corresponding author: Zhang Songbai; E-mail: yangtze2008@ 126. com

Identification of *Cucurbit chlorotic yellows virus* P4. 9 as a possible movement protein

Wei Ying, Chen Siyu, Li Honglian, Chen Linlin, Sun Bingjian, Shi Yan*
(*College of Plant Protection*, *Henan Agricultural University*, *Zhengzhou* 450002, *China*

Abstract: *Cucurbit chlorotic yellows virus* (CCYV) is a bipartite cucurbit-infecting crinivirus within the family *Closteroviridae*. The crinivirus genome varies among genera. P4. 9 is the first protein encoded by CCYV RNA2. P5, which is encoded by LIYV, is necessary for efficient viral infectivity in plants; however, it remains unknown whether CCYV P4. 9 is involved in movement. In this study, we used green fluorescent protein (GFP) to examine the intracellular distribution of P4. 9-GFP in plant cells, and observed fluorescence in the cytoplasm and nucleus. Transient expression of P4. 9 was localized to the plasmodesmata. Co-infiltration of agrobacterium carrying binary plasmids of P4. 9 and GFP facilitated GFP diffusion between cells. Besides P4. 9 was able to spread by itself to neighboring cells, and co-localized with a marker specific to the endoplasmic reticulum, HDEL-mCherry, but not with the Golgi marker Man49-mCherry. Together, these results demonstrate that CCYV P4. 9 is involved in cell-cell movement.

Key words: CCYV; P4. 9; movement

* Corresponding author: Yan Shi; E-mail: shiyan00925@126. com

Development of a sensitive and reliable reverse transcription droplet digital PCR assay for the detection of *Citrus tristeza virus* *

Wang Yingli[1,2**], Wang Qin[1,2], Yang Zhen[1,2], Li Ruhui[3], Liu Yingjie[1,2], Li Jifen[4], Li Zhengwen[4], Zhou Yan[1,2***]

(1. *National Citrus Engineering Research Center*, *Citrus Research Institute*, *Southwest University*, *Chongqing* 400712, *China*; 2. *Academy of Agricultural Sciences*, *Southwest University*, *Chongqing* 400715, *China*; 3. *USDA-ARS*, *National Germplasm Resources Laboratory*, *Beltsville* 20705, *MD*; 4. *Xinping Yi-Dai Autonomous County Agricultural Bureau*, *Yunnan* 653400, *China*)

Abstract: *Citrus tristeza virus* (CTV) is one of the most important viruses infecting citrus in the world. Droplet digital PCR (ddPCR) is a novel sensitive, accurate method that enables absolute quantitation without the need for a standard curve. The aim of this study was to develop a reverse transcription droplet digital PCR (RT-ddPCR) method to detect CTV, and to compare its quantitative linearity, sensitivity and accuracy with RT-qPCR. The results indicated that both methods showed a high degree of linearity (R^2 = 0.991) and quantitative correlation. The assays suggested that the detection limit for RT-ddPCR was approximately 4.0×10 copies/μL, a 100-fold greater sensitivity than RT-qPCR. The detection results for heat-treatment citrus samples also showed that the positive detection rate of RT-ddPCR (73.2%) was higher than that of RT-qPCR (53.6%). In summary, the results demonstrated that RT-ddPCR may contribute to improve CTV diagnosis by its higher sensitivity and specificity.

Key words: RT-ddPCR; CTV; detection

---

* Funding: Intergovernmental International Science, Technology and Innovation (STI) Collaboration Key Project of China's National Key R&D Programme (NKP) (2017YFE0110900), Overseas Expertise Introduction Project for Discipline Innovation (111 Center) (B18044) and Fundamental Research Funds for the Central Universities (XDJK2018AA002)

** First author: Wang Yingli, major in plant pathology; E-mail: 1409916703@ qq. com

*** Corresponding author: Zhou Yan, research on citrus virus diseases and construction of virus-free breeding system; E-mail: zhouyan@ cric. cn

Distribution and Molecular Characterization of *Citrus yellow vein clearing virus* in Yunnan Province of China*

Wang Qin[1**], Wang Yingli[1], Yang Zhen[1], He Shaoguo[2], Wu Qiang[2], Li Jifen[3], Li Zhengwen[3], Zhou Yan[1***]

(1. *National Citrus Engineering Research Center*, *Citrus Research Institute*, *Southwest University*, *Chongqing* 400712, *China*; 2. *Anyue Lemon Science and Technology Institute*, *Sichuan* 642300, *China*; 3. *Xinping Yi-Dai Autonomous County Agricultural Bureau*, *Yunnan* 653400, *China*)

Abstract: In 2009, a new citrus viral disease caused by *Citrus yellow vein clearing virus* (CYVCV) was first discovered in Yunnan province of China. In this study, a survey was conducted in 27 orchards from Yunnan province from April 2017 to September 2018. In all, 45 of a total of 513 citrus samples were tested positive for CYVCV by RT-PCR. Furthermore, the complete genome sequences of 6 CYVCV isolates from different hosts were sequenced. Comparisons of the whole genome sequences of these 6 CYVCV isolates as well as 34 isolates previously reported from around the world revealed the sequence identity ranged from 96.9%-99.8% at nucleotide level, indicating that there is a very low level of sequence heterogeneity among CYVCV isolates of different hosts in Yunnan province.

Key words: *Citrus yellow vein clearing virus*; distribution; molecular characterization

* Funding: Intergovernmental International Science, Technology and Innovation (STI) Collaboration Key Project of China's National Key R&D Programme (NKP) (2017YFE0110900), Overseas Expertise Introduction Project for Discipline Innovation (111 Center) (B18044) and Fundamental Research Funds for the Central Universities (XDJK2018AA002)

** First author: Wang Qin, major in plant pathology; E-mail: 1367272851@qq.com

*** Corresponding author: Zhou Yan, research on citrus virus diseases and construction of virus-free breeding system; E-mail: zhouyan@cric.cn

Evidence for non-transmission of *Citrus yellow vein clearing virus* by seed[*]

Wang Qin[**], Wang Yingli, Yang Zhen, Cao Mengji,
Liu Yingjie, Hu Wenzhao, Zhou Yan[***]
(*National Citrus Engineering Research Center*, *Citrus Research Institute*,
Southwest University, *Chongqing* 400712, *China*)

Abstract: In 2009, a new citrus viral disease caused by *Citrus yellow vein clearing virus* (CYVCV) was first discovered in China and now CYVCV is considered to be the most serious disease of lemon production in China. Despite CYVCV was frequently detected in the seed tissues, there is a lack of evidence that CYVCV could be transmitted through seed. Because in previous study, the numbers of seeds tested were low, we further investigated the possibility that CYVCV transmit through seed. In November 2016, healthy appearing seeds were collected from mature fruit of 17 CYVCV-infected Eureka lemon (*Citrus limon*) plants at Citrus Research Institute, Southwest University, China. Approximately 1 500 seeds were disinfected with 1% 8-hydroxyquinoline, thoroughly washing with distilled water, dried at room temperature. And then the seeds were planted in plastic pots with sterilized soil mix (1/3 sand, 1/3 chaff, 1/3 peat) and seedlings were maintained in an insect-proof greenhouse at 20-25℃. Six and 12 months after germination, total RNAs were obtained from young shoot of 1267 Eureka lemon seedlings with Trizol reagent and detected by RT-PCR with a specific primer pair (sense: 5′-TACCGCAGCTATCCATTTCC-3′ and antisense: 5′-GCAGA AATCCCGAACCACTA-3′) designed in the coat protein gene. However, CYVCV was not detected in all the seedlings and none of the seedlings showed symptoms. In fact, these seedlings looked like and grew at the same rate as control seedlings from healthy mother plants. The present study confirmed that CYVCV couldn't transmit through seed.

---

* Funding: This work was supported in part by the Intergovernmental International Science, Technology and Innovation (STI) Collaboration Key Project of China's National Key R&D Programme (NKP) (2017YFE0110900), and Fundamental Research Funds for the Central Universities (XDJK2018AA002)

** First author: Wang Qin, major in plant pathology; E-mail: 1367272851@ qq. com

*** Corresponding author: Zhou Yan, research on citrus virus diseases and construction of virus-free breeding system; E-mail: zhouyan@ cric. cn

Endoplasmic reticulum remodeling induced by *Wheat yellow mosaic virus* in wheat (*Triticum aestivum* L.) *

Xie Li [1,2], Song Xijiao [1], Liao Zhenfeng[1], Lv Mingfang [1], Wu Bin [3], Yang Jian [1], Hong Jian [2**], Zhang Hengmu [1**]

(1. *Institute of Virology and Biotechnology*, *Zhejiang Academy of Agricultural Sciences*, *Hangzhou* 310021, *China*; 2. *Analysis Center of Agrobiology and Environmental Sciences*, *Zhejiang University*, *Hangzhou* 310058, *China*; 3. *Institute of Plant Protection*, *Shandong Academy of Agricultural Sciences*, *Jinan* 250100, *China*)

Abstract: Viralinfection usually induces remodeling of host cellular membranes and organelles, especially endoplasmic reticulum (ER) for generating membrane packed viral factory, in which viruses replicate and transcribe its hereditary nucleic acids, translate into proteins, and then assemble proteins and nucleic acids into virion. During the infection of bymoviruses, a kind of membranous bodies (MBs) are generated in host cells, which have thought to be as ER aggregates (Hibino *et al.*, 1981). So far, studies on bymovirus-induced MBs usually focused on their viral components (Schenk *et al.*, 1993; Sun *et al.*, 2014), but less on its membrane origination and structure. In this study, MBs induced by *Wheat yellow mosaic virus* (WYMV) were intensively investigated with several transmission electron microscopy (TEM) -based methods, including cytological observation, component analysis by immuno-gold labeling and structural analysis by electron tomography (ET). WYMV infection induced at least two morphologies of MB, including the lamella dominated morphology (lamella-MB) looked like sprawling cirrus, and the tubule dominated morphology (tubule-MB) looked like latticed network. MBs were verified composing of ER as revealed by immuno-gold labeling by antibody against endoplasmic reticulum (ER) retention signal as well as by detailed observation of MB construction modules as double layer membrane. By immuno-gold labeling, both two MB morphologies (lamella-MB and tubule-MB) had same components in viral derived protein and membrane origination (from ER). Structural analysis by ET reconstruction revealed the organization of ER in MBs. Lamella-MBs were composed of cesER like structures arranged irregularly whereas tubule-MBs were composed of tubER like structures arranged regularly. This study provided insights into the structural details in how Bymovirus utilizing host membrane system.

* Funding: Natural Science Foundation of China (grant no. 31600123, 31070129)

** Corresponding authors: Zhang Hengmu; E-mail: zhhengmu@tsinghua.org.cn
Hong Jian; E-mail: jhong@zju.edu.cn

Chinese wheat mosaic virus-induced gene silencing at low temperature in plant *

Yang Jian[1,2], Zhang Tianye [1,2], Liao Qiansheng[3],
Peng Qiqi [1,3], He Long [1,2], Li Jing [1], Chen Jianping [2**], Zhang Hengmu [1**]
(1. *Institute of Virology and Biotechnology*, *Zhejiang Academy of Agricultural Sciences*, *Hangzhou* 310021, *China*; 2. *Institute of Plant Virology*, *Ningbo University*, *Ningbo* 315211, *China*; 3. *College of Life Science*, *Zhejiang SCI-Tech University*, *Hangzhou* 310021, *China*)

Abstract: Virus induced gene silencing (VIGS) is an important tool for functional genomics, especially for those organisms where genetic transformation is difficult. With this method, it is possible to target most endogenous genes and downregulate the messenger RNA (mRNA) in a sequence-specific manner. Low temperature is a key environmental factor for plant growth and development. However, there has been no effective VIGS vector demonstrated for gene silencing at low temperature. *Chinese wheat mosaic virus* (CWMV), a member of the genusFurovirus, family Virgaviridae, naturally infects cereal plants, in which infected cells produced numerous virus-derived interference RNAs (Yang *et al.* 2014; Guo *et al.*, 2019), suggesting that CWMV could induce active RNA silencing in infected plants at low temperature. In our latest study, its infectious full-length cDNA clones were developed, which provided us a key platform for its reverse genetics (Yang *et al.* 2016). To assess its potential for gene silencing at low temperature, a fragment of the N. benthamiana or wheat phytoene desaturase (PDS) gene was expressed from a modified CWMV RNA2 clone. Photobleaching phenotype and quantitative reverse-transcriptase polymerase chain reaction (qRT-PCR) assays consistently showed that the CWMV vector was effective in the down-regulating endogenous PDS in the N. benthamiana and wheat plants. In experiments using fragments of PDS ranging from 500 to 1 500 nucleotides, insert length influenced the stability and the efficiency of VIGS. The silencing system was also used to suppress miR165/166 and miR3134a through expression of short tandem miRNA-target mimics (STTM). The relative expression levels of mature miR165/166 and miR3134a decreased whereas the transcript levels of their target genes increased. When compared with those of *Barley stripe mosaic virus* (BSMV) or *Foxtail mosaic virus* (FoMV) -based vectors in wheat or *Tobacco rattle virus* (TRV) -based vector, the CWMV induced silencing system appeared to be much more efficient at low temperature. These results suggested that the CWMV vector could be used as a powerful tool for silencing endogenous genes and miRNAs in both monocot and dicot plants at low temperature.

---

* Funding: Natural Science Foundation of China (31501604) and National Key Project for Research on Transgenic Biology (2016ZX08002-001)

** Corresponding authors: Zhang Hengmu; E-mail: zhhengmu@ tsinghua. org. cn
Chen Jianping; E-mail: jpchen2001@ 126. com

百香果夜来香花叶病毒 RT-qPCR 体系的建立*

谢慧婷**，崔丽贤，李战彪，秦碧霞，陈锦清，蔡健和***
（广西壮族自治区农业科学院植物保护研究所，广西作物病虫害生物学重点实验室，南宁　530007）

摘　要：夜来香花叶病毒（*Telosma mosaic virus*，TeMV）为正义 ssRNA，Potyvirus 病毒，粒体呈线状，侵染百香果主要引起花叶症状。传播媒介目前尚未明确。广西百香果病毒病发生重、蔓延快，严重影响百香果产量和品质，其中 TeMV 的发生频率最高。本研究根据百香果 TeMV 外壳蛋白的保守序列设计引物建立 RT-qPCR 检测体系，为该病毒在百香果中的含量测定奠定基础，为百香果健康种苗繁育和综合防控提供技术支撑。

利用重组质粒 TeMV-1 作为标准品建立 SYBR Green Ⅰ 实时荧光定量方法。针对引物浓度、退火温度、灵敏度、特异性、重复性和稳定性等进行条件优化。试验结果：熔解曲线为单一峰，说明所用引物扩增效果好，特异性强，无非特异性扩增；引物在终浓度为400nmol/mL 时 Ct 值最小为 19.61，且荧光信号最强，为最适引物浓度；引物在 51℃时 Ct 值最小为 14.24，且荧光信号最强，为最适退火温度；重组质粒 TeMV-1 的浓度为 2.485×10^{10}copies/μL，选取 $2.485\times(10^5\sim10^9)$ copies/μL 进行梯度稀释建立的标准曲线方程为：$y=-4.427\log(x)+38.12$，$R^2=0.999$，标准曲线中样品的 Ct 值与浓度的 Log 值呈良好的线性关系；常规 PCR 只能检测到 2.485×10^5copies/μL，而 RT-qPCR 能够检测到 2.485×10^3copies/μL，RT-qPCR 比常规 PCR 检测的灵敏度高 100 倍。且 RT-qPCR 方法具有良好的重复性和稳定性。本研究建立的基于 SYBR Green Ⅰ 的 RT-qPCR 技术用于检测 TeMV，具有特异性强、速度快、灵敏度高、重复性好等特点，适合用于检测 TeMV 在百香果中的含量等研究。

关键词：百香果夜来香花叶病毒；SYBR Green Ⅰ；RT-qPCR；病毒检测

---

* 基金项目：广西农业科学院基本科研业务专项项目（桂农科 2017JM28）；广西自然科学基金项目（2018JJB130097）；广西创新驱动发展专项项目（AA17204041）；广西农业科学院基本科研业务专项项目（2015YT42）

** 第一作者：谢慧婷，助理研究员，主要从事植物病理学研究；E-mail：huitingx@163.com

*** 通信作者：蔡健和，研究员，主要从事植物病毒学研究；E-mail：caijianhe@gxaas.net

雀麦花叶病毒外壳蛋白基因原核表达及抗血清制备*

甘海锋，陈　雯，陈夕军，张　坤，贺　振**
（扬州大学园艺与植物保护学院，扬州　225009）

摘　要：雀麦花叶病毒（*Brome mosaic virus*，BMV）寄主范围广泛，多为害禾本科、豆科等植物，因此，建立快速有效的检测方法对于 BMV 的防控有着重要意义。本研究依据 BMV 外壳蛋白（Coat protein，CP）基因序列合成一对引物，扩增获得大小为 570 bp 的目的基因，将目的基因与原核表达载体 pET28a 连接，获得 pET28a-CP^{BMV}。将正确的重组质粒转化大肠杆菌 Rosetta 菌株，经 IPTG 诱导后，SDS-PAGE 电泳检测显示在分子量约为 26 ku 处有目的蛋白带，与预期的 BMV CP 大小一致。融合蛋白主要是以可溶性蛋白的形式存在。利用镍柱亲和纯化重组的 BMV CP 蛋白，并免疫健康新西兰大白兔，制备兔抗血清。Western blot 分析显示，在对多个样品进行检测时发现制备的抗血清能特异性地识别 BMV 的 CP 蛋白，在受 BMV 侵染的烟草植株中能检测到 CP 蛋白的表达，而在健康植株中未能检测到。对制备的抗血清按照比例稀释至 1∶20 000时，仍能特异地检测到目的蛋白条带。这说明通过大肠杆菌表达 CP 制备的 BMV 抗血清特异性强、效价高，为 BMV 的快速检测提供了有利条件。

关键词：雀麦花叶病毒；外壳蛋白；原核表达；抗血清

---

* 基金项目：国家自然科学基金（31601604）

** 通信作者：贺振，副教授，主要从事分子植物病毒学研究；E-mail：hezhen@ yzu. edu. cn

ASSVd田间扩展趋势及组培条件下的传播方式*

郗娜娜**，赵 坷，杨金凤，王亚南***，曹克强***
（河北农业大学植物保护学院，保定 071000）

摘 要：苹果锈果类病毒（*Apple scar skin viroid*，ASSVd）引起的苹果花脸病在苹果生产中危害严重，在我国北方苹果产区普遍发生，近年病情有逐年加重的趋势，目前，尚未找到有效的治疗措施。为明确ASSVd在田间的扩展趋势以及传播方式，2012—2015年对顺平南神南村3个果园进行持续调查和检测，果园A、B、C的显症率分别由3%增加至10%、13.98%增加至24.73%、10.75%增加至22.58%，带毒率由11%增加至23%、16.1%增加至26.9%、15.1%增加至25.8%。锈果病树田间分布有明显的发病中心；以携带ASSVd的组培苗和无毒组培苗为试材，通过模拟田间不同传播方式，通过RT-PCR检测，明确了ASSVd可通过汁液传染、根系接触、含毒基质、组培剪污染进行传播，其中根接传毒是风险最高的传播途径。研究结果可为生产中减少ASSVd的传播提供理论依据，为有效预防苹果锈果病的发生奠定基础。

关键词：苹果锈果类病毒；RT-PCR；扩展趋势；传播方式

* 基金项目：国家重点研发计划（2016YFD0201110）；河北省引进留学人员资助项目（C201839）；河北省高等学校优秀青年基金项目（YQ2014023）

** 第一作者：郗娜娜，在读硕士研究生；主要从事果蔬病毒与寄主的互作及其综合防治研究；E-mail：985372535@qq.com
*** 通信作者：王亚南，博士，教授，主要从事果蔬病毒与寄主的互作及其综合防治研究；E-mail：wyn3215347@163.com
曹克强，博士，教授，主要从事植物病害流行与综合防控研究；E-mail：caokeqiang@163.com

我国甘蔗主要育种材料的病原病毒检测*

郭　枫[1]，邹承武[2,3]，姚姿婷[2,3]，蒙姣荣[2,3]，温荣辉[1,3]，陈保善[1,2]**
（1. 广西大学生命科学与技术学院，南宁　530004；2. 广西大学农学院，南宁　530004；
3. 亚热带农业生物资源保护与利用国家重点实验室，南宁　530004）

摘　要：由于甘蔗以无性繁殖方式进行生产，容易积累多种病毒，导致品种退化，产量和含糖量降低，给甘蔗产业造成重大的经济损失。广西甘蔗种植面积占全国甘蔗种植面积的七成左右，主要受到甘蔗杆状病毒（*Sugarcane bacilliform virus*，SCBV）、甘蔗线条病毒（*Sugarcane streak virus*，SSV）、高粱花叶病毒（*Sorghum mosaic virus*，SrMV）、甘蔗花叶病毒（*Sugarcane mosaic virus*，SCMV）和甘蔗黄叶病毒（*Sugarcane yellow leaf virus*，ScYLV）等 5 种病毒病侵染。本研究利用多重 RT-PCR 检测方法对广西大学农科基地甘蔗种质资源圃中来自国内各地的 155 份甘蔗育种材料进行病原病毒检测。结果表明所采集的 80% 以上的育种材料感染病毒，其中 SCBV、SSV、SCMV、ScYLV 和 SrMV 的检出率分别为 41.3%（64/155）、4.5%（7/155）、10.3%（16/155）、31.6%（49/155）和 40.6%（63/155）。在 155 个甘蔗样品中，感染 SrMV 的甘蔗材料在田间表现出较严重的花叶症状。存在多种病毒复合侵染的情况，SCBV 和 SrMV 复合侵染率为 10.3%（16/155），SCBV 和 ScYLV 复合侵染率为 12.3%（19/155），SCBV、ScYLV 和 SrMV 复合侵染率为 3.9%（6/155）。通过评估甘蔗育种材料的带毒情况，可为甘蔗抗病毒育种提供参考。

关键词：甘蔗病毒；甘蔗育种材料；病原病毒检测

---

* 基金项目：广西蔗糖产业协同创新中心项目（桂教科研〔2014〕13 号）资助

** 通信作者：陈保善，教授；E-mail：chenyaoj@gxu.edu.cn

栗疫菌弱毒病毒 CHV1 的基因组序列多样性与重组分析

林　媛*，周　旋，杜亚楠，方守国，章松柏，邓清超**
（长江大学农学院，荆州　434025）

摘　要：CHV1 病毒的群体多样性研究对于栗疫病的生物防控具有巨大的理论和实践价值，中国是推测的 CHV1 病毒起源地，但关于中国 CHV1 病毒群体多样性的相关研究却相对匮乏。本研究对来自中国东部各地和日本的 30 个 CHV1 病毒进行了多区段的部分基因组序列测定和系统发育分析，发现中国的 CHV1 病毒具有较欧洲和日本更为丰富的遗传多样性，在之前已鉴定的 6 个欧洲 CHV1 病毒亚型的基础上，从亚洲东部鉴定出 8 个新的 CHV1 病毒亚型。挑选分属 8 个新鉴定亚型的 15 个代表性菌株进行了基因组全长测序，结合已报道的 6 个欧洲亚型的 12 个 CHV1 病毒的全长基因组序列，进行了系统发育和重组分析。依据全长基因组序列信息所绘制的系统发育树表明，东亚地区的 CHV1 病毒与欧洲 CHV1 病毒遗传差异明显，且多样性更为丰富；欧洲的 6 个亚型在系统发育树上分为 2 簇且聚集在一起，提示其具有较近的亲缘关系；只来自中国的亚型 CN6 与欧洲 CHV1 病毒的亲缘关系相对较近，但仍位于欧洲整体分支的外部，其他 7 个东亚亚型与欧洲 CHV1 病毒亲缘关系相对较远。基于 27 个 CHV1 病毒全长基因组序列的重组分析表明，CHV1 病毒间存在着广泛的基因组重组现象。部分重组事件涵盖一个或多个亚型，提示这些重组事件可能发生在亚型分化之前，部分重组事件只发生在某些亚型内的部分成员上，提示是较近时期才发生的重组事件。基于基因组核酸序列的病毒进化分析可能会受到重组所造成的信号干扰，造成分析结果的错误和不准确。本研究中，去除各个病毒基因组中的重组区段之后，用各个病毒的剩余主要基因组序列成分，重新进行了系统发育分析，发现此系统发育树与基于全长基因组序列所绘制的系统发育树相比，其拓扑结构未发生根本性的变化，但各分支更加明晰，支持度大为提高。

关键词：栗疫病；真菌病毒；CHV1；弱毒力；遗传多样性；遗传重组

* 第一作者：林媛，在读硕士研究生；E-mail：1160712993 @ qq. com
** 通信作者：邓清超，讲师；E-mail：Dengqingchao@ yangtzeu. edu. cn

稻瘟菌中一种弱毒相关真菌病毒 *Magnaporthe oryzae polymycovirus* 1 的基因组序列测定与生物学性状分析

周 旋*，林 媛，徐 炀，游 江，方守国，章松柏，邓清超**
（长江大学农学院，荆州 434025）

摘 要：在分离自湖北省天门市的稻瘟菌菌株 TM02 中发现了一种新型 dsRNA 病毒，暂命名为 *Magnaporthe oryzae polymycovirus* 1（MoPmV1）。MoPmV1 基因组包含四条独立的 dsRNA，通过克隆测序获得了四条 dsRNA 的全长核酸序列信息，并上传至 GenBank，其登录号依次为 MH231406.1、MH231407.1、MH231408.1 和 MH231409.1。其中 dsRNA1 全长 2401bp，推测其 36~2 334区段编码一个多肽 a1，包含一个 RNA 依赖的 RNA 聚合酶（RDRP）结构域；dsRNA2 全长 2 233bp，推测其 72—22 163区段编码一个功能未知的假定蛋白 b1；dsRNA3 全长 1 963bp，推测其 55~1 891区段编码一个功能未知的假定蛋白 c1；dsRNA4 全长 1 324bp，推测其 159~950 区段和 1 040~ 1 324区段分别编码两个功能未知的假定蛋白 d1 和 d2。Blast 序列比对和系统进化发育分析表明，MoPmV1 与一些新近报道的未分类的真菌 dsRNA 病毒，如 *Beauveria bassiana polymycovirus* 2、*Aspergillus fumigatus tetramycovirus*-1 和 *Botryosphaeria dothidea virus* 1 等具有较近的亲缘关系。dsRNA 提取克隆和 RT-PCR 分析在分离自湖南、贵州、云南、海南、浙江、福建、江苏、安徽和湖北等各省的稻瘟菌菌株中都发现了 MoPmV1 的存在，提示了该病毒在稻瘟菌群体中分布的广泛性。多次的病毒粒体提取实验表明 MoPmV1 在稻瘟菌细胞中不形成病毒粒子。通过对 TM02 菌株的药物处理和单分生孢子分离，获得了多个脱除 MoPmV1 病毒的单孢后代菌株。通过脱毒菌株与 TM02 和含有 MoPmV1 的单孢后代菌株的对比发现，MoPmV1 可显著降低稻瘟菌的生长速率和分生孢子产量；离体大麦叶片与感病品种水稻叶片接种实验表明，MoPmV1 可显著降低病斑大小；活体感病品种水稻接种实验表明，MoPmV1 能显著降低病斑形成的数量和大小。综合分析表明，MoPmV1 是一种新型的真菌 dsRNA 病毒，能够在稻瘟菌中造成弱毒现象，具有潜在的生物防治应用价值。

关键词：稻瘟菌；dsRNA 真菌病毒；弱毒力；生物防治

---

* 第一作者：周旋，在读硕士研究生；E-mail：1293179793@ qq. com
** 通信作者：邓清超，讲师；E-mail：Dengqingchao@ yangtzeu. edu. cn

侵染甘蔗的玉米黄花叶病毒分子检测及基因组序列分析*

陈建生**，孙生仁**，黄小聪，黄美婷，傅华英，高三基***
（福建农林大学甘蔗国家工程研究中心，福州　350002）

摘　要：玉米黄花叶病毒（*Maize yellow mosaic virus*，MaYMV）是近年来在玉米上发现的一个新的病毒，引起玉米叶片出现黄化褪绿的症状。最近，MaYMV 被确定为黄症病毒科 Luteoviridae 马铃薯卷叶病毒属 *Polerovirus* 成员。为确定我国蔗区甘蔗是否存在受到 MaYMV 侵染，本研究收集了 315 份甘蔗和 4 份玉米叶片样品，以已报道的 MaYMV-F 和 MaYMV-R 为引物进行 RT-PCR 分子检测，结果表明在广西、四川和福建省（自治区）共检测到 37 个阳性样品，其中甘蔗 36 个、玉米 1 个，阳性检出率平均为 11.6%。随后，通过病毒基因组的分段克隆，共获得了 3 条 MaYMV 全基因组序列，分别来自四川内江的玉米寄主（MaYMV-SC1）和甘蔗寄主（MaYMV-SC2）以及福建福州的甘蔗寄主（MaYMV-FZ1）分离物，核苷酸大小均为 5 642 bp，它们之间的基因组核苷酸一致性为 99.9%，与其他已报道的 MaYMV 基因组核苷酸一致性为 92.4%~98.7%。系统进化分析显示这 3 个分离物与全球其他国家的 MaYMV 分离物聚在同一进化分支上。基因组序列的重组分析发现 MaYMV 分离物之间存在遗传重组现象，表明遗传重组也是 MaYMV 遗传进化的主要动力之一。

关键词：甘蔗；玉米黄花叶病毒；基因组序列；系统进化树；重组分析

---

* 基金项目：国家现代农业产业技术体系（糖料）建设专项（CARS-170302）

** 第一作者：陈建生，硕士研究生，主要从事甘蔗病害检测与抗病分子育种；E-mail：js875325785@163.com
孙生仁，博士研究生，主要从事甘蔗病害检测与抗病分子育种；E-mail：ssr03@163.com

*** 通信作者：高三基，研究员，主要从事甘蔗病害检测与抗病分子育种；E-mail：gaosanji@yahoo.com

携带 GFP 的甜瓜坏死斑点病毒侵染性克隆载体构建*

吴会杰**，彭　斌，康保珊，刘丽锋，刘莉铭，古勤生***
（中国农业科学院郑州果树研究所，郑州　450009；中国农业科学院
郑州果树研究所瓜类重点实验室，郑州　450009）

摘　要：甜瓜坏死斑点病毒（*Melon necrotic spot virus*，MNSV）是中国新近报道为害甜瓜的病毒种。本研究首先从田间采集确认是 MNSV 侵染的甜瓜叶片，采用分段扩增的方式，获得了该病毒的 cDNA 全长基因。随后以双元载体 pXT1 为骨架质粒，利用 In-Fusion HD Cloning Kit 重组酶连接 pXT1 和 MNSV 全长 cDNA 片段，获得含 MNSV cDNA 全长基因的 pXT1-MNSV 克隆，接种甜瓜，通过症状观察及 PCR 检测鉴定其致病性。取接种 pXT1-MNSV 载体后发病明显的甜瓜叶片，经汁液摩擦接种到健康的甜瓜上，验证其是否能通过汁液摩擦接种。在此基础上，利用同源重组的方法将 GFP 去掉终子密码子的全长基因重组到 MNSV *cp* 基因后面，接种甜瓜后，通过观察接种植株的症状及 PCR 检测其侵染性，并利用激光共聚焦观察甜瓜根部及叶片，均出现 GFP 荧光。本研究首次在国内构建了携带 GFP 的 MNSV 侵染性克隆载体，为解析病毒的分子机理以及在侵染过程中病毒的分布与运动奠定了基础。

关键词：甜瓜；甜瓜坏死斑点病毒；侵染性克隆；重组载体；致病性

---

* 基金项目：国家自然基金项目（31701941，31071811）；河南省基础与前沿技术研究项目（152300410231）；中国农业科学院创新团队西甜瓜病虫害防控（CAAS-ASTIP-2018-ZFRI-08）；国家西甜瓜产业体系病毒防控岗（CARS-26-13）

** 第一作者：吴会杰，副研究员，主要从事瓜类病害研究；E-mail：wuhuijie@ caas. cn

*** 通信作者：古勤生，研究员，主要从事瓜类病害及其防控研究；E-mail：guqinsheng@ caas. cn

农杆菌介导的南瓜花叶病毒 cDNA 侵染性克隆的构建*

刘莉铭**，解昆仑，彭 斌，刘 美，吴会杰，古勤生***

（中国农业科学院郑州果树研究所，郑州 450009；中国农业科学院郑州果树研究所瓜类重点实验室，郑州 450009）

摘 要：南瓜花叶病毒（*Squash mosaic virus*，SqMV）属于豇豆花叶病毒属的成员，能够侵染甜瓜、南瓜、西瓜、黄瓜等葫芦科作物，可通过机械方法、种子和介体昆虫进行传播。由于该病毒的种传率较高，且该病毒的侵染可以引起植株叶片花叶、皱缩，开花果实期时病害更为严重，对葫芦科作物的种植生产造成了极大的威胁。本研究利用 RT-PCR 对南瓜花叶病毒 CH 99/211 分离物的两条基因组分别进行了分段扩增，并利用同源重组技术将 2 条基因组序列分别构建到植物表达载体 pXT1 中，获得该病毒的 cDNA 侵染性克隆 pSqMV-RNA1 和 pSqMV-RNA2。利用农杆菌介导的方式将 pSqMV-RNA1 和 pSqMV-RNA2 同时接种西葫芦，在接种初期，西葫芦新叶产生褪绿斑点，之后随着接种时间的延长，花叶症状逐渐明显。将发病叶片研磨成汁液，经机械摩擦方式进一步接种西瓜、甜瓜、黄瓜和本生烟，通过观察接种植株症状的变化和 RT-PCR 检测，发现它可以侵染甜瓜和黄瓜，但不能侵染西瓜和本生烟。以上结果说明该侵染性克隆产生的病毒后代具有生物学活性，南瓜花叶病毒的 cDNA 侵染性克隆构建成功。本研究为 SqMV 侵染性克隆成功获得的首次报道，为该病毒各基因功能的研究及致病机制的揭示奠定了基础，也为该病毒种传机制的研究提供了新方法。

关键词：南瓜花叶病毒；侵染性克隆

---

* 基金项目：国家西甜瓜产业体系（CARS-26-13）；中国农业科学院科技创新工程项目（CAAS-ASTIP-2018-ZFRI）；中央级公益性科研院所基本科研业务费专项（1610192019208）

** 第一作者：刘莉铭，研究实习员，主要从事瓜类病毒病及其防控研究；E-mail：liuliming@caas.cn

*** 通信作者：古勤生，研究员，主要从事瓜类病害及其防控研究；E-mail：guqinsheng@caas.cn

南瓜蚜传黄化病毒侵染百香果在中国的首次报道*

张绍康**，刘锦涛，宇良语，王　颖，张宗英，李大伟，于嘉林，韩成贵***
（中国农业大学植物病理学系，农业生物技术国家重点实验室，北京　100193）

摘　要：百香果（Passion fruit），学名西番莲，源自南美洲的热带雨林，是西番莲属多年生的草质藤本植物。20 世纪初百香果被引入中国，在我国的广东、广西、云南、福建、海南等温暖湿润的地区广泛栽培。近年来，因百香果具有多种生理疗效和治疗作用深受人们喜爱，使得百香果产业发展迅速，然而，病毒病的广泛发生严重影响了百香果的产量和质量。南瓜蚜传黄化病毒（*Cucurbit aphid-borne yellows virus*，CABYV）是侵染葫芦科作物的重要病毒，属于黄症病毒科（Luteoviridae）马铃薯卷叶病毒属（*Polerovirus*）。该病毒为正义单链 RNA（+ssRNA）病毒，局限在寄主植物韧皮部组织，主要通过蚜虫（*Myzus persicae* 和 *Aphis gossypii*）以持久循回型和非增殖方式传播，典型症状是引起叶片的黄化和增厚，造成田间植株约减产 50%。该病毒于 2018 年在巴西首次报道能够侵染百香果，本研究通过对中国百香果叶片中 CABYV 的检测与鉴定，以期为国内百香果病毒病的防治提供参考依据。

本研究于广西壮族自治区梧州市和云南省保山市采集 108 份具有黄化增厚症状的百香果叶片，提取植物组织总 RNA 后利用 CABYV 运动蛋白（CABYV-MP）特异性引物进行 RT-PCR 扩增，获得 CABYV-MP 片段。将其克隆至 pMD19-T（simple）载体后进行测序分析，共检测到 4 种核苷酸存在差异的 CABYV-MP 基因序列。提取百香果叶片的总蛋白后进行 Western Blot 验证，实验结果表明，CABYV-MP 抗血清可以在 20 ku 处检测到特异性条带。这是在中国 CABYV 侵染百香果的首次报道，为深入研究 CABYV 与百香果的互作、病害的有效防控以及提高百香果的产量打下了工作基础。

关键词：南瓜蚜传黄化病毒；百香果；首次报道

致谢：感谢王献兵教授和张永亮副教授对本研究的指导和建议。

---

* 基金项目：国家自然科学基金项目（31671995）部分资助

** 第一作者：张绍康，硕士研究生，主要从事植物病毒与寄主的分子互作研究；E-mail：zhangshaokang@ cau. edu. cn

*** 通信作者：韩成贵，教授，主要从事植物病毒病害及抗病转基因作物研究；E-mail：hanchenggui@ cau. edu. cn

南瓜蚜传黄化病毒运动蛋白的原核表达、纯化和抗血清的制备*

张绍康**，赵添羽，左登攀，时 兴，王 颖，
张宗英，李大伟，于嘉林，韩成贵***
（中国农业大学植物病理学系，农业生物技术国家重点实验室，北京 100193）

摘 要：南瓜蚜传黄化病毒（*Cucurbit aphid-borne yellows virus*，CABYV）是侵染葫芦科作物的重要病毒，属于黄症病毒科（Luteoviridae）马铃薯卷叶病毒属（*Polerovirus*）。该病毒于1992年在法国首次报道，局限在寄主植物韧皮部组织，主要通过两种蚜虫（*Myzus persicae* 和 *Aphis gossypii*）以持久循回型和非增殖方式传播。近年来，由CABYV侵染引起的葫芦科作物病毒病在我国发生面积不断扩大，严重影响了葫芦科作物的产量及品质，造成严重经济损失，CABYV既可以单一侵染，也可以复合侵染，典型症状是引起叶片的黄化和增厚，造成田间植株约减产50%。

CABYV为正义单链RNA（+ssRNA）病毒，具有直径为25~30 nm的正二十面体球形病毒粒子。病毒粒子较稳定，对氯仿和非离子去垢剂不敏感，但在高盐条件下长时间处理会被破坏。CABYV基因组全长约5.7 kb，共含有7个开放阅读框（ORFs），编码7个蛋白，前3个ORF通过基因组RNA表达，后4个通过亚基因组RNA表达，其中ORF4编码南瓜蚜传黄化病毒运动蛋白（CABYV-MP蛋白），通过渗透扫描表达，可定位在胞间连丝，在病毒复制和移动过程中发挥重要调节作用。

本研究利用实验室保存的CABYV全长cDNA克隆扩增获得CABYV-MP片段，成功构建到PDB-his-MBP原核表达载体上转化大肠杆菌菌株Rosetta，经0.2 mmol/L IPTG诱导后通过镍柱亲和层析并使用不同浓度的咪唑洗脱液进行洗脱。SDS-PAGE检测选择最优浓度浓缩获得CABYV-MP纯化蛋白。将此蛋白作为抗原免疫新西兰大白兔，制备获得多克隆抗血清。Western Blot结果表明，所制备的抗血清效价高达1∶1 000 000，稀释倍数为1∶100 000时在显色效果和经济角度较为理想。此抗血清可以检测到稀释128倍的感病样品，且不与同属内的其他病毒发生血清学交叉反应，为CABYV的检测以及MP功能的深入研究提供材料基础。

关键词：南瓜蚜传黄化病毒；运动蛋白；原核表达；抗血清制备；Western Blot检测
致谢：感谢王献兵教授和张永亮副教授对本研究的指导和建议。

---

* 基金项目：国家自然科学基金项目（31671995）部分资助
** 第一作者：张绍康，硕士研究生，主要从事植物病毒与寄主的分子互作研究；E-mail：zhangshaokang@cau.edu.cn
*** 通信作者：韩成贵，教授，主要从事植物病毒病害及抗病转基因作物研究；E-mail：hanchenggui@cau.edu.cn

甜瓜蚜传黄化病毒运动蛋白的原核表达、纯化和抗血清的制备*

时　兴**，张绍康，左登攀，姜　宁，张晓艳，
王　颖，于嘉林，李大伟，韩成贵***
（中国农业大学植物病理学系，农业生物技术国家重点实验室，北京　100193）

摘　要：甜瓜蚜传黄化病毒（*Melon aphid-borne yellows virus*，MABYV）属于黄症病毒科（Luteoviridae）马铃薯卷叶病毒属（*Polerovirus*），由蚜虫以持久非增殖型方式传播。MABYV 既可以单一侵染，也可以和马铃薯卷叶病毒属的其他病毒复合侵染，病毒侵染局限于寄主植物韧皮部，被侵染叶片发生黄化和增厚，影响田间植株产量和品质。近年来，葫芦科作物因感染 MABYV 而发病的面积不断扩大，造成严重经济损失。

MABYV 为正义单链 RNA（+ssRNA）病毒，病毒粒子正二十面体球形，由 180 个蛋白亚基按照 T=3 排列形成，直径为 25~30nm。病毒基因组全长约 5.7kb，共含有 7 个开放阅读框（ORFs），编码 7 个蛋白。运动蛋白 MP（191 aa）由 ORF4（3 544~4 119nt）通过渗透扫描表达，在 MABYV 传播、复制和表达过程中发挥重要作用。对于 MABYV 的检测，主要利用 RT-PCR 技术，而有关于 MABYV 的血清学检测尚未见正式报道。

为了制备抗血清检测甜瓜蚜传黄化病毒及其运动蛋白，首先将 MABYV 的 MP 构建到原核表达载体 pDB. His. MBP 上，通过转化大肠杆菌 BL21（DE3），利用 0.1 mmol/L 的 IPTG 在 18℃条件下诱导培养 16 h，经超声破碎和 Ni 亲和层析柱纯化，得到高浓度和纯度的 MP 融合蛋白。然后将总量约为 3mg 的表达蛋白作为抗原免疫健康的新西兰大白兔，制备并获得抗血清。最后，利用该抗血清进行了 Western-blot 检测，发现该抗血清能够检测到在本生烟接种叶上瞬时表达的 MABYV 运动蛋白，效价、灵敏度高且特异性好。

关键词：甜瓜蚜传黄化病毒；运动蛋白；原核表达；抗血清；检测

致谢：感谢刘俊峰、王献兵教授和张永亮副教授对本研究的指导和帮助。

---

* 基金项目：国家自然科学基金项目（31671995）

** 第一作者：时兴，硕士研究生，主要从事植物病毒检测研究；E-mail：shixing@ cau. edu. cn

*** 通信作者：韩成贵，教授，主要从事植物病毒学与抗病毒基因工程；E-mail：hanchenggui@ cau. edu. cn

大麦黄矮病毒 PAV 运动蛋白的原核表达、纯化和抗血清的制备

胡汝检[*]，赵添羽，王　颖，张宗英，韩成贵[**]

（中国农业大学植物病理学系，农业生物技术国家重点实验室，北京　100193）

摘　要：大麦黄矮病毒（*Barley yellow dwarf viruses*，BYDVs），自 1950 年在美国加利福尼亚首次发现该病毒后，许多国家相继对 BYDVs 进行了报道，目前 BYDVs 已经广泛分布于世界各地，其寄主植物种类可达到 150 余种，如小麦、大麦、玉米、水稻等单子叶植物。

BYDVs 病毒粒子外壳为二十面体（T=3），无包膜，是一类球形病毒。随着分子生物学技术的进步，BYDVs 的分类地位也越发清楚，ICTV 最新分类报告将 BYDVs 划分到黄症病毒属（*Luteovirus*，包括 PAV、PAS、MAV、Ker Ⅱ 和 Ker Ⅲ 五个种）以及马铃薯卷叶病毒属（*Polerovirus*，包括 RPS、RPV 和 RMV 三个种），另外还有一些株系还未正式分属，如 GPV 和 SGV 株系。国内报道的 BYDVs 有 GAV、PAV、RMV 和 GPV 四种，其中研究较多的 BYDV-GAV 已制备了相应的单克隆抗体，可应用于转基因植物的检测以及田间大规模样品检测。黄症病毒属和马铃薯卷叶病毒属的 BYDVs 基因组均编码 7 个 ORFs，区别在于马铃薯卷叶病毒属的 BYDVs 其 5′端区域具有一个 ORF0，而黄症病毒属病毒在 3′端区域有一个 ORF6。此外，大麦黄矮病毒 ORF4 编码的运动蛋白一般具有细胞间运动蛋白的生化特性，与病毒的致病性和寄主范围密切相关，其中 BYDV-PAV 的 ORF4 编码的运动蛋白是 RNA 沉默抑制因子，瞬时表达能够引起本生烟坏死反应。

本文使用 RT-PCR 技术扩增获得 BYDV-PAV 的运动蛋白（MP）基因，随后将 MP 基因构建到 pDB-His-MBP 表达载体上，转化大肠杆菌 BL21（DE3），挑取阳性菌落过夜培养，经 IPTG 诱导，MP 融合蛋白在上清液中大量表达，进一步利用镍柱亲和纯化蛋白并用不同浓度梯度咪唑洗脱，得到目的蛋白，纯化后的 MP 融合蛋白用于多克隆抗体的制备。实验结果表明，所制备的抗血清效价约为1∶32 000，当稀释倍数为1∶1 000时，检测效果更为理想，为检测 BYDV-PAV 和探究其 MP 功能打下了基础。

关键词：大麦黄矮病毒 PAV；运动蛋白；原核表达；抗血清制备

致谢：感谢于嘉林、李大伟、王献兵教授和张永亮副教授对本研究的建议

---

* 第一作者：胡汝检，硕士研究生，主要从事植物病毒与寄主的分子互作研究；E-mail：Hurujian@ cau. edu. cn

** 通信作者：韩成贵，教授，主要从事植物病毒病害及抗病转基因作物研究；E-mail：hanchenggui@ cau. edu. cn

海南省番茄黄化曲叶病病原的鉴定*

汤亚飞[1,2**]，张　丽[1]，李正刚[1]，佘小漫[1]，于　琳[1]，
蓝国兵[1]，邓铭光[1]，何自福[1,2***]
（1. 广东省农业科学院植物保护研究所，广州　510640；
2. 广东省植物保护新技术重点实验室，广州　510640）

摘　要：番茄黄化曲叶病是番茄生产上的一种毁灭性病害，已给我国很多地区的番茄生产造成了严重经济损失。2019 年 2 月，调查海南省三亚市、东方市、陵水县三市县蔬菜病害过程中发现番茄黄化曲叶病的发生非常严重，尤其东方市和陵水县的千禧番茄田间发病率达到 100%。同时从三市县采集典型症状病样 30 份，利用菜豆金色花叶病毒属病毒通用简并引物 AV494/CoPR，对采集的病样总 DNA 进行 PCR 检测，30 份病样中均能扩增出一条预期 570bp 大小条带，进一步应用滚环扩增（Rolling circle amplification，RCA）方法，对各地阳性病样代表分离物的病毒全基因组进行克隆，获得了侵染海南三亚、东方、陵水番茄的双生病毒 3 个分离物全基因组，它们大小均为 2 781 nt，编码 6 个 ORF，其中病毒链上编码 AV1 和 AV2，互补链上编码 AC1、AC2、AC3 和 AC4。同源性比较结果表明，3 个海南分离物基因组序列两两间同源性为 98.5%；与已报道的 TYLCV 各分离物同源性在 91%以上，而与来自中国不同地区的 TYLCV 分离物的同源率均在 98%以上。由此可知，引起海南三亚、东方、陵水番茄黄化曲叶病的病毒应属 TYLCV 分离物。本研究首次在海南番茄上检测到 TYLCV。

关键词：番茄黄化曲叶病；病原；鉴定

* 基金项目：国家重点研发计划（2018YFD0201200）；广州市科技计划项目（201901010173）

** 第一作者：汤亚飞，副研究员，研究方向植物病毒；E-mail：yf. tang1314@ 163. com
*** 通信作者：何自福，研究员；E-mail：hezf@ gdppri. com

柑橘衰退病毒 CP 与寄主互作蛋白的筛选及验证*

张永乐[1]，杨作坤[1]，王国平[1,2]，洪　霓[1,2]**
（1. 华中农业大学植物科技学院，湖北省作物病害监测与安全控制重点实验室，武汉　430070；2. 华中农业大学，农业微生物学国家重点实验室，武汉　430070）

摘　要：柑橘衰退病毒（*Citrus tristeza virus*，CTV）是长线性病毒科（Closteroviridae）长线性病毒属（*Closterovirus*）成员，所编码的外壳蛋白（coat protein，CP）在该病毒侵染中具有多重功能，参与病毒粒子组装且抑制细胞间 RNA 沉默效应。本研究以 CTV 侵染的墨西哥莱檬植株为材料，采用免疫共沉淀联合质谱分析（CoIP/MS）的方法，对与 CTV 的 CP 互作的柑橘蛋白进行了筛选，鉴定到与 CP 共沉淀的柑橘蛋白 25 个，同时鉴定出 CTV 自身编码的 3 个蛋白。KEGG 通路分析发现，这些柑橘蛋白主要注释到代谢和光合作用相关通路，少数蛋白注释到蛋白质加工、三羧酸循环和植物—病原互作等通路。蛋白质互作网络分析发现，多数鉴定到的柑橘蛋白之间存在关联，互作网络的中心节点蛋白多与光合作用相关。从中选取了柑橘 α-微管蛋白 CsTUA 进行下一步验证，从柑橘中克隆到该基因，经过双分子荧光互补和酵母双杂交试验确认 CsTUA 与 CP 之间存在互作。亚细胞定位分析发现，CP 定位于细胞核、细胞质和胞间连丝，CsTUA 与微管标记物 MAP-mCherry 共定位于细胞的微管丝；而当 CP 与 CsTUA 共注射表达时，CsTUA 与 CP 在细胞质和微管丝存在共定位。进一步通过双分子荧光互补分析发现 CP 和 CsTUA 均可以与自噬途径核心蛋白 CsATG6 互作，亚细胞定位分析发现 CsATG6 定位于细胞质中，当 CP 与 CsATG6 共注射表达时，CP 被 CsATG6 招募至其在细胞质中的点状结构中。qRT-PCR 检测 CTV 侵染后的本氏烟叶片发现，在接种叶和上部叶片中自噬相关基因 *NbPI3K*、*NbATG5*、*NbATG6*、*NbATG8f* 和 *NbATG9* 均显著上调表达，推测 CTV 侵染会诱导植物自噬途径，但其具体机制和作用仍需进一步研究。另外，其他筛选到的潜在互作蛋白有待进一步验证，该研究可为了解 CTV CP 与寄主蛋白之间的互作机制打下一定基础。

---

* 基金项目：国家自然科学基金（31870145）

** 通信作者：洪霓；E-mail：whni@ mail. hzau. edu. cn

苹果茎痘病毒编码的 TGBp 及 CP 蛋白互作和亚细胞定位分析*

李　柳[1,2]，王国平[1,2]，洪　霓[1,2]**

（1. 华中农业大学植物科学技术学院，湖北省作物病害监测与安全控制重点实验室，武汉　430070；2. 华中农业大学，农业微生物学国家重点实验室，武汉　430070）

摘　要：苹果茎痘病毒（*Apple stem pitting virus*，ASPV）是一种正义单链 RNA 病毒，其基因组包含 5 个开放阅读框（Open reading frame，ORF），其中 ORF2、ORF3 和 ORF4 构成的三基因盒编码病毒运动蛋白（Triple gene block protein，TGBp），共同介导病毒的胞间运动及长距离运输；ORF5 编码病毒的外壳蛋白（Coat protein，CP）。在编码 TGBp 的植物病毒中，根据 TGBp 在初级结构和功能上的差异，可分为大麦病毒属型（hordei-1ike）和马铃薯 X 病毒属型（potex-like）。在病毒的运动中，这两种类型之间的区别是 CP 是否参与病毒的胞间移动。目前对 ASPV 的 TGBp 和 CP 生物学功能尚缺乏研究，根据 ASPV 的 TGBp 的二级结构及分类地位，将该病毒归为 potex-like 型。本研究对 ASPV 分离物 HB-HN6 的 TGBp 和 CP 的亚细胞定位特点及蛋白间互作特点进行了研究。采用酵母双杂交核系统对 ASPV 分离物 HB-HN6 的 3 个 TGBp 及 CP 蛋白间互作进行了分析，结果显示 TGBp1 和 CP 有较强的自身互作，CP 可与 3 个 TGBp 蛋白发生互作，TGBp1 与 TGBp3 存在互作。该结果表明 CP 可能通过与 3 个 TGBp 互作而形成复合体，并在病毒的运动中行使功能。进一步通过农杆菌浸润接种本氏烟（*Nicotiana benthamiana*），在激光共聚焦显微镜下观察这些蛋白的亚细胞定位特点，发现该病毒的 TGBp1 在细胞膜及类似内质网处有明显的定位信号，该亚细胞定位特点与已报道的 potex-like 型病毒竹花叶病毒（*Bamboo mosaic virus*，BaMV）和马铃薯 X 病毒（*Potato virus X*，PVX）的 TGBp1 的定位不同，这两种病毒的 TGBp1 主要定位在胞质与胞核；TGBp2 定位于内质网，而 TGBp3 主要定位于周质内质网，偶有定位于核膜，该定位特点与 BaMV 与 PVX 的 TGBp2 和 TGBp3 的亚细胞定位相似；CP 定位于核质和细胞膜。研究结果为进一步阐明这些蛋白在 ASPV 侵染及移动中的功能提供了参考信息。

关键词：苹果茎痘病毒；运动蛋白；外壳蛋白；蛋白互作；亚细胞定位

---

* 基金项目：政府间国际科技创新合作重点专项（2017YFE0110900）；梨现代农业技术产业体系（CARS-28-15）

** 通信作者：洪霓，教授，植物病毒学；E-mail：whni@mail.hzau.edu.cn

刺盘孢菌真菌病毒的研究进展*

李春霞[1**]，李　敏[1]，高兆银[1]，弓德强[1]，周子骞[2]，洪小雨[1]，胡美姣[1***]
（1. 中国热带农业科学院环境与植物保护研究所，海口　571101；
2. 华中农业大学植物科学技术学院，武汉　430070）

摘　要：刺盘孢菌（*Colletotrichum* spp.）是禾本科、果树、蔬菜等经济作物的寄生菌，可引起严重的炭疽病，造成经济损失巨大。部分弱毒相关真菌病毒（Hypovirulence-associated mycovirus）可使植物病原真菌的致病力发生衰退，具有防治植物炭疽病的潜力。截至目前为止，分别在尖孢刺盘孢（*C. acutatum*）、山茶刺盘孢（*C. camelliae*）、镰形刺盘孢（*C. falcatum*）、果生刺盘孢（*C. fructicola*）、盘长孢状刺盘孢（*C. gloeosporioides*）、禾生刺盘孢（*C. graminicola*）、希金斯刺盘孢（*C. higginsianum*）、豆刺盘孢（*C. lindemuthianum*）和平头刺盘孢（*C. truncatum*）等9种刺盘孢菌中发现真菌病毒的存在，且基因组类型均为dsRNA，绝大部分病毒为直径为25~50 nm的球状病毒颗粒，此外，还发现目前唯一一例线状真菌病毒 *Colletotrichum camelliae filamentous virus* 1（CcFV-1）。其中 *Colletotrichum acutatum partitivirus* 1（CaPV1）、*Colletotrichum higginsianum Non-segmented dsRNA virus* 1（ChNRV1）、*Colletotrichum truncatum partitivirus* 1（CtParV1）、*Colletotrichum gloeosprioides chrysovirus* 1（CgCV1）、*Colletotrichum fructicola chrysovirus* 1（CfCV1）和CcFV-1的病毒序列已知。目前，在刺盘孢菌中仅发现2例能使寄主致病力衰退的弱毒相关病毒：CcFV-1能损害山茶刺盘孢（*C. camelliae*）细胞的动态平衡，引发弱毒现象，影响寄主形态，如色素生产和分配、生长率和菌丝环带，但被病毒感染后的真菌仍能维持成长活动，这一点与其他真菌病毒严重影响寄主生长活动的特征明显不同；而CfCV1虽然对果生刺盘孢（*C. fructicola*）菌落形态无明显影响，但对离体果实的弱毒力作用较无CfCV1菌株强。弱毒相关病毒的发现为植物炭疽病的防治和研究提供了新契机，但病毒传播能力有限，用于田间炭疽病的生物防治还有一定距离，因此，可利用宏转录组技术在刺盘孢菌中快速挖掘具有生防潜力的弱毒相关病毒，促进植物炭疽病的生防进展。目前，刺盘孢中真菌病毒的报道大部分是对不同种类真菌病毒的描述，对病毒与寄主真菌间的互作研究不够深入。建立真菌遗传操作系统和病毒反向遗传学系统，是研究寄主真菌与真菌病毒互作机制的首要条件。迄今已有27种刺盘孢菌基因组（https：//www. ncbi. nlm. nih. gov/genome）被测定，同时，盘长孢状刺盘孢菌、瓜类刺盘孢菌、禾生刺盘孢菌等多种刺盘孢菌的遗传转化体系已成功构建，这为进一步研究刺盘孢菌与病毒的分子互作提供有力支持。本文归纳总结了刺盘孢菌真菌病毒的种类和特征，分析弱毒相关病毒对刺盘孢菌的影响，指出利用弱毒相关病毒防治植物炭疽病仍存在不足，根据刺盘孢菌真菌病毒的研究现状提出未来研究方向，为刺盘孢菌真菌病毒的深入研究和利用提供参考。

关键词：刺盘孢菌；真菌病毒；炭疽病；生物防治

---

* 基金项目：海南省自然科学基金项目（319MS095）；中国热带农业科学院基本科研业务费项目（1630042019041）

** 第一作者：李春霞，博士，助理研究员，从事热带果蔬采后保鲜与病害防治研究；E-mail：lichunxia2012@ foxmail. com

*** 通信作者：胡美姣，博士，研究员，从事热带果蔬采后保鲜与病害防治研究；E-mail：humeijiao320@ 163. com

电光叶蝉鸟氨酸脱羧酶抗酶参与水稻条纹花叶病毒侵染的机制*

李　盼**，赵　萍，李光军，周燕燕，张晓峰，王宗文，
毛倩卓，陈红燕，魏太云，贾东升***
（福建农林大学植物病毒研究所，福建省植物病毒学重点实验室，福建　350002）

摘　要：水稻条纹花叶病毒（*Rice stripe mosaic virus*，RSMV）属于弹状病毒科胞质型弹状病毒属的一个新种，由电光叶蝉以持久增殖型方式传播。目前已明确其在介体内的侵染途径，但其增殖机制尚不清楚。本研究通过酵母双杂交技术筛选到电光叶蝉中与 RSMV-M 互作的候选互作蛋白鸟氨酸脱羧酶抗酶Ⅰ（Ornithine decarboxylase antizyme 1，OAZ1），并利用 GST-pull down 证实了 OAZ1 与 RSMV-M 之间存在特异性互作。OAZ1 是细胞内多胺代谢通路中鸟氨酸脱羧酶（ornithine decarboxylase，ODC）的抑制酶，主要通过反馈性负调控多胺的合成。本研究发现 RSMV 侵染可诱导电光叶蝉 OAZ1 蛋白上调表达，同时 ODC 的表达也上调。在叶蝉体内利用 dsRNA 诱导的 RNA 干扰技术抑制 OAZ1 的表达，可以显著降低 RSMV 的增殖，而抑制 ODC 的表达，对病毒的增殖无显著影响。进一步通过免疫荧光标记和免疫电镜发现 OAZ1 和 RSMV-M 在 RSMV 侵染的昆虫体内共定位，两者均标记在病毒原质周围的病毒粒体上，推测 OAZ1 与 RSMV 的侵染有关。利用杆状病毒表达系统发现 M 蛋白单独表达形成小管结构，OAZ1 单独表达形成颗粒状结构，当两者共表达时 OAZ1 与 M 共定位形成小管结构。基于 M 是病毒粒体的基质蛋白并参与粒体的装配，推测 OAZ1 蛋白也参与了 RSMV 粒体的装配。在叶蝉细胞内抑制 OAZ1 的表达，阻碍 RSMV 向邻近细胞的再侵染，进一步表明抑制 OAZ1 阻碍了病毒粒体的装配而无法再侵染邻近细胞。综合以上结果，本研究解析了 OAZ1 在 RSMV 侵染过程中的重要功能，揭示了 RSMV 与介体叶蝉互作的新机制。

关键词：水稻条纹花叶病毒；鸟氨酸脱羧酶抗酶；多胺通路；装配

---

* 基金项目：国家自然科学基金（31730071，31770166）

** 第一作者：李盼，硕士研究生，从事水稻病毒与介体昆虫互作机制研究；E-mail：xpmpyt@163. com

*** 通信作者：贾东升，副研究员；E-mail：jiadongsheng2004@163. com

番木瓜畸形花叶病毒 VPg 基因的遗传变异分析*

莫翠萍**，李华平***

（华南农业大学农学院，广州　510642）

摘　要：番木瓜畸形花叶病毒（*Papaya leaf distortion mosic virus*，PLDMV）是近几年在中国番木瓜局部产区新发现的一种病毒病原，能够侵染目前大面积商品化种植的转基因番木瓜‘华农 1 号’，导致局部地区番木瓜生产的重大损失。笔者在海南和广东该病害发病区进行了广泛调查，采集了不同年份和地区的 112 个感病样品进行了病毒全序列基因组测定。通过 PLDMV 基因组连接蛋白基因（Viral genome-linked protein，VPg）（广东 26 个，海南 86 个，从 NCBI 上下载其他国家和地区 8 个）的序列比对，笔者发现广东省 26 个和海南省 86 个各自 VPg 基因之间的核苷酸同一性分别为 98.5%～100%和 97.5%～100%；广东和海南省的 112 个 VPg 基因序列与日本 PLDMV 分离物 J56P、中国台湾 PLDMV 分离物 CZ_TW 和先前报道的海南 PLDMV 分离物 DF_HN的核苷酸同一性分别为 93.9%～94.8%、93.5%～94.8%和 99.4%～99.8%，这表明广东和海南的分离物与日本和中国台湾的分离物存在一定的差异，而与先前报道的海南分离物较为一致。选择 5 个海南（HA60，HA26，HA8，SD12，S27）和 5 个广东（GZ20，FM26，NSA3，NK8，GZ30）的分离物分别与日本 PLDMV J56P、台湾 CZ_TW 和先前报道的海南 DF_HN 进行 VPg 氨基酸序列比对发现，在 187 个氨基酸中仅有 6 个发生了变化（T5D，Q46H，D106N，R112K，L116I，I146V）。基于 NJ 法构建的 120 个分离物的 VPg 核苷酸序列系统进化树的结果显示，目前所发现的世界上的 PLDMV 可被分为 3 个大族。其中 Group Ⅰ主要为中国台湾分离物，Group Ⅱ为日本分离物，Group Ⅲ为 2 个已报道的海南分离物及本研究中所有的海南分离物和广东分离物。进一步的氨基酸差异的功能分析正在进行中。

关键词：番木瓜；畸形花叶病毒；基因组连接蛋白；遗传变异

---

* 基金项目：现代农业产业技术体系建设专项（CARS-31-09）

** 第一作者：莫翠萍，博士研究生，植物病理学；E-mail：cuiping2018@126.com

*** 通信作者：李华平，教授；E-mail：huaping@scau.edu.cn

利用 GFP 报告系统筛选抗 CGMMV 的高效人工 miRNA

苗　朔*，梁超琼，李晓宇，罗来鑫，李健强**
（中国农业大学植物病理学系，种子病害检验与防控北京市重点实验室，北京　100193）

摘　要：人工 microRNA（artificial microRNA，amiRNA）是一种利用植物内源 miRNA 的前体作为骨架，替换成熟 miRNA 序列以获得新的具有靶向能力的一段人工合成的碱基序列。通过设计多组含有靶向病毒不同保守结构域的 amiRNA 的前体，可以有效降低因病毒进化导致的 amiRNA 介导的病毒抗性丧失等问题，从而达到防治病毒病的目的。利用该技术，已开发出抗芜菁黄花叶病毒（*Turnip yellow mosaic virus*，TYMV）、芜菁花叶病毒（*Turnip mosaic virus*，TuMV）、黄瓜花叶病毒（*Cucumber mosaic virus*，CMV）、马铃薯 X 病毒（*Potato virus* X，PVX）和马铃薯 Y 病毒（*Potato virus* Y，PVY）的拟南芥转基因植株；抗黄瓜花叶病毒（*Cucumber mosaic virus*，CMV）的烟草、番茄转基因植株；抗小麦矮缩病毒（*Wheat dwarf virus*，WDV）的大麦转基因植株以及抗木薯褐色线条病毒（*Cassava brown streak virus*，CBSV）的木薯转基因植株等。

本研究以拟南芥 miRNA 前体 ath-miR156a、ath-miR164a 和 ath-miR171a 为骨架，设计了 6 条分别靶向黄瓜绿斑驳花叶病毒（*Cucumber green mottle mosaic virus*，CGMMV）外壳蛋白、运动蛋白和复制酶基因保守区域的 amiRNA 前体，将其合成后前体序列构入植物双元表达载体 pEarleyGate100（pEG100）中，同时将 amiRNA 的靶基因构入在 GFP CDS 区 3′端，并构入同样的双元表达载体，两个双元表达载体按 1∶1 的浓度比例混合，利用农杆菌介导的遗传转化法浸润 5~6 周大小的本生烟植株，以只包含 GFP 的双元表达载体作为对照，在手持紫外灯（365mm）的照射下，通过荧光强弱来筛选抗病毒效率较高的 amiRNA（荧光强度越强，amiRNA 对病毒基因的沉默效果越弱；荧光强度越弱则表明 amiRNA 的沉默效果越强）。

研究结果显示：靶向沉默 CGMMV 复制酶基因的 3 个 amiRNA 中，amiRNA2、amiRNA3 的效果优于 amiRNA1；靶向沉默 CGMMV 运动蛋白的 2 个 amiRNA 中，amiRNA4 的效果优于 amiRNA5；靶向沉默 CGMMV 外壳蛋白的 amiRNA 沉默效果明显；通过 CGMMV 挑战接种试验检测浸润不同 amiRNA 的烟草叶片中的 CGMMV 基因表达量，显示不同 amiRNA 对 CGMMV 的抑制效果与其 GFP 报告系统的试验结果一致。本研究为快速、精准地筛选抗病毒的高效人工 miRNA 提供了参考方法，为作物抗病毒研究提供了新思路。

关键词：人工 miRNA；GFP 报告系统；黄瓜绿斑驳花叶病毒

---

* 第一作者：苗朔，博士研究生；E-mail：ms_0825@163.com
** 通信作者：李健强，教授，主要从事种子病理学研究；E-mail：lijq231@cau.edu.cn

新疆紫花苜蓿病毒分子鉴定与多重 RT-PCR 体系的建立*

阿孜古丽·木汗买提，热甫卡提·雪合拉提，李克梅**
（新疆农业大学农学院，乌鲁木齐　830052）

摘　要：紫花苜蓿因其高蛋白含量而被称为“牧草之王”，在世界各国广泛种植。病毒病对苜蓿的生产造成严重的危害，近年来，苜蓿病毒病在我国新疆地区盛行，我国研究者对苜蓿病毒病的研究相对较少，对病毒种类缺乏了解。2017 年，课题组借助 Si-RNA 高通量测序初步判断分析出疑似紫花苜蓿病毒病病株中可能携带有菜豆卷叶病毒（BLRV）、菜豆黄花叶病毒（BYMV）、苜蓿花叶病毒（AMV）、豇豆花叶病毒（CPMV）、白三叶草花叶病毒（WCMV）等病毒。本研究通过对北疆地区紫花苜蓿疑似病毒病病株的采集，采用分子检测技术对新疆北疆地区的紫花苜蓿病毒病害进行了分子检测和鉴定。现报告研究结果如下：

（1）2017—2018 年对新疆北疆乌鲁木齐市的乌拉泊、板房沟，昌吉的呼图壁、昌吉阿什里乡以及阿勒泰地区的北屯市、布尔津县、富蕴县等 7 个区县共采集 219 个苜蓿疑似病毒病毒株样品，并对此样品进行 RT-PCR 的检测与鉴定。检测结果表明：供试样品中含有菜豆卷叶病毒（*Bean leafroll virus*，BLRV）、菜豆黄花叶病毒（*Bean yellow mosaic virus*，BYMV）以及苜蓿花叶病毒（*Alfalfa mosaic virus*，AMV）的三种病毒类型。检出率分别为 BLRV 为 25.1%、BYMV 为 6.8%、AMV 为 53.4%；样品同时受 3 种病毒的复合侵染率为 2.3%。根据检测结果可知 AMV 是我国新疆北部地区紫花苜蓿病毒病的优势病原，且有较为明显的复合侵染现象。

（2）扩增并克隆了苜蓿的 BLRV 外壳蛋白基因，运用生物信息学手段对 BLRV 所对应的蛋白的理化特性以及结构域等进行分析，结果表明，分离得到的 955 个核苷酸序列位于 BLRV 基因组的 ORF3 区域，为衣壳蛋白 P3。该序列编码 328 个氨基酸，分子量（Mw）为 22.04 ku，理论等电点（pI）为 11.48%。其蛋白为亲水性蛋白。预测三维结构表明，BLRV 衣壳蛋白 P3 无规则卷曲使蛋白倾向于球状构象，这种构象具有高度的特异性，与蛋白质的生物活性密切相关。

（3）建立多重 RT-PCR 的方法可同时检测 BLRV、BYMV、AMV 三种病毒。通过对多重 RT-PCR 的退火温度、循环参数、引物用量以及灵敏度等方面进行优化设计，建立了可同时检测 2~3 种病毒的双重以及多重 RT-PCR 方法。多重 RT-PCR 反应条件优化结果显示，在 25 μL 反应体系中理想反应条件为退火温度 54℃、循环参数 35 个循环、引物浓度 10 μmol/L 时用到的量为 1 μL；灵敏度试验表明，3 种病毒的最低检测浓度为 9.5 ng/μL。建立的该多重 RT-PCR 体系可以准确、快速、灵敏的同时检测出 3 种病毒。

关键词：紫花苜蓿；病毒病；分子检测；RT-PCR；生物信息学；多重 RT-PCR

---

* 基金项目：国家自然科学基金（31760708）

** 通信作者：李克梅；E-mail：835004213@qq.com

Banana bunchy top virus (BBTV) nuclear shuttle protein interacts and re-distributes BBTV coat protein in *Nicotiana benthamiana*[*]

Yu Naitong[1,2**], Li Weili[1], Liu Zhixin[1,2***]

(1. *Key Laboratory of Biology and Genetic Resources of Tropical Crops, Ministry of Agriculture and Rural Affairs, Institute of Tropical Bioscience and Biotechnology, Chinese Academy of Tropical Agricultural Sciences, Haikou*, 571101, *China*; 2. *Hainan Provincial Key Laboratory of Microbiology, Haikou*, 571101, *China*)

Abstract: *Banana bunchy top virus* (BBTV) is a circular single-stranded DNA virus with multi-components. The knowledge about interaction between viral proteins and pathogenesis mechanism of BBTV are remains unclear. In this study, the coat protein gene (*CP*, ORF 516 bp) and nuclear shuttle protein gene (*NSP*, ORF 465 bp) from BBTV B2 isolate of the Southeast-Asia group were cloned. The intracellular localization analysis showed the CP locates in the cell nucleus of tobacco cells, while the NSP distributes in the cell nucleus and cytoplasm. Co-localization analysis indicated the NSP does not change itself distribution, but CP re-distributes to the cell nucleus and cytoplasm, suggesting that NSP interacts with CP and re-locates the CP in the cell. The interaction between CP and NSP was further verified by co-immunoprecipitation (Co-IP) in tobacco protoplasts. The study will help us to understand the interaction between viral proteins and pathogenesis mechanism of BBTV in host plants.

Key words: *Banana bunchy top virus*; CP; NSP; Intracellular localization; Interaction; Co-IP

* 基金项目：国家自然科学基金青年项目（31401709）；中国热带作物学会青年人才托举工程项目（CSTC-QN201704）

** 第一作者：余乃通，博士，助理研究员，研究方向：病毒及分子生物学；E-mail：yunaitong@163.com

*** 通信作者：刘志昕；E-mail：liuzhixin@itbb.org.cn

含酰腙结构化合物Ⅶ-6对烟草花叶病毒的抑制效果及机理*

吕 星[1**]，向顺雨[1]，袁梦婷[1]，刘昌云[1]，李 斌[2]，汪代斌[3]，
陈海涛[3]，徐 宸[3]，张 帅[3]，陈德鑫[4]，孙现超[1***]
（1. 西南大学植物保护学院，重庆 400716；2. 中国烟草总公司四川省公司，成都 610000；
3. 中国烟草公司重庆市公司烟草科学研究所，重庆 400716；
4. 中国农业科学院烟草研究所，青岛 266101）

摘 要：迄今为止，烟草花叶病毒（*Tobacco mosaic virus*，TMV）已经在农业生产中造成了巨大损失，且没有高效的抗病毒药剂。本研究分析了苯环上含有4-氯酰腙化合物Ⅶ-6，的抗烟草花叶病毒活性及机制。结果发现，在浓度为100 μg/mL、200 μg/mL、400 μg/mL、800 μg/mL、1 600 μg/mL梯度下对药剂进行浓度筛选时，在100μg/mL浓度下具有较好的抗病毒活性，在400 μg/mL浓度时抗病毒率达到50%以上，抗病毒活性显著高于香菇多糖处理组和其他药剂处理组，后用RT-qPCR和ELISA分别从核酸和蛋白水平对抗病毒结果进行验证，结果与荧光斑结果一致。对其抗病毒机理进行研究，结果表明，该化合物在浓度为400μg/mL时能够在2 d、4 d、6 d分别显著提高SOD、POD、CAT活性。进一步研究发现该化合物能够明显诱导水杨酸、茉莉酸、乙烯途径抗性基因（*NPR*1，*PR*1，*PR*2，*CO*Ⅰ1等）表达，且400μg/mL时效果最佳。生测试验初步证明了含酰腙结构化合物Ⅶ-6具有较高的抗烟草花叶病毒病活性，适合被开发为新型抗植物病毒制剂。

关键词：烟草花叶病毒；抗病毒活性；酶活性；抗性基因

---

* 基金项目：中国烟草公司四川省公司科技项目（SCYC201703）；中国烟草公司重庆市公司技项目（NY20180401070010，NY20180401070001，NY20180401070008）

** 第一作者：吕星，硕士研究生，从事植物病理学研究

*** 通信作者：孙现超，博士，研究员，博士生导师，主要从事植物病毒学及植物病害控制研究；E-mail：sunxianchao@163.com

番茄 SYTA 与 Fd I 及 DCL 1 相互作用的研究*

罗　可**，韦学峰，张　坚，邹艾红，樊光进，孙现超***
（西南大学植物保护学院，重庆　400715）

摘　要：实验室前期研究显示番茄 SYTA（*S. lycopersicum* SYTA，S.l SYTA）与铁氧还蛋白Ⅰ（ferredoxin，FdⅠ）与 DCL 1（Dicer-like，DCL 1）存在相互作用。为进一步明确 *S. l* SYTA 为中心的互作链 CP_{TMV}-FdⅠ-*S. l SYTA*-MP_{TMV}中各蛋白是如何相互作用进而影响寄主本身抗病及感病能力。本实验根据番茄基因组含有的 SYTA 同源基因序列，利用 Primer premier 5.0 软件设计克隆引物采用 RT-PCR 技术克隆 *S. l SYTA* 全长序列，利用在线 TMHMM Server2.0 分析各蛋白结构域。然后，设计引物，PCR 缺失突变技术分别获得各 *S. l* SYTA，Fd I 和 DCL1 蛋白的缺失突变体。分别构建进入双分子荧光互补所用的 pCV-nYFP-C1 和 pCV-cYFP-C1 载体中，构建相应缺失突变体的重组载体。通过热激法将融合表达的植物载体转化至农杆菌 EHA105，共聚焦显微镜下观察蛋白互作情况。利用 TRV 病毒载体在番茄植株中沉默 *S. l SYTA* 接种 TMV，紫外灯下观察并利用实时荧光定量 PCR（qRT-PCR）检测 *S. l SYTA*，FdⅠ和 DCL1 及 TMV-GFP 的积累和移动情况。结果显示克隆得到1 620bp的 *S. l SYTA* 基因开放阅读框全长，通过序列比对及蛋白结构域分析表明，其编码的氨基酸序列具有 SYT 家族的典型特征，含有 N 端的跨膜区和两个 C2 结构域。C2A 和 C2B 结构域分别编码 100 个和 99 个氨基酸。在共聚焦显微镜下显示，*S. l SYTA* 与FdⅠ和 DCL1 存在相互作用，且 C2A 结构与 Fd 1 在共聚焦显微镜下发现荧光，C2B 与 FdⅠ不存在相互作用，证明 C2A 为主要行使功能的结构域。通过紫外灯和 qRT-PCR 观察和检测 TMV-GFP 的累积和移动情况显示，在 SYTA 沉默番茄植株中，TMV 的侵染速度以及病毒累积情况相较于空载体对照明显缓慢和下降，证明 *S. l SYTA* 对 TMV 的侵染和移动有促进作用，同时比较 *Fd*Ⅰ和 *DCL* 1 的表达量发现，番茄 Fd Ⅰ的表达量与 *S. l SYTA* 呈负相关，而番茄 *DCL* 1（Dicer-like 1）基因的表达量与 *S. lSYTA* 呈正相关。

关键词：番茄 SYTA；FdⅠ；DCL 1

---

* 基金项目：国家自然科学基金（31670148，31870147）；重庆市社会事业与民生保障创新专项（cstc2016shmszx0368）

** 第一作者：罗可，硕士研究生，从事植物病理学研究
*** 通信作者：孙现超，博士，研究员，博士生导师，主要从事植物病毒学及植物病害防控研究；E-mail：sunxianchao@163.com

辣椒抗性相关基因 *CaNHL* 家族的全基因组鉴定与分析*

刘昌云**，陈雯镜，刘朝龙，李欣羽，向顺雨，薛　杨，樊光进，孙现超***
（西南大学植物保护学院，重庆　400615）

摘　要：*NHL*（NDR1/ Hin1-like）是从拟南芥基因组数据库中发掘出来的与拟南芥 *NDR*（Non-race-specific disease resistance gene）基因和烟草 *HIN*1（Harpininduced gene 1）基因序列同源的基因，是一些抗病基因（Resistance genes）产物诱导的拟南芥对细菌和真菌病原体的抵抗所必须的基因。该家族在多种植物防卫、免疫反应中起到重要作用。本研究充分利用生物信息学手段与分子生物学手段，以辣椒为对象，系统探究辣椒 *CaNHL* 基因家族在辣椒抗病防御中的作用。我们利用生物信息学手段，以本实验室前期已克隆得到的 CaHIN1（AB162221）为诱饵，从辣椒全基因组数据库中共筛选出 15 个 *CaNHL* 家族基因，其编码的蛋白被分为 4 个亚组，且每个亚组间具有相似的蛋白结构和 Motif 分布，并且该家族共享两个高度保守的 Motif 基序（Motif 1：NPNKRIGIYYD；Motif 2：NPNKRIGIYYD）；染色体定位分析发现该家族基因随机分布在 7 条染色体上；顺势元件分析得出该家族基因启动子存在多个重复茉莉酸甲酯（MeJA）、水杨酸（SA）、脱落酸（ABA）、耐低温等激素与抗逆响应反应元件。随后，我们通过分析转录组数据，构建该家族基因的组织表达与胁迫表达模式，发现除 *CaNHL*3 外，其他基因都在各个组织中有较高表达，并且有 *CaNHL*1、*CaNHL*4、*CaNHL*6、*CaNHL*10、*CaNHL*11、*CaNHL*12 等基因在 SA、JA、ABA 处理下呈现特异高表达，揭示这些基因可能参与上述抗病激素的响应。通过 qRT-PCR 分析发现烟草花叶病毒胁迫下，上述激素处理下的高表达基因均出现差异表达，揭示上述基因可能参与辣椒的抗病毒反应。综合上述分析，笔者推测 CaNHL 家族可能在辣椒抗病毒防御中起到了非常关键的作用，揭示 NHL 不仅在植物抵御细菌和真菌，甚至在抵御病毒中发挥重要作用，为辣椒的抗病育种等提供理论基础。

关键词：辣椒；*CaNHL*；生物信息学；组织表达分析；胁迫表达分析

---

* 基金项目：国家自然科学基金（31670148，31870147）；重庆市社会事业与民生保障创新专项（cstc2016shmszx0368）

** 第一作者：刘昌云，本科生，从事植物病理学研究

*** 通信作者：孙现超，博士，研究员，博士生导师，主要从事植物病毒学及植物病害防控研究；E-mail：sunxianchao@163. com

IP-L 相关本氏烟基因对 TMV 侵染的影响*

陈　雪**，刘昌云，韦学峰，张　坚，樊光进，马冠华，孙现超***
（西南大学植物保护学院，重庆　400715）

摘　要：IP-L（Interaction Protein L）是从烟草 cDNA 文库中筛选出的与 ToMV 外壳蛋白（coat protein，CP）互作的寄主蛋白，两者共定位于本氏烟细胞表皮质膜。NbLSP 是从烟草 cDNA 文库筛选到的 IP-L 的互作蛋白，ToMV 侵染本氏烟诱导了 IP-L 及其编码蛋白表达量显著上调，IP-L 沉默本氏烟接种 TMV 后，IP-L 的表达量上调，同时 IP-L 过表达会促进 PVX 的表达，从 IP-L 沉默植株转录组筛选出植物与病原互作通路的差异表达基因 RIN4、CML、LHY 及 CNGCS。对筛选出的差异表达基因构建 VIGS 沉默载体，通过农杆菌浸润接种本氏烟，沉默后 8d 通过摩擦接种的方法用绿色荧光蛋白（Green Fluorescence Protein，GFP）标记的 TMV 侵染本氏烟，在接种后 2d、3d、4d、5d 在紫外灯下观察病毒侵染移动情况，并利用实时荧光定量 PCR（RT-qPCR）定量 TMV-GFP 的积累量。结果显示，CML 沉默本氏烟新叶在接种 TMV-GFP 后 3d 出现绿色荧光，RIN4 沉默本氏烟新叶在接种 TMV-GFP 后 5d 出现绿色荧光，而对照组在接种 TMV-GFP 后 4d 出现绿色荧光，说明 RIN4 沉默可能对 TMV 侵染本氏烟有一定的抑制作用，CML 沉默则可能对 TMV 侵染本氏烟有促进作用。实时荧光定量也表现出相同的结果，接种 TMV-GFP 后 3d 后，CML 沉默本氏烟新叶 TMV-GFP 积累量显著高于对照，而 RIN4 沉默本氏烟接种后 5d 在新叶中检测到 TMV-GFP 的积累，并且显著低于对照组。LHY 及 CNGCS 沉默对 TMV 侵染本氏烟没有明显的影响。

关键词：IP-L；本氏烟；TMV；基因沉默

* 基金项目：国家自然科学基金（31670148，31870147）；重庆市社会事业与民生保障创新专项（cstc2016shmszx0368）

** 第一作者：陈雪，硕士研究生，从事植物病理学研究
*** 通信作者：孙现超，博士，研究员，博士生导师，主要从事植物病毒学及植物病害防控研究；E-mail：sunxianchao@163. com

马铃薯合作 88 抗卷叶病毒基因的 QTL 定位*

梁静思[1,2**]，陶　宇[1]，汤淑丽[1]，张　佩[1]，张荣英[1]，唐　唯[1,2***]
（1. 云南师范大学生命科学学院，昆明　650500；
2. 云南师范大学马铃薯科学研究院，昆明　650500）

摘　要：病毒病是马铃薯生产中的主要病害之一。在侵染马铃薯的病毒中，发生较为普遍是卷叶病毒（*Potato leaf roll virus*，PLRV）引起的卷叶病。本研究利用西南地区马铃薯四倍体主栽品种合作 88 的 F_2代自交分离群体，通过田间性状调查得到抗 PLRV 性状分离群体共 50 株。首先利用 DAS-ELISA 检测侵染病毒的类型，随机选取 5 株抗病群体和 5 株感病极端群体，利用 Illumina 重测序，以马铃薯双单倍体 DM 为参考基因组进行单倍体组装和注释。通过 MISA 程序进行全基因组 SSR 挖掘和比对，共筛选出 277 个与抗卷叶病相关的特异性 SSR 标记；用 Primer 3.0 设计引物、PCR 扩增后在 LabChip GX Touch 平台检测得到 232 个等位位点，其中多态性位点 152 个。利用 JoinMap 4.0 构建了合作 88 抗 PLRV 连锁的遗传图谱；遗传图谱共包含 8 个连锁群，总长度为2 684.3 cM，标记间平均距离 11.57 cM。最后用 Tetraploid Map 检测到 5 个与抗卷叶病相关的 QTL，其遗传贡献率分别为 4.31%、8.70%、10.89%、13.52%、7.82%。本研究可为高通量开发马铃薯抗卷叶病分子标记提供技术参考，也可为精细定位马铃薯优良性状相关基因提供理论依据。

关键词：马铃薯卷叶病毒；SSR；单倍体组装；遗传图谱；QTL

* 基金项目：国家自然科学基金项目（31660503）；云南师范大学大学生科研训练基金项目（ky2018-136）
** 第一作者：梁静思，硕士研究生，研究方向：植物病理；E-mail：569020457@ qq. com
*** 通信作者：唐唯，讲师，研究方向：植物病理；E-mail：4311@ ynnu. edu. cn

樱桃小果病毒 1（LChV-1）LAMP 检测方法的建立与应用*

刘雅馨[1**]，齐志彦[1]，曹欣然[2]，李庆亮[1]，丁诚实[1]，王甲威[3]，王德亚[1***]
（1. 枣庄学院生命科学学院，病毒基因工程泰山学者工作站，枣庄　277000；
2. 山东农业大学植物保护学院植物病理系，山东省农业微生物重点实验室，泰安　271018；
3. 山东省农业科学院果树研究所，泰安　271018）

摘　要：樱桃小果病毒 1（*Little cherry virus* 1，LChV－1）属于长线形病毒科（Closteroviridae）长线病毒属（*Closteriovirus*），侵染樱桃可引起果实缩小、口味下降等，果实品质和产量明显下降，造成了巨大的经济损失。根据 NCBI 数据库中 LChV-1 的 CP 基因序列设计了 3 对引物用于 RT-LAMP 检测，筛选并获得 1 组特异性引物。以感染 LChV-1 的甜樱桃叶片总 RNA 为模板，建立并优化了一步法 RT-LAMP（One-step reverse transcription loop-mediated isothermal amplification）检测方法。实验表明反应温度为 57℃、甜菜碱浓度为 2mmol/L、Mg^{2+}浓度为 8mmol/L、dNTPs 为 2.5mmol/L、酶活力单位为 8U、内外引物浓度之比为 1：6 反应时间为 60min 时为最优反应体系，扩增产物利用琼脂糖电泳分析，同时对其特异性进行测定。结果表明，建立的 RT-LAMP 检测方法能特异性扩增 LChV-1。用 RT-LAMP 方法对 35 个疑似樱桃小果病的田间样品进行检测，发现 13 个样品感染了 LChV-1，其检测结果与 RT-PCR 法一致。RT-LAMP 法具有特异性强、操作简单、快速等特点，适合对 LChV-1 样品的快速检测与鉴定。

关键词：樱桃小果病毒 1；RT-LAMP；快速检测

---

* 基金项目：山东省自然科学基金项目（ZR2019PC011）；枣庄学院博士启动基金（2018BS040）

** 第一作者：刘雅馨，本科生，主要从事分子植物病毒学研究；E-mail：1224853754@qq.com
*** 通信作者：王德亚，讲师，主要从事分子植物病毒学研究；E-mail：wangdeyasdny@163.com

酵母双杂交技术筛选与甜菜坏死黄脉病毒 p14 蛋白互作的寄主因子[*]

刘 唱[**]，姜 宁，张宗英，韩成贵，王 颖[***]
（中国农业大学植物病理学系，农业生物技术国家重点实验室，北京 100193）

摘 要：甜菜坏死黄脉病毒（*Beet necrotic yellow vein virus*，BNYVV）引起的甜菜丛根病是甜菜上的一种重要病害。p14 蛋白是 BNYVV RNA 2 链编码的最后一个蛋白质，具有 RNA 沉默抑制功能。因此可以通过酵母双杂交技术探究 p14 与植物蛋白的互作关系，进而筛选到一些相关基因，有助于阐明 BNYVV 病毒的致病机理以及与植物的互作机制，为病害的综合防治提供理论依据。

将 p14 构建到膜蛋白酵母双杂交的诱饵载体 pDHB1 上，首先验证了诱饵克隆 pDHB1-p14 未表现出自激活活性，因此可作为后续双杂交实验的诱饵克隆。之后将诱饵克隆转入酵母菌株 NMY51 中并制作酵母感受态，再将植物 cDNA 文库（购于上海海科公司）转入酵母感受态中。最后用 1×TE 悬浮菌体，并且涂布于四缺营养缺陷型培养基上，将在四缺培养基上长出的单菌落重新划线到含有 X-gal 的四缺培养基上，筛选与 p14 互作的寄主因子。通过获取测序以及在 NCBI 数据库中的比对，初步筛选到 45 个互作因子，主要是与泛素化通路有关的蛋白。

关键词：BNYVV；酵母双杂交；p14；互作因子

致谢：感谢于嘉林、李大伟、王献兵教授和张永亮副教授对本研究的建议。

* 基金项目：国家自然科学基金项目（31872921）
** 第一作者：刘唱，硕士研究生，主要从事植物病毒与寄主互作研究；E-mail：changchang@ cau. edu. cn
*** 通信作者：王颖，副教授，主要从事植物病毒与寄主互作研究；E-mail：yingwang@ cau. edu. cn

电光叶蝉 siRNA 抗病毒途径介导水稻条纹花叶病毒与水稻瘤矮病毒在昆虫体内的协生关系[*]

刘 烨[**]，赵 萍，贾东升，魏太云，卫 静[***]
（福建农林大学植物病毒研究所，福建省植物病毒学重点实验室，福州 350002）

摘 要：由介体电光叶蝉（*Recilia dorsalis*）以持久增殖型方式传播的水稻条纹花叶病毒（Rice stripe mosaic virus，RSMV）和水稻瘤矮病毒（*Rice gall dwarf virus*，RGDV）目前在我国南方稻区有蔓延暴发的趋势。近期研究发现二者存在协生关系，RGDV 的侵染促进了 RSMV 在介体电光叶蝉体中肠内的增殖。通过对叶蝉中肠小 RNA 测序发现，RSMV 源的 siRNA 在与 RGDV 复合侵染的叶蝉中肠内的数量低于其单独侵染的数量，表明虫体抗 RSMV 的 RNAi 通路被抑制，进而促进 RSMV 的积累。同时发现 RGDV 源的 siRNA 在复合侵染叶蝉肠道内的数量高于在单独侵染的肠道内的数量，表明 RGDV 可能通过吸引虫体的 RNAi 反应进而减少该途径对 RSMV 的防御，从而促进 RSMV 的增殖。通过干扰 RNAi 通路关键基因 dicer2 的表达，发现 RSMV 和 RGDV 在叶蝉肠道内的积累均增加，表明 RGDV 可能通过调控昆虫 RNAi 免疫通路促进 RSMV 的增殖。因此本研究建立 RGDV 通过调控叶蝉 RNAi 通路介导其与 RSMV 协生关系的模型，为探索新的病毒病防控技术提供理论依据。

关键词：水稻条纹花叶病毒；水稻瘤矮病毒；协生关系；RNAi

---

* 基金项目：国家自然科学基金（31870148）
** 第一作者：刘烨，硕士研究生，从事水稻病毒与介体昆虫互作机制研究；E-mail：liuye092793@163.com
*** 通信作者：卫静，副教授；E-mail：weijing0306@163.com

电光叶蝉化感蛋白与水稻条纹花叶病毒的互作关系研究*

梁启福**，陈曼尼，霍晨阳，贾东升，魏太云***

（福建农林大学植物病毒研究所，福州　350002）

摘　要：化感蛋白（Chemosensory proteins，CSPs）是一类小的可溶性蛋白，与气味结合蛋白（Odorant-binding proteins，OBPs）相似，CSPs 涉及信息素在昆虫血淋巴中溶解和转运，且影响昆虫行为及发育。有研究表明虫媒病毒可通过血淋巴突破宿主免疫屏障，推测水稻条纹花叶病毒可能通过化感蛋白突破介体昆虫血淋巴免疫屏障。本研究在介体电光叶蝉（*Recilia dorsalis*）中分离到 9 个 CSP 基因，利用酵母双杂交系统发现只有 RdCSP9 蛋白与 RSMV 的 G 蛋白互作，且不与 RSMV 的 M 和 N 蛋白互作。杆状病毒表达系统研究显示 RdCSP9 与 G 在 Sf9 细胞质中有部分共定位。我们还发现 *RdCSP9* 在成虫中高表达，雄虫中显著高于雌虫，并且 RSMV 侵染后电光叶蝉体内 *RdCSP9* 表达显著上调。RNAi 抑制 *RdCSP9* 表达后 RSMV 在介体内的增殖减少，且雌虫卵巢发育受到影响，推测 RdCSP9 与 RSMV 在电光叶蝉体内的侵染有关。本研究结果从一个新的角度揭示了虫媒病毒与介体昆虫互作的关系。

关键词：水稻条纹花叶病毒；化感蛋白；电光叶蝉

---

* 基金项目：国家自然科学基金（31730071，31770166）

** 第一作者：梁启福，博士研究生，从事水稻病毒与介体昆虫互作机制研究；E-mail：lqf5687@ 126. com

*** 通信作者：魏太云，研究员；E-mail：weitaiyun@ 163. com

侵染电光叶蝉的一种新病毒的发现和特性研究

王子尧*，施夏敏，毛倩卓，张晓峰，吴　维，魏太云**
（福建农林大学植物保护学院，福州　350002）

摘　要：电光叶蝉（*Recilia dorsalis*）属于半翅目（Homoptera）叶蝉科（Cicadellidae），是一种重要的农业害虫。电光叶蝉在取食水稻的同时，能够传播水稻病毒病，严重威胁农业生产安全。像其他昆虫一样，电光叶蝉体内存在许多病毒或类病毒粒子，这些内共生病毒与昆虫的关系是复杂多样的。本研究报道了一种电光叶蝉体内的正单链 RNA（+ssRNA）病毒，经转录组测序拼接后得到的片段全长为15 974 bp，含有 7 个互相不重叠的开放阅读框，BLAST 比对结果显示该病毒属于 Virgaviridae 科，基因结构与同科的烟草花叶病毒（*Tobacco mosaic virus*）相似。该病毒在电光叶蝉中侵染较为普遍，PCR 检测结果显示，实验室培养的电光叶蝉种群带毒率为 29%～50%，不同地域田间采样的电光叶蝉种群带毒率为 10%～80%。在若虫期和成虫期的电光叶蝉中，以及在电光叶蝉的唾液腺、消化道、生殖系统和神经系统等部位均能检测到病毒的存在，说明病毒在叶蝉体内能持久存在。然而，该病毒在水稻中仅能存在 3～5 d，因此推测病毒不能在水稻中增殖，只能短暂停留，水稻是病毒在昆虫间传播的媒介。生物学交配实验结果发现，在带毒虫产下的卵中能检测到病毒，说明病毒能够垂直传播给后代，单管实验证明病毒能够从父本也能够从母本向子代传递。此外，携带该病毒的电光叶蝉种群寿命和产卵量均显著高于不带毒种群，因此该病毒的侵染有利于电光叶蝉在自然环境下的种群繁衍。

关键词：内共生病毒；电光叶蝉；传播方式

* 第一作者：王子尧，硕士研究生，从事病毒与介体昆虫互作机制研究；E-mail：ziyaowang9601@163.com
** 通信作者：魏太云，研究员；E-mail：weitaiyun@fafu.edu.cn

中国南瓜曲叶病毒编码的 AC5 基因功能分析*

吴会杰**，彭　斌，康保珊，刘丽锋，刘莉铭，古勤生***
（中国农业科学院郑州果树研究所，中国农业科学院
郑州果树研究所瓜类重点实验室，郑州　450009）

摘　要：中国南瓜曲叶病毒（*Squash leaf curl china virus*，SLCCNV）属双生病毒科（Geminiviridae）菜豆金色花叶病毒属（*Begomovirus*），是一类具有孪生颗粒形态的植物单链 DNA 病毒。自然条件下主要侵染葫芦科作物，由烟粉虱以持久方式传播。侵染后植株整株矮化，叶片皱缩，产量下降品质变劣。

该属病毒基因组大多含 DNA-A 和 DNA-B 双组分，2.5~3.0 kb，编码 6~7 个完整的开放阅读框。国内外对 SLCCNV 病毒编码的蛋白功能及其致病性的研究甚少。

本研究探讨了 SLCCNV AC5 基因的功能。应用 SLCCNV 侵染性克隆构建了定点突变 AC5 基因的突变体，通过接种试验证实了 AC5 基因是该病毒的致病因子。此外，利用 p35S-AC5 与 p35S-GFP 共浸润接种 16c 本氏烟，接种 5 d 后接种部位出现明显的绿色荧光信号，表明了 AC5 基因可抑制局部沉默。利用 p35S-GFP、p35S-dsFP 和 p35S-AC5 共浸润野生本氏烟，接种 5 d 后在接种部位未发现绿色荧光信号，表明 AC5 不抑制双链 GFP（IR-PTGS）诱导的沉默。在 AC5 基因 ORF 之前引入终止密码子，明确了 AC5 是在蛋白水平发挥作用的。利用 AC5 核定位信号缺失突变体与 p35S-GFP 共浸润 16c 本氏烟，发现该突变体失去了沉默抑制作用，表明 AC5 蛋白的核定位信号与沉默抑制活性相关。

关键词：中国南瓜曲叶病毒；AC5 基因；编码蛋白；致病性

---

* 基金项目：国家自然基金项目（31701941）；河南省基础与前沿技术研究项目（152300410231）；中国农业科学院创新团队西甜瓜病虫害防控（CAAS-ASTIP-2018-ZFRI-08）；国家西甜瓜产业体系病毒防控岗（CARS-26-13）

** 第一作者：吴会杰，副研究员，主要从事瓜类病害研究；E-mail：wuhuijie@ caas. cn

*** 通信作者：古勤生，研究员，主要从事瓜类病害及其防控研究；E-mail：guqinsheng@ caas. cn

尖孢镰刀菌甜瓜专化型 T-FJ019 携带真菌病毒的研究[*]

吴思颖[**]，华晖晖，张小芳，吴学宏[***]
（中国农业大学植物保护学院，北京 100193）

摘 要：由尖孢镰刀菌甜瓜专化型（*Fusarium oxysporum* f. sp. *melonis*）引起的甜瓜枯萎病被称为甜瓜的“癌症”，它造成甜瓜产量降低和品质下降，导致了严重的经济损失。真菌病毒是一种可以侵染真菌并在其体内进行复制增殖的病毒。世界范围内已在多种镰刀菌中发现真菌病毒，其中来源于尖孢镰刀菌的仅有两种，一种是 *Fusarium oxysporum chrysovirus* 1（FocV1），存在于尖孢镰刀菌甜瓜专化型（*F. oxysporum* f. sp. *melonis*）；另一种是 *Fusarium oxysporum* f. sp. *dianthi virus* 1（FodV1），存在于尖孢镰刀菌康乃馨专化型（*F. oxysporum* f. sp. *dianthi*）。本研究针对来源于我国 19 个城市 35 个采样点的甜瓜枯萎病病害样品中分离得到的尖孢镰刀菌甜瓜专化型，进行高通量测序、RT-PCR 及病毒双链 RNA 提取，从菌株 T-FJ019 中检测到一种真菌病毒；该病毒与 FodV1 的同源性较高，将其暂命名为 *Fusarium oxysporum* f. sp. *melonis virus* 1（FomV1）。将真菌病毒 FomV1 的 cDNA 序列进行克隆、分析，并将其与其他相近病毒的依赖于 RNA 的 RNA 聚合酶（RNA-dependent RNA polymerase，RdRp）构建系统发育树；结果表明，该病毒与产黄青霉科的病毒成员聚在一起，属于产黄青霉病毒科。真菌病毒 FomV1 共有 4 条链，每条链均存在一个开放阅读框（ORF），分别编码病毒的 RdRp、假定蛋白 2（HP2）、外壳蛋白（CP）和假定蛋白 4（HP4）。

关键词：甜瓜枯萎病；尖孢镰刀菌甜瓜专化型；真菌病毒；序列分析

---

* 基金项目：国家公益性行业（农业）科研专项（编号：201503110-14）

** 第一作者：吴思颖，硕士研究生，主要从事甜瓜枯萎病菌对杀菌剂的敏感性及其真菌病毒的研究；E-mail：799650234@qq.com

*** 通信作者：吴学宏，教授，主要从事植物病原真菌种类鉴定及其遗传多样性研究；E-mail：wuxuehong@cau.edu.cn

尖孢镰刀菌甜瓜专化型真菌病毒多样性分析*

吴思颖**，张小芳，梁芷健，吴学宏***
（中国农业大学植物保护学院，北京 100193）

摘 要：甜瓜是热带和亚热带地区广泛种植的重要园艺水果作物，作为甜瓜生产和消费大国，我国甜瓜种植总面积大约为40万公顷。作为甜瓜生产上的重要病害，甜瓜枯萎病常导致甜瓜植株死亡，严重影响其产量和品质，其病原菌为尖孢镰刀菌甜瓜专化型（*Fusarium oxysporum* f. sp. *melonis*）。世界范围内已在镰刀菌中检测到多种真菌病毒，其中大多数真菌病毒（已报道全基因组序列）主要来源于禾谷镰刀菌（*F. graminearum*）。本研究以从甜瓜枯萎病病害样品中分离鉴定得到的135株尖孢镰刀菌甜瓜专化型为对象，利用高通量测序分析其中真菌病毒的种类，以期探索其生防潜力，为防治甜瓜枯萎病提供生防资源。高通量测序数据处理得到61个病毒相关Contigs，这些信息涵盖了8个不同的科以及一些未明确分类的病毒。按照病毒核酸类型，这61个病毒序列包括39个正单链RNA病毒（占64%）、10个双链RNA病毒（占16%）、7个未分类的病毒（12%）、3个未分类RNA病毒（5%）以及2个双链DNA病毒（3%）。其中，正单链RNA病毒主要分布在马铃薯Y病毒科（41%）（主要为甘蔗花叶病毒，SCMV）中，这是首次在尖孢镰刀菌甜瓜专化型中检测到疑似跨界侵染的甘蔗花叶病毒。

关键词：甜瓜枯萎病；尖孢镰刀菌甜瓜专化型；高通量测序；真菌病毒；多样性

---

* 基金项目：国家公益性行业（农业）科研专项（编号：201503110-14）

** 第一作者：吴思颖，硕士研究生，主要从事甜瓜枯萎病菌对杀菌剂的敏感性及其真菌病毒的研究；E-mail：799650234@qq. com

*** 通信作者：吴学宏，教授，主要从事植物病原真菌种类鉴定及其遗传多样性研究；E-mail：wuxuehong@ cau. edu. cn

尖孢镰刀菌西瓜专化型携带 Contg 1267 真菌病毒分析[*]

韩　涛[**]，梁芷健，陈垦西，吴学宏[***]
（中国农业大学植物保护学院，北京　100193）

摘　要：西瓜枯萎病（Fusarium wilt of watermelon）由尖孢镰刀菌西瓜专化型（*Fusarium oxysporum* f. sp. *niveum*）引起，并在西瓜上普遍发生的一种危害极大的真菌性病害。世界范围内已在多种镰刀菌中发现真菌病毒，其中大多数真菌病毒（已报道全基因组序列）主要来源于禾谷镰刀菌（*F. graminearum*）。本研究通过高通量测序对 90 株尖孢镰刀菌西瓜专化型进行宏转录组检测，共获得 58 个病毒相关信息，并发现 Contig 1267 在大部分菌株中存在。为了探究含有 Contig 1267 真菌病毒在哪些菌株中存在，根据 RNA 病毒在复制过程中都会产生 dsRNA 复制体这一特性，通过提取并检测尖孢镰刀菌西瓜专化型的 dsRNA，发现 2018 年从北京顺义地区采集的西瓜枯萎病病害样品中分离到一株尖孢镰刀菌西瓜专化型（编号为 4-8），其携带 3 条 dsRNA 片段，大小为2 300～4 000 bp。通过在 NCBI 网站上进行 BLAST 比对分析，3 条 dsRNA 的部分 cDNA 序列分别与 *Alternavirus* 的真菌病毒相似性为 30%～68%，但其分类地位尚未明确。选取含有 Contig 1267 真菌病毒的 7 株尖孢镰刀菌西瓜专化型，采用浸根法进行致病性试验，其平均发病率和平均病情指数均高于未含 Contig 1267 真菌病毒的菌株，但其致病力升高是否可能与 Contig 1267 的存在有关还有待进一步研究进行确认。

关键词：西瓜枯萎病；尖孢镰刀菌西瓜专化型；Contig 1267；致病力

---

* 基金项目：国家公益性行业（农业）科研专项（编号：201503110-14）

** 第一作者：韩涛，硕士研究生，主要从事西瓜枯萎病菌对杀菌剂的敏感性及其真菌病毒的研究；E-mail：1131140436@qq. com

*** 通信作者：吴学宏，教授，主要从事植物病原真菌种类鉴定及其遗传多样性研究；E-mail：wuxuehong@ cau. edu. cn

尖孢镰刀菌西瓜专化型真菌病毒多样性分析*

韩　涛**，梁佳媛，陈垦西，吴学宏***
（中国农业大学植物保护学院，北京　100193）

摘　要：由尖孢镰刀菌西瓜专化型（*Fusarium oxysporum* f. sp. *niveum*）侵染引起的西瓜枯萎病是西瓜上发生最为广泛、最为严重的真菌病害之一。真菌病毒是指存在于真菌并能够在其中稳定复制的一类病毒，此类病毒主要分布在包括植物病原真菌在内的丝状真菌中，部分真菌病毒导致宿主植物病原真菌的致病力降低，具有一定的生防潜力。已有研究报道表明，尖孢镰刀菌甜瓜专化型（*F. oxysporum* f. sp. *melonis*）中存在真菌病毒 *Fusarium oxysporum chrysovirus* 1（FocV1），尖孢镰刀菌康乃馨专化型（*F. oxysporum* f. sp. *dianthi*）中存在真菌病毒 *Fusarium oxysporum* f. sp. *dianthi virus* 1（FodV1）。本研究采用高通量测序技术对90株尖孢镰刀菌西瓜专化型进行宏转录组检测，获得58个病毒相关信息，表明尖孢镰刀菌西瓜专化型中存在一定数量的真菌病毒资源；这些病毒的核酸类型呈现多样性，其中unclassified ssRNA占48.30%，+ssRNA占17.20%，dsRNA占13.80%，Retro-transcribing viruses占6.90%，dsDNA占6.90%，unclassified viruses占6.90%。通过序列比对和保守结构域的进化分析，明确这58个Contigs中有裸露病毒科（Narnaviridae）、双分病毒科（Partitiviridae），以及植物病毒，如马铃薯Y病毒科（Potyviridae）、帚状病毒科（Virgaviridae）（主要为烟草花叶病毒属 *Tobamovirus*）；此外还有许多病毒的分类地位尚未明确。

关键词：西瓜枯萎病；尖孢镰刀菌西瓜专化型；高通量测序；真菌病毒；多样性

---

* 基金项目：国家公益性行业（农业）科研专项（编号：201503110-14）

** 第一作者：韩涛，硕士研究生，主要从事西瓜枯萎病菌对杀菌剂的敏感性及其真菌病毒的研究；E-mail：1131140436@qq.com

*** 通信作者：吴学宏，教授，主要从事植物病原真菌种类鉴定及其遗传多样性研究；E-mail：wuxuehong@cau.edu.cn

引起甜瓜叶部病害的链格孢菌携带一种 dsRNA 病毒*

刘　泉**，李郁婷，马国苹，吴学宏***

（中国农业大学植物保护学院，北京　100193）

摘　要：本研究从北京地区甜瓜感病叶片上分离到一株 *Alternaria tenuissima*（命名为 21-5），其中检测到一种 dsRNA 病毒，通过随机克隆和末端克隆拼接得到该病毒的 cDNA 序列全长。经研究发现该病毒有 2 条 dsRNA，各编码一个开放阅读框（ORF）。dsRNA 1 全长 6 188 bp，预测编码 1 个开放阅读框（ORF 1），其氨基酸序列经 BLASTp 比对后并没有发现保守区域；dsRNA 2 全长 5 903 bp，预测编码 1 个开放阅读框（ORF 2），其氨基酸序列经 BLASTp 比对后发现含有 1 个保守的 RdRp 结构域（RdRp_4）。该病毒与 *Alternaria botybirnavirus* 1（ABRV1）相似性极高，且认定为同一种病毒，但是两者来自于不同种类的链格孢菌，因此将该病毒命名为 *Alternaria botybirnavirus* 1-T1（ABRV1-T1）。采用邻接法对 ABRV1-T1 预测的 RdRp 氨基酸序列构建系统发育树，结果表明 ABRV1-T1 与 Botybirnaviridae 的病毒聚在一起一支，属于 Botybirnaviridae。对 ABRV1-T1 的宿主真菌 21-5 进行致病性研究发现，其致病力显著下降，但是其致病力下降是否与其携带这种真菌病毒相关还需要进一步的研究。

关键词：甜瓜叶部病害；链格孢菌；dsRNA 病毒；cDNA 克隆

---

* 基金项目：北京市西甜瓜产业创新团队（BAIC10-2019）

** 第一作者：刘泉，硕士研究生，主要从事引起甜瓜叶部病害的链格孢菌对杀菌剂的敏感性及其真菌病毒的研究；E-mail：2235262089@ qq. com

*** 通信作者：吴学宏，教授，主要从事植物病原真菌种类鉴定及其遗传多样性研究；E-mail：wuxuehong@ cau. edu. cn

引起甜瓜叶部病害的链格孢菌真菌病毒多样性分析*

刘 泉**，李郁婷，陈垦西，吴学宏***
（中国农业大学植物保护学院，北京 100193）

摘 要：链格孢菌（*Alternaria* spp.）是一种较为常见的真菌，它不仅是植物病原菌，而且还能危害人和动物。农业生产上，防治由链格孢菌引起的真菌病害主要通过施用化学杀菌剂的方法，但由此引起的的环境污染、病原菌抗药性和农药残留等问题促使人们积极寻找可以替代部分化学杀菌剂的生防资源。随着真菌病毒 *Cryphonectria parasitica hypovirus* 1（CHV1）在控制欧洲板栗疫病上的成功应用，使得真菌病毒作为一种生防资源引起了广大科技工作者的高度重视。本研究利用高通量测序技术对引起甜瓜叶部病害的链格孢菌中的真菌病毒资源进行挖掘。测序结果在NCBI non-redundant protein 数据库中共比对得到 29 种病毒信息，其中+ssRNA 病毒数量最多，所占比例为 49%，dsRNA 病毒、未分类的病毒及-ssRNA 病毒所占比例分别为 19%、23%和 7%，仅有一条 Contig 比对结果为逆转录病毒，但并未比对到 DNA 病毒的序列信息。这 29 种病毒分别属于 9 个不同的病毒科、属以及未分类病毒，其中 *Ourmiavirus*、*Narnaviridae* 和 *Chrysoviridae* 所占比例较大。通过对链格孢菌中真菌病毒的深入研究可以为真菌病毒的起源和进化以及真菌病毒与寄主之间的相互关系提供理论依据，为利用真菌病毒防治链格孢菌引起的病害提供潜在的生防资源。

关键词：甜瓜叶部病害；链格孢菌；真菌病毒；高通量测序

* 基金项目：北京市西甜瓜产业创新团队（BAIC10-2019）

** 第一作者：刘泉，硕士研究生，主要从事引起甜瓜叶部病害的链格孢菌对杀菌剂的敏感性及其真菌病毒的研究；E-mail：2235262089@qq.com

*** 通信作者：吴学宏，教授，主要从事植物病原真菌种类鉴定及其遗传多样性研究；E-mail：wuxuehong@cau.edu.cn

CGMMV 的 CP 嵌合突变体构建以及致病性研究

周　涛*，毕馨月，安梦楠**，吴元华**
（沈阳农业大学，沈阳　110866）

摘　要：黄瓜绿斑驳花叶病毒（*Cucumber green mottle mosaic virus*，CGMMV）属于烟草花叶病毒属（*Tobamovirus*）成员之一，该病毒是葫芦科作物中一种重要的检疫性病害，严重威胁着黄瓜、西瓜、甜瓜等葫芦科作物的生产。辽宁地区 CGMMV 侵染西瓜引起血瓤症状，常年造成严重危害。侵染性克隆技术是研究病毒致病机理的重要工具，可以对全长克隆进行缺失、突变或嵌合等操作，为我们利用反向遗传学分析法开展对病毒的研究奠定了基础。本研究以该辽宁分离物侵染性克隆 pCB-CGMMV-LN 为材料构建嵌合突变表达载体，分别根据辣椒轻斑驳病毒（*Pepper mild mottle virus*，PMMoV）和（*Tobacco mosaic virus*，TMV）病毒的外壳蛋白（coat protein，CP）基因序列设计相应的 vector 和 insert 特异性扩增引物。

vector 引物：

CG-TCP+ TGCAACTTGATTTCGAGGGTCTTCTGATGG；

CG-TCP-CTGTAAGACATCTTCAAAAGAAACAGAAC；

CG-PCP+ AACTCCTTAATTTCGAGGGTCTTCTGATGG；

CG-PCP-TGTGTAAGCCATCTTCAAAAGAAACAGAAC。

insert 引物：C-TMV-CP+ TCTTTTGAAGATGTCTTACAGTATCACTAC；

C-TMV-CP-CCTCGAAATCAAGTTGCAGGACCAGAGGTC；

C-PMV-CP+ TCTTTTGAAGATGGCTTACACAGTTTCCAG；

C-PMV-CP-AAGACCCTCGAAATTAAGGAGTTGTAGCCC。

采用同源重组的克隆方式，以 PMMoV 和 TMV 的 DNA 序列为模板通过 C-PMV-CP+& C-PMV-CP-，C-TMV-CP+& C-TMV-CP-扩增得到的 PMMoV 和 TMV 的 CP 基因片段连接到以 pCB-CGMMV-LN 为模板采用 CG-PCP+& CG-PCP-，CG-TCP+& CG-TCP-扩增后用 DMT 酶处理的 vector 片段中，即将 CGMMV-ZJ 的 CP 蛋白质基因序列分别替换为 PMMoV 和 TMV 的 CP 基因序列，最终成功构建了 CGMMV-TCP 和 CGMMV-PCP 嵌合突变体的侵染性克隆，然后利用农杆菌浸润法接种本生烟，初步结果显示接种 CGMMV-TCP 的突变体的本生烟有明显的花叶皱缩等现象，确认其有侵染性，而接种 CGMMV-PCP 突变体的本生烟发病症状并不明显。本研究预期采用不同的葫芦科作物进行野生型和突变体 CGMMV 的接种，通过继续构建新的表达载体和突变体，以及植株症状观察、病毒检测、病毒对寄主的影响几方面揭示不同突变体在致病过程中的关键因子，从而阐明 CGMMV 以及其他两种病毒的 CP 蛋白在植物寄主致病性以及传播机制方面的作用。

关键词：CGMMV；侵染性克隆；CP；嵌合突变体；致病性

* 第一作者：周涛，主要从事植物病毒学研究；E-mail：zt13332465967@163.com
** 通信作者：安梦楠，讲师，主要从事病毒学研究；E-mail：anmengnan1984@163.com
吴元华，教授，主要从事病毒学研究；E-mail：wuyh7799@163.com

四川烟草辣椒脉斑驳病毒的分离鉴定和 RPA 检测方法的建立*

焦裕冰**，李嘉伦，徐传涛，夏子豪***，吴元华***
（沈阳农业大学植物保护学院，沈阳 110866）

摘 要：辣椒脉斑驳病毒（*Chilli veinal mottle virus*，ChiVMV）是马铃薯 Y 病毒属（*Potyvirus*）的正单链 RNA 病毒，能够侵染烟草、辣椒和番茄等茄科植物，并造成严重经济损失，我国四川、贵州、湖南等辣椒产区已有报道，云南烟草产区 2010 年也发现了该病毒在烟草上零星为害。四川是我国烟草和辣椒的主产区之一，ChiVMV 在烟草和辣椒之间可相互传播，加重了病害的发生和为害，给农业生产造成了严重经济损失。因此，研究不同作物上 ChiVMV 的生物学特性、基因组结构和功能，以及建立新的快速、灵敏、特异性强的检测方法，对于 ChiVMV 的防控具有重要理论和实践意义。

本研究中，从中国四川西南部泸州地区感病烟草中分离得到 ChiVMV-LZ，根据 NCBI 数据库中已报道 ChiVMV 的全基因序列设计引物，采用分段克隆法和 RACE 扩增法获得了 ChiVMV-LZ 基因组全序列。该序列全长为 9 742 nt，包含 1 个长 9 270 nt 的开放阅读框，编码 1 个长度为 3 089 aa 的多聚蛋白，具有马铃薯 Y 病毒属病毒的典型结构特点。利用 DNAMAN 和 MEGAX 等软件与 GenBank 上已经登陆的分离物序列进行序列比对与分析。结果显示 ChiVMV-LZ 与 ChiVMV 四川辣椒分离物 ChiVMV-Yp8（KC711055.1）的基因序列和氨基酸序列的同源性分别为 98.76%和 98.55%，亲缘关系最近，而与广东（KU987835.1）和海南（GQ981316.1）分离物亲缘关系较远。与云南烟草上的 ChiVMV 分离物（JX088636.1）同源性仅有 83%，表明 ChiVMV 存在着遗传分化现象，并且具有一定的地理差异。与此同时，我们建立了一种新型恒温 DNA 扩增检测技术-重组酶聚合酶扩增（RPA）技术用于快速检测 ChiVMV。结果表明，RPA 法可在 38℃ 条件下恒温反应 30 min 完成扩增反应，且与其他主要烟草病毒无交叉反应，特异性良好。灵敏度分析表明，RPA 法的最低检测限度是普通 PCR 方法的 10 倍，灵敏度高。此外，该方法已成功应用于田间采集的烟草和辣椒的检测。本研究建立的 RPA 检测方法为在田间和基层实验室中快速检测 ChiVMV 提供了可靠、灵敏和有效的方法。

关键词：辣椒脉斑驳病毒（ChiVMV）；系统发育分析；重组酶聚合酶扩增（RPA）

---

* 基金项目：四川省烟草公司计划管理类项目（2018-510500-2-4-024）
** 第一作者：焦裕冰，博士研究生，主要从事植物病毒学研究；E-mail：417783085@qq.com
*** 通信作者：夏子豪，讲师，主要从事植物病毒学研究；E-mail：zihao8337@syau.edu.cn
吴元华，教授，主要从事植物病理学研究；E-mail：wuyh09@syau.edu.cn

烟草花叶病毒外壳蛋白抑制寄主抗性研究

于　曼[*]，安梦楠[**]，吴元华[**]
（沈阳农业大学，沈阳　110866）

摘　要：烟草花叶病毒（*Tobacco mosaic virus*，TMV）属于烟草花叶病毒属（*Tobamovirus*），寄主范围广，对作物造成严重危害。TMV 的外壳蛋白（CP）在病毒侵染过程中发挥着多功能作用，除了形成病毒粒子，外壳蛋白还参与病毒翻译、复制、细胞间运输以及系统运输。然而，CP 在抑制寄主抗性方面的机制尚不清楚。在本研究中，基于 TMV 侵染性克隆 pCB-TMV-SY 利用同源重组方法构建了 CP 替换黄瓜绿斑驳花叶病毒（*Cucumber green mottle mosaic virus*，CGMMV）和辣椒轻斑驳病毒（*Pepper mild mottle virus*，PMMoV）的两个嵌合突变体 pCB-TMV-C-CP 和 pCB-TMV-P-CP。将含有野生型和突变体克隆的农杆菌浸润本生烟，5d 后发现突变体与野生型 TMV 接种的本生烟都表现出皱缩以及花叶症状。提取病毒粒子经透射电镜检测，突变体可以形成与野生型 TMV 形态相同的杆状病毒粒子。将上述发病的本生烟汁液分别摩擦接种普通烟 NC89，野生型 TMV 接种的烟草表现出明显的花叶症状，相比之下，接种突变体的烟草表现出明显的矮化症状。笔者利用转录组测序技术比较了接种突变体病毒 TMV-C-CP 和 TMV-P-CP 相对于野生型 TMV 烟草中的基因表达调控变化。结果表明接种 TMV-C-CP vs TMV 和接种 TMV-P-CP vs TMV 的烟草 NC89 中显著差异表达基因（DEGs）分别为 4 860个和 3 751个。KEGG 通路分析结果表明嵌合突变体 TMV 侵染诱导多个 DEG 在烟草与病原体相互作用和植物激素信号转导途径中显著富集；诱导了多个参与生长素以及细胞分裂素信号转导的基因的显著下调；可诱导上调表达参与水杨酸、乙烯和赤霉素信号传导途径的关键基因。此外，结果表明 WRKY、PR-1、LRR-RLK 和钙调节相关基因也显著上调，这些基因与植物与病原体相互作用有关。上述结果揭示 TMV-CP 可能在抑制寄主抗性以及躲避植物免疫反应攻击中起到重要作用，该假说仍需要后续试验进行充分验证。

关键词：烟草花叶病毒；外壳蛋白；嵌合突变体；转录组分析

* 第一作者：于曼，硕士研究生，主要从事植物病毒学研究；E-mail：ym960527@163.com
** 通信作者：安梦楠，讲师，主要从事病毒学研究；E-mail：anmengnan1984@163.com
吴元华，教授，主要从事病毒学研究；E-mail：wuyh7799@163.com

呼长孤病毒 SsMYRV4 单个基因对核盘菌生物学特性的影响*

庞茜丹**，吴 吞，程家森，付艳苹，姜道宏，谢甲涛***
（农业微生物学国家重点实验室，湖北省作物病害监测和
安全控制重点实验室，华中农业大学，武汉 430070）

摘 要：核盘菌（*Sclerotinia sclerotiorum*）所引起的油菜菌核病在全球各大产区均有发生，影响油菜整个生长发育周期，为害茎、叶、花和角果。一些真菌病毒能引起寄主弱毒现象，具有防治菌核病的潜力，然而核盘菌复杂的营养体不亲和性（Vegetative incompatibility）限制了真菌病毒通过菌丝间融合的水平传播，影响低毒相关真菌病毒的实践应用。在前期研究中，发现感染 SsMYRV4 的菌株与不同营养体亲和群的菌株对峙培养时，SsMYRV4 能抑制在非我识别初期的信号转导蛋白 G 蛋白基因的表达，抑制 *het*（Heterokaryon）基因及活性氧（Reactive Oxygen Species，ROS）相关基因的表达，最终导致营养体不亲和性反应受到抑制。这种由 SsMYRV4 介导的对核盘菌非我识别的抑制作用，能作为异源病毒传播的桥梁，突破营养体不亲和性的限制进行更广泛的水平传播。为了解析呼肠孤病毒 SsMYRV4 基因组中抑制寄主营养体不亲和性反应的分子机制，发掘可用于克服 VIC 反应的有益病毒或核盘菌基因资源，进行以下研究。根据病毒全长序列信息，从呼长孤病毒 SsMYRV4 基因组的 12 条 dsRNA 片段中克隆出 S1～S7 全长 cDNA 序列，构建了 S1～S7 单一基因的超表达载体 pCHEF-S1～S7。利用农杆菌介导法转化核盘菌 1980，通过潮霉素抗性筛选和 PCR 证明病毒片段整合到核盘菌 1980 并表达，获得超表达病毒片段 S1、S2、S4、S5、S6、S7 的核盘菌转化子的数量各为 4 株、1 株、6 株、2 株、16 株和 6 株。对核盘菌转化子进行了生物学特性研究，发现超表达 S6 基因的转化子在 PDA 固体培养基上的菌落边缘出现明显扇变，气生菌丝发达，菌核形成时间推迟，与 1980 相比致病力显著下降。超表达 S7 基因的转化子菌落形态和菌核形态正常，气生菌丝发达，其生长速度略慢于菌株 1980，菌丝尖端和产酸能力正常，在油菜上能形成侵染垫，但致病力显著降低，在水中易发生原生质体渗漏，在含有 1mol/L NaCl 和 1.4mol/L Sorbitol 的 PDA 培养基中生长受到的抑制率为 61%～67%、61%～64%，均高于对照菌株 1980 的 45%和 53%。关于超表达 SsMYRV4 基因片段对寄主生物学特性的影响及其对真菌病毒水平传播的影响，正在进行相关研究。

关键词：核盘菌；呼肠孤病毒；超表达；生物学特性

---

* 基金项目：国家自然科学基金（31722046）

** 第一作者：庞茜丹，硕士研究生，主要从事真菌病毒与寄主互作研究；E-mail：pxd2018wuhan@163.com

*** 通信作者：谢甲涛，教授，主要从事生物防治相关研究；E-mail：jiataoxie@mail.hzau.edu.cn

烟草丛顶病毒非翻译区远距离互作的分子开关定位*

窦宝存**，王德亚，于成明，原雪峰***
（山东农业大学植物保护学院植物病理系，山东省农业微生物重点实验室，泰安 271018）

摘　要：烟草丛顶病毒（*Tobacco bushy top virus*，TBTV）属于番茄丛矮病毒科幽影病毒属，基因组由一条正义单链 RNA（+ssRNA）病毒组成，全长 4 152 nt。5′端没有帽子结构（m7GPPPN），3′端也没有 poly（A）尾序列。TBTV 基因组的 5′UTR 为 10 nt；3′UTR 长达 645 nt，存在一个不依赖帽子翻译调控元件——BTE 结构。BTE 结构中含有结合翻译起始因子 eIF4E 的关键区域，也还有与 5′UTR 存在远距离 RNA-RNA 互作的区域，此远距离 RNA-RNA 互作对不依赖帽子翻译至关重要。

研究发现 3 个 TBTV 分离物（TBTV-MDI、TBTV-MDII、TBTV-JC）的 p35 蛋白表达量明显低于野生型 TBTV。嵌合病毒的翻译分析表明，RI 区（TBTV 5′末端的 511 nt）和 RV 区（包括 BTE 在内的3138—3885区域）是影响蛋白表达的关键区域，推测 RI 区可能是通过干扰基因组 5′UTR 与 BTE 的远距离互作从而抑制了 p35 的表达。TB-MDI（1-511）嵌合病毒的基础上增加 5′UTR，发现增加两个 5′UTR 序列后蛋白表达量恢复到野生型的 50%以上。暗示在原有的 5′末端 511 nt 中存在突变位点，此突变位点通过与 5′UTR 的局部互作而影响 5′UTR 与 BTE 的远距离 RNA-RNA 互作，进而抑制不依赖帽子的翻译水平。比对 3 个新分离物 1-511 与 TBTV 野生型对应区域的序列，找到了在 3 个分离物中均存在的 9 个共有突变位点，其中关于 299 位、405 位和 464 位的突变可以使嵌合病毒 TB-MDI（1-511）的蛋白表达明显提高，说明这 3 个位点及其上下游序列的存在与 5′UTR 形成潜在的局部互作，在一定程度上阻止了 5′UTR 与 BTE 的远距离 RNA-RNA 互作，抑制了不依赖帽子的翻译水平。TBTV 基因组的 5′端存在分子开关，此开关控制 5′UTR 与 3′UTR 间的远距离 RNA-RNA 互作，通过调节 TBTV RNA 的环化状态来调控 TBTV 的不依赖帽子翻译。

关键词：烟草丛顶病毒；分子开关；定位

---

* 基金项目：国家自然科学基金（31872638，31670147）；山东省“双一流”奖补资金资助经费（SYL2017XTTD11）
** 第一作者：窦宝存，硕士研究生，从事植物病毒学研究；E-mail：baocun0106@163.com
*** 通信作者：原雪峰，教授，主要从事分子植物病毒学研究；E-mail：snowpeak77@163.com

山东大樱桃病毒病病原分析*

曹欣然[1,2]**，耿国伟[1]，张雅雯[1]，于成明[1]，李向东[1]，原雪峰[1]***
（1. 山东农业大学植物保护学院 山东省农业微生物重点实验室，泰安 271000；
2. 烟台市农业技术推广中心，烟台 264001）

摘 要：本研究的目的是调查山东大樱桃病毒病种类，掌握大樱桃病毒病发生和流行情况，为防治提供参考。2016—2018 年，共采集来自山东省 13 个市、62 个县区的 110 个具有代表性果园的大樱桃样品。110 个果园的樱桃感病样品中，针对 17 种候选病毒进行 RT-PCR 检测，共检测到 8 种病毒，分别为樱桃病毒 A（*Cherry virus A*，CVA）、樱桃绿斑驳病毒（*Cherry green ring mottle virus*，CGRMV）、樱桃小果病毒 1（*Little cherry virus* 1，LChV-1）、樱桃小果病毒 2（*Little cherry virus* 2，LChV-2）、李树皮坏死与茎痘伴随病毒（*Plum bark necrosis stem pitting-associated virus*，PBNSPaV）、李属坏死环斑病毒（*Prunus necrotic ringspot virus*，PNRSV）、李矮缩病毒（*Prune dwarf virus*，PDV）和黄瓜花叶病毒（*Cucumber mosaic virus*，CMV）。山东樱桃园病毒检出率高，110 个果园的样品中，有 93 个果园可检测到侵染樱桃的病毒，潜隐性病毒（CVA、CGRMV）普遍存在，CVA 检出率 96.8%，CGRMV 检出率 43%；造成严重危害的小果病毒（LChV-1、LChV-2）存在率低，LChV-1 检出率 11.8%，LChV-2 检出率 3.2%。93 个检测到病毒的果园中，多病毒侵染现象普遍，只有 4 个果园是 CVA 单独侵染，果园中存在两种及两种以上病毒占 95.7%，最高达 8 种病毒同时存在一个果园中。结果显示山东省引起樱桃病毒病病毒共有 8 种，分别为 CVA、CGRMV、LChV-1、LChV-2、PBNSPaV、PNRSV、PDV 和 CMV，其中 CMV 首次在山东省樱桃园上检测到。山东大樱桃主产区病毒病普遍存在，同一果园中存在两种及两种以上病毒概率高，不同地区间樱桃病毒病的发生情况呈现分布不均，烟台和泰安检测到樱桃病毒种类最多，可能与种苗调运频繁程度相关。

关键词：樱桃病毒病；山东省；反转录 PCR；鉴定

* 基金项目：国家自然科学基金（31872638，31670147）；山东“双一流”奖补资金资助经费（SYL2017XTTD11）；山东省农业微生物重点实验室开放课题资助（SDKL2017015）
** 第一作者：曹欣然，博士，从事植物病毒学研究；E-mail：xinran1001@163.com
*** 通信作者：原雪峰，教授，主要从事分子植物病毒学研究；E-mail：snowpeak77@163.com

小麦黄花叶病毒 RNA1 复制调控研究*

耿国伟**，于成明，原雪峰***
（山东农业大学植物保护学院植物病理系，山东省农业微生物重点实验室，泰安 271018）

摘 要：小麦黄花叶病毒（*Wheat yellow mosaic virus*，WYMV）是小麦土传花叶病的主要病原之一，对我国多个地区冬小麦的生长、发育造成严重危害。WYMV 为二分体基因组，由两条线性单链正义 RNA 组成。RNA1 全长约 7.6 kb，RNA2 全长约 3.6 kb，分别编码一个 270 ku 和 100 ku 的多聚蛋白。WYMV RNA 的 5′端有共价结合的 VPg，3′端有一个 poly（A）尾；NIb 蛋白具有依赖 RNA 的 RNA 聚合酶（RdRp）活性。

利用原核表达载体 pMAL-C2X 表达纯化 NIb 蛋白，通过体外复制实验发现 NIb 蛋白只能识别 WYMV 基因组的 3′UTR，在基因组的 5′UTR 与中间位置不能发挥作用；并且其他病毒（TBTV）的 RdRp 在 WYMV 的 3′UTR 同样不能发挥作用。随后通过原核表达了 WYMV 体内复制相关的蛋白：14K、P3 和 VPg，通过体外复制实验发现这 3 种蛋白单独均不能发挥作用，在复制过程中发挥核心作用的是 NIb 蛋白，建立了 WYMV NIb 介导的体外复制体系。

通过 In-line probing（RNA 体外结构分析技术）发现 WYMV RNA1 的 3′末端具有 5 个茎环结构。通过定点突变对这 5 个茎环进行突变即打破茎的稳定性和将环的碱基突变为互补碱基。通过体外复制实验发现对于第一个茎环和第三个茎环的突变不影响复制的效率，第二个茎环和第四个茎环是维持复制的最核心的结构，同时第五个茎环在复制过程中也发挥了不可替代的作用。

关键词：小麦黄花叶病毒；3′UTR；体外复制

---

* 基金项目：国家自然科学基金资助项目（31670147；31370179）；山东省“双一流”奖补资金资助经费（SYL2017XTTD11）

** 第一作者：耿国伟，博士研究生，主要从事分子植物病毒学研究；E-mail：guowgeng@163.com

*** 通信作者：原雪峰，教授，博士生导师，主要从事分子植物病毒学研究；E-mail：snowpeak77@163.com

菜豆普通花叶病毒花生株系蛋白翻译调控*

李 哲**，耿国伟，于成明，原雪峰***
（山东农业大学植物保护学院植物病理系，山东省农业微生物重点实验室，泰安 271018）

摘 要：花生条纹病是花生上分布最广泛的病毒病之一，引起该病的病原是菜豆普通花叶病毒（*Bean common mosaic virus*，BCMV）花生株系。BCMV 病毒粒体线状，是正单链 RNA 病毒，全长约 10kbp，包含一个开放阅读框，编码一个多聚蛋白，通过蛋白酶切割产生 10 个成熟的功能蛋白。BCMV 5′端没有帽子结构，但含有病毒基因组结合蛋白（VPg），3′端有 Poly（A）尾巴；Nib 蛋白具有 RNA 依赖的 RNA 聚合酶（RdRp）活性，负责基因组的复制。

本研究通过体外翻译系统，利用萤火虫荧光酶素（Fluc）载体报告基因研究 BCMV 非翻译区对蛋白表达的调控作用。研究发现含有 BCMV 5′UTR 的 Fluc 载体相对于空 Fluc 载体翻译效率可以提高 6 倍，含有 BCMV 3′UTR 的 Fluc 载体对翻译调控作用不明显，说明 5′UTR 可以正调控不依赖帽子翻译。同时在 5′UTR 存在的情况下 3′UTR 的翻译效率相对于单独 5′UTR 可以提高 3 倍，说明 3′UTR 可以协同提高 5′UTR 的正调控作用。本研究为 BCMV 不依赖帽子的翻译调控奠定基础，对进一步研究 BCMV 5′UTR 和 3′UTR 对蛋白表达的调控作用提供数据支撑。

关键词：菜豆普通花叶病毒；体外翻译系统；不依赖帽子翻译

---

* 基金项目：国家自然科学基金（31872638，31670147）；山东省“双一流”奖补资金资助经费（SYL2017XTTD11）
** 第一作者：李哲，在硕士研究生，主要从事分子植物病毒学研究；E-mail：17863801865@163.com
*** 通信作者：原雪峰，教授，博士生导师，主要从事分子植物病毒学研究；E-mail：snowpeak77@163.com

黄瓜花叶病毒（CMV）弱毒突变体构建的位置选择*

刘珊珊**，亓　哲，于成明，原雪峰***

（山东农业大学植物保护学院植物病理系，山东省农业微生物重点实验室，泰安　271018）

摘　要：黄瓜花叶病毒（*Cucumber mosaic vitus*，CMV）是目前所知寄主范围最多、分布最广、最具经济危害的植物病毒之一，对其弱毒疫苗的研制对植物预防 CMV 具有重要意义。本研究以 CMV Fny 株系侵染性克隆为研究对象，选取 CMV TLS（tRNA-liking-structure）结构和 2b 蛋白作为改造位点，对 CMV 的 TLS 结构与 2b 蛋白进行突变，构建 TLS 保守区域缺失的极端突变 pCB-Fny1-TLSm1、pCB-Fny2-TLSm1、pCB-Fny3-TLSm1；TLS 保守区域插入 18 个碱基型突变 pCB-Fny1-TLSm2、pCB-Fny2-TLSm2、pCB-Fny3-TLSm2；2b 蛋白提前终止型突变 pCB-Fny2-del2bN。将突变型侵染性克隆利用农杆菌浸润方法，接种 6 叶期本生烟，7d 后发现 TLS 缺失型突变与插入型突变均能引起严重的病毒病症状，感染植株均表现为叶片严重皱缩，植株矮化，伴随不同程度的花叶症状；接种 2b 蛋白提前终止型突变的植株未出现叶片皱缩、植株矮化等症状，与未接种的植株长势类似。对接种植株做 RT-PCR 检测，均能检测到 CMV 的存在，对 PCR 产物进行测序分析，发现 2b 突变位点仍存在，TLS 缺失型突变位点有不同程度的回复突变现象，TLS 插入型突变位点均回复突变。研究表明，TLS 结构保守区域的突变和 2b 蛋白提前终止的突变都不影响 CMV 的侵染性；TLS 突变不影响 CMV 的致病力，而对 2b 蛋白的突变能够造成 CMV 致病力的明显减弱。3 种不同类型的 CMV 突变体的致病力研究对 CMV 弱毒疫苗构建的突变位点选择提供了实验依据。

关键词：黄瓜花叶病毒；TLS 结构；2b 蛋白；病毒致病性

---

* 基金项目：国家自然科学基金（31670147，31370179）；山东省“双一流”奖补资金资助经费（SYL2017XTTD11）

** 第一作者：刘珊珊，博士研究生，从事植物病毒研究；E-mail：shansd1218@126.com

*** 通信作者：原雪峰，教授，主要从事分子植物病毒学研究；E-mail：snowpeak77@163.com

水稻条纹病毒的亚基因组表达调控的研究*

刘志菲**，杨 晨，原雪峰***
（山东农业大学植物保护学院植物病理系，山东省农业微生物重点实验室，泰安 271018）

摘 要：水稻条纹叶枯病，俗称水稻癌症，在生产中造成的损失严重。引起该病的病原是水稻条纹病毒。水稻条纹病毒（*Rice stripe tenuivirus*，RSV），是一种负单链 RNA 病毒（-ssRNA），为纤细病毒属的模式种，自然界中主要通过灰飞虱传播。RSV 基因组由 4 条 RNA 链组成，R1 链负义编码 RdRp 蛋白，其他 3 条链采用双义编码策略编码 6 个蛋白，病毒链与病毒互补链分别产生共 5′末端的亚基因组。笔者针对亚基因组的表达调控展开研究。

本研究通过 RACE 技术，完成 RSV 亚基因组 NSvc2-sgRNA、NS3-sgRNA、CP-sgRNA 末端定位。通过萤火虫荧光素酶（Fluc）载体的体外翻译分析，Fluc 读数结果显示单独亚基因组 CP-5U 的存在对翻译起正调控作用，单独亚基因组 CP-3U 的存在对翻译起抑制作用，亚基因组的 5U 和 3U 同时存在对翻译有更大的提高作用。表明 5U 与 3U 可能存在互作对翻译起正调控作用。在外翻译系统中以 VCR3 和 cp-sgRNA 为模板在加帽和不加帽的情况下分别来表达 CP 蛋白，通过同位素^{35}S 放射自显影的方法检测蛋白表达量，实验结果表明 RSV 共 5′末端亚基因组能够提高病毒 CP 蛋白表达量，暗示了共 5′末端亚基因组的产生的生物学意义及对病毒蛋白表达的必要性。

关键词：水稻条纹病毒；亚基因组；翻译

* 基金项目：国家自然科学基金（31872638，31670147）；山东省“双一流”奖补资金资助经费（SYL2017XTTD11）

** 第一作者：刘志菲，在读硕士研究生，主要从事分子植物病毒学研究；E-mail：wliuzhifei@163.com

*** 通信作者：原雪峰，教授，博士生导师，主要从事分子植物病毒学研究；E-mail：snowpeak77@163.com

番茄斑萎病毒（TSWV）亚基因组非翻译区的研究*

杨　晨**，窦宝存，于成明，原雪峰***

（山东农业大学植物保护学院植物病理系，山东省农业微生物重点实验室，泰安　271018）

摘　要：番茄斑萎病毒病的病原番茄斑萎病毒（*Tomato spotted wilt virus*，TSWV）属于布尼亚病毒科（Bnuyaviridae）番茄斑萎病毒属（*Tospovirus*），该病毒寄主植物广泛，主要侵染番茄、烟草、花生、辣椒、大豆等蔬菜作物和数以千计的观赏植物，其基因组包含有 3 条 RNA 链，分别为 L（8.9kb），M（4.8kb），S（2.9kb），共编码 5 个蛋白，其中 M 和 S 均为双义 RNA，分别编码两个非重叠的蛋白。

本研究主要通过 RACE 技术和体外翻译体系研究 TSWV 亚基因组的 5′和 3′末端序列及其对翻译调控的作用。通过 RACE 实验分别定位了 4 条亚基因组的 5′和 3′末端序列，同时 5′RACE 发现在 TSWV 中存在 4 种不同长度的 5′UTR 序列。通过报告基因萤火虫荧光酶素（Fluc）的体外翻译系统发现单独的 5′UTR 对翻译调控有增强作用，并且 4 种不同类型的 5′UTR 对增强翻译能力是有差别的；单独的 3′UTR 对翻译调控有抑制作用；当 5′UTR 和 3′UTR 同时存在时，对翻译调控作用明显增强并且远远强于单独 5′UTR 对翻译的提高能力。

本研究表明了亚基因组产生的必要性，同时暗示了通过亚基因组表达蛋白的高效性。

关键词：番茄斑萎病毒；亚基因组；RACE；蛋白翻译

---

* 基金项目：国家自然科学基金资助项目（31670147；31370179）；山东省“双一流”建设资助经费（SYL2017XTTD11）

** 第一作者：杨晨，硕士，主要从事分子植物病毒学研究；E-mail：yang1993_chen@yeah.net

*** 通信作者：原雪峰，教授，博士生导师，主要从事分子植物病毒学研究；E-mail：snowpeak77@163.com

烟草丛顶病毒-1位移码机制调控元件的研究[*]

于成明[**]，耿国伟，刘珊珊，原雪峰[***]
（山东农业大学植物保护学院植物病理系，山东省农业微生物重点实验室，泰安 271018）

摘　要：烟草丛顶病毒（*Tobacco bushy top virus*，TBTV）属于番茄丛矮病毒科幽影病毒属。TBTV 基因组是一条（+）ssRNA，全长由 4 152个核苷酸组成，有 4 个开放阅读框（ORF），编码 4 个蛋白，5′端缺乏帽子结构（m7GPPPN），3′末端也不带 poly（A）尾。其中 TBTV 的 RNA 依赖的 RNA 聚合酶（RdRp）是 ORF2 通过-1 位移码机制翻译表达的以 ORF1/ORF2 融合蛋白的形式存在。

本研究主要对参与调控-1 位移码的 4 个层次的调控元件展开研究，确定并发现了影响移码的关键性元件。通过序列和结构预测对比分析确定了 TBTV 中影响-1 位移码的七核苷酸滑动序列以及七核苷酸滑动序列下游的二级茎环结构；通过突变以及 Western Blot 确定了影响 TBTV-1 位移码的七核苷酸滑动序列为 946—952 位的 GGATTTT，该序列与下游二级颈环结构之间的距离为 6~9nt；通过 EMSA 实验发现移码区与全基因组的 3′末端 200nt 处存在两处远距离互作，并且这两处远距离互做均参与调控移码；通过 In-line probing 和 SHAPE 技术分别对移码区及发生远距离互作的 3′末端区域进行了结构解析，进一步确定了发生远距离互作的互作区域为移码区第一个颈环的侧环以及第三个颈环的顶环与全基因组 3′末端的 Pr 环发生互作；同时通过 SHAPE 技术和 EMSA 以及 Western 实验发现在 TBTV 移码区有一处假节点结构的存在，该结构对于移码作用发挥着至关重要的作用。本研究通过对 TBTV 移码现象的研究有多个重大发现：同时发现 4 个层次的移码调控元件；首次发现假节点结构存在并对移码效率的调控有重要作用；首次发现参与移码的两处远距离互做。本研究成功的揭示了 TBTV RdRp 表达机制的调控作用，对于烟草病毒病的防控有一定的指导作用。

关键词：烟草丛顶病毒；移码；调控元件；RNA 结构

---

* 基金项目：国家自然科学基金（31872638，31670147）；山东省“双一流”奖补资金资助经费（SYL2017XTTD11）
** 第一作者：于成明，在读博士生，主要从事分子植物病毒学研究；E-mail：ycm2006. apple@ 163. com
*** 通信作者：原雪峰，教授，博士生导师，主要从事分子植物病毒学研究；E-mail：snowpeak77@ 163. com

基于黄瓜花叶病毒（CMV）多联弱毒疫苗的制备*

张雅雯**，刘珊珊，于成明，原雪峰***
（山东农业大学植物保护学院植物病理系，山东省农业微生物重点实验室，泰安 271018）

摘　要：黄瓜花叶病毒（*Cucumber mosaic virus*，CMV）是雀麦花叶病毒科（Bromoviridae）黄瓜花叶病毒属（*Cucumovirus*）的典型成员，为三组分 RNA 病毒，基因组由 RNA1、RNA2、RNA3 三条正义单链 RNA 组成。CMV 寄主范围广泛，分离物多，可人为获得组合型分离物，为容纳更多的外源序列提供可能性。

本研究以 CMV Fny 株系 RNA2 为基础，选取 2b 蛋白编码框作为弱毒突变体的改造位点进行改造，构建了 2b 蛋白提前终止型突变。在 CMV RNA2 中插入大小不同的其他病毒片段，构建携带其他病毒片段的 2b 蛋白缺陷型突变体。将该弱毒突变体通过农杆菌注射接种 6 叶期本氏烟，3d 后注射接种 Fny 强毒株系来测试弱毒突变体对 CMV 的防治效果和稳定性。初步实验结果显示该弱毒突变体能够明显减轻 CMV 造成的花叶皱缩等症状，表明该弱毒突变体对 CMV 强毒有一定程度的防治效果。下一步将进一步构建携带多种病毒片段的 2b 蛋白缺陷型突变体，从弱毒疫苗的稳定性以及应用广泛性进行研究，从而实现多联弱毒疫苗的制备及强毒可调型弱毒疫苗的应用。

关键词：黄瓜花叶病毒；弱毒疫苗；防治

* 基金项目：国家自然科学基金（31872638，31670147）；山东省“双一流”奖补资金资助经费（SYL2017XTTD11）
** 第一作者：张雅雯，在读硕士生，主要从事分子植物病毒学研究
*** 通信作者：原雪峰，教授，博士生导师，主要从事分子植物病毒学研究；E-mail：snowpeak77@163.com

建立樱桃中快速高效检测樱桃病毒 A 的重组酶聚合酶扩增技术体系*

陈 玲**，段续伟，张开春***，张晓明，王 晶，闫国华，周 宇
（北京市林业果树科学研究院，农业农村部华北地区园艺作物生物学与种质创制重点实验室，北京市落叶果树工程技术研究中心，北京 100093）

摘　要：樱桃病毒 A（*Cherry virus* A，CVA）是 β 线形病毒科（Betaflexiviridae）发型病毒属（*Capillovirus*）成员，分布较广，在我国已广泛传播。CVA 寄主范围广泛，常见于甜樱桃和酸樱桃，前期研究发现 CVA 通常与侵染樱桃的一至多种病毒存在复合侵染。针对 CVA 的高发性，本研究从 NCBI 上收集了位于不同地理位置，不同进化组的 24 个 CVA 全序列，根据 CVA 外壳蛋白保守区域设计特异性引物（F：5′-GATGAAGATATTGTCAATCCTCCAA CTGTG-3′，R：5′-Biotin-CTGTCTCTTGCTTCTG GCATGATACCTTCT-3′）和探针（Probe：5′-FAM-TGGCACTCATT-GACGTCAGTGTTGACCAGAG-THF-TTCAGGAAAGGTGGTTTC-PO_4-3′），建立了在樱桃中检测 CVA 的等温 DNA 扩增技术，即重组酶聚合酶扩增（recombinase polymerase amplification，RPA）技术。利用 TwistAmp™nfo 试剂盒建立的 RPA 技术在恒温 38℃的条件下反应 40min 完成，经 PCR 产物纯化试剂盒回收后电泳，正向和反向引物扩增获得 138 bp 的扩增子，TA 克隆后测序正确，加探针后获得 110 bp 的扩增子。灵敏性试验表明该 RPA 技术的灵敏性是 PCR 技术的 10 倍，并且与国内侵染樱桃的其他 6 种病毒原（PNRSV、PDV、CGRMV、CNRMV、PBNSPaV、LChV2）和 1 种类病毒（HSVd）无交叉反应。此外，获得的具有标记的扩增子可通过侧向流试纸条检测进行直观观察，便于 CVA 检测的高通量应用。田间样品检测进一步证实了该方法的实用性。以上结果表明本研究开发的 CVA RPA 检测技术具有简单、快速、灵敏及特异等特点，能用于田间任何生长季的樱桃样品是否被 CVA 侵染的常规检测。

关键词：樱桃病毒 A；重组酶聚合酶扩增；樱桃；检测

---

* 基金项目：北京市博士后基金；北京市农林科学院博士后基金

** 第一作者：陈玲；E-mail：chenlingvip1988@ 126. com
*** 通信作者：张开春；E-mail：kaichunzhang@ 126. com

北京地区樱桃病毒原检测初报*

陈 玲**，段续伟，张晓明，闫国华，王 晶，周 宇，张开春***
（北京市林业果树科学研究院，农业农村部华北地区园艺作物生物学与种质创制重点实验室，北京市落叶果树工程技术研究中心，北京 100093）

摘 要： 樱桃病毒感染在现阶段樱桃生产中普遍存在，其主要通过嫁接相互传播，造成树势衰弱，降低果实产量和品质。明确樱桃病毒原种类，对于建立樱桃病毒检测体系及樱桃植株无病毒化技术，获得具有优良性状的甜樱桃品种和砧木的无病毒原种以及樱桃病毒病的防控具有重要的理论意义和应用价值。

为了明确北京地区现有樱桃种质资源中的病毒原种类，从海淀区北京市林业果树科学研究院樱桃资源圃以及通州区樱桃基地共采集 100 棵砧木盆栽苗、24 棵采穗圃砧木、94 棵甜樱桃品种的样品，通过 RT-PCR 和小 RNA 测序的方法进行樱桃病毒原的检测鉴定。结果在被检测樱桃种质资源中共发现了 9 种病毒和 1 种类病毒。包括李矮缩病毒（*Prunus dwarf virus*，PDV）、李属坏死环斑病毒（*Prunus necrotic ring spot virus*，PNRSV）、樱桃绿环斑驳病毒（*Cherry green ring mottle virus*，CGRMV）、樱桃坏死锈斑驳病毒（*Cherry necrotic rusty mottle virus*，CNRMV）、樱桃病毒 A（*Cherry virus* A，CVA）、黄瓜花叶病毒（*Cucumber mosaic virus*，CMV）、李树皮坏死茎痘伴随病毒（*Plum bark necrosis stem pitting-associated virus*，PBNSPaV）、樱桃小果病毒-1（*Little cherry virus*-1，LChV-1）以及啤酒花矮化类病毒（*Hop stunt viroid*，HSVd）。RT-PCR 检测以上病毒均获得特异性条带，测序后比对，序列正确。此外，在 3 棵甜樱桃品种上发现樱桃扭叶伴随病毒（*Cherry twisted leaf associated virus*，CTLaV），设计特异性鉴定引物（F：5′-TCAGCAAGATTAAG-GAGGTTG-3′，R：5′-ATNGGTTGAATTTGGCCAGT-3′）进行 RT-PCR 检测，检测结果显示特异的 CTLaV 条带，测序验证无误。CTLaV 在国内尚未报道，下一步计划获得其全序列进行进一步分析。

关键词： 樱桃病毒原；RT-PCR；小 RNA 测序；检测；鉴定

---

* 基金项目：北京市博士后基金；北京市农林科学院博士后基金

** 第一作者：陈玲；E-mail：chenlingvip1988@126.com

*** 通信作者：张开春；E-mail：kaichunzhang@126.com

内蒙古自治区马铃薯病毒病的检测

魏　瑶[*]，图门白拉，胡　俊，赵明敏[**]
（内蒙古农业大学园艺与植物保护学院，植物病毒研究室，呼和浩特　010000）

摘　要：马铃薯病毒病是我国马铃薯作物常见病害之一，造成大量减产，严重时可减产70%~80%。马铃薯病毒病是引起马铃薯退化的根本原因。目前，仍不清楚内蒙古自治区马铃薯病毒病是由哪些病毒引起的。因此，马铃薯病毒病的检测对保证马铃薯质量和产量具有重要意义。课题组利用RT-PCR技术在内蒙古自治区的马铃薯主要种植产区进行了采样和几种病毒病的检测研究。研究表明，27份样品中，11份样品为复合侵染，16份为单独侵染。其中，6份样品为马铃薯Y病毒（*Potato virus* Y，PVY）和马铃薯卷叶病毒（*Potato Leaf Roll virus*，PLRV）的复合侵染；2份样品为PVY和马铃薯M病毒（*Potato virus* M，PVM）的复合侵染；1份样品为PVY、PLRV和马铃薯S病毒（*Potato virus* S，PVS）的复合侵染；1份样品为PVY、PLRV和PVM的复合侵染；1份样品为PVY、PLRV、PVM和马铃薯A病毒（*Potato virus* A，PVA）的复合侵染。16份单独侵染的样品中均检测到PVY。27份样品中未检测到马铃薯X病毒（*Potato virus* X，PVX）。由此可见，无论复合侵染还是单独侵染均检测到了PVY，说明PVY仍然是引起当地马铃薯病毒病的主要病原之一。

关键词：马铃薯病毒病；反转录-聚合酶链式反应；植物病毒检测

---

* 第一作者：魏瑶，硕士研究生，主要从事植物病毒相关研究；E-mail：weiyao2018@163.com
** 通信作者：赵明敏，教授，研究方向为分子植物病毒学，主要从事植物病毒致病分子机制与绿色防控技术研究；E-mail：mingminzh@163.com

基于小 RNA 深度测序技术的吉林省马铃薯病毒鉴定与分析*

王永志[1]，马俊丰[1]，李小宇[1]，苏　颖[1]，张春雨[1]，王忠伟[1]，
李　闯[1]，徐　飞[2]，张胜利[2]，韩忠才[2]，刘婷婷[1]，周雪平[3**]
（1. 吉林省农业科学院，公主岭　136100；2. 吉林省蔬菜花卉科学研究院，
长春　130031；3. 中国农业科学院植物保护研究所，北京　100193）

摘　要：本研究对吉林省马铃薯病毒进行首次普查，确定了吉林省马铃薯病毒名录，为吉林省马铃薯病毒防控工作提供理论依据。2017 年在吉林省马铃薯主要产区，采集带有典型病毒症状的马铃薯样品，每份样品拿出一部分，组成混合样品，使用小 RNA 深度测序技术对混合样品进行检测，再根据检测结果对混合样品进行 RT-PCR 验证，最后进行部分病毒的生物信息学分析。结果显示，在 106 份样品中检测出马铃薯 Y 病毒（*Potato virus* Y，PVY）、马铃薯 S 病毒（*Potato virus* S，PVS）、马铃薯 M 病毒（*Potato virus* M，PVM）、马铃薯 H 病毒（*Potato virus* H，PVH）、马铃薯 X 病毒（*Potato virus* X，PVX）、马铃薯卷叶病毒（*Potato leafroll virus*，PLRV）共 6 种病毒和马铃薯纺锤块茎类病毒（*Potato spindle tuber viroid*，PSTVd），其中 PVH、PVX 为吉林省首次发现。通过全基因序列扩增与分析，明确了 PVH 吉林分离物的基因结构；通过 PVX CP 基因扩增与分析，确定吉林省 PVX 病毒的发生与种薯调运有关；通过 PVY VPg 基因扩增与分析，发现吉林省 PVY 具有株系多样。研究表明，近些年随着吉林省马铃薯种植面积的扩大，马铃薯病毒病的发生也逐年增多，因此完善大田防控措施，加强种薯原产地的检疫，严控无毒种薯的选择与培育流程，是解决当下问题的有效之策。

关键词：小 RNA 深度测序；马铃薯病毒；系统发育

* 项目基金：科技部国家重点研发项目（2017YFD0201604）；吉林省科学技术厅重点科技研发项目（20180201013NY）

** 通信作者：周雪平，博士，教授，主要从事植物病理学研究；E-mail：zzhou@ zju. edu. cn

重组酶聚合酶扩增技术检测番茄黄化曲叶病毒*

周　莹**，刘　梅，乔广行，黄金宝

（北京市农林科学院植物保护环境保护研究所，北京　100097）

摘　要：番茄黄化曲叶病毒（*Tomato yellow leaf curl virus*，TYLCV）是番茄上发生很普遍的双生病毒之一，隶属于双生病毒科（Geminiviridae）菜豆金色花叶病毒属（*Begomovirus*），主要通过烟粉虱（*Bemisia tabaci*）传播，可以侵染茄科、豆科等多种植物，基因组只含单组分的DNA-A，大小约2.8 kb，共编码6个开放阅读框（ORF）。随着全球气候变暖和耕作制度的改变以及国际贸易的增加，番茄黄化曲叶病毒（TYLCV）在全球范围传播。该病毒引起的为害与侵染番茄植株的早晚直接相关，植株越早感染造成的损失越大，因此做好病毒检测是病害防控工作中的一项重要环节。

等温核酸体外扩增是20世纪90年代出现的一种新型技术，RPA（Recombinase polymerase amplifcafion）是近几年发展起来最接近等温的扩增技术。RPA技术在25~42℃范围内的等温条件下，重组酶与引物DNA紧密结合，形成酶和引物的聚合体，当引物在模板DNA上搜索到与之完全互补的序列时，在单链DNA结合蛋白SSB的作用下，使模板DNA解链，并在DNA聚合酶的作用下，形成新的DNA互补链。RPA产物与普通PCR的产物一样，扩增出来的是单一产物。目前RPA技术在医学病原物的快速诊断中得到了一些应用，在农业领域中主要用于转基因作物的检测，植物病原物的检测方面有若干报道。国内尚未有RPA技术在番茄病毒检测方面的报道。

本研究依据TYLCV DNA-A基因保守序列，设计了8对RPA引物并进行了筛选，获得1对用于RPA检测的特异性扩增引物，建立了TYLCV-RPA检测方法，并分析了其特异性和灵敏度。该方法在37℃下反应40min即可达到高于普通PCR 10倍的扩增结果，目的条带为343 bp大小的特异性条带；仪器要求简单，仅需要一个恒温装置即可满足扩增要求，而且方法特异性好，与番茄其他常见病毒无交叉反应。研究表明，该方法适合基层单位和现场检测使用，为TYLCV的病害调查、田间诊断以及无毒苗木的生产提供简单高效的技术方法。

关键词：番茄黄化曲叶病毒；重组酶聚合酶扩增；等温扩增；检测方法

---

* 基金项目：北京市农林科学院青年科研基金（QNJJ1201705）

** 通信作者：周莹，高级农艺师；E-mail：zhouying16_ 2013@ 163. com

Function analysis of vacuolar protein sorting *vps*26、*vps*29 and *vps*35 in *Aspergillus flavus*

Wang Sen[1], Wang Yu[1], Liu Yinghang[1,2], Wang Shihua[1] *

(1. *Key Laboratory of Pathogenic Fungi and Mycotoxins of Fujian Province*, *Key Laboratory of Biopesticide and Chemical Biology of Education Ministry*, *and School of Life Sciences*, *Fujian Agriculture and Forestry University*, *Fuzhou* 350002, *China*; 2. *State Key Laboratory of Microbial Technology*, *Shandong University*, *Jinan* 250100, *China*)

Abstract: In eucaryon, the Retromer complex are responsible for retrograding transport of receptors from endosomes to the trans-Golgi network, which play vital roles in regulation of growth and development process. Here, we study the functions of the subunits of Retromer complex, Cargo - Selective subunit in fungal development, pathopoiesity and aflatoxin biosynthesis in *A. flavus*. We identified Cargo-Selective subunit that is a retromer including *Vps*26, *Vps*29 and *Vps*35 via bioinformatics method. These disruption mutants exhibited dramatic reduction in hyphal growth, conidiation, sclerotia production, and also displayed remarkable reduction in aflatoxin production. The stress experiments revealed that the *vps* genes deletion mutants were more sensitive to the stress factors when compared to the wild-type. We also found that *vps* gene deleted strains were reduced in the ability of infecting crop seeds, while the aflatoxin yield was increased when infecting peanuts. Further study on proteins' subcellular location showed that *Vps*26, *Vps*29 and *Vps*35 were all compartmentalized punctuated structures near vacuole. The interaction pattern among *Vps*26, *Vps*29 and *Vps*35 were also performed by Yeast two-hybrid technology.

In conclusion, *Vps*26, *Vps*29 and *Vps*35 included in theRetromer complex all take part in the process of growth and development, pathopoiesis and aflatoxin biosyntheses in *A. flavus*. This study is useful for understanding the function of the Retromer complex, and would provide valuable imformation for the controling of *A. flavus*.

Key words: *Aspergillus flavus*; Retromer complex; Vacuolar protein sorting; Aflatoxin

* Corresponding author: Wang Shihua; E-mail: wshyyl@ sina. com. cn

稻瘟病菌核转运蛋白 PoKap123 调控生长发育、致病性和氧化胁迫

曹雪琦[1]，张丽梅[1]，齐　敏[1,2]，梁　楠[1]，郑华坤[1]，王宗华[1,3*]，鲁国东[1*]
（1. 福建农林大学闽台作物有害生物生态防控国家重点实验，福州　350002；2. 福建农林大学生命科学学院，福州　350002；3. 闽江大学海洋研究院，福州　350108）

摘　要：核孔转运蛋白（Karyopherins，Kaps）通过介导大分子物质的核质穿梭过程，在许多细胞生物学过程中均具有重要作用。Kap123 在酵母和动物细胞中可以介导组蛋白 H3 和 H4 的入核过程，但是在稻瘟病菌（*Pyricularia oryzae*；syn. *Magnaporthe oryzae*）等丝状真菌中其同源蛋白功能仍不清楚。为了探究稻瘟病菌中 Kap 蛋白的功能，我们通过同源比对鉴定了一个稻瘟病菌 Kap123 同源蛋白的编码基因，并构建了该基因的敲除突变体进行表型分析。氨基酸序列比对结果表明，稻瘟病菌 PoKap123 与粗糙脉孢菌同源蛋白亲缘关系较近，而与酵母和人类同源蛋白的亲缘关系较远。表型分析结果表明，*PoKAP*123 基因的敲除导致稻瘟病菌丝生长速率、产孢量和致病性显著下降，对氧化胁迫的敏感性显著增强。综上所述，PoKap123 可能通过介导组蛋白等大分子物质的入核过程，调控稻瘟病菌的生长发育、胁迫响应和致病性。

关键词：稻瘟病菌；核孔转运蛋白；Kap123；致病性；生长发育

* 通信作者：王宗华，博士，研究员，主要从事真菌功能基因组学研究；E-mail：zonghuaw@163. com
鲁国东，博士，研究员，主要从事真菌功能基因组学研究；E-mail：gdlufafu@163. com

Endoplasmic reticulum (ER) membrane protein MoScs2 is required for asexual development and pathogenesis of *Magnaporthe oryzae*

Chen Xuehang [1], Wang Min [1], Zheng Qiaojia[1], Tang Wei [1*], Wang Zonghua [1,2*]

(1. *State Key Laboratory of Ecological Pest Control for Fujian and Taiwan Crops*, *College of Plant Protection*, *Fujian Agriculture and Forestry University*, *Fuzhou* 350002, *China*
2. *Institue of Ocean Science*, *Minjiang University*, *Fuzhou* 350108, *China*)

Abstract: Rice blast is one of the most destructive diseases, caused by the hemibiotrophic filamentous ascomycete *Magnaporthe oryzae* (syn. *Pyricularia oryzae*), seriously threatened rice production. Previous studies revealed that ER membrane protein Scs2 involved in regulate phospholipid metabolism, connect ER to plasma membrane (PM), and control PI4P levels in *Saccharomyces cerevisiae*. However, the molecular mechanism and physiological roles in *M. oryzae* are still unclear. In this study, we identified and characterized the Scs2 orthologous protein MoScs2 by generating its knock-out mutants in the rice blast fungus. Similarly, MoScs2 was localized to the ER. Compared to the wild type Guy11, deletion of *MoSCS2* resulted in significant defects in hyphae growth, sporulation and pathogenesis. The conidium morphology of the Δ*Moscs2* mutant is abnormal. Additionally, the Δ*Moscs2* mutant is defective in appressorium development. Furthermore, MoScs2 is involved in the dithiothreitol-induced or tunicamycin-induced ER stress response. Briefly, these results showed that the ER membrane protein MoScs2 is necessary for vegetative growth, asexual development, pathogenesis, and ER stress response of *M. oryzae*, but further investigation of MoScs2 and its molecular mechanism about pathogenesis is warranted.

Key words: *Magnaporthe oryzae*; ER; MoScs2; pathogenesis

* Corresponding authors: Tang Wei; tangw@fafu.edu.cn
Wang Zonghua; wangzh@fafu.edu.cn

The arabidopsis CtBP/BARS homolog angustifolia as a crucial negative regulators in pattern-triggered immunity

Gao Xiuqin*, Lai Wenyu, Tian Shifu, He Dou,
Guo Cuiting, Li Yalin, Peng Changlin, WangAirong**
(*College of Plant Protection, Fujian Agriculture and Forestry University, Fuzhou* 350002, *China*)

Abstract: Plasma membrane-localized pattern recognition receptors (PRRs) such as FLAGELLIN SENSING2 (FLS2) recognize microbe-associated molecular patterns (MAMPs) to activate immune signaling pathway, this response was called pattern-triggered immunity (PTI). The ANGUSTIFOLIA (AN) gene, encodes a homolog of the animal C-terminal binding proteins (CtBPs) in Arabidopsis. In contrast to animal CtBPs, AN seems unable to function as a transcriptional co-repressor and instead functions outside nucleus where it might be involved in Golgi-associated membrane trafficking and the formation of stress granules (SGs). AN T-DNA mutant can cope better with abiotic stress and are more resistant to bacterial infection, correlating with increased accumulation of reactive oxygen species (ROS). In this study, we report AN as a critical PTI negative regulator in *Arabidopsis thaliana*. Knock-out (KO) mutants of AN not only were remarkable resistance to *Pst* DC3000 *and Psm* ES4326, but significantly insensitive to necrotrophic fungal, such as *Sclerotinia sclerotiorum* and *Botrytis cinerea*. Accordingly, loss-of-function mutations *an-t*1 and *an-t*2 showed a distinct PTI responses, including increase ROS accumulation and callose deposition upon flg22 and chitin treatment. Moreover, Arabidopsis lines overexpressing AN-OE were hypersusceptible to bacteria and fungi and showed a defective PTI response, including the reduce of ROS accumulation and callose deposition triggered by MAMP. This work reveals AN as a novel negative regulatory for PTI activation when involve in biotic stress.

Key words: Angustifolia; *Arabidopsis thaliana*; PTI; ROS; Biotic stress

* First author: Gao Xiuqin, Master Degree Candidate; E-mail: 1107710370@qq.com
** Corresponding author: Wang Airong; E-mail: airongw@126.com

禾谷镰刀菌中假定 GTP 酶激活蛋白 FgMsb3 的功能研究*

李玲萍[1]，苗鹏飞[2]，余 芝[2,3]，林 梅[2]，郑华伟[1,3**]，王宗华[1,3**]
（1. 福建农林大学植物保护学院，福州 350002；2. 福建农林大学生命科学学院，福州 350002；3. 闽江学院海洋研究院，福州 350108）

摘 要：禾谷镰刀菌（*Fusarium graminearum*）引起的小麦赤霉病（Wheat scab）是世界小麦生产上的主要病害。前期研究表明，禾谷镰刀菌 Rab GTP 酶介导的囊泡运输对禾谷镰刀菌的生长发育及侵染致病起着非常重要的作用，然而关于禾谷镰刀菌 Rab GTP 酶的调控机制尚不明确，对其调控机制展开研究对小麦赤霉病的防治具有非常重要的意义。GTP 酶激活蛋白（GAPs）作为 Rab GTP 酶的负调控因子，在哺乳动物及酵母中，已经鉴定的 Rab GTP 酶的 GAPs 均含有保守的 TBC（Tre2/Bub2/Cdc16）结构域。为研究 Rab GTP 酶调控网络，从 SMART 数据库中鉴定了禾谷镰刀菌所有含有 TBC 结构域的 12 个假定 GAPs，对这个蛋白家族的基因进行敲除和表型分析。初步结果显示，其中一个基因 *FgMSB3*（FGSG_04033）缺失导致突变体生长变慢，对小麦麦穗的致病力基本丧失，表明该基因对禾谷镰刀菌的生长发育及对宿主的侵染至关重要。同时我们在遗传上证实了 FgMsb3 可作为 FgRab8 的 GAP 发挥功能。此外，笔者发现 FgMsb3 主要定位于菌丝顶端，能够和顶体（Spitzenkörper，SPK）Marker（FM4-64）部分共定位。后续将进一步对 *FgMSB3* 基因缺失突变体和野生型 PH-1 的差异分泌蛋白质组及 FgMsb3 的互作蛋白质组展开研究，解析 FgMsb3-FgRab GTP 酶可能通过极性胞吐或分泌来调控致病性机制。

关键词：小麦赤霉病；致病机理；禾谷镰刀菌；FgMsb3；Rab GTP 酶激活蛋白

* 基金项目：国家自然科学基金青年基金（31701742）
** 通信作者：郑华伟；王宗华

稻瘟病菌中与NDR激酶Dbf2相互作用的Momob1蛋白的功能研究*

刘 丹，张 君，赵倩倩，李明阳，李昀溪，王宗华**
（福建农林大学生物农药与化学生物学教育部重点实验室，福州 350002）

摘 要：蛋白激酶Dbf2属于NDR（Nuclear Dbf2-Related）激酶家族，现有研究表明，真核生物中的NDR激酶序列具有高度保守性，并调控细胞的有丝分裂、细胞生长和发育等过程。真菌的胞质分裂主要包括3个阶段：胞质分裂起始位点的选择；调节蛋白的有序激活和定位；肌动球蛋白环的收缩。该过程需要一个非常保守的信号网络的激活，该信号网络在芽殖酵母中称为有丝分裂退出网络（Mitotic Exit Network，MEN），Dbf2和Mob1是信号通路的一部分。Dbf2与Mob1会形成复合体，并且Mob1为Dbf2的共激活因子。本实验对稻瘟病菌中*Mob*1基因的功能进行了初步的探索，基于反向遗传学的策略，通过基因敲除技术得到*Mob*1缺失突变体，再通过对该突变体表型变化的观察来推断该基因的功能。

表型分析发现，与野生型Ku80相比，*Momob*1的缺失使得菌落的生长速率受到明显抑制、分生孢子的形态发生严重畸形、分生孢子及隔膜的数量明显减少、附着胞的形成明显延迟、致病菌的侵染能力显著下降。亚细胞定位表明Momob1定位于菌丝的隔膜处，这与Modbf2在菌丝中的亚细胞定位相同。以上结果表明*Momob*1对稻瘟病菌的发育和致病侵染过程至关重要。综上所述，稻瘟病菌中*Mob*1基因可与*Dbf*2基因相互作用，调控稻瘟病菌的产孢，隔膜和附着胞的形成以及胞质分裂等相关进程，其具体调控机制有待深入研究。

关键词：稻瘟病菌；Dbf2；Mob1；胞质分裂；隔膜

---

* 基金项目：国家青年自然科学基金（31770156）

** 通信作者：王宗华，研究员，博士，主要从事植物与微生物相互作用分子生物学；E-mail：wangzh@ fafu. edu. cn

稻瘟病菌蛋白激酶 CK2a 磷酸化水平相关的磷酸酶基因的功能分析

谢雨漫，张连虎，张　甜，蔡　燕，王　倩，王宗华，张冬梅
（福建农林大学闽台作物有害生物生态防控国家重点实验室，福州　350002）

摘　要：在真核细胞中，蛋白激酶控制的磷酸化作用及蛋白磷酸酶控制的去磷酸化作用共同调控着蛋白质的磷酸化水平，进而调控细胞增殖与分化、胚胎发育、信号转导和胞内转运等生理过程。目前，蛋白激酶和蛋白磷酸酶的研究成为了真核功能基因研究的热点之一。已有的稻瘟病菌蛋白激酶和蛋白磷酸酶功能基因研究的结果表明，它们在稻瘟病菌的生长发育、隔膜的形成、分生孢子的形成和对宿主的致病性方面发挥着重要作用。本研究通过对稻瘟病菌丝/苏氨酸蛋白激酶 CK2 催化亚基 MoCKa pull-down 实验获得的一系列蛋白进行整理分析，以获得的丝/苏氨酸蛋白磷酸酶 MoPpz1、MoSki2、MoPph22 和 MoPtc5 为对象进行功能研究和分析。首先通过酵母双杂实验对 MoPpz1、MoPph22、MoPtc5 与 MoCk2 之间的关系进行了研究。酵母双杂实验结果发现 MoPpz1、MoPph22 和 MoPtc5 与 MoCKa 没有直接互作，它们可能是通过 MoCK2 全酶或其他组分蛋白或某个接头蛋白与 MoCKa 间接互作。后续酵母双杂实验验证了 MoPpz1、MoPph22 和 MoPtc5 与 MoCK2 两个调节亚基（MoCKb1 和 MoCKb2）没有直接互作。同时，对 MoPpz1、MoSki2 和 MoPph22 的功能研究发现，MoPpz1 均定位在细胞质中；MoSki2 在稻瘟病菌的生长发育、分生孢子产生和侵染致病等生理过程中发挥重要作用。

关键词：稻瘟病菌；丝/苏氨酸蛋白磷酸酶；基因敲除；功能分析

Roles of SNARE protein MoSnc1 in effector secretion of the rice blast fungus

Zhang Jin, Yang Piao, Wu Huiming, Fang Wenqin, Zheng Wenhui*

(*State Key Laboratory of Ecological pest Control for Fujian and Taiwan Crops, College of Plant Protection, Fujian Agriculture and Forestry University, Fuzhou* 350002, *China*)

Abstract: SNAREs (soluble N-ethylmaleimide-sensitive factor attachment protein receptors) belong to a superfamily of transmembrane proteins locating on organelles and vesicular membranes. Vesicular anchoring and fusion mediated by SNAREs are the main ways how materials and information are exchanged between organelles in endocytic secretory pathways of eukaryotic cells. *Saccharomyces cerevisiae* involves two v-SNAREs (Snc1, Snc2), but only one v-SNARE exists in the rice blast fungus *Magnaporthe oryzae* (named MoSnc1) and has not been systematically characterized so far. In this study, we generated *MoSNC*1 gene knockout mutant in the *M. oryzae*. Phenotypic analyses revealed that the fungal growth, conidiation and pathogenicity of *ΔMosnc*1 all decreased to some extents compared with that of the wild type Guy11. Additionally, we observed that MoSnc1-GFP localized in a bright fluorescent punctum close to the tips of growing and invasive hyphae. We expressed the plasmid vector containing both of cytoplasmic effector Pwl2-mCherry and apoplastic effector Bas4-GFP in the *ΔMosnc*1 and checked their localizations using a high-resolution laser confocal microscopy. We found that in *ΔMosnc*1, Bas4-GFP was not completely located within the extra-invasive hyphal membrane (EIHM), but partially mistakenly located into the vacuoles, and Pwl2-mCherry showed several biotrophic interfacial complex (BICs) -like localizations, suggesting that MoSnc1 is required for correct localizations of cytoplasmic and apoplastic effectors in *M. oryzae*. In the future, the mechanism of how SNARE proteins function in secretion of effector proteins of *M. oryzae* will be further studied, so as to pursue a clearer understanding of effector's translocation in the rice blast fungus.

Key words: *Magnaporthe oryzae*; MoSnc1; SNARE; effector

* Corresponding author: Zheng Wenhui

Genome data of *Fusarium oxysporum* f. sp. *cubense* race 1 and tropical race 4 isolates using long-read sequencing

Chen Shasha, Yun Yingzi, Song Aixia, Lu Songmao,
Wang Zonghua, Zhang Liangsheng
(*State Key Laboratory of Ecological Pest Control for Fujian and Taiwan Crops*,
FujianAgriculture and Forestry University, *Fuzhou* 350002, *China*)

Abstract: Fusarium wilt of banana is caused by the soil-borne fungal pathogen *Fusarium oxysporum*f. sp. *cubense* (*Foc*) and is one of the most catastrophic plant diseases in the world. *Foc* race 1 (*Foc*1) nearly destroyed global banana trade in the 19th century before the development of the *Foc*1-resistant cultivar Cavendish. *Foc* tropical race 4 (*Foc*TR4) is a new serious threat to banana cultivation due to its strong pathogenicity affecting nearly all banana cultivars including the Cavendish. To study the differences in their genomes and pathogenicity, we generated two chromosome-level assemblies of *Foc*1 and *Foc*TR4 strains using single-molecule real-time sequencing technology. The *Foc*1 and *Foc*TR4 assemblies had 35 and 29 contigswith contig N50 lengths of 2. 08 Mb and 4. 28 Mb, respectively, and the contigs were further anchored and oriented onto 11 chromosomes. These two new references genomes represent a greater than 100-fold improvement over the contig N50 statistics of the previous short read-based *Foc* assemblies. Orthology groups and phylogenetic analyses revealed that *Foc*TR4 had more specific genes and a longer branch length compared to the analyzed *Foc*1 strains. Transposable element (TE) analysis revealed that the *Foc*TR4 genome had many more TEs (7. 09% of the assembly) than the *Foc*1 genome (4. 22% of the assembly). We also found a closer relationship between the TE distribution and the location of putative effector genes in *Foc*TR4 than in *Foc*1. Compared to *Foc*1 strains, the findings indicate that the faster evolution of *Foc*TR4 and the greater number of TEs in the *Foc*TR4 genome may have contributed to its higher pathogenicity. The two high-quality assemblies reported here will be a valuable resource for the comparative analysis of *Foc*races at the pathogenic, ecological, and evolutionary levels.

Key words: banana; *Foc*; single-molecule real-time sequencing (SMRT)

A *Barley stripe mosaic virus*-based guide RNA delivery system for targeted mutagenesis in wheat and maize*

Hu Jiacheng[1], Li Shaoya[2], Li Zhaolei[1], Li Huiyuan[2], Song Weibin[3], Zhao Haiming[3], Lai Jinsheng[3], Xia Lanqin[2], Li Dawei[1], Zhang Yongliang[1]**

(1. *State Key Laboratory of Agro-Biotechnology and Ministry of Agriculture Key Laboratory of Soil Microbiology*, *College of Biological Sciences*, *China Agricultural University*, *Beijing* 100193, *China*; 2. *Institute of Crop Sciences*, *Chinese Academy of Agricultural Sciences*, *Beijing* 100081, *China*; 3. *State Key Laboratory of Agrobiotechnology and National Maize Improvement Center*, *Department of Plant Genetics and Breeding*, *China Agricultural University*, *Beijing* 100193, *China*)

Abstract: Plant RNA virus-based gRNA delivery has substantial advantages compared to that of the conventional constitutive promoter-driven expression due to the rapid and robust amplification of gRNAs during virus replication and movement. Although several plant viruses have be developed to deliver gRNAs for targeted mutagenesis in plants, to the best of our knowledge, there are currently no virus-induced genome editing (VIGE) tools for economically important crops like wheat and maize. In this study, we have engineered a *Barley stripe mosaic virus* (BSMV) -based gRNA delivery system for CRISPR/Cas9-mediated targeted mutagenesis in wheat and maize. BSMV-based delivery of single guide (g) RNAs for targeted mutagenesis was first validated in *Nicotiana benthamiana*. To extend this work, we transformed wheat and maize with the Cas9 nuclease gene, and selected the wheat *TaGASR*7 and maize *ZmTMS*5 genes as targets to assess the feasibility and efficiency of BSMV-mediated mutagenesis. Positive targeted mutagenesis of the *TaGASR*7 and *ZmTMS*5 genes was achieved with for wheat and maize with efficiencies of up to 78% and 48%. In addition to wheat and maize, BSMV strains are able to infect barley (*Hordeum vulgare* L.), *Brachypodium distachyon*, oat (Pacak *et al.*, 2010), millet (*Setaria italic*), sorghum (*Sorghum bicolor*) and several other cereals. Our proof-of-concept study of BSMV-based VIGE in *N. benthamiana*, wheat and maize provides the basis for expanded use in other economically important monocot plants.

---

* Funding: This work is supported by National Key R & D Program of China (2016YFD0100502) and the Transgenic Research Program of China (2016ZX08010-001)

** Corresponding author: Zhang Yongliang, Ph. D & Associate Professor; E-mail: cauzhangyl@ cau. edu. cn

A Barley stripe mosaic virus-based guide RNA delivery system for targeted mutagenesis in wheat and maize

第四部分 细 菌

Integrated transcriptomics and secretomics approaches reveal critical pathogenicity factors in *Pseudofabraea citricarpa* inciting citrus target spot*

Yang Yuheng[1**], Fang Anfei[1], Yu Yang[1], Bi Chaowei[1], Zhou Changyong[2]

(1. *College of Plant Protection*, *Southwest University*, *Chongqing* 400715, *China*;
2. *Citrus Research Institute*, *Southwest University*, *Chongqing* 400712, *China*)

Abstract: Target spot is a newly emerging citrus disease caused by *Pseudofabraea citricarpa*. Outbreaks of this disease result in massive economic losses to citrus production. Here, an integrated study involving comparative transcriptomic and secretomic analyses was conducted to determine the critical pathogenicity factors of *P. citricarpa* involved in the induction of citrus target spot. A total of 701 transcripts and their cognate proteins were quantified and integrated. Among these transcripts and proteins, 99 exhibited the same expression patterns. Our quantitative integrated multi-omics data highlight several potentially pivotal pathogenicity factors, including 16 unigenes that were annotated as plant cell-wall-degrading enzymes, 13 unigenes homologous to virulence factors from various fungi, and one unigene described as a small cysteine-rich secreted protein, were screened and analyzed. The screening of differentially expressed genes that encode secondary metabolism core enzymes implicated terpene metabolism in the pathogenicity of *P. citricarpa*. Overall, results indicated that plant cell wall degradation, plant-pathogen protein/polyribonucleotide interaction, and terpene biosynthesis have critical roles in the pathogenicity of *P. citricarpa*. This work demonstrated that integrated-omic approaches enable the identification of pathogenicity/virulence factors and provide insights into the mechanisms underlying the pathogenicity of fungi. These insights would aid the development of effective disease management strategies.

Key words: *Pseudofabraea citricarpa*; RNA sequencing; secretory proteome; integrated analysis; pathogenicity factors

* 基金项目：国家重点研发计划（2018YFD0200500）；重庆市基础科学与前沿技术研究专项（cstc2016jcyjA0316）；重庆市博士后科研项目特别资助（Xm2016124）

** 第一作者：杨宇衡，主要从事植物病理学研究；E-mail：yyh023@swu.edu.cn

Petroleum ether fraction of *Polygonum orientale* seeds acts as an antimicrobial agent against *Clavibacter michiganensis* subsp. *michiganensis* by damaging the cell membrane*

Cai Jin[1**], Shi Xiaojing[2], Gao Yichen[3], Du Beibei[3]

(1. *Institute of Applied Chemistry*, *Shanxi University*, *Taiyuan* 030006, *China*; 2. *Xinzhou Teachers University*, *Xinzhou* 034000, *China*; 3. *School of Life Science*, *Shanxi University*, *Taiyuan* 030006, *China*)

Abstract: Tomato is an important crop. Bacterial canker of tomato disease, caused by *Clavibacter michiganensis* subsp. *michiganensis* (Cmm), is known to cause significant economic losses. This study was to investigate the antimicrobial activity of petroleum ether fraction from *Polygonum orientale* fruits (PPF) against Cmm. Active compounds of PPF were analyzed by gas chromatography-mass spectrometry (GC-MS). Twenty-one active compounds in PPF were reported in our work, and the major compounds were (Z, Z) -9, 12-octadecadienoic acid (54.55%), oleic acid (17.66%), linoleic acid ethyl ester (8.85%), n-hexadecanoic acid (5.83%), ethyl oleate (5.66%), hexadecanoic acid, ethyl ester (2.3%), and octadecanoic acid (1.53%). Scanning electron microscopy (SEM) and transmission electron microscopy (TEM) analyses showed that the PPF treatment resulted in damaged cell envelopes and vacuole formation. An Annexin V-FITC/PI double-staining test confirmed the disruptive action of PPF on cell membrane permeability. Cmm cells treated with PPF showed decreased membrane potential, and a significant accumulation of reactive oxygen species (ROS), which damaged cell membrane integrity. Together, these results indicate that cell membrane of Cmm is the primary target of PPF antimicrobial action, and PPF could be used as a promising alternative to chemicals for inhibiting Cmm.

Key words: antimicrobial activity; *Polygonum orientale*; *Clavibacter michiganensis* subsp. *michiganensis*; active compounds; antimicrobial mechanism

* 基金项目：国家自然科学基金青年基金项目（31601677）；山西省应用基础研究青年基金面上项目（201701D221179）；山西省高等学校科技创新项目（2016117）

** 第一作者及通信作者：蔡瑾，副教授，主要研究方向为植物病害生物防治；E-mail：caijin@sxu.edu.cn

Improved primers for the specific detection of *Leifsonia xyli* subsp. *xyli* in sugarcane using a conventional PCR assay*

Sun Shengren[1**], Chen Junlü[1,2], Duan Yaoyao[1], Chu Na[1], Huang Meiting[1], Fu Huaying[1], Gao Sanji[1***]

(1. *National Engineering Research Center for Sugarcane*, *Fujian Agriculture and Forestry University*, *Fuzhou*, *Fujian* 350002, *China*; 2. *Guangdong Provincial Bioengineering Institute* (*Guangzhou Sugarcane Industry Research Institute*), *Guangzhou*, *Guangdong* 510316, *China*)

Abstract: Ratoon stunting disease (RSD), one of the most important diseases of sugarcane, is caused by the bacterium *Leifsonia xyli* subsp. *xyli* (*Lxx*). *Lxx* infects sugarcane worldwide and RSD results in high yield losses and varietal degeneration. It is highly challenging to diagnose RSD based on visual symptomatology because this disease does not exhibit distinct external and internal symptoms. In this study, a novel *Lxx*-specific primer pair *Lxx*-F1/*Lxx*-R1 was designed to detect this pathogen using a conventional PCR assay. These primers were then compared with four published *Lxx*-specific primers and one universal *Leifsonia* generic primer pair *Lay*F/*Lay*R. Sugarcane leaf samples were collected from *Saccharum* spp. hybrids in commercial fields (315 samples) and from wild germplasm clones of five *Saccharum* species and *Erianthus arundinaceus* (216 samples). These samples were used for comparative field diagnosis with six conventional PCR assays. Sensitivity tests suggested that the PCR assay with primers *Lxx*-F1/*Lxx*-R1 had the same detection limit (1 pg of *Lxx* genomic DNA) as the primer pairs *Cxx*1/*Cxx*2 and *Cxx*ITSf#5/*Cxx*ITSr#5 and had 10-fold higher sensitivity than the primer pairs Pat1-F2/Pat1-R2, *Lay*F/*Lay*R, and C2F/C2R. Comparison of PCR assays revealed that natural *Lxx*-infection incidence (6.1%) in field sample evaluation identified by *Lxx*-F1/*Lxx*-R1 primers was higher than incidences (0.7% - 3.0%) determined by other primer pairs. Moreover, no non-specific DNA amplification occurred within these field samples with *Lxx*-F1/*Lxx*-R1 primers, unlike with the primer pairs *Cxx*1/*Cxx*2 and *Lay*F/*Lay*R. Diverse *Leifsonia* strains were identified by PCR detection with *Lay*F/*Lay*R primers in the field samples, whereas whether these *Leifsonia* strains were pathogenic to sugarcane requires further research. Our investigations revealed that the PCR assay with the newly designed primers *Lxx*-F1/*Lxx*-R1 could be widely used for RSD diagnosis and *Lxx*-pathogen detection with satisfactory sensitivity and specificity.

Key words: *Leifsonia xyli* subsp. *Xyli*; Polymerase chain reaction (PCR) Molecular detection; Ratoon stunting disease (RSD); Sugarcane

---

* 基金项目：国家现代农业产业技术体系（糖料）建设专项（CARS-170302）

** 第一作者：孙生仁，博士研究生，主要从事甘蔗病害检测与抗病分子育种；E-mail：ssr03@163.com

*** 通信作者：高三基，研究员，主要从事甘蔗病害检测与抗病分子育种；E-mail：gaosanji@yahoo.com

Molecular identification of *Xanthomonas albilineans* infecting elephant grass (*Pennisetum purpureum*) in China*

Meng Jianyu[1**], Mbuya sylvain Ntambo[1,2], Luo Linmei[1,2], Huang Meiting[1], Fu Huaying[1], Gao Sanji[1***]

(1. *National Engineering Research Center for Sugarcane, Fujian Agriculture and Forestry University, Fuzhou, Fujian* 350002, *China*; 2. *College of Crop Science, Fujian Agriculture and Forestry University, Fuzhou, Fujian* 350002, *China*)

Abstract: Elephant grass (*Pennisetum purpureum*) is an important fodder crop worldwide. Numerous plant pathogens have been found to infect this grass, which plays a major role as a reservoir for important pathogens of other grass crops including sugarcane. In this study, we collected 36 leaf samples and 108 juice samples from six symptomatic plants and 30 asymptomatic plants of elephant grass in Fuzhou, China for molecular detection of *Xanthomonas albilineans*, the putative causal agent of scald disease in grasses. *X. albilineans* was detected by PCR with specific primers XAF1/XAR1 in 25.0% (9/36) of the leaf and 5.6% (6/108) of the juice samples, while 55.6% (20/36) of the leaf and 81.5% (88/108) of the juice samples tested positive using a real-time quantitative PCR (qPCR) assay. PCR-amplified DNA fragments of two housekeeping genes, *abc* and *gyrB*, were cloned and sequenced from 15 positive samples, and the nucleotide sequences were 100% identical for both genes. Based on the concatenated sequences of the two genes (1 512 nucleotides), phylogenetic analysis revealed that the *X. albilineans* strains infecting elephant grass clustered in the PFGE-B group with six reference strains from sugarcane that originated from Guadeloupe and Martinique (France), Florida (USA), and Fuzhou (China); All nucleotide sequences in this group were 100% identical. Additionally, two strains of *X. albilineans* isolated from elephant grass were identified as being highly pathogenic in young plants of elephant grass and sugarcane by artificial inoculation. The findings presented here attempt to address a gap in the literature, particularly with regard to molecular identification of *X. albilineans* naturally infecting elephant grass.

Key words: Molecular detection; phylogenetic analysis; *Pennisetum purpureum*; *Xanthomonas albilineans*

---

* 基金项目：国家现代农业产业技术体系（糖料）建设专项（CARS-170302）

** 第一作者：孟建玉，硕士研究生，主要从事甘蔗病害检测与抗病分子育种；E-mail：jian1yu2@163.com

*** 通信作者：高三基，研究员，主要从事甘蔗病害检测与抗病分子育种；E-mail：gaosanji@yahoo.com

A convenient gene deletion toolbox for genetic analysis in *Ralstonia solanacearum*

Yan Jinli*, Zhang Lianhui

(*Guangdong Province Key Laboratory of MicrobialSignals and Disease control*, *Integrative Microbiology Research Centre*, *South China Agricultural University*, *Guangzhou* 510642, *China*)

Abstract: *Ralstonia solanacearum*, a soilborne bacterial pathogen, is the causal agent of bacterial wilt and infects over 200 plant species in 50 families. The pathogen is lethal to many solanaceous species, including eggplant, tobacco and peanut. The pathogen is notorious for its genetic variations, and the regulatory mechanism that governs the pathogenicity of *R. solanacearum* is not well-understood. In this study, we set to develop a convenient and reliable delivery method for stable and directed deletion of gene or DNA sequence in the genome of *R. solanacearum*. The method is modified from the suicide vector pRC, which was developed previously for gene knock-in at a permissive site downstream of *glmS* in the genome of *R. solanacearum* strain GMI1000 through homologous recombination. The modified gene deletion system consists of a fusion DNA fragment containing the upstream and downstream region of a target gene and a gentamycin resistance gene as a selection marker, as well as a growth medium which primes *R. solanacearum* to a state prone to receiving DNA fragments for integration. The feasibility of this gene deletion system was tested by targeting a regulatory gene in *R. solanacearum* EP1, which is a virulent strain isolated from the diseased eggplant in Guangdong Province of China. The results showed that the resultant mutants displayed much reduced EPS production. Development of such a convenient gene deletion system would facilitate further investigations of the pathogenic mechanisms of this important bacterial pathogen.

Key words: *Ralstonia solanacearum*; gene deletion; toolbox; genetic analysis

* Corresponding author: Yan Jinli; E-mail: 2417617672@qq.com

The VfmIH two-component system modulates multiple virulence traits in *Dickeya zeae*

Lv Mingfa[1], Hu Ming[1], Li Peng[2], Jiang Zide[1], Zhou Jianuan[1], Zhang Lianhui[1*]

(1. *Guangdong Province Key Laboratory of Microbial Signals and Disease Control*, *Integrative Microbiology Research Centre*, *South China Agricultural University*, *Guangzhou* 510642, *China*; 2. *Ministry of Education Key Laboratory for Ecology of Tropical Islands*, *College of Life Sciences*, *Hainan Normal University*, *Haikou* 571158, *China*)

Abstract: Pathogenic bacteria *Dickeya zeae* strain EC1 produces a family of antibiotics-like phytotoxins named zeamines, which are encoded by the *zms* gene cluster and critical for bacterial virulence. In this study, we identified a zeamine deficient mutant with Tn5 inserted in a gene encoding a two-component system (TCS) sensor histidine kinase (HK). Bioinformatics analysis showed that this kinase gene is the homologue of the *vfmI* gene of *D. dadantii* strain 3 937. Similar to the case of *D. dadantii* 3 937, a homologue of *vfmH* encoding a response regulator (RR) is located at the vicinity of *vfmI* and sharing the same operon. Deletion in-frame of either *vfmI* or *vfmH* resulted in significantly reduced production of zeamines and cell wall degrading enzymes (CWDEs), and mitigated the bacterial virulence on rice seeds and potato tubers. RNA-seq and Q-RT PCR analysis of *D. zeae* EC1 and the mutant of VfmH showed that the TCS positively modulated the transcriptional expression of a range of virulence genes, including *zms*, T1SS, T2SS, T6SS, flagellar and CWDEs genes. In *D. dadantii* 3937, VfmH protein was shown to interact with the promoters of *vfmA* and *vfmE* to affect the transcription. This study found that the VfmH protein not merely could directly interact with the promoters of *vfmA* and *vfmE* but also interact with the promoters of *vfmF* to affect the transport of Vfm signals.

Key words: *Dickeya zeae*; VfmIH two-component system; virulence traits

---

* Corresponding author: Zhang Lianhui; E-mail: lhzhang01@ scau. edu. cn

Stress resistance of VBNC cells in *Clavibacter michiganensis* subsp. *michiganensis*[*]

Chen Xing[**], Xu Xiaoli, Bai Kaihong, Jiang Na, Li Jianqiang, Luo Laixin[***]

(*Department of Plant Pathology, China Agricultural University/ Beijing Key Laboratory of Seed Disease Testing and Control, Beijing* 100193, *China*)

Abstract: Bacterial canker of tomato is an important seedborne bacterial disease caused by the Gram-positive bacterium *Clavibacter michiganensis* subsp. *michiganensis* (Cmm). It was indicated that Cu^{2+} and low pH can induce Cmm enter a viable but non-culturable (VBNC) state in previous studies. Most researchers believed that VBNC state is a common survival strategy when bacteria are exposed to harsh environment stresses. The VBNC Cmm cells, which induced by 50 μmol/L Cu^{2+}, was tested the characteristics of resistance to stress conditions, and the culturable bacterial cells in log-phase were used as a control. The stress conditions included heat challenge (55℃ for 30 min), oxidative stress (H_2O_2 at a final concentration of 50 mmol/L) and antibiotic challenge (streptomycin and penicillin at a final concentration of 50 μg/mL). The results showed that the population of culturable log-phase Cmm cells decreased from $\log_{10}$ CFU/mL of 7.0 to 0.8, 1.6, 3.3 and 4.1 when treated by heat challenge, oxidative stress (50 mmol/L of H_2O_2 treatment) and antibiotic challenge (50 μg/mL of streptomycin or penicillin), respectively. However, there were no significant decline of VBNC Cmm cells when exposed to the same stresses, just decreased from $\log_{10}$ CFU/mL of 7.0 to 5.0, 6.5, 4.5 and 5.5, respectively. We concluded that the VBNC Cmm cells significantly enhanced its tolerance to heat, H_2O_2, streptomycin and penicillin than log-phase culturable Cmm cells. Further studies will be focused on the resistance mechanism of VBNC cells and the functions of related genes in Cmm.

Key words: Cmm; VBNC; resistance

* Funding: This research was supported by the National Natural Science Foundation of China (No. 31571972)

** First author: Chen Xing, PhD student; E-mail: chenxing2028@163.com

*** Corresponding author: Luo Laixin; E-mail: luolaixin@cau.edu.cn

Resuscitating *Acidovorax citrulli* cells from VBNC state is different from the un-induced cells according to the proteomic analysis*

Kan Yumin, Lv Qingyang, Jiang Na, Li Jianqiang, Luo Laixin**

(*Department of Plant Pathology, China Agricultural University / Beijing Key Laboratory of Seed Disease Testing and Control, Beijing* 100193, *China*)

Abstract: *Acidovorax citrulli* is the causal agent of bacterial fruit blotch (BFB), which is a serious threat to cucurbits production worldwide. In a previous study, we confirmed that *A. citrulli* can enter into a viable but nonculturable (VBNC) state following exposure to copper sulfate and copper-induced VBNC cells can be resuscitated by several strategies including addition of EDTA solution. Moreover, the resuscitated cells had an equivalent virulence to watermelon seedlings with the un-induced cells in log phase, thus making resuscitated *A. citrulli* cells a great threat to the cucurbits production. A good knowledge of mechanism in *A. citrulli* VBNC state resuscitation is quite important.

In this study, we compared the protein profiling of resuscitated cells in four stages with the untreated cells in log phase using isobaric tags for relative and absolute quantification (iTRAQ) approach. A total of 459 proteins were identified differentially expressed between four stages of resuscitated cells and un-induced log-phase cells, with 127 upregulated and 332 proteins downregulated. Expression levels of 8 in 11 selected genes tested by qPCR (Aave_0233, Aave_0312, Aave_0415, Aave_0846, Aave_0854, Aave_1832, Aave_1897, Aave_3650, Aave_3692, Aave_4073 and Aave_4272) were consistent with the protein levels by iTRAQ analysis. Many differentially expressed proteins were associated to the general metabolism. While in the resuscitating process, the results showed that energy production played an important role, as well as the stress response, especially the oxidative stress response proteins. We identified five upregulated proteins involved in type Ⅵ secreted system and type Ⅳ pili, which were showed associated to virulence of *A. citrulli*, thus indicating the great risk of resuscitated cells to the cucurbits production.

Key words: *Acidorvorax citrulli*; Bio-PCR; primer; specificity; sensitivity

* Funding: This research was supported by the National Key Research and Development Program of China (Grant No. 2017YFD0201601)

** Corresponding author: Luo Laixin; E-mail: luolaixin@cau.edu.cn

Effect of L-Proline on Colonization Characteristics of *Bacillus subtilis* Strain NCD-2*

Zhao Weisong**, Dong Lihong, Wang Peipei, Guo Qinggang,
Zhang Xiaoyun, Su Zhenhe, Lu Xiuyun, Li Shezeng, Ma Ping***
(*Institute of Plant Protection, Hebei Academy of Agriculture and Forestry Sciences, Integrated Pest Management Center of Hebei Province, Key Laboratory of IPM on Crops in Northern Region of North China, Ministry of Agriculture, Baoding* 071000, *China*)

Abstract: *Bacillus subtilis* NCD-2 was isolated from cotton rhizosphere soil and screened as a biocontrol agent against Verticillium wilt in cotton. Previous studies have shown that the biocontrol efficiency of strain NCD-2 was related to its root colonization, and there was a significant difference in the ability of strain NCD-2 to colonization rhizosphere of different cotton varieties. Biofilm formation plays an very important role in root colonization and Verticillium wilt control by *B. subtilis* strain NCD-2. Small-molecular-weight substances, such as organic acids, sugars, amino acids, and secondary metabolites were produced by plant roots. Microorganisms can use root exudates as carbon or energy substances for growth and development. In addition to serving as nutrient, certain substances in root exudates can attract bacteria to colonize in plant rhizosphere and promote biofilm formation, or act as signal molecules to change the expression of specific genes in microorganism. In order to clarify the effect of L-proline on the colonization characteristics of strain NCD-2, the effects of L-proline on the growth, biofilm formation, swimming and swamming of strain NCD-2 were analyzed. The results showed that the three-dimensional structure of biofilm is more complex, the degree of folds is enhanced, but also the ability of strain NCD-2 in biofilm formation is increased in the presence of L-proline. In order to gain new insights into molecular mechanisms of L-proline in strain NCD-2, transcriptone sequencing of L-proline-treated and non-treated strain NCD-2 was performed via Illumina Hiseq technology. Differentially-expressed genes (DEGs) were then sorted using GO terms and KEGG pathways analyses. As a result, 602 genes were up-regulated and 469 genes were down-regulated in a total 1071 DEGs. Some DEGs were selected randomly for further analyses using quantitative real-time PCR (qRT-PCR). The study will be explained the colonization difference of strain NCD-2 in different cotton varieties, and will be provided a reference for genetic improvement and scientific application of strain NCD-2.

Key words: L-proilne; *Bacillus subtilis*; biofilm formation; colonization

* Funding: This work was funded by the earmarked fund for National Natural Science Foundation (31801786) and Financial Special Program of Hebei Academy of Agriculture and Forestry Sciences (201820101)

** First author: Zhao Weisong; E-mail: zhaoweisong1985@163.com

*** Corresponding authors: Ma Ping; E-mail: pingma88@126.com

Three Monoclonal Antibodies of Membrane Proteins for Detecting Citrus HLB

Wu Xiaoyan

(*Integrative Microbiology Research Centre*, *South China Agricultural University*, *Guangzhou* 510642, *China*)

Abstract: Huanglongbing (HLB), caused by the pathogen *Candidatus* Liberibacter asiaticus (*C*Las), is one of the most destructive disease of citrus. Transmitted by the Asian citrus psyllid, *C*Las is the phloem-residing bacteria and could not be cultured under *in vitro* conditions. Current integrated control methods for HLB depend on early detection and eradication of HLB-infected citrus plants. To identify infected citrus and to interdict the spread of disease timely are the key points to limit HLB. Currently, methods for detecting HLB include electron microscopy, PCR, and qPCR, but these conventional detection methods are time consuming, expensive, and inconvenient for farmers. With an aim to develop a convenient and rapid hands-on machine to detect HLB, we set to generate HLB-specific monoclonal antibody. Through bioinformatics analysis, we selected three genes encoding outer membrane proteins as potential targets. The proteins encoded by these genes were purified by affinity chromatography, and designated as Q1G, Q2G and Q3G, respectively. The anti-Q1G, anti-Q2G and anti-Q3G monoclonal antibodies will be prepared soon.

Key words: huanglongbing; membrane proteins; monoclonal antibody

不同环境温度下烤烟漂浮育苗营养液中的细菌多样性研究*

孟颢光[1**]，常 栋[2]，李 豪[1]，周 博[1]，张琳婧[1]，胡登辉[1]，
崔江宽[1***]，蒋士君[1***]
（1. 河南农业大学植物保护学院，郑州 450002；
2. 河南省烟草公司平顶山市公司，平顶山 467000）

摘 要：烤烟是我国重要的经济作物之一，培育无病壮苗是优质烟叶生产的重要基础之一。烟草漂浮育苗具有出苗齐、成苗快、抗病毒病等优点，应用漂浮育苗已经成为黄淮烟区有效控制烟草病毒病的关键技术之一。2010 年以来河南省烟草漂浮育苗的推广覆盖度达 100%。但由于苗床期 3—4 月气温多变，育苗大棚内的温湿度忽高忽低，烟苗根系健康问题日趋严重，移栽隐性感染的亚健康烟苗已成为大田根茎病害多发重发的重要诱因。为探究漂浮育苗根系隐性感染与环境温度的关系，本研究系统分析了四连体大棚中在 20℃、25℃、30℃、35℃和 40℃条件下，育苗营养液中细菌多样性的动态变化。提取的样品总 DNA，用带测序接头的 16S rDNA 保守区引物进行 PCR 扩增引物，并对扩增产物进行纯化、定量和均一化形成测序文库，质检合格的文库用 Illumina HiSeq 2500 进行测序、测序拼接、过滤，共产生1 589 868条 Clean tags，每个样品至少产生52 119条 Clean tags，平均产生66 245条 Clean tags。将 OTU 的代表序列与微生物参考数据库进行比对共鉴定到细菌物种 102 纲、185 目、292 科、182 属，主要包括：黄单胞菌属、假单胞杆菌属、土壤杆菌、鞘脂单胞菌、酸杆菌和芽单胞菌等多种致病性细菌。研究还利用 QIIME 软件分析了不同分类水平上的物种丰度表，利用 R 语言工具绘制各分类学水平下的群落结构图。研究结果对于探明烟苗亚健康的发生机制及其控制方法，提供了直接的科学依据。

关键词：漂浮育苗；烟草；细菌；隐性感染

* 基金项目：河南省烟草公司平顶山市公司科技项目（PYKJ201803）；河南农业大学高层次人才引进项目（30500663）
** 第一作者：孟颢光，博士，讲师，主要从事烟草病害防控技术研究；E-mail：menghaoguang@ henau. edu. cn
*** 通信作者：崔江宽，博士，讲师，主要从事烟草土传病病害防控技术研究；E-mail：jiangk. cui@ henau. edu. cn
蒋士君，硕士，副教授，主要从事烟草病害防控技术研究；E-mail：Jiangsj001@ 163. com

甜瓜自毒物质降解细菌的筛选及发酵条件优化研究*

唐爽爽**，张照然，马周杰，何世道，高增贵***
（沈阳农业大学植物保护学院，沈阳 110866）

摘 要：甜瓜的自毒物质是导致连作障碍的重要原因之一，其中酚酸类物质是甜瓜自毒物质的主要成分，对甜瓜植株的生长和发育均具有抑制作用，降解自毒物质是缓解甜瓜自毒作用行之有效的方法之一。

本试验通过外源添加自毒物质的筛选培养基对甜瓜连作土壤细菌和甜瓜内生细菌进行初筛选，分离得到234株细菌。通过复筛及HPLC降解率验证，得到9株可稳定生长于筛选培养基中并有明显降解效果的菌株。9株菌株经过16SrRNA扩增产物测序，blast比对后最终划分为4种降解细菌，T58、T194为*Burkholderia cepacia*（洋葱伯克霍尔德菌）；T49、T147为*Burkholderia gladioli*（唐菖蒲伯克霍尔德菌）；T15、T31、T113、T182为*Burkholderia phytofirmans*；H16为*Pseudomonas oryzihabitans*（栖稻假单胞菌）。

筛选到的4种降解菌对自毒物质具有显著的降解作用，所有菌株对没食子酸20 d后的降解率均为100%，栖稻假单胞菌和唐菖蒲伯克霍尔德菌对阿魏酸和对香豆酸20 d后的降解率均为100%，*Burkholderia phytofirmans*对香兰素20 d后的降解率为100%，洋葱伯克霍尔德菌和唐菖蒲伯克霍尔德菌对芥子酸20 d后的降解率均为100%。

通过对4种降解细菌的最优发酵条件研究，以菌体生物量为指标，采用单因素方法对菌株的发酵条件进行优化。结果表明，洋葱伯克霍尔德菌在NA培养基中生长量最大，唐菖蒲伯克霍尔德菌在MD培养基中生长量最大，栖稻假单胞菌在PDA培养基中生长量最大，*Burkholderia phytofirmans*在LB培养基中生长量最大，4种降解细菌的最佳发酵条件为温度37 ℃，pH值=6.0，装液量50/250 mL，接种量为10%，转速为180 r/min。为进一步研究其降解机理及研发其生防制剂提供了理论基础。

关键词：甜瓜；自毒物质；降解细菌；筛选；HPLC；降解作用；发酵条件

---

* 基金项目：公益性行业（农业）科研专项经费资助项目（201503110-04）

** 第一作者：唐爽爽，博士研究生，主要从事甜瓜病害研究

*** 通信作者：高增贵，研究员，博士生导师；E-mail：gaozenggui@ sina. com

枯草芽孢杆菌 L1-21 全基因组重测序研究*

李咏梅[1**]，Shahzad Munir[1]，何鹏飞[1,3]，吴毅歆[2,3]，何月秋[2,3***]

（1. 云南农业大学植物保护学院，昆明　650201；2. 云南农业大学农学与生物技术学院，昆明　650201；3. 微生物菌种筛选与应用国家地方联合工程研究中心，昆明　650217）

摘　要：枯草芽孢杆菌 L1-21 是一株从健康柑橘叶片分离的内生细菌，能够有效地防治柑橘黄龙病。此外，枯草芽孢杆菌 L1-21 可定殖于多种植物体内，并且对常见的植物病原细菌和真菌均表现出较强的抑菌活性，具有广泛应用前景。为进一步解析枯草芽孢杆菌 L1-21 防治柑橘黄龙病的潜在作用机理，笔者利用重测序方法比对了 L1-21 与生防枯草芽孢杆菌 XF-1 的基因组。用 Illumina 平台进行全基因组测序，使用 BWA 软件将 clean reads 比对到参考基因组上，bedtools 软件作覆盖度统计，GATK 软件检测并校正单核苷酸多态性位点（single nucleotide polymorphism，SNP）及插入缺失位点（InDel），ANNOVAR 软件对检测出的变异进行功能注释。结果表明：共获得 831.4 万条 reads，在参考基因组的覆盖率为 95.01%，完全匹配率为 84.35%；检测到45 831个 SNP，978 个 InDel；L1-21 菌株基因组内还含有大量的次生代谢产物合成相关基因，如嗜铁素（siderophore）、溶杆菌素（bacilysin）、杆菌烯（bacillaene）和大环内酯（macrolactin）等聚酮类化合物，以及表面活性素（surfactin）和芬荠素（fengycin）等脂肽类抗生素。本研究结果可为探究柑橘内生芽孢杆菌 L1-21 与寄主植物以及黄龙病菌的互作规律提供信息支持。

关键词：内生菌；枯草芽孢杆菌；柑橘黄龙病；重测序

---

* 基金项目：科技部重点研发计划项目（2018YFD0201500）

** 第一作者：李咏梅，在读博士生，主要从事植物病害生物防治相关研究；E-mail：kala.111@163.com

*** 通信作者：何月秋，从事功能微生物开发与植物病理学研究；E-mail：ynfh2007@163.com

南瓜青枯菌 RSCM 菌株 *RipG* 效应子家族基因克隆分析*

佘小漫[1,2**]，蓝国兵[1]，汤亚飞[1]，于　琳[1]，李正刚[1]，邓铭光[1]，何自福[1***]
（1. 广东省农业科学院植物保护研究所，广州　510640；
2. 广东省植物保护新技术重点实验室，广州　510640）

摘　要： 青枯菌 RipG 家族效应蛋白是一类含有富亮氨酸重复序列（leucine-rich repeat，LRR）的蛋白。青枯菌 GMI1000 菌株中 RipG 家族效应蛋白 7 个基因共同决定 GMI1000 的致病力，并且不同 RipG 基因互作影响 GMI1000 对不同寄主致病力；*RipG*2、*RipG*3、*RipG*6 和 *RipG*7 是决定寄主特异性的候选基因。南瓜青枯病是新发生病害，RSCM 菌株除了能侵染南瓜外，还能侵染番茄、茄子、辣椒及烟草等作物。本研究对菌株 RSCM 的 *RipG* 家族效应子基因进行克隆分析，结果表明，菌株 RSCM 基因组缺少 *RipG*1 基因，存在 6 个 *RipG* 效应子家族基因，分别是 *RipG*2、*RipG*3、*RipG*4、*RipG*5、*RipG*6 和 *RipG*7，其基因大小分别为 3 108 bp，1 851 bp，1 389 bp，1 674 bp，1 866 bp 和 1 941 bp。与 GMI1000 菌株的 *RipG* 效应子家族比对，基因序列同源性均在 99%以上。利用 SMART 工具对 RSCM 的 6 个 *RipG* 效应子家族蛋白结构预测，结果显示，*RipG*2、*RipG*3、*RipG*4 和 *RipG*5 与 GMI1000 菌株的 *RipG*2、*RipG*3、*RipG*4 和 *RipG*5 结构基本一致；RSCM 菌株与 GMI1000 菌株的 *RipG*6 和 *RipG*7 蛋白结构存在较大差异。南瓜青枯菌 RSCM 菌株与 GMI1000 菌株的 *RipG*6 和 *RipG*7 蛋白基因功能是否有差异还有待下一步研究。

关键词： 南瓜青枯病；*RipG* 效应子；基因克隆

---

* 基金项目：广东省自然科学基金项目（2018A030313566）

** 第一作者：佘小漫，研究员，博士，研究方向植物细菌；E-mail：lizer126@ 126. com
*** 通信作者：何自福，研究员；E-mail：hezf@ gdppri. com

枯草芽孢杆菌 Czk1 全基因组测序研究*

梁艳琼**，吴伟怀，谭施北，习金根，李　锐，郑金龙，
陆　英，黄兴，贺春萍***，易克贤***
（中国热带农业科学院环境与植物保护研究所，农业部热带农林有害生物入侵检测与控制重点开放实验室，海南省热带农业有害生物检测监控重点实验室，海口　571101）

摘　要：枯草芽孢杆菌（*Bacillus subtilis*）是一种能够产生很多抗菌谱广、可用于农业和药物的具有抗菌活性物质（包括聚酮化合物和脂肽类等次级代谢天然产物）的微生物。枯草芽孢杆菌 Czk1 是本实验室前期在橡胶树根木质部分离到的一株细菌，对橡胶树根病、炭疽病以及西瓜枯萎病等有良好的抑制作用，其可产生脂肽类抗生素对病原菌产生拮抗作用。为进一步研究枯草芽孢杆菌 Czk1 的代谢机理，深入探明其抑菌机制，进而为今后对该菌株进行改造提供理论基础，因此利用 Illumina 测序技术对 Czk1 菌株的基因组进行扫描测序，构建了 Illumina PE 文库，使用 ABySS 拼接软件和 GapCloser 软件组装，使用 GeneMarkS 软件进行编码基因预测，将预测基因的蛋白序列分别与 Nr、Genes、eggNOG 和 GO 数据库进行 blastp 比对，从而获得预测基因的注释信息。结果发现，Czk1 菌株基因总长度为3 648 858，获得4 251个蛋白序列，其中 Nr 数据库为4 078个，GO 数据库为1 651个，eggNOG 为3 399个，KEGG 为2 121个，注释功能涉及次级代谢产物的合成、转运和代谢，氨基酸的转运、合成和代谢，ATP 的产生与转换，信号传导机制，DNA 模板复制和转录调控等，涉及的代谢产物为 Bacillibactin、Fengycin、Bacillaene、Surfactin、Bacilysin、Subtilosin 等。通过在线软件 anti SMASH 对 Czk1 全基因组中的可能产生的次级代谢产物合成基因簇进行挖掘，结果共注释到 7 种，18 个次级代谢基因簇，注释到的基因簇有 7 个微菌素（Microcin）、5 个 Nrps（nonribosomal peptides）、2 个萜烯（Terpene）、2 个第三类聚酮类化合物（T3pks）、1 个 Transatpks-Nrps-Otherks、1 个 Sactipeptide 和一个尚不明确的基因簇。

关键词：枯草芽孢杆菌；Czk1；测序；基因

---

* 基金项目：国家天然橡胶产业技术体系建设项目（No. nycytx-34-GW2-4-3；CARS-34-GW8）；海南省科协青年科技英才学术创新计划项目（QCXM201714）

** 第一作者：梁艳琼，助理研究员，研究方向：植物病理；E-mail：yanqiongliang@ 126. com

*** 通信作者：贺春萍，研究员，研究方向：植物病理；E-mail：hechunppp@ 163. com
易克贤，研究员，研究方向：分子抗性育种；E-mail：yikexian@ 126. com

小麦内生细菌在不同器官和生长阶段的多样性及作为生物肥料潜力的研究*

庞发虎**，Camilo Ayra-Pardo，王 坦，余自伟，黄思良***
（南阳师范学院农业工程学院，南阳 473061）

摘 要：内生细菌在提高寄主植物的环境适应能力、抗逆性和产量方面发挥着不可或缺的作用。虽然从小麦或其他植物的特定部位分离内生细菌并用于生物防治有所研究，但目前关于内生细菌在不同植物器官和生长阶段中作为生物肥料的潜在分布和生物多样性还鲜有报道。本研究于2012—2013 年，在小麦的不同生长阶段，从小麦植株的不同器官（根、茎、叶和种子）中共分离出 127 株内生细菌。通过对菌株的形态、生化特征和 16S rDNA 序列分析，将其鉴定为 10 个属，14 个特定种和 8 个未确定种。香农多样性指数和辛普森多样性指数表明，根是内生细菌群落最为丰富的器官，其次是茎、叶和种子。随着小麦的生长发育，从分蘖期到种子灌浆期分离出的内生细菌属的数量不断增加，但有些属具有器官特异性，只能在小麦的特定生长阶段被分离出来。从分离频率来看，蜡样芽孢杆菌（*Bacillus cereus*）和枯草芽孢杆菌（*B. subtilis*）是优势种群。嗜麦芽窄食单胞菌（*Stenotrophomonas maltophilia*）和葡萄球菌（*Staphylococcus* sp.）被鉴定为小麦新的内生细菌。在促生方面，分别有 45%、29%和 37%的菌株产吲哚-3-乙酸和溶无机磷和有机磷。用内生细菌接种生长在贫瘠土壤中的小麦植株，与未接种的对照组相比，内生细菌可使小麦根系干重、地上部分干重和株高分别增加到 1. 9%~61. 3%（平均 27. 7%）、1. 6%~54. 2%（平均 28. 5%）和 0. 3%~54. 9%（平均 22. 3%）。研究结果揭示了内生细菌类群在小麦各器官和生长阶段中的分布的多样性，其中根系是主要的分布器官，是筛选内生细菌促进小麦生长的首选组织。

关键词：分离频率；种群多样性；促生；内生细菌

---

* 基金项目：河南省科技厅重大科技专项（182102110468）

** 第一作者：庞发虎，副教授，博士研究生，主要从事植物病害生物防治研究；E-mail：pangfahu@ 163. com

*** 通信作者：黄思良，教授，主要从事植物病害生物防治研究；E-mail：silianghuang@ 126. com

胡萝卜细菌性心腐病的病原菌鉴定*

郑雪玲[1**]，杨 迪[1,2]，孙静怡[1]，王 潞[1]，庞发虎[1]，陶爱丽[1]，王 坦[1]，黄思良[1***]
（1. 南阳师范学院农业工程学院，南阳 473061；
2. 广西农业科学院植物保护研究所，南宁 530007）

摘 要：2017 年 7 月，在河南省南阳市当地超市购买的胡萝卜（*Daucus carota* L. var. *sativa* Hoffm.）中发现一种呈中空腐烂的未知心腐病。发病胡萝卜初期外观症状不明显，后期外观颜色暗淡，切开发病胡萝卜的肉质根，可见病部呈水渍状，组织软腐，后期内部组织部分中空，肉质根中部组织病害扩展速度最快。挑取发病组织镜检，观察到明显的细菌性病害常见的“喷菌”现象。发病组织无传统的 *Erwinia carotovora* subsp. *carotovora* 引起的胡萝卜软腐病组织特有的恶臭味。取染病胡萝卜样品，在超净工作台中依次用 75%酒精浸泡 15s、0. 1%升汞表面消毒处理 3 min，经无菌水冲洗 5 次后，用灭菌解剖刀切开胡萝卜样品，取内部绿豆颗粒大小的染病组织，于无菌水中捣烂病组织，制成细菌悬浮液，将细菌悬浮液稀释 10×，原液和稀释液各 3 个重复涂布于 NA 平板上，将涂菌平板于 28℃恒温箱中培养 3 d 后分离获得形态一致的单菌落细菌，随机选取两株细菌（H1 和 SP-2），通过刺伤接种试验依据柯赫法则验证了它们对胡萝卜的致病性。对菌株 H1 和 SP-2 的生理生化特性进行了检测以及基于 16S rDNA 和 *gyr*B 基因序列的系统发育分析。结果显示：供试菌株为革兰氏阴性，具周身鞭毛，在牛肉膏蛋白胨琼脂（NA）平板上菌落为乳白色不透明状，光滑圆球形，凸起，边缘整齐，黏稠，表面湿润，易挑取，28℃培养 3 d 的单菌落直径达 4~5 mm；接触酶、甲基红、V-P、葡萄糖氧化、丙二酸盐利用、酯酶、20℃明胶水解、七叶灵、硫化氢、亚硝酸盐利用、脱氧核糖核酸酶、吲哚试验、糖发酵等为阳性反应，而氧化酶、淀粉水解、果胶酸盐利用、脲酶等为阴性反应。分子鉴定结果表明：菌株 H1 和 SP-2 的 16S rDNA 序列（GenBank 登录号分别为 MK177601 和 MH490936）与普利茅斯沙雷氏菌 *Serratia plymuthica*（GenBank 登录号 CP012097）的最大相似性（MS）分别达 99. 65%和 99. 7%，其 *gyr*B 序列（GenBank 登录号分别为 MK511449 和 MK511451）与 *S. plymuthica*（登录号 CP012097）的 MS 均达 100%，基于 16S rDNA 和 *gyr*B 序列的系统发育分析结果均一致支持菌株 H1 和 SP-2 属 *S. plymuthica*。综合形态学与生理生化特性，将引起胡萝卜心腐病的病原细菌鉴定为普利茅斯沙雷氏菌 *S. plymuthica*。为了解该菌的寄主范围，在室内用刺伤接种法就菌株 H1 和 SP-2 对不同蔬菜作物的致病力进行了测定。结果表明，供试菌株对白菜［*Brassica pekinensis*（Lour.）Rupr.］、甘薯［*Ipomoea batatas*（L.）Lam.］和萝卜（*Raphanus sativus* L.）有较强致病力，对青椒（*Capsicum annuum* L.）的致病力较弱，而对芹菜（*Apium graveolens* L.）、马铃薯（*Solanum tuberosum* L.）和山药（*Dioscorea oppositifolia* L.）无致病力。一般认为 *S. plymuthica* 是生活在植物根际的细菌，可产生具有较宽抗真菌谱的代谢产物，常作为生防菌被用于辣椒、黄瓜、草莓等多种作物真菌性病害的生物防治。此外，该菌不同菌株间有较大的遗传多样性，除了植物病害生防菌株外，有的可致洋葱球茎腐烂。由 *S. plymuthica* 引起的胡萝卜心腐病为首次报道。

关键词：胡萝卜；细菌性心腐病；普利茅斯沙雷氏菌；生理生化特性；寄主范围

---

* 基金项目：河南省高校科技创新团队支持计划项目（2010JRTSTHN012）
** 第一作者：郑雪玲，硕士，研究方向为植物病害生物防治；E-mail：1203017121@ qq. com
*** 通信作者：黄思良；E-mail：silianghuang@ 126. com

西瓜食酸菌调控因子 CueR 的生物信息学分析

颉兵兵[1]，刘　君[2]
（1. 新疆农业大学农学院，乌鲁木齐　830052；
2. 新疆农业大学林学与园艺学院，乌鲁木齐　830052）

摘　要：CueR 被证实在细菌中参与抗铜组分的转录调控，但在西瓜食酸菌（*Acidovorax citrulli*）中还未见报道。通过生物信息学手段对西瓜食酸菌及其他菌中的 CueR 蛋白进行类比分析，为进一步验证西瓜食酸菌 CueR 蛋白在铜稳态机制中的功能提供参考。本研究以已经鉴定的大肠杆菌等 4 个模式细菌中的 CueR 为参照，对西瓜食酸菌的 CueR（AcCueR）与大肠杆菌的 EcCueR、铜绿假单胞菌的 PaCueR、沙门氏菌的 SeCueR、霍乱弧菌的 VcCueR 蛋白进行一级结构、聚类分析、氨基酸理化性质、亚细胞定位、铜离子结合位点、保守结构域、二级及三级结构等特征分析。研究结果显示：5 种蛋白一级结构序列、铜离子结合位点、二级结构百分比、三级结构等相似但并不完全相同；西瓜食酸菌的 AcCueR 与霍乱弧菌的亲缘关系较近；5 种蛋白均为亲水性的、细胞内任何位置定位的、不稳定蛋白质，且 N 端均含有蛋氨酸残基；除 EcCueR 为酸性蛋白外，AcCueR 等其他蛋白均为碱性蛋白；5 种蛋白均属于 HTH_MerR-SF 超家族，三级结构主要由 α-螺旋和无规则线圈构成；AcCueR 可以与西瓜食酸菌中 CopA 产生互作，且在 *copA* 基因上游存在一个 *copA* 启动子序列。对西瓜食酸菌 CueR 蛋白的生物信息学分析表明：推测 AcCueR 蛋白与已鉴定的大肠杆菌等菌中的 CueR 具有相似的结构与功能，显示西瓜食酸菌中可能存在类 CueR 蛋白。

关键词：西瓜食酸菌；CueR 蛋白；生物信息学；类比分析

生防芽孢杆菌 Bs916 中 RNA 结合蛋白 Hfq 编码基因的突变及其功能初步研究*

乔俊卿[1**]，衡　阳[2]，刘永锋，陈志谊[1]，刘邮洲[1***]
（1. 江苏省农业科学院植物保护研究所，南京　210014；
2. 扬州大学园艺与植物保护学院，扬州　225009）

摘　要：由真菌引起的水稻纹枯病、稻瘟病和稻曲病对水稻的高产、稳产造成严重的威胁。笔者所在的江苏省农业科学院植物保护研究所水稻病害研究室，于20世纪90年代筛选获得一株对水稻纹枯病和稻曲病均具有较好防效的生防芽孢杆菌 Bs916，并成功创制为水稻纹枯病和稻曲病的生物杀菌剂——纹曲宁，在全国各稻区大面积推广应用，为高产、优质水稻的产业化生产提供了核心保障技术。

笔者研究团队在分析生防芽孢杆菌 Bs916 定殖植物根部的转录组数据中发现 RNA 结合蛋白 Hfq 编码基因 *hfq* 发生了高表达，为了明确 *hfq* 参与定殖过程中的作用，本研究利用分子生物学技术构建芽孢杆菌 Bs916 的 *hfq* 突变体 Bs916Δ*hfq*，并检测了该突变体的生物膜形成能力和群集游动能力，初步明确了 *hfq* 基因参与定殖过程的方式。本研究首先基于芽孢杆菌 Bs916 和枯草芽孢杆菌 Bs168 菌株的全基因组序列，查找 *hfq* 基因及其侧翼序列，并进行比对分析，结果显示：Bs916 中 *hfq* 及其相邻基因 *ymaF*、*miaA*、*ymzC* 四个基因总长为2 091bp，而 Bs168 中的全长为2 187bp；两个菌株的 *hfq* 基因大小都是222bp，核苷酸序列相似性为87%，氨基酸序列同源性为96%。笔者利用重叠 PCR 方法构建了用于突变生防芽孢杆菌 Bs916 中 *hfq* 基因的单交换同源重组载体 pT-hfqL-cm，并进行了芽孢杆菌 Bs916 的化学转化，经 PCR 验证，最终获得3个正确转化子。随后，进一步检测了 *hfq* 突变株的生物膜形成能力和群集游动能力（Swarming）。结果显示：菌株 Bs916 和 Bs916Δ*hfq* 在 MSgg 培养基中都能够形成质地均匀、厚实多皱褶的生物膜；菌株 Bs916 在0.7%的 LB 半固体培养基中迅速扩展，而突变体 Bs916Δ*hfq* 相较野生型扩展缓慢，群集游动能力显著下降。以上研究初步表明，*hfq* 基因不能显著影响芽孢杆菌 Bs916 的生物膜形成，但通过调控群集游动能力参与了芽孢杆菌定殖植物根部的过程。后续笔者将深入研究 RNA 结合蛋白 Hfq 调控芽孢杆菌 Swarming 过程的具体分子机制。

关键词：芽孢杆菌 Bs916；*hfq* 基因；同源重组；生物膜；群集游动能力

---

* 基金项目：国家自然科学基金青年科学基金（31660543）；江苏省农业科学院院基金（6111615）；苏州市科技计划项目（No. SNG2018095）

** 第一作者：乔俊卿，副研究员，主要从事植物病害及生物防治研究；E-mail：junqingqiao@ hotmail. com

*** 通信作者：刘邮洲，研究员，主要从事植物病害及生物防治研究；E-mail：shitouren88888@ 163. com

柑橘溃疡病菌多位点序列分型方法的建立*

许晓丽[1**]，张丹丹[1]，李健强[1]，冯建军[2,3]，高瑞芳[2,3]，章桂明[2,3]，罗来鑫[1***]

（1. 中国农业大学植物保护学院，种子病害检验与防控北京市重点实验室，北京 100193；
2. 深圳海关动植物检验检疫技术中心，深圳 518045；
3. 深圳市外来有害生物检测技术研发重点实验室，深圳 518045）

摘　要：多位点序列分型法（multilocus sequence typing，MLST）是一种基于持家基因序列测定的分析方法，用于菌株之间基因组相关性及遗传多样性研究。柑橘溃疡病由柑橘黄单胞菌柑橘亚种（*Xanthomonas. citri* subsp. *citri*）侵染引起，是目前芸香科水果生产中一大毁灭性病害，其病原菌是国内外重要的检疫性有害生物。研究柑橘溃疡病菌的遗传多样性，对于了解和掌握病原菌进化、传播及病害发生流行都具有极其重要的意义。本研究共收集 53 株柑橘溃疡病菌株，其中包括从浙江、湖南、广西、广东、福建等地采集发病样品，通过分离培养以及对纯培养物进行特异性引物 JYF5（5′-TTCGGCGTCAACAAAATG-3′）和 JYR5（5′-AACTCCAGCACATACGGGTC-3′）鉴定为阳性，以及来自印度、日本、巴西、美国等产区菌株。根据 NCBI 中 *Xanthomonas. citri* subsp. *citri* 306、AW13 和 *Xanthomonas fuscans* subsp. *aurantifolii* FDC 1561 等菌株全基因组序列，通过 ART+ Bwa 0. 7. 17+ Samtools 1. 9 软件组合分析基因组之间的单核苷酸多态性差异，结合部分代表不同生化功能的基因，共筛选出 6 个适用于柑橘溃疡病菌 MLST 分析的基因，分别是 *filL*、*pilY*1、*egl*、*copB*、*yapH* 及 *hrpA*。START v2. 0 软件分析结果表明，这些基因在进化过程中较为保守，可为运用多位点序列分型研究柑橘溃疡病菌的种群分化、遗传多样性提供数据来源。

关键词：柑橘溃疡病菌；看家基因；多位点序列分型（MLST）

---

* 基金项目：国家重点研发计划课题（2016YFF0203204）

** 第一作者：许晓丽；E-mail：xuxl1123@ 126. com

*** 通信作者：罗来鑫；E-mail：luolaixin@ cau. edu. cn

同时检测两种瓜类种传细菌的微滴数字 PCR 检测技术*

赵子婧[1,2**]，芦 钰[2]，李健强[1]，罗来鑫[1***]，徐秀兰[2***]

（1. 中国农业大学植物保护学院，种子病害检验与防控北京市重点实验室，北京 100193；
2. 北京市农林科学院蔬菜研究中心，农业部华北地区园艺作物生物学与种质创制重点实验室，农业部都市农业（北方）重点实验室，北京 100097）

摘 要：西瓜嗜酸菌（*Acidovorax citrulli*，简称 Ac）和瓜类角斑病菌（*Pseudomonas syringae* pv. *lachrymans*，简称 Psl）是引起瓜类细菌性果斑病和细菌性角斑病的重要种传细菌。带菌种子、种苗是病害长距离传播的主要途径，种子检测是预防和控制该病害的首要环节。本实验基于微滴数字聚合酶链式反应（droplet digital polymerize chain reaction，简称 ddPCR）技术，建立了同时检测种子携带 Ac 和 Psl 两种病原细菌的方法。实验中平行测试比较了 ddPCR 和实时荧光定量 PCR（real-time PCR）两种检测技术对两种细菌的检测灵敏度与稳定性。对于等比例混合的两种细菌菌悬液和 DNA 样品，ddPCR 的最低检测限分别为 10^3 CFU/mL 和 2.28 fg/μL，其检测灵敏度均为 real-time PCR 的 10 倍。非等比混合菌悬液和 DNA 样品检测结果表明，ddPCR 可以同时检出两种病原菌的混合比例分别为 1：100（10^4：10^6 CFU/mL）和 1：10 000（2.28 fg/μL：22.8 ng/μL），检测灵敏度同样是 real-time PCR 检测的 10 倍。此外，ddPCR 方法在人工接菌种子测试中可检测到带菌种子比例均为 0.2%（$n=500$）的西瓜种子样品，同时在两种细菌带菌种子比例为 1：10（带菌率分别为 0.2%和 2%）时可以准确检出两个目标菌；而对于 real-time PCR 检测方法则只能在带菌率为 1%（$n=100$）的接种样品中检出阳性。综上所述，本实验基于 ddPCR 技术建立了一种同时检测 Ac 和 Psl 两种重要葫芦科种传细菌的种子检测方法，为提高种传病原菌的检测灵敏度提供了新思路。

关键词：ddPCR；real-time PCR；种子带菌检测；西瓜嗜酸菌；瓜类角斑病菌

---

* 基金项目：北京市农林科学院科技创新能力专项（KJCX20180203）

** 第一作者：赵子婧，在读硕士研究生；E-mail：zhaozijing94@163.com
*** 通信作者：罗来鑫，副教授；E-mail：08030@cau.edu.cn
徐秀兰，副研究员；E-mail：xuxiulan@nercv.org

受 DegU 调控且与 fengycin 合成相关基因的功能分析*

王培培[1**]，郭庆港[1]，李社增[1]，鹿秀云[1]，赵卫松[1]，张晓云[1]，苏振贺[1]，马　平[1***]

（1. 河北省农林科学院植物保护研究所，河北省农业有害生物综合防治工程技术研究中心，保定　071000）

摘　要：Fengycin 是枯草芽孢杆菌 NCD-2 菌株产生的主要抑菌活性物质。DegU/DegS 是枯草芽孢杆菌 NCD-2 菌株中一个重要的双因子调控系统，笔者采用基因缺失突变的分子遗传学方法，获得了 NCD-2 菌株 *degU* 基因缺失突变子。初步研究发现，*degU* 突变子的 fengcyin 合成能力显著下降，对番茄灰霉病菌的抑菌能力也明显降低。随后采用 RNA-seq 和蛋白质组学研究技术并结合生物信息学的方法，从 mRNA 水平和蛋白质水平上分析受 DegU 调控且与 fengycin 合成相关的目的基因。转录组测序结果表明差异表达的基因共 566 个，其中相对于野生型菌株，在突变子中表达量下调的基因 295 个，表达量上调的基因 271 个；差异基因大部分与生物进程相关；蛋白质组测序结果表明差异表达蛋白基因共 250 个，其中相对于野生型菌株 NCD-2，突变子中蛋白表达量下调的对应基因 133 个，表达量上调蛋白对应的基因 117 个。对转录组结果和蛋白质组结果进行联合分析后结果显示，转录和蛋白表达量同时下调的基因 37 个，转录和蛋白表达量同时上调的基因 38 个，转录和蛋白差异表达的基因 7 个。随机挑选了 20 个基因进行 RT-qPCR 验证，验证结果与组学结果一致。对获得的这些基因进行分析后结果显示，从中挑选出感兴趣的基因作为目的基因，目前正在进行目的基因功能研究，预期结果将明确 DegU 对 fengycin 合成的调控机理，为构建 NCD-2 菌株的 fengycin 合成网络奠定基础，为更加合理的应用 NCD-2 菌株提供科学参考依据。

关键词：枯草芽孢杆菌；NCD-2 菌株；DegU 调控；fengycin 合成；基因缺失突变

---

* 基金项目：国家自然科学基金（31601680）；河北省农林科学院财政项目（F17E10003）

** 第一作者：王培培，博士，副研究员，主要从事植物病害的生物防治研究；E-mail：wangpeipei0010@163.com

*** 通信作者：马平，研究员，博士生导师，主要从事棉花病害和植物病害生物防治研究；E-mail：pingma88@126.com

水稻细菌性条斑病菌中 PilZ 结构域蛋白的功能分析*

魏　超[1**]，王黎锦[1**]，刘朋伟[2]，施菲菲[1]，王冠群[2]，孙文献[1,2***]
（1. 中国农业大学植物保护学院，农业部作物有害生物监测与绿色防控重点实验室，
北京　100193；2. 吉林农业大学植物保护学院，长春　130118）

摘　要：环二鸟苷酸（c-di-GMP）是一类普遍存在于细菌内的第二信使分子，参与调控多种细胞代谢过程。细菌体内存在多种类型 c-di-GMP 的受体，其中包括含退化的 GGDEF-EAL 结构域蛋白和 PilZ 结构域蛋白等。这些受体蛋白负责感知细胞内 c-di-GMP 浓度的变化，并启动或抑制下游特定基因的表达。

水稻细菌性条斑病（bacterial leaf streak，BLS）是由稻生黄单胞菌条斑致病变种（*Xanthomonas. oryzae* pv. *oryzicola*，*Xoc*）引起的一种细菌性病害。在柑橘溃疡病菌中，FimX-PilZ-PilB 形成蛋白复合体调控四型纤毛生物合成。本研究成功构建了 *Xoc* 中 3 个编码 PilZ 结构域蛋白的基因敲除突变体 $\Delta pilZ_{Xoc}$-1、$\Delta pilZ_{Xoc}$-2、$\Delta pilZ_{Xoc}$-3，对其生物膜、胞外多糖、蛋白酶分泌、游动性、滑动性及致病性进行了检测。发现 $\Delta pilZ_{Xoc}$-1、$\Delta pilZ_{Xoc}$-2、$\Delta pilZ_{Xoc}$-3 生物膜形成、胞外多糖合成、蛋白酶分泌以及游动性与野生型菌株相比均无明显差异，但 $\Delta pilZ_{Xoc}$-1 滑动性上升，致病力下降；ITC 实验显示 $PilZ_{Xoc}$-2 和 $PilZ_{Xoc}$-3 在体外均能与 c-di-GMP 结合，$PilZ_{Xoc}$-1 则不能与 c-di-GMP 结合；通过细菌双杂交与离体 GST pull-down 技术，也证实了 *Xoc* 中 $FimX_{Xoc}$-$PilZ_{Xoc}$-1-$PilB_{Xoc}$ 三元复合体的存在；$PilZ_{Xoc}$-1 的 104—114 位氨基酸是 $PilZ_{Xoc}$-1 与 $PilB_{Xoc}$ 互作的关键区域，且在滑动性的调控上发挥重要作用；此外，透射电镜观察到 $\Delta pilZ_{Xoc}$-1 菌体表面的 T4P 数量与野生型相比明显减少。以上结果为研究 c-di-GMP 信号调控途径提供了重要信息。

关键词：水稻细菌性条斑病菌；c-di-GMP；PilZ

---

* 基金项目：作物细菌性病害防控技术研究与示范（201303015）

** 第一作者：魏超，博士研究生，研究方向：分子植物病理学；E-mail：478608841@ qq. com
王黎锦，硕士研究生，研究方向：分子植物病理学；E-mail：929466449@ qq. com

*** 通信作者：孙文献，教授，主要从事植物与病原细菌、真菌互作；E-mail：08042@ cau. edu. cn

水稻细菌性条斑病菌三型效应蛋白 AvrBs2 致病机理研究*

徐嘉擎**，王善之，李　帅，孙文献***

（中国农业大学植物保护学院，农业部作物有害生物监测与绿色防控重点实验室，北京　100193）

摘　要：稻生黄单胞菌条斑致病变种（*Xanthomonas oryzae* pv. *oryzicola*，*Xoc*）引起的细菌性条斑病是水稻上最重要的细菌病害之一，简称细条病菌。然而，对于该病菌如何克服水稻免疫系统成功侵染水稻的致病机制知之甚少。解析细条病菌关键致病效应蛋白的毒性机理将有助于揭示该病菌的致病机制。

前期研究表明，水稻细菌性条斑病菌 *Xoc* 菌株 RS105 的 23 个 non-TAL 效应蛋白基因单敲除体接菌水稻后，仅有 *avrBs*2 突变体的致病能力显著降低，其余单敲除体致病力均未减弱，表明 AvrBs2 是 *Xoc* 侵染水稻过程中的重要毒力因子。在异源表达 *avrBs*2 的转基因水稻中研究发现，AvrBs2 抑制由 flg22 和几丁质等病原相关分子模式（PAMPs）诱导的活性氧暴发和 *PR* 基因上调表达，并且促进 *Xoc* 对水稻的致病能力，表明 AvrBs2 进入植物细胞后抑制了植物的免疫反应，在致病性中起关键作用。本研究对 *avrBs*2 截断体、酶活点突变回补菌株致病力分析发现，截断体回补菌株发病病斑长度较野生型菌株（RS105）显著降低，在 8 个点突变回补菌株中，仅有 Δ*avrBs*2（*avrBs*2-*E304A*，*D306A*、*avrBs*2-*H319A*）的病斑长度较野生型菌株（RS105）相比显著降低，其余点突变回补菌株致病性与野生型菌株（RS105）相似。实验结果表明，N 端、C 端和 GDE 上的两个酶活位点对 AvrBs2 发挥毒性均有重要作用。此外，体外分泌实验发现，Δ*avrBs*2（*avrBs*2-*E304A*，*D306A*、*avrBs*2-*H319A*）菌株均可分泌相应 AvrBs2 突变体蛋白。在此研究基础上，将进一步探寻 AvrBs2 在植物细胞中的毒力机制，研究结果将加深对细条病菌致病分子机制的理解。

关键词：稻生黄单胞菌条斑致病变种；三型效应蛋白；AvrBs2；水稻

---

* 基金项目：国家自然科学基金（3180110557）

** 第一作者：徐嘉擎，硕士研究生，研究方向：水稻与病原细菌互作分子机理研究；E-mail：136810185@ qq. com

*** 通信作者：孙文献，教授，主要从事水稻与病原细菌、真菌互作分子机理研究；E-mail：wxs@ cau. edu. cn

旱柳细菌性溃疡病病原的鉴定*

李　永，朴春根，薛　寒，郭民伟，汪来发
（中国林业科学研究院森林生态环境与保护研究所，北京　100091）

摘　要：旱柳（*Salix matsudana*）是中国北方常用的庭荫树、行道树，常栽培在河湖岸边或孤植于草坪，对植于建筑两旁，亦用作公路树、防护林及沙荒造林等。2016 年 5 月在山东省菏泽市赵王河公园内的柳树上首次发现一种枝干溃疡病，是柳树上新发生的溃疡病症状的病害，其症状是发病初期流出大量酸臭的汁液，随后出现树皮开裂、韧皮部局部坏死、坏死处组织肿胀的症状，病斑长可达 0.5m。目前该病害主要分布在山东省菏泽市和宁夏回族自治区中卫市，主要为害旱柳主干和侧枝导致枝干死亡。

本研究利用组织分离的方法分离获得细菌优势种群 3 个、真菌优势种群 2 个。人工接种结果显示，一种透明的细菌人工接种的情况下可以造成类似自然发病的溃疡症状，而且通过了柯赫氏法则验证，证实该菌是旱柳溃疡病的病原菌。然后我们利用 16S rDNA 序列析、多基因序列分析（MLSA）、生理生化分析、脂肪酸分析、基因组同源性比较等方法对分离的菌株 hezel2-1-2，L2-3，L6-4B 进行了分类地位的研究。16S rDNA 序列析显示，这些菌株与 *Lonsdalea populi* LMG 27349^{T}16Sr RNA 基因的同源性为 99.8%～100%。16S rDNA 和多基因序列系统进化结果显示，hezel2-1-2、L2-3、L6-4B 与 *Lonsdalea populi* 聚在同一个进化枝之上。基因组同源性分析结果显示，菌株 hezel2-1-2、L2-3、L6-4B 与 *Lonsdalea populi* LMG 27349^{T}基因组同源性为 99.3%～99.4%，高于种间基因组差异的阈值 95%～96%。16S rDNA 序列析、多基因序列分析（MLSA）、生理生化分析、脂肪酸分析、基因组杂交等结果一致将菌株 hezel2-1-2、L2-3、L6-4B 鉴定为 *Lonsdalea populi*，因此，笔者最终确定旱柳细菌性溃疡病是由 *Lonsdalea populi* 侵染导致的。

关键词：旱柳；细菌性溃疡病；病原鉴定；*Lonsdalea populi*

* 基金项目：国家重点研发计划资助项目（2018YFC1200400）

生防蜡样芽孢杆菌 905 菌株突变体文库的构建及 MnSOD2 调控基因的筛选*

高坦坦[1,2]**，丁明政[2]，李　燕[2]，王　琦[2]***
（1. 农业农村部华北都市农业重点实验室，北京农学院生物与资源环境学院，北京　102206；
2. 中国农业大学植物保护学院，北京　100193）

摘　要：生防蜡样芽孢杆菌 905（*Bacillus cereus* 905）是本实验室前期从小麦植株内分离获得的一株有益芽孢杆菌，温室和田间应用结果显示，该细菌可以在小麦根围（内）稳定定殖，且表现出良好的促生效果。研究发现 *B. cereus* 905 的基因组上包含 *sodA*1 和 *sodA*2 基因，分别编码锰超氧化物歧化酶 MnSOD1 和 MnSOD2，其中后者在 *B. cereus* 905 定殖于植物根内及抵御纳米 TiO_2 的光催化氧化毒性中发挥更为重要的作用。目前在生防芽孢杆菌中，*sodA*2 基因的调控机制尚不清楚。本研究通过构建 *B. cereus* 905 转座子随机插入突变体文库，筛选调控 MnSOD2 表达的基因，为深入研究 *B. cereus* 905 定殖的分子机制奠定基础。

本研究首先将利用 *sodA*2 启动子指导 GFP 表达的质粒 pGFPsodA2 转入 *B. cereus* 905 菌株中，获得 GFP 标记的报告菌株，以 *gfp* 基因的表达水平指示 *sodA*2 启动子的转录活性。然后将携带转座元件 Tn*YLB*-1 的温度敏感型质粒 pMarA 通过电击转化转入报告菌株中。利用 37℃ 诱导转座，46℃ 高温消除质粒技术使转座子 Tn*YLB*-1 随机插入到 *B. cereus* 905 的基因组上，成功构建了 *B. cereus* 905 的转座子随机插入突变体文库。通过流式细胞仪监测 GFP 的荧光强度，共完成约 2 000个突变子的筛选，获得 26 个 GFP 荧光强度发生显著变化的突变子。以 Tn*YLB*-1 中部分卡那霉素编码基因片段（约 380 bp）为模板合成 DNA 探针进行 Southern 杂交，确定转座子插入的拷贝数；采用反向 PCR 克隆突变子中 Tn*YLB*-1 插入位点的侧翼序列，测序确定转座子插入的基因位点。本研究共获得 12 个插入基因的序列，经分析，正调控 *sodA*2 基因表达的为：磷酸烯醇丙酮酸蛋白磷酸转移酶、磷酸转移载体蛋白、热诱导转录抑制因子、重组酶 A、氯离子通道蛋白和一个未知功能蛋白；负调控 *sodA*2 基因表达的为：2 个亚铁离子运输蛋白 B、核苷二磷酸激酶、组氨酸解氨酶、苏氨酸脱水酶和 RNA 聚合酶 sigma-70 亚基。

本研究通过转座子插入突变体文库的构建及目标基因的筛选，为开展 *sodA*2 调控途径的后续研究奠定了基础。解析生防菌 *B. cereus* 905 中 MnSOD 的调控途径，不仅有助于揭示生防菌定殖及耐受外界环境胁迫的分子机制，并且利于探寻提高生防菌功能稳定性的途径。

关键词：生防蜡样芽孢杆菌；pMarA；突变体文库；MnSOD；调控基因

---

* 基金项目：国家自然科学基金（31701860，31171893）
** 第一作者：高坦坦，博士，讲师；E-mail：gaotantan0537@ 163. com
*** 通信作者：王琦，博士，教授；E-mail：wangqi@ cau. edu. cn

枯草芽孢杆菌 9407 基因组中生防相关抗生素基因簇的分析

顾小飞[*]，曾庆超，李　蓉，杨攀雷，李　燕，王　琦[**]
（中国农业大学植物保护学院，北京　100193）

摘　要： 枯草芽孢杆菌 9407 是一株分离自苹果果实，对苹果轮纹病具有良好防效的生防菌，并对多种植物病原真菌和细菌有良好的抑制作用。菌株 9407 对各种病原菌起主要作用的代谢物质尚不明确。本研究通过基因组学和分子生物学对菌株 9407 进行深入的研究。研究结果表明 9407 基因组中包含 surfactin、fengycin、subtilocin A、bacilycin、bacillaene 等抗生素合成基因簇，其中 fengycin、subtilocin A、bacilycin、bacillaene 抗生素基因簇与已报道菌株的基因簇的相似度为 100%，surfactin 基因簇的相似度也高达 82%。进一步分析发现菌株 9407 中只含有 2 种脂肽类抗生素基因簇，即 surfactin 和 fengycin。功能分析表明 surfactin 和 fengycin 是抑制瓜类细菌性果斑病菌（*Acidovorax citrulli*）和苹果轮纹病菌（*Botryosphaeria dothidea*）的主要活性物质。在其他类抗生素中，subtilocin A 是一种细菌素，通过改变细胞膜通透性发挥作用；bacilycin 是溶杆菌素，通过破坏细胞壁的合成对细菌和真菌发挥作用；bacillaene 是一种多烯抗生素，通过抑制蛋白合成发挥作用，对细菌和真菌都有作用。因此，笔者推测其他类的抗生素在菌株 9407 的抑菌作用中可能发挥一定的作用，后续对其他类抗生素进行功能分析。为深入了解生防菌株 9407 的作用机理奠定基础，并为菌株 9407 的田间应用提供理论依据。

关键词： 枯草芽孢杆菌；抗生素；基因簇

---

* 第一作者：顾小飞，博士研究生；E-mail：gxfmail2013@ 163. com
** 通信作者：王琦，教授，主要从事植物病害生物防治与微生态研究；E-mail：wangqi@ cau. edu. cn

水稻细菌性谷枯病的发生与病原鉴定*

袁 斌**，张 舒***，刘友梅，黄 薇，吕 亮，常向前，杨小林，王佐乾
（湖北省农业科学院植保土肥研究所，农业部华中作物有害生物综合治理重点实验室，
农作物重大病虫草害防控湖北省重点实验室，武汉 430064）

摘 要：水稻细菌性谷枯病（Bacterial grain rot of rice，BGRR）又称水稻穗枯病（Bacterial panicle blight of rice，BPBR），该病于 1956 年 Goto 等在日本发现，由颖壳伯克氏菌（*Burkholderia glumae*）引起。在我国周边国家严重发生，一般造成 15%～20%的损失，在越南，严重田块产量损失高达 75%。2007 年我国尚未发现该病有发生时，国内科研人员对水稻细菌性谷枯病对我国水稻生产的潜在威胁进行了风险评估，认为水稻细菌性谷枯病属于高度危险性有害生物，国内有可能严重发生。本研究室 2016—2019 年从广西、广东及湖北等多个省份采集到水稻疑似病穗，采用稀释平板法分离到 12 个来源不同地方的病原细菌。将这 12 个菌株的 3×10^8 CFU/mL 菌悬液分别注射接种本生烟叶片，结果表明 12 个菌株均能引起烟草叶片发生过敏性反应；利用特异性引物对 12 个菌株进行检测，扩增出特异条带，表明分离得到的菌株为细菌性谷枯病菌；用 3×10^8 CFU/mL 菌悬液喷雾接种水稻品种 IR24 的四叶一心期幼苗，引起植株茎基部腐烂，与该病菌导致的苗期病状一致；采集接种发病的样品重新分离病原物，PCR 鉴定表明与接种病菌为同一种病菌；为了进一步鉴定该病原物，利用 16S rDNA 通用引物扩增目标区段并送样测序，结果表明 12 个菌株的 16S rDNA 与 NCBI 数据库中细菌性谷枯病菌的 16S 序列 100%同源，所以本研究室从水稻病穗分离的病原菌确定为水稻细菌性谷枯病菌。周求根等（2014）在江西南昌分离到该菌；徐以华等（2018）在浙江、福建、广东等全国各地的水稻病穗上分离到该病原物。这表明国内已经广泛发生水稻细菌性谷枯病，且呈上升趋势，随着全球气候变暖，有可能成为水稻新的主要病害。

关键词：水稻；细菌性谷枯病；病原鉴定

---

* 基金项目：国家自然科学基金（31772154）；国家重点研发计划（2016YFD0200807）；湖北省技术创新专项重大项目（2017ABA146）；湖北省农业科技创新中心项目（2016-620-000-001-017）

** 第一作者：袁斌，博士，副研究员，研究方向为分子植物病理学；E-mail：yuanbin2000@139.com
*** 通信作者：袁斌，博士，副研究员，研究方向为分子植物病理学；E-mail：yuanbin2000@139.com
张舒，硕士，研究员，研究方向为植物保护学；E-mail：ricezs6410@163.com

甘蓝黑腐病菌 *Xanthomonas campestris* pv. *campestris topoIB* 基因的克隆及功能分析

李珅瑀*，张　迎，王　诚，孔维文，张清霞**
（扬州大学园艺与植物保护学院，扬州　225009）

摘　要：野油菜黄单胞菌野菜致病变种（*Xanthomonas campestris* pv. *campestris*，*Xcc*）是严重威胁十字花科植物的重要病原真菌之一，可造成全世界范围内农业的严重减产。DNA 拓扑异构酶是通过调节 DNA 的拓扑结构进而影响 DNA 复制、重组和转录过程。目前尚未见文献报道有关病原细菌中拓扑异构酶功能研究。为分析拓扑异构酶是否影响野油菜黄单胞菌野菜致病变种的致病性，本实验以 *Xcc* 8004 中 *Xcc*_0034 基因，预测编码的 TopoIB 蛋白作为研究对象，通过对 TopoIB 蛋白的理化性质进行初步的生物信息学分析，表明 TopoIB 蛋白含有核定位信号和信号肽。在农杆菌介导烟草叶片中瞬时表达，发现 p1305-GFPTopoIB 融合蛋白定位于烟草叶片的细胞核上。构建 *topoIB* 基因的原核表达载体，纯化的 TopoIB 重组蛋白反应 15 min 可以完全解旋超螺旋 DNA，表明 TopoIB 蛋白具有拓扑异构酶活性。将 *topoIB* 基因突变后得到突变体 ΔTopoIB，表型检测结果显示，ΔTopoIB 菌株与野生型 *Xcc* 8004 在植物体内和培养基中的生长状况无明显差异，且胞外多糖产量基本相同，此外胞外蛋白酶、胞外纤维素酶和胞外淀粉酶产量也与野生型 *Xcc* 8004 基本一致。在拟南芥和甘蓝（‘京华一号’）上致病性检测显示 ΔTopoIB 菌株致病性并没有受到影响。因此表明 *topoIB* 与 *Xcc* 8004 的致病性无明显的直接关系。

关键词：*Xcc* 8004；*topoIB*；缺失突变；亚细胞定位；原核表达

---

* 第一作者：李珅瑀，硕士研究生，资源利用与植物保护专业；E-mail：2248863592@ qq. com
** 通信作者：张清霞，副教授，从事植物病害生物防治研究；E-mail：zqx817@ 126. com

PMA-qPCR 检测两种丁香假单胞菌活性研究*

张丹丹[1**]，许晓丽[1]，李健强[1]，罗来鑫[1]，冯建军[2,3]，高瑞芳[2,3]，章桂明[2,3***]

(1. 中国农业大学植物保护学院，北京 100193；2. 深圳海关动植物检验检疫技术中心，深圳 518045；3. 深圳市外来有害生物检测技术研发重点实验室，深圳 518045)

摘 要：由丁香假单胞菌豌豆致病变种（*Pseudomonas syringae* pv. *pisi*）引起的豌豆细菌性疫病和由丁香假单胞菌斑点致病变种（*Pseudomonas syringae* pv. *maculicola*）引起的十字花科细菌性黑斑病既是世界性病害，也是我国进境植物检疫性有害生物。前者在我国尚未见报道，后者近年来在我国发生趋势有所上升。如进境种子携带该 2 种病原菌传入我国将会对我国寄主作物造成较大经济损失。目前国内外对于这两种病原菌的死活均无快速检测方法。本研究利用 PMA（叠氮溴化丙锭）这一新型核酸结合染料，可以穿过受损的细胞膜进入细胞内，导致 DNA 在 PCR 过程中无法打开双链而不能被扩增的原理，将 PMA 结合实时荧光定量 PCR，通过筛选的特异性引物以及探针进行扩增检测，建立一种用于细菌活性的快速、高灵敏度、定性、定量检测方法。本研究结果表明，未经 PMA 处理的活菌菌液在实时荧光扩增时 Ct 值与菌液浓度呈良好的线性关系，即 $y=-3.0809x+43.07$（$R^2=0.9971$），向 80℃ 水浴处理 20min 后的死菌菌液中添加终浓度为 10μmol/L 的 PMA 后，Ct 值较活菌菌液显著增加，表明 PMA 可以抑制热灭活死菌的 DNA 扩增，在此基础上，本研究将对 PMA 添加量、黑暗孵育时间、曝光时间等进行优化并开展应用研究。

关键词：丁香假单胞菌；PMA-qPCR；活性检测

---

* 基金项目：国家重点研发计划课题（2016YFF0203204）

** 第一作者：张丹丹，在读硕士研究生；E-mail：15237373080@163.com

*** 通信作者：章桂明；E-mail：zgm2001cn@163.com

番茄溃疡病菌微滴式数字 PCR 检测方法的建立*

王　丽[1**]，周　佩[2,3]，田　茜[2]，孙现超[1]，赵文军[2***]
（1. 西南大学植物保护学院，重庆　400715；2. 中国检验检疫科学研究院，
北京　100176；3. 中国农业大学植物保护学院，北京　100193）

摘　要：由 *Clavibacter michiganensis* subsp. *michiganensis* 引起的番茄溃疡病是茄科作物生产中最具毁灭性的种传病害之一，被列为我国检疫性有害生物。带菌番茄种子是该病害的主要初侵染源，是病原菌远距离传播的最主要方式，因此，对该病原菌尤其是种子上该病原菌的准确检测鉴定是防止病害传播为害的关键。本研究根据番茄溃疡病菌 *pat*-1 基因序列，筛选设计了其特异性检测引物（CMM CF/CR）及探针（CMM CP），并在此基础上建立了相应的数字 PCR 检测技术。利用 5 株番茄溃疡病菌、18 株同属其他菌株及 5 株其他属病原菌，对所建立的检测体系进行了测试，并采用人工接种及自然带菌番茄种子进行验证。测试结果显示，所建立的数字 PCR 检测方法可以特异性检测出番茄溃疡病菌，而其他近缘菌株的检测结果均为阴性；该方法对番茄溃疡病菌的检测下限达到 3.8 CFU/mL，高于实时荧光 PCR 检测方法；3 次重复试验拷贝数的相对标准偏差小于 25%，证明实验数据可以较为精确的反应样品浓度，且结果重复性良好。此外，所建立的数字 PCR 检测方法在模拟带菌种子及自然种子样品上也得到了成功验证，该方法可以准确检测出所有带菌种子。总的来说，本研究所建立的番茄溃疡病菌数字 PCR 检测方法具有特异性好，灵敏度高，重复性好等优点，能够满足病害实际检测中的需求。

关键词：番茄溃疡病菌；数字 PCR；检测

---

* 基金项目：中国检验检疫科学研究院基本科研业务费项目（2018JK004）
** 第一作者：王丽，硕士研究生；植物病理学；E-mail：492321854@ qq. com
*** 通信作者：赵文军；E-mail：wenjunzhao@ 188. com

云南省烟草青枯病病原多样性初探*

卢灿华[1**]，刘俊莹[2]，马俊红[1]，盖晓彤[1]，余　清[1]，雷丽萍[1]，莫笑晗[1]，夏振远[1***]
（1. 云南省烟草农业科学研究院，昆明　650021；2. 玉溪师范学院
化学生物与环境学院，玉溪　653100）

摘　要：烟草青枯病是由假茄科雷尔氏菌（*Ralstonia pseudosolanacearum*）侵染引起的维管束病害。该病的病原菌最早被 Smith（1896）命名为 *Bacillus solanacearum*，1996 年由 IJSB 更名为茄科雷尔氏菌（*R. solanacearum* E. F Smith），2014 年 Safni 等根据 16S-23S rRNA ITS、部分内源葡聚糖酶（endoglucanase，egl）基因序列及 DNA-DNA 杂交结果将烟草青枯菌（演化型 Ⅰ）并入假茄科雷尔氏菌（*R. pseudosolanacearum*）。1880 年，烟草青枯病首先发现于美国北卡罗来纳州格兰维尔（Granville），也称作“Granville wilt”，而后在印度尼西亚、日本、澳大利亚和韩国等国逐渐演变为烟草上的重要病害。该病害在我国长江流域及其以南各烟区普遍发生，其中广西、广东、福建、湖南、浙江、安徽、四川及贵州等地危害严重。云南省早在 1987 年就有烟草青枯病的报道，但危害较轻。2002 年以来，云南省南部旱地烟区青枯病危害逐渐加重，局部地块发病率达到 80%以上。目前，该病在云南省的文山、保山、临沧、红河、昆明、玉溪、曲靖、昭通、大理、丽江、楚雄和德宏等州市均有发生，其中文山、临沧、保山和德宏的部分地区发病较为严重。虽然烟草青枯病在云南省发病较为严重，但对该病的研究相对不足，尤其在病原学方面的研究尚少。2018 年，笔者从文山、临沧、玉溪、昆明、曲靖、普洱、德宏等 7 个州市采集分离 60 份烟草青枯菌，经序列分析发现分离自玉溪龙树的 1 株菌（LLRS-1）的 16S rDNA 序列与蒲桃雷尔氏菌（*R. syzygii* subsp. *syzygii*）R001 菌株一致，其他 59 株青枯菌的 16S rDNA 序列与 *R. solanacearum* GMI1000 相似性最高；经复合 PCR 检测，发现除 LLRS-1 菌株为演化型 Ⅳ 外，其他菌株均为演化型 I；分析青枯菌对 3 种糖、3 种醇的利用情况，初步研究结果表明分离的 60 株青枯菌种存在 3、2 生化型，但也存在 5 种生化型（1、2、3、4、5）以外的菌株。综上所述，对 2018 年获得的 60 株烟草青枯菌的研究结果表明除假茄科雷尔氏菌外，蒲桃雷尔氏菌（*R. syzygii* subsp. *syzygii*）也能侵染引起烟草青枯病，病原菌在种类、演化型、生化型方面存在分化说明云南省不同烟区烟草青枯菌间可能存在差异。因此，有必要深入研究云南省各植烟州市烟草青枯菌的遗传多样性与生理分化情况，为该病的生防资源筛选、抗病资源评价、抗病育种及品种合理布局提供重要参考。

关键词：烟草；青枯病；病原菌；遗传多样性

---

* 基金项目：中国烟草总公司云南省公司科技计划重点项目（2018530000241006，2017YN08）

** 第一作者：卢灿华，助理研究员，从事植物病原细菌相关分子生物学与生物化学研究；E-mail：lucanhua1985@163.com

*** 通信作者：夏振远，研究员，主要从事烟草病虫害绿色防控研究；E-mail：648778650@qq.com

新疆田旋花黄化病植原体16SrRNA基因的序列分析*

王帅杰**，胡惠清，王浩东，都业娟***
（石河子大学农学院，新疆绿洲农业病虫害治理与植保资源利用重点实验室，石河子　832002）

摘　要：植原体侵染植物后可导致衰退、叶片黄化、畸形等症状，冬季主要在多年生植物的根部越冬。田旋花（*Convolvulus arvensis* L.）为旋花科多年生植物，农业生产中作为田间有害杂草，广泛存在于农田、果园的田边及田间，争夺农作物养分的同时，还是多种病原物的越冬寄主。调查中发现田旋花除表现出丛簇等植原体症状外，还会表现全株黄化的症状，为明确采集到的表现黄化症状的田旋花植株是否感染植原体及所感染的植原体的种类，本研究利用植原体16S rRNA基因通用引物P1（5′-AGAGTTTGATCCTGGCTCAGGA-3′）和P7（5′-CGTCCTCATCG-GCTCTT-3′）、R16F2n（5′-GAAACGACTGCTAAGACTGG-3′）和R16R2（5′-TGACGGGCGGT-GTGTACAAACCCCG-3′）对新疆石河子地区采集的表现黄化症状的田旋花植株总DNA进行巢式PCR扩增，并对扩增片段进行克隆和测序。结果表明：表现黄化症状的田旋花病株利用巢氏PCR可扩增到与阳性对照一致的特异性目标片段，克隆测序表明该16S rRNA基因片段长1 244 bp，NCBI blast结果表明与16SrⅫ-A亚组植原体的同源性高于99.6%，构建系统进化树表明其与16SrⅫ-A亚组植原体聚于同一个亚分支上，明确新疆田旋花黄化病植原体属于16SrⅫ-A亚组。

关键词：田旋花；黄化病；植原体；分子鉴定

---

* 基金项目：兵团区域创新计划项目（2018BB403）

** 第一作者：王帅杰，主要从事植原体病害研究；E-mail：2472921879@qq.com

*** 通信作者：都业娟，主要从事植物病毒及植原体病害研究；E-mail：dyjagr@sina.com

几株 16SrV-B 亚组植原体的鉴定与比较分析*

任争光[1**]，杨　静[1]，廖亚军[2]，宁钧辉[1]，董雅容[1]，李梦涵[1]，王进忠[1]，王　合[3***]
（1. 北京农学院生物与资源环境学院，农业农村部华北都市农业重点实验室，北京　102206；
2. 北京农学院后勤基建处，北京　102206；3. 北京林业保护站，北京　100029）

摘　要：笔者课题组在北京市昌平区北京农学院校园及附近分别发现了患有典型植原体病害症状的刺槐、红花槐、丝绵木和暴马丁香植株。这些植株共同的症状特征为小叶和丛枝，其中丝绵木病株还表现花变叶的症状，而刺槐、红花槐和暴马丁香病株不再有花蕾和开花现象。为了检测这些患病植物中的植原体存在与否，本实验对上述 4 种病株进行了采样，并以枣疯病（Jujube witches' broom，JWB）植物样品为对照，分别提取了样品的总 DNA。利用特异性引物 R16mF2/R16mR2 对样品中植原体的 16S rDNA 部分序列进行扩增，结果在这些样品中都能直接 PCR 扩增到目的片段（1.4 kb），经过测序和比对分析发现刺槐丛枝（*Robinia pseudoacacia* witches' broom，RpWB）、红花槐丛枝（*R. pseudoacacia* var. *decaisneana* witches' broom，RpdWB）、丝绵木丛枝（*Euonymus bungeanus* witches' broom，EbWB）和暴马丁香（*Syringa reticulate* witches' broom，SrWB）丛枝植原体与 16SrV-B 亚组的 JWB 植原体序列相似率均达到了 99% 以上。利用 MEGA5.0 软件构建系统进化树，结果发现这些植原体与枣疯植原体聚在同一分支。同样的利用特异性引物对这些植原体的 *rp*、*secY* 和 *tuf* 基因序列进行了扩增、序列比对分析和系统进化树构建，结果都显示它们与枣疯植原体的序列有较高相似率和最近亲缘关系，因此将以上 4 种植原体暂划定为 16SrV-B 组成员。

作为 16SrV 组最典型的代表，枣疯植原体及其引起的枣疯病已经给我国大枣产业造成了巨大的损失。在北京地区，无论是山区种植，公园栽植，还是庭院养护的枣树都经常发生枣疯病。本研究中发现的 RpWB、RpdWB、EbWB 和 SrWB 植株周围也都有枣疯病树，而相对于枣疯病，这些丛枝病的发生面积和发生率极小，因此，结合分子生物学鉴定结果，推测它们中的植原体来源应和枣疯植原体有关，即这 4 种植物是枣疯植原体的潜在寄主。当然要证明这 4 种植物是枣疯植原体的新寄主，还有很多工作要做，比如传播介体叶蝉的鉴定和传毒实验，更多基因或者全基因组序列的比对分析等。

虽然上述 4 种病害极少大面积发生，但是近几年调查发现这些病害有加重的趋势，特别是刺槐丛枝病在北京地区的发病率逐年升高，而且对植株的破坏性极大。以本实验发现的刺槐丛枝病树为例，从发现丛枝的枝条到整树死亡不到 5 年时间，而红花槐丛枝病树更是两年之内就死亡。因此加强新植原体病害的监测，及时制定防控措施，对保护首都农林植物健康具有重要的意义。

关键词：16SrV-B 亚组；植原体；分子生物学；鉴定

---

* 基金项目：北京市自然科学基金（6182002）；北京农学院大北农青年教师科研基金（15ZK005）

** 第一作者：任争光，讲师，主要从事原核生物病害及其生物防治研究；E-mail：zgren2005@126.com

*** 通信作者：王合，高级工程师，主要从事林业有害生物防控研究；E-mail：lbz6329@sina.com

南瓜饲养对不同龄期剑麻新菠萝灰粉蚧体内植原体的脱毒效果初探[*]

王桂花[1**]，吴伟怀[2]，莫秀芳[2]，郑金龙[2]，黄　兴[2]，
贺春萍[2]，习金根[2]，梁艳琼[2]，谭施北[2]，易克贤[2***]
（1. 海南大学热带林学院，海口　570228；2. 中国热带农业科学院环境与植物保护研究所，农业农村部热带作物有害生物综合治理重点实验室，海南省热带农业有害生物监测与控制重点实验室，海口　571101）

摘　要：利用植原体 16S rRNA 基因的通用引物 R16mF2/R16mR1 和 R16F2n /R16R2，对来自剑麻田间的不同龄期剑麻新菠萝灰粉蚧，以及经南瓜饲养 10 代后，其总 DNA 进行巢式 PCR 扩增，以了解南瓜饲养对剑麻新菠萝粉蚧体内植原体的脱毒效果。结果表明，来自剑麻紫色卷叶病株 1 龄（30 头）、2 龄（30 头）、3 龄雌（30 头）或雌成虫（30 头），其植原体检出率分别为 10%、20%、13.3%和 26%。比较而言，同批次的新菠萝灰粉蚧经南瓜室内饲养 10 代后，无论是 1 龄（30 头）、2 龄（30 头），还是 3 龄其雌（30 头）或雌成虫（30 头）的 DNA 均检测不出植原体。由此表明，南瓜饲养对新菠萝粉蚧体内植原体具有良好的脱毒效果。

关键词：剑麻；新菠萝；灰粉蚧；植原体

---

* 基金项目：国家麻类产业技术体系剑麻生理与栽培岗位（CARS-16-E16）；中央级公益性科研院所基本科研业务费专项（1630042019012；1630042019028）

** 第一作者：王桂花，在读博士，研究方向：农业生态学；E-mail：wgh8308@ 163. com

*** 通信作者：易克贤，博士，研究员，研究方向：热带作物真菌病害及其抗性育种；E-mail：yikexian@ 126. com

第五部分　线　虫

Effects of α-pinene on the pinewood nematode (*Bursaphelenchus xylophilus*) and its symbiotic bacteria

Wang Xu[1,2], Yu Yanxue[2], Ge Jianjun[2], Cheng Xinyue[1]

(1. *College of Life Sciences*, *Beijing Normal University*, *Beijing* 100875, *China*;

2. *Chinese Academy of Inspection and Quarantine*, *Beijing* 100176, *China*)

Abstract: The pinewood nematode (PWN), *Bursaphelenchus xylophilus*, is an important plant-parasitic nematode that can cause severe mortality of pine trees. This PWN-induced harm to plants may be closely related to the abundance and diversity of the symbiotic microorganisms of the parasitic nematode. In this study, nematodes were divided into untreated and antibiotic-treated groups. Nematodes were treated by fumigation with different amounts of α-pinene, and the resultant mortality rates were analyzed statistically. Concentrations of symbiotic bacteria were calculated as colony-forming units/mL. High-throughput sequencing was used to investigate the bacterial community structure. The results showed that the mortality of nematodes increased slightly with increasing concentration of α-pinene, and nematodes untreated with antibiotics were more sensitive to α-pinene than those treated with antibiotics. The highest abundance of symbiotic bacteria was obtained via medium and low levels of α-pinene, but for which community diversity was the lowest (Shannon and Simpson indexes). The proportion of *Pseudomonas* spp. in the symbiotic bacteria of nematodes without antibiotics was relatively high (more than 70%), while that of *Stenotrophomonas* spp. was low (6%-20%). However, the proportion of *Stenotrophomonas* spp. was larger than that of the *Pseudomonas* spp. in the symbiotic bacteria associated with the antibiotic-treated nematodes. *Pseudomonas* sp. increased after pinene treatment, whereas *Stenotrophomonas* spp. decreased. These results indicate that although α-pinene has low toxicity to PWNs over a short time period, α-pinene ultimately influences the abundance and community diversity of the symbiotic bacteria of these nematodes; this influence may potentially disturb the development and reproduction of nematodes in the process of infecting pine trees.

Key words: pinewood nematode (PWN); *Bursaphelenchus xylophilus*; α-pinene; symbiotic bacteria

黄淮麦区重要小麦品种对菲利普孢囊线虫（*Heterodera filipjevi*）的抗性鉴定[*]

周　博[**]，任豪豪，胡登辉，吴佳欢，曹梦园，刘马鑫芝，
徐玲玲，张梦宁，白慧敏，何宇豪，崔江宽[***]
（河南农业大学植物保护学院，郑州　450002）

摘　要：小麦（*Triticum aestivum* L.）是世界上最重要的粮食作物之一，是全世界分布范围最广的作物种类，其播种面积和产量约占世界粮食总量的 1/3。中国是世界上最大的小麦生产国，河南是国内最大的小麦生产基地。河南省总耕地面积 693.33 万 hm^2，常年种植小麦 487 万 hm^2 左右，约占全国小麦种植面积的 1/4。小麦年产量占全国总产的 24% 左右，每年供应的小麦商品粮约占全国的 25%~30%，播种面积、总产和供应商品粮总量均位居全国第一（王绍中 等，2004）。菲利普孢囊线虫（*Heterodera filipjevi*），是一种发生分布范围迅速蔓延的禾谷类作物孢囊线虫。目前在俄罗斯、塔吉克斯坦、挪威、伊朗、北美、土耳其、印度和中国等世界上 20 多个国家报道发生（Subbotin 等，2010）。Peng 等（2010）和 Li 等（2010）首次在我国的河南省许昌地区报道发现了菲利普孢囊线虫（*H. filipjevi*）。此后，Peng 等（2015）在我国的安徽省宿州地区再次报道发现了菲利普孢囊线虫（*H. filipjevi*）。最近，Peng 等（2018）在我国的新疆和山东等地首次报道发现了菲利普孢囊线虫（*H. filipjevi*）。

本实验收集了黄淮麦区主栽品种共计 243 个，其中审定品种 59 个，区试品种 184 个。供试菲利普孢囊线虫（*H. filipjevi*）群体采自河南省许昌市。成熟褐色孢囊经形态学鉴定后，再辅助 SCAR 标记分子鉴定，从而确定供试线虫群体为菲利普孢囊线虫（*H. filipjevi*）。将分离出的成熟孢囊，消毒后置于 4℃冰箱内冷藏催化 2 个月。然后，移入温度为（16±2）℃的人工气候培养箱中进行二龄幼虫（J2）的孵化，每天收集新孵化的幼虫。在显微镜下统计线虫悬浮液中的线虫数量，并取 3~5 次重复，计算线虫悬浮液平均值（浓度）。将小麦种子进行催芽，在芽长至 0.5~1 cm 时即可进行播种。

根据每株上的白雌虫与孢囊总和来划分小麦抗感性，免疫（M）：根系不形成白雌虫；高抗（HR）：根系形成白雌虫数 0.1~5.0；中抗（MR）：白雌虫数 5.1~10.0；中感（MS）：白雌虫数 10.1~20.0；高感（HS）：白雌虫数 > 20。统计结果显示：获得免疫品种 1 个，高抗品种 14 个，中抗品种 55 个，中感品种 57 个，高感品种 116 个。从鉴定结果看出我国黄淮地区推广的重要小麦品种中感病品种占有的比例远远高于抗性品种。同时，发现免疫品种并不能阻止二龄幼虫（J2）对根尖的侵入，因此尚需进行多批次的验证和田间鉴定试验，确保免疫品种的抗性表现稳定可靠；对于抗性品种来说，二龄幼虫（J2）的侵染数量明显增加很多；而高感品种二龄幼虫（J2）的侵染数量激增，远远高于免疫和抗性品种。而这些品种的田间抗性表现尚需进行多批次、多地点的鉴定试验。

关键词：菲利普孢囊线虫；抗性；黄淮麦区

---

* 基金项目：国家自然科学基金（31801717）；河南省高等学校重点科研项目计划（19A210017）

** 第一作者：周博，硕士研究生，主要从事植物病理学研究；E-mail：1635101947@qq.com

*** 通信作者：崔江宽，博士，讲师，主要从事植物线虫致病机理和防控技术研究；E-mail：jiangk.cui@henau.edu.cn

谷子种子带线虫检测*

白　辉[1**]，宋振君[1,2]，全建章[1]，王永芳[1]，马继芳[1]，李志勇[1***]，董志平[1***]
（1. 河北省农林科学院谷子研究所，国家谷子改良中心，河北省杂粮重点实验室，
石家庄　050035；2. 河北师范大学生命科学学院，石家庄　050024）

摘　要：为了快速检测谷子种子携带线虫情况，利用 PCR 技术对采集自中国谷子主产区不同区域内的 99 份谷子种子的带线虫情况进行检测。结果表明，根据贝西滑刃线虫核糖体 28S rRNA-D2/D3 片段设计的引物对谷子线虫具有高度的特异性和灵敏性，扩增出 245bp 的特异性目的片段，对谷子线虫的检测浓度最低为 0.125ng/μL。在 99 份谷子种子 DNA 中，有 33 份种子检测到特异性扩增条带，测序结果表明所有扩增条带对应序列与线虫 28S rRNA 序列的相似性达 97%以上，进一步说明 33 份种子携带线虫。重复实验结果表明，阳性种子的带线虫频率介于 33.3%~100%，且带线虫的谷子种子主要来自春谷区。本研究建立了一种特异性好，灵敏度高的检测谷子种子带线虫的一步 PCR 方法；谷子线虫病是夏谷区的主要病害，随着谷子春夏谷区的交流，线虫病成为春谷区病害，种子带病可能是主要初侵染源。

关键词：谷子；贝西滑刃线虫；种子带菌；聚合酶链式反应

---

* 基金项目：国家自然科学基金项目（31872880）；河北省农林科学院创新工程（2019-4-2-3）；国家现代农业产业技术体系（CARS-07-13.5-A8）；河北省优秀专家出国培训项目

** 第一作者：白辉，博士，副研究员，从事谷子抗病分子生物学研究；E-mail：baihui_mbb@126.com
*** 通信作者：李志勇，博士，研究员，从事谷子病害研究；E-mail：lizhiyongds@126.com
董志平，硕士，研究员，从事谷子病虫害研究；E-mail：dzping001@163.com

拟禾本科根结线虫的水稻取食位点中蔗糖的胞内运输机制研究*

许立鹤**，肖立英，肖雪琼，王高峰，肖炎农***
（华中农业大学植物科学技术学院，湖北省植物病害监测与安全控制重点实验室，武汉 430070）

摘　要：拟禾本科根结线虫（*Meloidogyne graminicola*，简称为 MG）是为害水稻的根结线虫主要类群之一，在我国水稻的主要种植区均有分布，已成为威胁我国水稻生产安全的重要土传病原。MG 从水稻根尖侵入并在根组织中建立取食位点细胞——巨型细胞（Giant Cell），从中摄取糖类等营养物质。前人研究发现，根结线虫与孢囊线虫，以及不同根结线虫间，其取食位点细胞中蔗糖的胞内运输机制存在差异。本研究以 MG 为研究对象，解析在 MG 诱导形成的水稻根部巨型细胞中，蔗糖的胞内运输机制。首先，通过 CFDA（Carboxyfluoresceine-diacetate）叶片加载试验，分析在巨型细胞中胞间连丝介导的蔗糖被动运输。试验结果显示，在 MG 侵染后的二龄幼虫和雌虫成虫中，以及在相应的巨型细胞中均检测到 CFDA 荧光信号。这表明，胞间连丝介导的蔗糖被动运输，可能参与了取食位点细胞中蔗糖的胞内运输。在此基础上，以实验室所培育的水稻胼胝质降解酶基因 *OsGN*1 的超表达转基因水稻为植物材料，进行 MG 接种试验。试验结果表明，与野生型水稻相比，超表达 *OsGN*1 可显著提高 MG 的侵染率，并促进 MG 在水稻根组织中的发育。据此推测，OsGN1 通过降解胼胝质，提高了 MG 取食位点细胞中胞间连丝介导的蔗糖被动运输效率，进而提高了水稻对 MG 的易感性。然后，应用定量 PCR 技术检测了在 MG 诱导形成的水稻根结（Galls）组织中，水稻 *OsSUTs* 基因家族成员 *OsSUT*1-5 的基因表达模式。试验结果显示，*OsSUT*1-5 不受 MG 诱导表达。同时，以实验室所培育的 4 个水稻遗传材料 $\text{Promoter}_{OsSUT2-5}$：*GUS* 为植物材料，通过 MG 接种及 GUS 组织染色分析试验，进一步证实了在 MG 诱导形成的水稻取食位点细胞中，*OsSUT*2-5 这 4 个基因不受 MG 诱导表达。由此可见，*OsSUTs* 介导的蔗糖主动运输途径，未参与巨型细胞中蔗糖的胞内运输。综合上述研究结果可知，在 MG 诱导形成的水稻根部巨型细胞中，蔗糖可能通过胞间连丝介导的被动运输途径向胞内运输，而非 *OsSUTs* 介导的蔗糖主动运输途径。本研究为进一步解析 MG 对水稻根部巨型细胞中碳水化合物供应的调控机制奠定了基础。

关键词：拟禾本科根结线虫；蔗糖运输；取食位点细胞；胞间连丝；*OsSUTs*

---

* 基金项目：国家自然科学基金（31501613）

** 第一作者：许立鹤，博士研究生，植物病理学；E-mail：248760951@ qq. com
*** 通信作者：肖炎农，博士，教授，植物病原线虫致病机制及绿色防控；E-mail：xiaoyannong@ mail. hzau. edu. cn

第六部分
抗病性

Antifungal effects of dimethyl trisulfide against *Colletotrichum gloeosporioides* infection on mango*

Tang Lihua[1,2**], Mo Jianyou[1,2], Guo Tangxun[1,2],
Huang Suiping[1,2], Li Qili[1,2***], Ning Ping[3], Tom Hsiang[4]
(1. *Institute of Plant Protection*, *Guangxi Academy of Agricultural Sciences*, *Nanning* 530007, *China*; 2. *The Key Lab for Biology of Crop Diseases and Insect Pests of Guangxi*, *Nanning* 530007, *China*; 3. *Guangxi Agricultural Vocational College*, *Nanning* 530007, *China*; 4. *School of Environmental Sciences*, *University of Guelph*, *Guelph*, *ON*, *N1G 2W1*, *Canada*)

Abstract: *Colletotrichum gloeosporioides*, one of the main agents of mango anthracnose, causes latent infections in unripe mango, and can lead to huge losses during fruit storage and transport. Dimethyl trisulfide (DMTS) is an antifungal agent produced by several microorganisms or plants, but its effects on the infection process of *C. gloeosporioides* have not been well characterized. A histological investigation demonstrated that DMTS exhibits strong inhibitory effects on the infection process of *C. gloeosporioides* in planta by inhibiting the germination of conidia and formation of appressoria, damaging cytoplasm to cause cells to vacuolate, and contributing to deformation of appressoria prior to penetration. This is the first study to demonstrate antifungal activity of DMTS against *C. gloeosporioides* on mango by suppression of the infection process, thus providing a novel postharvest biorational control for mango anthracnose.

Key words: dimethyl trisulfide, *Colletotrichum gloeosporioides*, mango, infection process

1 INTRODUCTION

Mango (*Mangifera indica* L.) is one of the most important tropical fruits in the world because of its desirable taste and flavor (Sherman *et al.*, 2015). After India, China is the second-largest world-wide mango producer, concentrated in the provinces of Hainan, Guangxi, Yunnan, Sichuan, Guangdong, and Fujian (Wu *et al.*, 2014; Mo *et al.*, 2018). *Colletotrichum gloeosporioides* has been reported as one of the most important pathogens worldwide that can infect at least 1 000 plant species, such as avocado (*Persea americana*), chili (*Capsicum annuum*), coffee (*Coffea arabica*), papaya (*Carica papaya*), strawberry (*Fragria frageriae*), and mango (*Mangifera indica*) (Phoulivong *et al.*, 2010). Mango anthracnose, caused by the fungus *C. gloeosporioides*, has become a major destructive disease both in the field and at postharvest (Bautista-Rosales *et al.*, 2014; Xu *et al.*, 2017). Typical symptoms on ripe fruit are sunken, prominent, dark brown to black decay spots. The spots on fruit usually do coalesce and are associated with extensive fruit rot. On unripe fruit, infections remain latent and with no macroscopic symptoms until fruit ripen and environmental factors favour the development

* 基金项目：国家自然科学基金（31560526，31600029，31860482）；广西自然科学基金（2016GXNSFCB380004，2018GXNSFBA281088）；广西作物病虫害生物学重点实验室基金（17-259-47-ST-3）
** 第一作者：唐利华，助理研究员，主要研究方向为果树病害及其防治；E-mail：654123597@qq.com
*** 通信作者：李其利，副研究员，主要研究方向为果树病害及其防治；E-mail：65615384@qq.com，liqili@gxaas.net

of the disease (Arauz 2000; Nelson 2008; De Souza *et al.*, 2013; Bautista-Rosales *et al.*, 2014). Currently, effective management of mango anthracnose mainly relies on chemical control. However, applying chemical fungicides to control mango anthracnose can result in environmental and food pollution, as well as a build-up of fungicide resistance (Rezende *et al.*, 2015). Biocontrol agents to control plant diseases include nonpathogenic microorganisms and antifungal compounds from plants or microorganisms, and these can serve as alternatives to chemical fungicides leading to their reduced use (Živković *et al.*, 2010; Xu *et al.*, 2017).

Dimethyl trisulfide (DMTS), a novel antifungal compound, can be produced by Chinese leek (*Allium tuberosum*), onion (*Allium cepa*), cabbage (*Brassica oleracea*), and some bacteria and fungi (Zhang *et al.*, 2013; Li *et al.*, 2014). DMTS has shown strong inhibitory effects on the growth of *Penicillium italicum* and *Fusarium oxysporum* f. sp. *cubense* tropical race 4 (Li *et al.*, 2010; Zuo *et al.*, 2017). However, the toxicity of DMTS to *C. gloeosporioides* has not been previously demonstrated. In the present study, we examined the inhibitory effects of DMTS *in planta* on the infection process of *C. gloeosporioides* on mango to better understand the mechanism by which DMTS can affect fungal pathogens and reduce disease.

2 MATERIALS AND METHODS

2.1 Dimethyl trisulfide (DMTS) preparation

DMTS (W327506-100G-K, Sigma-Aldrich) was prepared to a concentration of 100 μL/L DMTS of air as follows: a small 1 mL container holding a piece of filter paper (15 mm×15 mm) containing 400 μL liquid DMTS was placed in the centre of the porcelain shelf of a clear glass desiccator (volume 4 L) (Lee *et al.*, 2003). And then the glass desiccator was well sealed using parafilm to prevent dissipation.

2.2 Pathogen isolation and conidial suspension preparation

Colletotrichum gloeosporioides isolate TD3 was collected from a lesion on a symptomatic mango leaf from Pingma town, Tiandong County, Guangxi Province of China. In previous work, isolate TD3 has been confirmed to be pathogenic to leaves and fruit (Mo *et al.*, 2018), as well as producing abundant conidia in culture. Before inoculation, conidial suspensions were made by growing the fungus of TD3 on potato dextrose agar (PDA) at 25℃ for 10 days to produce conidia, and then washing conidia and diluting to 5×10^5 conidia/mL in sterile water.

2.3 Mango fruit preparation

Healthy-looking fruits of susceptible mango "Tainong 1" were obtained from a local mango market, in Tiandong County, Guangxi province, China. Mango fruits were surface sterilized in 2.5% sodium hypochlorite for 3 min, rinsed three times using sterile water, and then surface air dried (Li *et al.*, 2010). After that, six mango fruits were placed in each 4 L glass desiccator for inoculation and treatment, with separate dessicator for each sampling time.

2.4 Inoculation and treatments

Preventive and curative tests involved histological observations of the infection process on detached mango fruit surfaces. For preventive tests (Table 1), mangos were first fumigated for 24 hours before inoculation (HBI) using 100 μL/L DMTS of air. For inoculation, three spots on the surface of each mango fruit were each wounded with six punctures using a sterilized needle, and each spot inoculated with 2 μL droplets of a conidial suspension (5×10^5 conidia/mL). All treatments were incubated at 25℃

in 12 h darkness/12 h light cycle with 100% relative humidity, and then sampled at 3, 6, 9, 12, 24, 48, 72, and 96 hours after inoculation (HAI) (Table 1). For curative tests (Table 1), DMTS exposure started at several different times after inoculation (3, 12, 24 or 48 HAI) and continued until the fruits were sampled (up to 96 HAI).

Table 1　Schedule for fumigation treatments with DMTS, and sampling time in the histological investigation of infection process of *C. gloeosporioides*

DMTS treatment time	Sampling time (hours after inoculation)
Fumigation for 24 h before inoculation (preventive)	3, 6, 9, 12, 24, 48, 72, 96
Fumigation starting at 3 h after inoculation (curative)	6, 9, 12, 24, 48, 72, 96
Fumigation starting at 12 h after inoculation (curative)	24, 48, 72, 96
Fumigation starting at 24 h after inoculation (curative)	48, 72, 96
Fumigation starting at 48 h after inoculation (curative)	72, 96

2.5　Sampling, staining and investigation

The sampling timesare listed in Table 1. At each sampling time, 5 mm × 5 mm mango skin slices that had been inoculated were sampled and stained with ethanol-lactophenol trypan blue following Koch and Slusarenko (1990). At least 100 conidia and associated appressoria per sample were observed by microscopy (Nikon, Ni-E), and the number of conidia that germinated (or not) were counted as well as the incidence of appressoria, and their morphology recorded. This was done in triplicate for each sampling time, and the experiment was repeated three times.

2.6　Statistical analysis

Statistical analysis was performed using Data Processing System software (DPS7.0) for Windows (Zhejiang University, Hangzhou, China) (Tang and Zhang, 2013), and data were subjected to analysis of variance (ANOVA) followed by Duncan's multiple range tests for means separation. An arcsine transformation was applied to percentage data of conidial germination and appressorial formation prior to ANOVA.

3　RESULTS

3.1　Effects of DMTS on the rate of conidial germination and appressorial formation of *C. gloeosporioides*

In thenon-treated control group, conidia could normally germinate starting at 6 hours after inoculation (HAI), and form appressoria starting at 24 HAI. In contrast, when DMTS was applied for 24 hours before inoculation (HBI) or if DMTS treatment began at 3 HAI with subsequent samplings from 6 to 96 HAI, there was no conidial germination, which was significantly different than the inoculated control ($P \leq 0.01$). For DMTS fumigation which began at 12 HAI and continued through sampling periods at 24 to 96 HAI, the rates of conidial germination at each sampling time were above 70%, and there was a significant difference from the inoculated control at 48 HAI sampling time ($P \leq 0.05$), but not at other sampling times of 24 HAI, 72 HAI, and 96 HAI. For DMTS fumigation which began at 24 HAI or 48 HAI, the rates of conidial germination were all above 95% at each sampling time, with no significant differences from the respective inoculated controls (Table 2). For appressorial formation, when DMTS was applied for 24 HBI and started at 3 HAI, 12 HAI, or 24 HAI, the rate was zero at every sampling time, which was significant different than inoculated control ($P \leqslant 0.01$). For DMTS treatment which began at 48 HAI, the rates of appressorial formation at sampling times of 72 HAI and 96 HAI were significantly less than the inoculated control ($P \leqslant 0.01$) (Table 3).

Table 2 Effects of 100 μL/L DMTS at 25℃ on the rate of conidial germination of *C. gloeosporioides* on mango fruit *in planta*

Treatment [A]	Conidial germination (%) at several sampling times (hours after inoculation, HAI) [B]							
	3 h	6 h	9 h	12 h	24 h	48 h	72 h	96 h
Control	0	5.7±0.6	29.0±3.6	57.7±1.5	73.3±2.5	94.3±3.8	97.0±1.0	97.7±1.5
24 HBI	0	0**	0**	0**	0**	0**	0**	0**
3 HAI	_	0**	0**	0**	0**	0**	0**	0**
12 HAI	_	_	_	_	72.7±4.2	87.7±4.7*	96.7±1.5	96.0±1.7
24 HAI	_	_	_	_	_	95.3±0.6	95.3±1.5	97.3±1.2
48 HAI	_	_	_	_	_	_	95.7±0.6	96.7±1.2

[A] DMTS treatments were applied for 24 h before inoculation (24 HBI) or starting at 3 h after inoculation (3 HAI), 12 h after inoculation (12 HAI), 24 h after inoculation (24 HAI), or 48 h after inoculation (48 HAI). The control was an inoculated check without DMTS treatment.

[B] Mean±standard deviations are based on 3 replicates, and at least 100 conidia per replicate were observed. Asterisks indicate significant differences versus the control, *, $P < 0.05$, **, $P < 0.01$, "–" not tested.

Table 3 Effects of 100 μL/L DMTS at 25℃ on the rate of appressorium formation of *C. gloeosporioides* on mango fruit *in planta*

Treatment [A]	Appressorium formation (%) at several sampling times (hours after inoculation, HAI) [B]							
	3 h	6 h	9 h	12 h	24 h	48 h	72 h	96 h
Control	0	0	0	0	5.3±1.5	33.3±4.2	40.3±2.5	69.7±2.1
24 HBI	0	0	0	0	0**	0**	0**	0**
3 HAI	_	0	0	0	0**	0**	0**	0**
12 HAI	_	_	_	_	0**	0**	0**	0**
24 HAI	_	_	_	_	_	0**	0**	0**
48 HAI	_	_	_	_	_	_	36.7±1.5**	35.3±2.5**

[A] DMTS treatments were applied for 24 h before inoculation (24 HBI) or starting at 3 h after inoculation (3 HAI), 12 h after inoculation (12 HAI), 24 h after inoculation (24 HAI), or 48 h after inoculation (48 HAI). The control was an inoculated check without DMTS treatment.

[B] Mean± standard deviations are based on 3 replicates, and at least 100 conidia and associated appressoria per replicate were observed. Asterisks indicate significant differences versus the control, *, $P < 0.05$, **, $P < 0.01$, "–" not tested.

3.2 Effects of DMTS on the morphology of conidia of *C. gloeosporioides* during the infection process

Morphology ofconidia of *C. gloeosporioides* with or without DMTS treatments during infection process was evaluated. In the inoculated control, the morphology of conidia at every sampling time was intact, and the conidia could germinate normally and form regular appressoria. However, with DMTS treatments for 24 HBI or which began at 3 HAI, no germinated conidia were observed, and conidial plasmolysis appeared at 96 HAI, with enhanced dissolution and decrease of cytoplasmic contents for conidia with the longest exposure (sampling at 96 HAI and exposed to DMTS for 93 hours). For DMTS treatments starting at 12 HAI or 24 HAI, germinated conidia were observed at each sampling time, but no appressoria were found. The conidia and germinated hyphae at 96 HAI were visibly altered with destruction and decrease of cytoplasmic contents as well. For DMTS treatment starting at 48 HAI, deformed appressoria were observed at 72 and 96 HAI, and the cytoplasm of conidia and germinated hyphae were obviously vacuolated (Fig. 1).

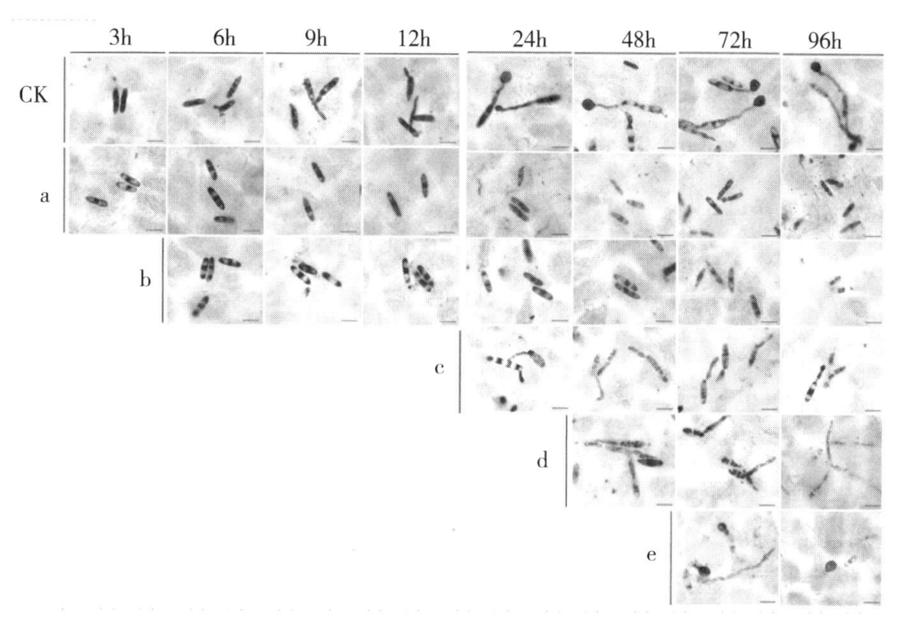

Fig. 1 Morphology of conidia of *C. gloeosporioides* treated with 0 or 100 μL/L DMTS at 25℃ on mango fruit during the infection process (Scale bar = 10 μm). Sampling times were 3, 6, 9, 12, 24, 48, 72, and 96 h after inoculation. Control (CK) was an inoculated check without DMTS treatment. (a) Treatment for 24 h before inoculation. (b) Treatment starting at 3 h after inoculation. (c) Treatment starting at 12 h after inoculation. (d) Treatment starting at 24 h after inoculation. (e) Treatment starting at 48 h after inoculation

4 DISCUSSION

Mango anthracnose begins as a quiescent infection on young fruit and can result in failure to producemature fruit. Conidial germination and appressorial formation on the fruit surface are the key events for infection by *C. gloeosporioides* (Estrada *et al.*, 2000). Antifungal substances from plants or microorganisms have been found to inhibit fungal development either *in vitro* or *in vivo* by inhibiting conidial germination or appressorial formation (Barrera-Necha *et al.*, 2008; Yang *et al.*, 2008; Li *et al.*, 2010; Li *et al.*, 2011; Li *et al.*, 2012). In our study, when DMTS was applied for 24 HBI or beginning at 3 HAI, no germinated conidia were detected. Previous studies showed that DMTS together with the volatiles from Chinese leek could inhibit conidial germination of *Fusarium oxysporum* f. sp. *cubense* tropical race 4 (*Foc* TR4), DMTS displayed the strongest inhibition among volatiles (Zhang *et al.*,

2013; Zuo *et al.*, 2015; Zuo *et al.*, 2017). Moreover, DMTS together with the volatiles emitted from *Streptomyces globisporus* JK-1 exhibited strong inhibitory effects on conidial germination of *P. italicum* and *Botrytis cinerea*, and DMTS acted as the main active compound (Li *et al.*, 2010; Li *et al.*, 2012). When DMTS was applied starting at 12 HAI, 24 HAI, or 48 HAI, DMTS exposure had no inhibitory effect on conidial germination of *C. gloeosporioides* compared to the inoculated control, likely since conidia normally germinate by 6 HAI. However, this post-inoculation exposure had strong effects on appressorial formation which normally start at 24 HAI. This is similar to the inhibitory effect of volatiles from *S. globisporus* JK-1 on the infection process of *B. cinerea* on tomato fruit (Li *et al.*, 2012), and DMTS was found in those volatiles (Li *et al.*, 2010). Similarly, the compound Physcion, which is also an antifungal agent from the roots of Chinese rhubarb, showed similar inhibitory effects on *Blumeria graminis* (Yang *et al.*, 2008).

Previous studies showed that thedevelopment of appressoria is linked to remodeling of the actin cytoskeleton, mediated by septin GTPases, and rapid cell wall differentiation (Deising *et al.*, 2000; Ryder and Talbot, 2015). All of above results indicated that DMTS mainly affected the early stages of the infection process of mango anthracnose by inhibiting conidial germination and appressorial formation. Early stages of infection by a variety of fungal phytopathogens, including *C. gloeosporioides*, is associated with an increase in peroxidase enzyme activity, and the increase of peroxidase enzyme activity is related to the inducible expression of peroxidase gene (Harrison *et al.*, 1995; Mir *et al.*, 2015). We speculate that the inhibitory effects of DMTS on the early stages of infection of *C. gloeosporioides* on mango may be associated with suppression of expression of the peroxidase gene, but further research is required.

Morphological observation demonstrated that DMTS had an impact on cytoplasm to cause cells become nearly fully vacuolated and contributed to deformation of appressoria in our study. Li *et al.* (2012) reported that volatiles from *S. globisporus* JK-1 caused the destruction of cytoplasmic contents to increase the vacuolation of conidia of *B. cinerea*. And Li *et al.* (2014) reported that garlic oil could firstly penetrate into hyphae cells and even their organelles, and then destroy the cellular structure, finally leading to the leakage of both cytoplasm and macromolecules. In these studies, DMTS was the active ingredient in the volatiles and garlic oil (Li *et al.*, 2012; Li *et al.*, 2014). Therefore, DMTS plays an important role in damaging cellular structures, but its deeper mechanism of action against plant pathogens need further investigation.

5 CONCLUSIONS

This study demonstrated for the first time that the volatile substanceDMTS affects the infection process of *C. gloeosporioides* by suppressing conidial germination and appressorial formation *in planta*, as well as damaging cytoplasm to cause cells to become vacuolated. Our findings suggest that the use of DMTS may be an efficacious alternative in the control of enclosed environment postharvest diseases. Although the study demonstrated effective suppression of infection of *C. gloeosporioides* on mango by DMTS, the molecular mechanisms of action and main target site are not well understood. Further studies are required to provide new insights into the mechanism of action of DMTS against *C. gloeosporioides*.

References

Arauz L F. 2000. Mango anthracnose: Economic impact and current options for integrated managaement [J]. Plant disease, 84 (6): 600-611.

Barreranecha L L, Bautistabanos S, Floresmoctezuma H F, *et al.* 2008. Efficacy of essential oils on the conidial germination, growth of *Colletotrichum gloeosporioides* (Penz.) Penz. and Sacc and control of postharvest diseases in papaya (*Carica papaya* L.) [J]. Plant Pathology Journal, 7 (2): 174-178.

Bautista-Rosales P U, Calderon-Santoyo M, Servín-Villegas R, *et al.* 2014. Biocontrol action mechanisms of *Cryptococcus laurentii* on *Colletotrichum gloeosporioides* of mango [J]. Crop Protection, 65 (4): 194-201.

Deising H B, Werner S, Wernitz, M. 2000. The role of fungal appressoria in plant infection [J]. Microbes & Infection, 2 (13): 1631.

De Souza A, Delphino Carboni R C, Wickert E, *et al.* 2013. Lack of host specificity of *Colletotrichum* spp. isolates associated with anthracnose symptoms on mango in Brazil [J]. Plant Pathology, 62 (5): 1038-1047.

Estrada A B, Dodd, J C, Jeffries P. 2000. Effect of humidity and temperature on conidialgermination and appressorium development of two Philippine isolates of the mango anthracnose pathogen *Colletotrichum gloeosporioides* [J]. Plant Pathology, 49 (5): 608-618.

Harrison S J, Curtis M D, Mcintyre C L, *et al.* 1995. Differential expression of peroxidase isogenes during the early stages of infection of the tropical forage legume *Stylosanthes humilis* by *Colletotrichum gloeosporioides* [J]. Molecular Plant-microbe Interactions, 8 (3): 398-406.

Koch E, Slusarenko A, *et al.* 1990. *Arabidopsis* is susceptible to infection by a downy mildew fungus [J]. Plant cell, 2: 437-445.

Lee B H, Annis P C, Tumaali F, *et al.* 2003. The potential of 1, 8-cineole as a fumigant for stored wheat [C] // *Proceedings of the Australian Postharvest Technical Conference*, Canberra: 25-27.

Li Q, Jiang Y, Ning P, *et al.* 2011. Suppression of *Magnaporthe oryzae* by culture filtrates of *Streptomyces globisporus* JK-1 [J]. Biological Control, 58 (2): 139-148.

Li Q, Ning P, Zheng L, *et al.* 2012. Effects of volatile substances of *Streptomyces globisporus* JK-1 on control of *Botrytis cinerea* on tomato fruit [J]. Biological Control, 61 (2): 113-120.

Li Q L, Ping N, Lu Z, *et al.* 2010. Fumigant activity of volatiles of *Streptomyces globisporus* JK-1 against *Penicillium italicum* on *Citrus microcarpa* [J]. Postharvest Biology & Technology, 58 (2): 157-165.

Li W R, Shi Q S, Liang Q, *et al.* 2014. Antifungal effect and mechanism of garlic oil on *Penicillium funiculosum* [J]. Applied Microbiology & Biotechnology, 98 (19): 8337-8346.

Mir A A, Park S Y, Sadat M A, *et al.* 2015. Systematic characterization of the peroxidase gene family provides new insights into fungal pathogenicity in *Magnaporthe oryzae* [J]. Scientific Reports, 5: 11831.

Mo J, Zhao G, Li Q, *et al.* 2018. Identification and characterization of *Colletotrichum* species associated with mango anthracnose in Guangxi, China [J]. Plant Disease, 102 (7): 1283-1289.

Nelson S C. 2008. Mango anthracnose (*Colletotrichum gloeosporiodes*). University of Hawai'i at Manoa [D]. College of Tropical Agriculture and Human Resources [D]. Cooperative Extension Service.

Phoulivong S, Cai L, Chen H, *et al.* 2010. *Colletotrichum gloeosporioides* is not a common pathogen on tropical fruits [J]. Fungal Diversity, 44 (1): 33-43.

Rezende D C, Fialho M B, Brand S C, *et al.* 2015. Antimicrobial activity of volatile organic compounds and their effect on lipid peroxidation and electrolyte loss in *Colletotrichum gloeosporioides* and *Colletotrichum acutatum* mycelia [J]. African Journal of Microbiology Research, 9 (23): 1527-1535.

Ryder L S, Talbot N J. 2015. Regulation of appressorium development in pathogenic fungi [J]. Current Opinion in Plant Biology, 26: 8-13.

Sherman A, Rubinstein M, Eshed R, *et al.* 2015. Mango (*Mangifera indica* L.) germplasm diversity based on single nucleotide polymorphisms derived from the transcriptome [J]. Bmc Plant Biology, 15 (1): 277.

Tang Q Y, Zhang C X. 2013. Data Processing System (DPS) software with experiental design, statistical analysis and data mining developed for use in entomological research [J]. Insect Science, 20 (2): 254-260.

Wu H X, Jia H M, Ma X W, *et al.* 2014. Transcriptome and proteomic analysis of mango (*Mangifera indica* Linn) fruits [J]. Journal of Proteomics, 105: 19-30.

Yang X J, Yang L J, Yu D Z, *et al.* 2008. Effects of physcion, a natural anthraquinone derivative, on the infection process of *Blumeria graminis* on wheat [J]. Canadian Journal of Plant Pathology, 30 (3): 391-396.

Xu X , Lei H, Ma X, *et al.* 2017. Antifungal activity of 1-methylcyclopropene (1-MCP) against anthracnose (*Colletotrichum gloeosporioides*) in postharvest mango fruit and its possible mechanisms of action [J]. International Journal of Food Microbiology, 241: 1.

Zhang H, Mallik A, Zeng R S. 2013. Control of Panama disease of banana by rotating and intercropping with Chinese chive (*Allium tuberosum* Rottler): role of plant volatiles [J]. Journal of Chemical Ecology, 39 (2): 243-252.

Živković S, Stojanović S, Ivanović Ž, *et al.* 2010. Screening of antagonistic activity of microorganisms against *Colletotrichum acutatum* and *Colletotrichum gloeosporioides* [J]. Archives of Biological Sciences, 62 (3): 611-623.

Zuo C, Li C, Li B, *et al.* 2015. The toxic mechanism and bioactive components of Chinese leek root exudates acting against *Fusarium oxysporum* f. sp. *cubense* tropical race 4 [J]. European Journal of Plant Pathology, 143 (3): 447-460.

Zuo C, Zhang W, Chen Z, *et al.* 2017. RNA sequencing reveals that endoplasmic reticulum stress and disruption of membrane integrity underlie dimethyl trisulfide toxicity against *Fusarium oxysporum* f. sp. *cubense* tropical race 4 [J]. Frontiers in Microbiology: 8.

*BnLPT*1 is a causal gene underlying a quantitative locus conferring resistance to *Sclerotnia sclerotiorum*

Zuo Rong*, Tang Minqiang, Bai Zetao, Li Yan, Huang Junyan,
Zhong Xue, He Yizhou, Cheng Xiaohui, Liu Yueying, Liu Shengyi**
(*Oil Crops Research Institute of Chinese Academy of Agricultural Sciences, Key Laboratory of Biology and Genetics Improvement of Oil Crops, Ministry of Agriculture and Rural Affairs, Wuhan* 430062, *China*)

Abstract: *Brassica napus* (oilseed rape) is one of the four major oil crops in the world and is the first source of China domestic edible oil. Sclerotinia stem rot (SSR), caused by *Sclerotinia sclerotiorum*, is reported to the most serious diseases in oilseed rape. In current agricultural production, the measures for controlling SSR mainly include chemical and biological control and field management, etc., which have certain effect, but the effect is not only relatively limited but also increases the production cost, and leads to environmental pollution. Consequently, selection and breeding disease-resistant varieties is the most economic and effective way to control SSR. In this study, a significant QTL, located on A04 chromosome, was identified by analyzing recombinant population and associated population combined with the phenotype data under natural field conditions and indoors inoculation which were related to SSR resistance, such as SSR incidence and disease index. The homologous gene in *Arabidopsis thaliana* is *AtLTP*1 (At2g38540). Therefore, we over-expressed *BnLTP*1 of *B. napus* and *AtLTP*1 in *A. thaliana* wild type, at the same time, RNAI interferences of *AtLTP*1 was constructed . We got stable integration and expression of transgenes in T0 and T2 generation plants through PCR and qRT-PCR analyses. The resistance to *S. sclerotiorum* of transgenic materials, which show inheritance in Mendelian fashion (3 : 1), were evaluated under greenhouse conditions. The results revealed that the lines of over-expression of *BnLTP*1 and *AtLTP*1 were more resistant to *S. sclerotiorum*, but the interference lines of *AtLTP*1 were more sensitive to *S. sclerotiorum* at 36h after inoculation and afterward. Furthermore, the crude proteins of BnLTP1and AtLTP1 expressed in prokaryotes could significantly inhibit the hyphal growth of *S. sclerotiorum* through *in vitro* bacteriostasis experiment. Therefore *BnLTP*1 can improve resistance to SSR in oilseed rape.

Key words: *Brassica napus*; *Sclerotinia sclerotiorum*; resistance; *BnLTP*1

* First author: Zuo Rong
** Corresponding author: Liu Shengyi; E-mail: liusy@ oilcrop. cn

Genetic analysis of quantitative resistance to stripe rust in wheat landrace "Wudubaijian" in multi-environment trials*

Chao Kaixiang, Li Juan, Wang Wenli, Zhang Jia, Li Qiang**, Wang Baotong**

(*State Key Laboratory of Crop Stress Biology for Arid Areas, College of Plant Protection, Northwest A&F University, Yangling* 712100, *China*)

Abstract: Stripe rust, caused by *Puccinia striiformis* f. sp. *tritici* (*Pst*), is one of the most important diseases of wheat (*Triticum aestivum* L.) and causes substantial yield losses in many wheat-growing regions of the world. Identifying and utilization of new resistance genes is the most effective way for achieving durable disease control. Wudubaijian, a wheat landrace derived from Gansu Province, exhibits adult-plant resistance to Chinese predominant *Pst* races. Mingxian169/Wudubaijian $F_{2:3}$ lines were evaluated for stripe rust reaction in the four environments of Yangling and Tianshui in 2015 and 2016. Analysis of the relative area under disease progress curve (rAUDPC) indicated that the resistance in Wudubaijian was controlled by more than one Quantitative trait loci (QTL). Combined with phenotypic data and molecular markers, two QTLs were identified with high Phenotypic variation explained (PVE) in two calculation methods. *QYrwdbj. nwafu-2B.* 1 with the phenotypic variance of 9. 54%-10. 40%, is located in the bin C-2BS1-0. 53 of chromosome 2BS; *QYrwdbj. nwafu-5A* with the phenotypic variance of 15. 02%-40. 26%, is located between 5AS1-0. 40-0. 75 and 5AS3-0. 75-0. 98 of chromosome 5AS. Molecular detection and epistasis analysis suggested that *QYrwdbj. nwafu-5A* may be a new major QTL that can be used in conjunction with other sites. This model provides new ideas for the realization of durable resistance in breeding design. The flanking markers *AX*-111500211 and *AX*-110411572 linked to *QYrwdbj. nwafu-2B.* 1 and the flanking markers *AX*-110458796, *AX*-109621625 and *AX*-111568277 linked to *QYrwdbj. nwafu-5A* can also be used in molecular marker assisted breeding.

* Funding: This research was supported by the National Key Research and Development Program of China (2016YFD0300705, 2018YFD0200404), the Natural Science Basic Research Plan in Shaanxi Province of China (2019JZ-17)

** Corresponding authors: Li Qiang; E-mail: qiangli@ nwsuaf. edu. cn

Wang Baotong; E-mail: wangbt@ nwsuaf. edu. cn

A Kind of Plant Small Molecular Chaperones Enhance the Tolerance to Biotic and Abiotic Stresses*

Guo Liuming[1,2], Li Jing [2], He Jing[1,2], Liu Han [1,2],
Yuan Zhengjie [1], Zhang Hengmu [1**]
(1. *Institute of Virology and Biotechnology*, *Zhejiang Academy of Agricultural Sciences*, *Hangzhou* 310021, *China*; 2. *College of Chemistry and Life Science*, *Zhejiang Normal University*, *Jinhua* 321004, *China*)

Abstract: Small heat shock proteins (sHSPs) have thought to function as chaperones, protecting their targets from denaturation and aggregation when organisms are subjected to various biotic and abiotic stresses (Haslbeck *et al.*, 2019). In our previous study, a sHSP from Oryza sativa (OsHSP20) and its homolog (NbHSP20) from *N. benthamiana* were cloned and found to self-interact and forms granules within the cytoplasm. Interestingly, they could interact *in vitro* and *vivo* with RNA-dependent RNA polymerase (RdRp) of *Rice stripe virus* (RSV), which causes a severe disease of rice in Eastern Asia. Further analysis showed that RSV infection could alter the expression pattern or sub-cellular distribution of HSP20s by their hetero-interaction in protoplasts of rice and epidermal cells of *N. benthamiana* (Li *et al.*, 2015). However, their functions remain largely unclear. To deep insight into their roles, here we investigated their expression pattern, molecular chaperone activity, and effects on tolerance to various biotic and abiotic stresses in *E. coli* (a prokaryote), yeast, and plants (eukaryotes) overexpressing the HSP20s. Their expressions were significantly up-regulated by viral infection, heat shock and high salinity. The purified recombinant proteins were shown to inhibit the thermal aggregation of the mitochondrial malate dehydrogenase (MDH) enzyme *in vitro*, suggesting their molecular chaperone activity. Heterologous expression of HSP20 in *Escherichia coli* or *Pichia pastoris* cells enhanced heat and salt stress tolerance compared with the control cultures. Transgenic plants had longer roots and higher germination rates than those of control plants when exposed to heat and salt treatments, consistently supporting the viewpoint that HSP20s confer heat and salt tolerance by its molecular chaperone activity in different organisms. On the other hand, the transgenic plants were artificially inoculated with *Tomato mosaic virus* (ToMV), *Turnip Mosaic Virus* (TuMV) and RSV, and were found to delay the symptom development, indicating that HSP20s may positively regulate the defense response of plants to viral diseases. The result could be helpful to elucidate the biological function of HSP20s during virus infection in plants.

* Funding: Natural Science Foundation of China (31501604) and the Key R&D Project of Zhejiang Province (2019C02018)
** Corresponding author: Zhang Hengmu; E-mail: zhhengmu@tsinghua.org.cn

Investigating the antifungal activity and mechanism of a microbial pesticide Ningnanmycin against *Phoma segeticola* var. *camelliae*[*]

Li Dongxue[**], Wang Xue, Yin Qiaoxiu, Ren Yafeng, Bao Xingtao, Dharmasena Dissanayake-saman-pradeep, Wu Xian, Song Baoan, Chen Zhuo[***]

(*State Key Laboratory Breeding Base of Green Pesticide and Agricultural Bioengineering, Guizhou University, Guiyang* 550025, *China*)

Abstract: *Phoma segeticola* var. *camelliae* was isolated from tea leaf spot disease in tea garden, Shiqian county, Guizhou province. Tea leaf spot disease affected seriously the quality and yield of tea. At present, there are no effective measures of prevention and control. Ningnanmycin (NNM) is a new cytosine nucleoside peptide antibiotic with high efficiency and low toxicity. The method of mycelial growth rate was used to determine the antifungal activity of NNM against *P. segeticola* var. *camelliae* strain GZSQ-4, with EC_{50} of 1 287. 54 U/mL. In order to further study the mechanism of NNM on GZSQ-4, the mycelium treated with NNM at the dosage of EC_{50} for 1 h and 14 h was observed by optical microscope, scanning electron microscopy and transmission electron microscopy, respectively. It was found that the mycelium swelled, the mycelial cytoplasmic density decreased, the granular content appeared and the number of mitochondria granules increased. After GZSQ-4 being treated by NNM at the dosage of EC_{50} for 1 h, RNA-seq showed that 1 363 differentially expressed genes (DEGs) were identified with 743 DEGs up-regulated and 620 DEGs down-regulated. By GO enrichment analysis, the DEGs were most enriched in the structural constituent of ribosome of molecular function, ribosome of cellular component and the oxidation-reduction process of biological process including 56, 38 and 116 DEGs, respectively ($P<0.05$). By KEGG enrichment analysis, the DEGs were most enriched in the pathway of ribosome including 58 DEGs. Ten selected randomly DEGs from treatment group of NNM for 1 h were verified by RT-qPCR. The results showed that the expression level of DEGs was consistent with the result of RNA-seq. These results indicated that NNM may affect the processes of translation of GZSQ-4. Our study will provide useful information for management of tea leaf spot caused by *P. segeticola* var. *camelliae* and increase our understanding of action mechanism of NNM against the pathogen.

Key words: tea disease; *Phoma segeticola* var. *camellia*; Ningnanmycin; antifungal activity; action mechanism

* Funding: This work was supported by National Key Research Development Program of China (2017YFD0200308) and its Post-subsidy project (2018-5262), and the Major Science and Technology Projects in Guizhou Province (No. 2012-6012)

** First author: Li Dongxue; E-mail: gydxli@ aliyun. com

*** Corresponding author: Chen Zhuo; E-mail: gychenzhuo@ aliyun. com

The transcriptome and ultrastructure uncover the anti-fungal mechanism of Ningnanmycin against *Pseudopestalotiopsis camelliae-sinensis* *

Wang Xue[1**], Wen Xiaodong[1,2], Song Xingchen[1,2], Yin Qiaoxiu[1],
Li Dongxue[1], Ren Yafeng[1], Wu Xian[1], Song Baoan[1], Chen Zhuo[1***]
(1. *State Key Laboratory Breeding Base of Green Pesticide and Agricultural Bioengineering, Guizhou University, Guiyang* 550025, *China*; 2. *College of Agriculture, Guizhou University, Guiyang* 550025, *China*)

Abstract: Tea grey blight is an important disease of tea plant (*Camellia sinensis*) all over the world, which leads to a huge loss of tea leaves. *Pseudopestalotiopsis camelliae-sinensis* is the causal agent of tea grey blight in tea garden in Guizhou Province, China. Ningnanmycin (NNM), which was known as an environmentally friendly microbial pesticide, can effectively prevent and control crop diseases caused by fungal pathogens. However, the anti-fungal mechanism of NNM for *Ps. camelliae-sinensis* is still unclear. In this study, we determined the antifungal activity of NNM on *Ps. camelliae-sinensis in vitro*, with EC_{50} of 75.92 U. NNM caused mycelial terminal deformity and swell, and induced to produce the granular contents in the mycelial tube. When the time of treatment was delayed to 14 h, the mycelial deformity was more obvious and the mycelial terminal represented obstructive of development, disorderly. The branches of the growing hyphae represented thinner and shorter using scanning electron microscope. The boundary structures of hypha represented unclear, organelles became atrophy using transmission electron microscope. A time-course transcriptome of analysis revealed that NNM can significantly up-regulated 615 and 865 genes ($P<0.05$) and down-regulated 314 genes and 827 genes ($P<0.05$) for 1 hr or 14 hr at the dosage of 500 U, respectively. Twelve differentially expressed genes (DEGs) were verified using qRT-PCR, and the results showed that these genes represent similar expression trends with RNA-seq data. DEGs were enriched in transmembrane transport (43), oxidation-reduction process (36) and mycelium development (30) at biological process, nucleus (48), integral component of membrane (48) and cytosol (39) at cellular component, as well as zinc ion binding (30), catalytic activity (26) and oxidoreductase activity (26) at molecular function. DEGs were further enriched in biosynthesis of amino acids (23), ABC transporters (19), Tryptophan metabolism (15) for KEGG. Our study will contribute to disease management of tea grey blight caused by *Ps. camelliae-sinensis*, and develop the microbial pesticide through its action mechanism.

Key words: *Camellia sinensis*; *Pseudopestalotiopsis camelliae-sinensis*; Ningnanmycin; anti-fungal activity; transcriptome

---

* Funding: This work was supported by National Key Research Development Program of China (2017YFD0200308) and its Post-subsidy project (2018-5262), and the Major Science and Technology Projects in Guizhou Province (No. 2012-6012)

** First author: Wang Xue; E-mail: gdwangxue@ aliyun. com

*** Corresponding author: Chen Zhuo; E-mail: gychenzhuo@ aliyun. com

Discovering responsive miRNAs in tea plant (*Camellia sinensis*) against leaf spot caused by *Lasiodiplodia theobromae* using high-throughput sequencing and prediction of their targets through degradome*

Jiang Shilong[1,2]**, Li Dongxue[2], Ren Yafeng[2], Bao Xingtao[2], Wu Xian[2], Jiang Xuanli[1]***, Chen Zhuo[2]***

(1. *College of Agriculture, Guizhou University, Guiyang* 550025, *China*;
2. *State Key Laboratory Breeding Base of Green Pesticide and Agricultural Bioengineering, Guizhou University, Guiyang* 550025, *China*)

Abstract: Tea leaf spot infected by *Lasiodiplodia theobromae* was firstly reported in tea plantation in Guizhou Province, which led to a huge loss of tea yield. The response mechanism of tea plant (*Camellia sinensis*) under the stress of the pathogen would contribute for the future research on host-pathogen interactions. Eight hundred and eighteen known miRNAs or conservative miRNAs and 4098 novel miRNAs were identified from tea leaves infected or un-infected by *L. theobromae*. Comparing with two treatment groups, the number of miRNAs with up- and down- regulated trends were 25 and 88 ($P< 0.01$), respectively. Eleven PC miRNAs, nine gma miRNAs, three mtr miRNAs, fve-miR156f, peu-MIR2916-p3_ 2ss18TC19GA, ptc-miR6427-3p_ L-1R-2_ 1ss16AT, rco-MIR403a-p5_ 1ss2TG and stu-miR156a were significantly up-regulated for pathogen infection ($P < 0.01$). Seventy-eight PC miRNAs, five gma miRNAs, seven mtr miRNAs, cas-miR11592_ R+5_ 1ss19AC, eun-miR535a-3p_ 2ss10AC14CT and ppe-miR166a were significantly down-regulated for pathogen infection ($P<0.01$). The Targetfinder software detected a total of 18967 potential targets, which were recognized and cleaved by conserved and novel tea miRNAs (total 3678) using the analysis of the degradome sequence data sets. Target sites (8773) was detected by target gene prediction and sequencing method, and 1747 and 7026 were up- and down- regulated, respectively. The target genes predicted using miRNAs and mRNAs from the degradation group density file were combined to calculate. The number of transcripts corresponding with interacted miRNAs was 4811 with the value of Degradome_ Valid_ miRNA being greater than 0. The target genes of miRNAs were classified at the level of BP, CC and MF. The number of the target genes of miRNAs at oxidation-reduction process, integral component of membrane, and ATP binding was 62, 133 and 107, respectively. The target genes of miRNAs were classified into 22 kinds by KOG. Beside of the kind of [[S], the number of target genes under the kind of [[K] toped list, with the value being 264, followed by [[O] and [[T], with the value being 239 and 224. The target genes of miRNAs were enriched by KEGG. For instance, 126 transcripts were enriched in pathway of Plant-pathogen

* Funding: This work was supported by National Key Research Development Program of China (2017YFD0200308) and its Post-subsidy project (2018-5262), and the Major Science and Technology Projects in Guizhou Province (No. 2012-6012)

** First author: Jiang Shilong; E-mail: jiangsl2003@gmail.com
*** Corresponding authors: Jiang Xuanli; E-mail: jxl32371@163.com
Chen Zhuo; E-mail: gychenzhuo@aliyun.com

interaction. Taking miRNAs ptc-miR6427-3p_ L-1R-2_ 1ss16AT, aly-miR858-5p_ L-1R-2, ath-miR858b_ 1ss21GT and PC-3p-3062093_ 3 as examples, they targeted and down-regulated transcription factors, such as disease resistance protein RPS2 - like (TEA031299. 1), MYBPAR (TEA002308. 1), transcription repressor MYB4 (TEA025886. 1) and disease resistance protein RPM1-like (TEA032387. 1) . Our study provided a significance data for the study of molecular mechanism of tea plant for leaf spot infected by *L. theobromae*.

Key words: miRNAs; degradome; target genes; molecular mechanism; plant-pathogen interaction

Molecular regulation of broad-spectrum resistance against rice blast disease

Chen Xuewei*

(*State Key Laboratory of Exploration and Utilization of Crop Genetic Resource in Southwest China* (*In preparation*), *State Key Laboratory of Hybrid Rice*, *Rice Research Institute*, *Sichuan Agricultural University at Wenjiang*, *Chengdu*, *Sichuan* 611130, *China*)

Abstract: Blastis the most devastating disease of rice. My laboratory has focused on characterization and creation of rice resources with broad-spectrum resistance, uncovering the underlying molecular mechanisms, and breeding disease resistant rice varieties. Through large-scale screening, we have identified 13 rice resources with broad-spectrum resistance, of which five are from natural resources and eight are from ethyl methanesulfonate-induced mutation. Employing these resources, we have discovered several truly novel mechanisms underlying plant broad-spectrum, durable disease resistance. More interestingly, we have uncovered a novel phospho-switch mechanism which allows rice to promote two important but competing biological processes, namely yield and blast resistance. By using these resistant rice resources and resistance alleles, we have generated nineteen elite hybrid rice varieties with broad-spectrum resistance through co-operation with breeders. These rice varieties have been widely utilized for production in China.

* Corresponding author: Chen Xuewei; E-mail: xwchen88@163.com

Post-transcriptionsl modifications play important roles in balancing rice yield and resistance

Wang Jing, Shi Hui, Yi Hong, Long Xiaoyu, Yin Junjie, Chen Xuewei
(*State Key Laboratory of Exploration and Utilization of Crop Genetic Resource in Southwest China (In preparation). State Key Laboratory of Hybrid Rice, Key Laboratory of Major Crop Diseases and Collaborative Innovation Center for Hybrid Rice in Yangtze River Basin, Rice Research Institute, Sichuan Agricultural University at Wenjiang, Chengdu, Sichuan* 611130, *China*)

Abstract: Post-translational modifications (PTMs) are versatile regulatory changes critical for plantdevelopment and immune response processes. Significantly, PTMs are involved in the crosstalk that serves as a fine-tuning mechanism to adjust cellular responses to various biotic and abiotic stress. Plant yield is often inhibited by an active immune response, resulting in yield penalties in crops when defending against various diseases. However, we have uncovered a novel plant disease-resistant mechanism whereby rice employs a single transcription factor IPA1 to enhance its immunity without reducing growth fitness. Phosphorylation of IPA1 at Serine 163 within its DNA binding domain occurs rapidly upon infection by the fungus *M. oryzae*. This phosphorylation then switches the DNA binding specificity of IPA1 towards the promoter of the pathogen defense gene *WRKY45*, leading to the activation of *WRKY45* expression and enhanced disease resistance. IPA1 returns to a nonphosphorylated state after 48 hours of infection, resuming support of the growth needed for high yield. In addition, we have found that ubiquitination is also an important PTM in balancing rice yield and immunity mediated by IPA1. IPI1 is RING-finger containing E3 ligase, contributing to IPA1-meidated ideal plant architecture generation, Recently, we also found that IPI1 was an important factor regulating rice immunity. However, the exact molecular mechanism remains to be elucidated. We will further study the IPA1 phosphorylation - and ubiquitination - dependent signaling pathways that rice employs to balance its development and immunity. With these conceptual innovations, we can develop practical strategies for breeding rice with excellent agronomic traits in high yield and resistance.

Key words: Balance between yield and resistance; Development; Immunity; Rice

Molecular mechanism of transcription factors regulating rice blast resistance

Li Weitao, Zhu Ziwei, Chen Xuewei

(*State Key Laboratory of Discovery and Utilization of Crop Genetic Resources in Southwest China* (*In preparation*), *State Key Laboratory of Hybrid Rice*, *Key Laboratory of Major Crop Diseases & Collaborative Innovation Center for Hybrid Rice in Yangtze River Basin*, *State Key Laboratory of Hybrid Rice*, *Hunan Hybrid Rice Research Center*, *Rice Research Institute*, *Sichuan Agricultural University at Wenjiang*, *Chengdu*, *Sichuan* 611130, *China*)

Abstract: Rice blast is the most devastating disease of rice. To effectively control this disease, it is necessary to elucidate molecular functions of resistance-related genes. We have compared the transcriptomic profile between the durably resistant rice variety Digu and the susceptible rice variety Lijiangxintuanheigu in response to infection by *Magnaporthe oryzae*, prior to full development of the appressorium. We identified 37 transcription factors (TFs) which regulate the blast response in the durably resistant rice Digu before the full maturation of the appressorium of *M. oryzae*. Meanwhile, we found that many biological processes were specifically activated in Digu shortly after infection, such as extracellular recognition, biosynthesis of antioxidants, terpenes and hormones. The results above indicated that contact and perception were present before the full maturation of the appressorium in the interaction of rice-*M. oryzae*. We also identified a natural allele of *bsr-d*1 that confers broad-spectrum resistance to blast disease. Mechanistically, this allele causes a single nucleotide change in the promoter of the *Bsr-d*1 gene, resulting in reduced expression of this gene due to the enhanced binding by its repressive TF MYBS1 and, consequently, an inhibition of H_2O_2 degradation and enhanced disease resistance. Collectively, our studies advance and deepen understanding of blast disease resistance.

Key words: Transcription factor; Resistance; Broad-spectrum; *Magnaporthe oryzae*; Rice

A rice TPR domain protein BSR-K1 negatively regulates broad-spectrum resistance via binding *OsPAL*1 mRNA

Zhou Xiaogang*, Liao Haicheng*, Chern Mawsheng*, Yin Junjie*, Chen Yufei, Wang Jianping, Zhu Xiaobo, Chen Zhixiong, Yuan Can, Zhao Wen, Wang Jing, Li Weitao, He Min, Ma Bingtian, Wang Jichun, Qin Peng, Chen Weilan, Wang Yuping, Liu Jiali, Qian Yangwen, Wang Wenming, Wu Xianjun, Li Ping, Zhu Lihuang, Li Shigui, Pamela C. Ronald, Chen Xuewei**

(*Key Laboratory of Major Crop Diseases*, *Sichuan Agricultural University*, *Wenjiang*, *Chengdu*, *Sichuan* 611130, *China*; *bState Key Laboratory of Hybrid Rice*, *Sichuan Agricultural University*, *Wenjiang*, *Chengdu*, *Sichuan* 611130, *China*)

Abstract: Crops carrying broad-spectrum resistance loci provide an effective strategy for controlling infectious disease because these loci typically confer resistance to diverse races of a pathogen or even multiple species of pathogens. Despite their importance, only a few crop broad-spectrum resistance loci have been reported. Here, we report the identification and characterization of the rice*bsr-k*1 (broad-spectrum resistance Kitaake-1) mutant, which confers broad-spectrum resistance against *Magnaporthe oryzae* and *Xanthomonas oryzae* pv. *oryzae* with no major penalty on key agronomic traits. Map-based cloning reveals that *Bsr-k*1 encodes a tetratricopeptide repeats (TPRs) -containing protein, which binds to mRNAs of multiple *OsPAL* (*OsPAL*1-7) genes and promotes their turnover. Loss of function of the *Bsr-k*1 gene leads to accumulation of *OsPAL*1-7 mRNAs in the *bsr-k*1 mutant. Furthermore, overexpression of *OsPAL*1 in wild-type rice TP309 confers resistance to M. oryzae, supporting the role of *OsPAL*1. Our discovery of the *bsr-k*1 allele constitutes a significant conceptual advancement and provides a valuable tool for breeding broad-spectrum resistant rice.

* For the authors contributed equally to this work

** Corresponding author: Chen Xuewei; E-mail: xwchen88@ 163. com

马铃薯 H 病毒的 p12 蛋白原核表达

郭志鸿[*]，李梦林，张宗英，韩成贵，王 颖[**]

（中国农业大学植物病理学系，农业生物技术国家重点实验室，北京 100193）

摘 要： 马铃薯 H 病毒（*Potato virus* H，PVH）是香石竹潜隐病毒属病毒（*Carlaviruses*）的成员之一，2012 年首次从内蒙古呼和浩特市采集的马铃薯植株上检测到 PVH。目前在中国的云南、广西、河北、辽宁、黑龙江、新疆以及内蒙古均检测到了 PVH。2019 年在孟加拉国的马铃薯产区首次检测到了马铃薯 H 病毒。PVH 具有特异的寄主范围，并能与多种马铃薯病毒（例如，马铃薯 X 病毒、马铃薯 Y 病毒）复合侵染马铃薯。PVH 的病毒粒子成丝状，微曲，长度约为 570 nm。PVH 单链基因组为 8 410 nt，其编码的 p12 蛋白具有锌指结构，是病毒编码的沉默抑制子（viral suppressor of RNA silencing，VSR），能够抑制植物的基因沉默。其抗血清的制备对后续研究 p12 与寄主的互作情况提供重要实验材料。将 p12 分别克隆到原核表达载体 PDB-His-MBP 与 PHAT2-His 上，得到阳性克隆 PDB-p12 与 pHAT2-p12，经小量诱导发现 p12 在 PDB-His-MBP 中表达量较高。经转化大肠杆菌 BL21（DE3），0. 2 mmol/L IPTG 诱导，18℃ 175 r/min 培养 18 h，预期大小为 56 ku 的 p12-MBP-12 * His 的融合蛋白在大肠杆菌中得到大量表达。随后，通过镍柱亲和层析并使用不同浓度的咪唑洗脱液进行洗脱，在 40~80 mmol/L 咪唑浓度下洗脱效果最好，浓缩纯化的蛋白用于多克隆抗血清的制备。

关键词： 马铃薯 H 病毒；p12 蛋白；原核表达

致谢： 感谢于嘉林、李大伟、王献兵教授和张永亮副教授对本研究的建议。

* 第一作者：郭志鸿，硕士研究生，主要从事植物病毒病害的研究；E-mail：guozhihong@ cau. edu. cn

** 通信作者：王颖，副教授，主要从事植物病毒与寄主互作研究；E-mail：yingwang@ cau. edu. cn

SAPK10-mediated phosphorylation onWRKY72 releases its suppression on jasmonic acid biosynthesis and bacteria blight resistance in rice

Hou Yuxuan, Wang Yifeng, Tang Liqun, Tong Xiaohong,
Wang Ling, Liu Lianmeng, Huang Shiwen, Zhang Jian
(*State Key Lab of Rice Biology*, *China National Rice Research Institute*, *Hangzhou* 311400, *China*)

Abstract: Bacterial blight caused by the infection of *Xanthomonas oryzae* pv. *oryzae* (*Xoo*) is a devastating disease that severely challenges the yield of rice. Here, we report the identification of a "SAPK10-WRKY72-AOS1" module, through which *Xoo* infection stimulates the suppression of JA biosynthesis to cause *Xoo* susceptibility. WRKY72 directly binds to the W-box in the promoter of JA biosynthesis gene *AOS*1, and represses its transcription *via* inducing DNA hypermethylation on the target site, which finally led to lower endogenous JA level and higher *Xoo* susceptibility. ABA-inducible SnRK2 type kinase SAPK10 phosphorylates WRKY72 at Thr 129. The SAPK10-mediated phosphorylation impairs the DNA binding ability of WRKY72, and releases its suppression on *AOS*1 and JA biosynthesis. Our work highlights a module of how pathogen stimuli lead to plant susceptibility, as well as a potential pathway for ABA-JA interplay with post-translational modification and epigenetic regulation mechanism involved.

Key words: Rice; bacterial blight; jasmonic acid; SnRK2; WRKY

Co-expression of antimicrobial peptide BnPRP1 and RsAFP2 in transgenic *Arabidopsis thaliana* confers enhanced resistance to *Sclerotnia sclerotiorum*

Li Yan*, Bai Zetao, Huang Junyan, Zuo Rong, He Yizhou, Cheng Xiaohui, Liu Yueying, Liu Shengyi**

(*Oil Crops Research Institute of Chinese Academy of Agricultural Sciences*, *Key Laboratory of Biology and Genetics Improvement of Oil Crops*, *Ministry of Agriculture and Rural Affairs*, *Wuhan* 430062, *China*)

Abstract: *Brassica napus* (oilseed rape) is an economically important oil crop all over the world. It not only provides vegetable oil for people, but also provides high quality fodder for animals. In addition, owing to its favorable agronomic properties, such as cultivation under different seasons and rotation with cereals, *B. napus* is preferred by farmers worldwide. However, Sclerotinia stem rot (SSR) is a devastating fungal disease and a major yield limiting factor in oilseed rape production. SSR is caused by *Sclerotinia sclerotiorum* (Lib.) de Bary, a cosmopolitan pathogen of many economically important crops. As a necrotrophic pathogen, it infects more than 600 plant species, including important oil crops such as oilseed rape, soybean, and sunflower. SSR not only deteriorates the quality of the seed, but also significantly reduces the oil content. To develop resistant materials against this fungal disease, the antimicrobial peptide gene *BnPRP*1 of *B. napus* and *RsAFP*2 from *Raphanus sativus* were expressed in *Arabidopsis thaliana* with a linker peptide. Stable integration and expression of transgenes in T0 and T2 generation plants were confirmed by PCR and qRT-PCR analyses. The crude proteins showed significant inhibition of *S. sclerotiorum* hyphal growth. The homozygous T2 plants, showing inheritance in Mendelian fashion (3 : 1), were further evaluated under greenhouse conditions for resistance to *S. sclerotiorum*. After vital leaf inoculation, restricted size and expansion of lesions were observed in transgenic plants compared with wild type. Consequently, co-expression of BnPRP1 and RsAFP2 in *Arabidopsis thaliana* provide subsequent protection against SSR disease and can be implemented in transgenic *B. napus*.

Key words: *Brassica napus*; *Sclerotinia sclerotiorum*; antimicrobial peptide; resistance

* First author: Li Yan

** Corresponding author: Liu Shengyi; E-mail: liusy@ oilcrop. cn

The study of proteomics of interaction between wheat and *Blumeria graminis* f. sp. *tritici* and *Puccinia striiformis* f. sp. *tritici* as well as the functional verification of the related resistant proteins*

Wang Qiao, Guo Jia, Jin Pengfei, Li Juan, Li Qiang**, Wang Baotong**

(*State Key Laboratory of Crop Stress Biology for Arid Areas, College of Plant Protection, Northwest A&F University, Yangling* 712100, *China*)

Abstract: To explore the similarities and differences of molecular mechanism of wheat defense against *Puccinia striiformis* f. sp. *tritici* (*Pst*) and *Blumeria graminis* f. sp. *tritici* (*Bgt*), quantitative proteomic analysis of Xingmin318 (XM318), a wheat cultivar highly resistant to both wheat stripe rust and powdery mildew, was performed by the TMT technology. A total of 741 proteins were significantly different and identified as differentially accumulated proteins (DAPs). Bioinformatics analysis indicated that all these DAPs relate to reactive oxygen species metabolic process, stimulus response, immune system process and other basal metabolism. In addition, results of bioinformatics showed that the main functional or metabolism categories for DAPs was identical between *Pst* and *Bgt* treatments, but some categories were existed obvious difference, such as process of detoxification. Intriguingly, only 42 DAPs responded to both *Pst* and *Bgt* treatments. Twelve proteins of DAPs were randomly selected and performed qRT-PCR analysis. The results showed that the mRNA expression levels of eleven proteins were consistent with the protein expression models. Furthermore, two proteins, member of Glutathione S-transferase family and member of ABA/WDS induced protein family, which was in response to both *Pst* and *Bgt* treatments, were performed gene silencing using the virus-induced gene silencing (VIGS) system. The results indicated that the Glutathione S-transferase protein mainly played an important role in anti-*Bgt* and had no function in resistance to *Pst*. As to the ABA/WDS induced protein, this protein was only reflected resistance in wheat-*Pst* interaction. This study revealed that wheat cultivar was likely to have different response mechanisms under the stress of *Pst* and *Bgt*. We also provide a new vision to better understand the complex regulation of the gene expression and numerous metabolic pathways between wheat and *Pst* or *Bgt* interaction.

Key words: Wheat; *Pst*; *Bgt*; VIGS; Glutathione S-transferase; ABA/WDS induced protein

* Funding: This research was supported by the National Key Research and Development Program of China (2018YFD0200403, 2016YFD0300705), the Natural Science Basic Research Plan in Shaanxi Province of China (2019JZ-17)

** Corresponding authors: Li Qiang; E-mail: qiangli@ nwsuaf. edu. cn

Wang Baotong; E-mail: wangbt@ nwsuaf. edu. cn

Paecilomyces variotii extracts (ZNC) enhance plant immunity and promote plant growth

Lu Chongchong[1*], Liu Haifeng[1*], Jiang Depeng[1],
Wang Lulu[1,3], Jiang Yanke[1], Tang Shuya[1], Hou Xuwen[1], Han Xinyi[1],
Liu Zhiguang[2], Zhang Min[2], Chu Zhaohui[1], Ding Xinhua[1**]

(1. *State Key Laboratory of Crop Biology, Shandong Provincial Key Laboratory for Biology of Vegetable Diseases and Insect Pests, Shandong Agricultural University, Taian* 271018, *China*; 2. *National Engineering Laboratory for Efficient Utilization of Soil and Fertilizer Resources, College of Recourses and Environment, Shandong Agricultural University, Taian* 271018, *China*; 3. *Shandong Pengbo Biotechnology Co., LTD, Taian* 271018, *China*)

Abstract: The crude extract of the endophyte*Paecilomyces variotii* known as ZhiNengCong (ZNC) has function of promoting plant growth and enhancing disease resistance and is widely used in China. Our study aims to evaluate the molecular mechanisms of plant growth promotion and disease protection. We generated transcriptome profiles from ZNC-treated seedlings using RNA sequencing. The function of salicylic acid (SA) in ZNC-mediated immunity was examined using SA biosynthesis and signaling pathway mutants. The concentrations of nitrogen (N) and phosphorus (P) in seedlings under ZNC treatment were measured. The effect of ZNC on the level of the hormone auxin in roots was tested using transgenic plants containing DR5 : GFP. ZNC exhibited ultrahigh activity in promoting plant growth and enhancing disease resistance, even at concentrations as low as 1-10 ng/mL. ZNC induced ROS accumulation, callose deposition, and expression of PR genes. SA biosynthesis and signaling pathways were required for the ZNC-mediated defense response. Moreover, in improving plant growth, ZNC increased the level of auxin in root tips and regulated the absorption of N and P. According to these results, ZNC is a highly effective plant elicitor that promotes plant growth by inducing auxin accumulation at the root tip at low concentrations and enhances plant disease resistance by activating the SA signaling pathway at high concentrations.

Key words: Plant elicitors; Plant growth promotion; Plant endophyte; Disease resistance; Auxin; Salicylic acid

* These authors: contributed equally to this work

** Corresponding author: Ding Xinhua; E-mail: xhding@ sdau. edu. cn

Study on *TaSBT*1 in wheat resistance against stripe rust*

Zhou Tianyu[1], Chen Fajing[2], Zhang Fengfeng[2], Fang Anfei[2],
Yu Yang[2], Bi Chaowei[2], Yang Yuheng[2]**
(1. *Citrus Research Institute*, *Southwest University*, *Chongqing* 400712, *China*;
2. *College of Plant Protection*, *Southwest University*, *Chongqing* 400715, *China*)

Abstract: Plant subtilases (SBTs) could induce hypersensitive response and systemic acquired resistance in plant-pathogen interactions, indicating that SBTs play important roles in plant disease resistance. However, study on plant SBTs against obligate biotrophic fungal pathogens has rarely been reported. Our previous study showed that a wheat subtilase TaSBT1 may be involved in the incompatible interaction between wheat and *Puccinia striiformis* f. sp. *tritici* (*Pst*). In this study, three homologous of *TaSBT*1 were cloned and thereafter named as TaSBT1a, TaSBT1b and TaSBT1d, which located on chromosomes 4AL, 4BS and 4DS of hexaploid wheat, respectively. Transient expression of TaSBT1 fused to GFP in the leaves of *Nicotiana benthamiana* showed that TaSBT1 proteins are distribute outside the plasma membrane, and their secretion function were also confirmed by the yeast signal peptide screen assay. In the incompatible interaction between wheat and avirulent *Pst* race CYR20, the transcripts of all three homologous were up-regulated between 12 hour-post inoculation (hpi) and 24 hpi compared with control treatment. The highest expression level of TaSBT1b was 7.2-fold at 12 hpi compared with the control, while TaSBT1a and TaSBT1d were 4.3-fold and 6.2-fold of the control, respectively. Subsequently, BSMV-VIGS technique was used to silence *TaSBT*1 in wheat leaves with two specific fragments of *TaSBT*1. Significant necrosis was observed on the control wheat leaves inoculated with CYR20, while amount of urediniospores were produced on the gene-silencing leaves. Moreover, TaSBT1 was only induced by Salicylic acid (SA) in wheat seedlings after treated by exogenous hormones SA, Methyl Jasmonate and Ethylene, suggesting that TaSBT1 may be involved in SA signal pathway. In summary, TaSBT1 is involved in the incompatible interactions between wheat and avirulent *Pst*, and regulates the wheat against wheat against *Pst*.

Key words: wheat; stripe rust; TaSBT1; BSMV-VIGS; disease resistance

---

* 基金项目：国家自然科学基金（31801719）；国家重点研发计划（2016YFD0101603）；重庆市基础科学与前沿技术研究专项（cstc2016jcyjA0316）

** 通信作者：杨宇衡，主要从事植物病理学研究；E-mail：yyh023@swu.edu.cn

Rice Black-Streaked Dwarf Virus (RBSDV) P5-1 Facilitates Viral Infection by Regulating the Ubiquitination Activity of SCF E3 Ligases and Jasmonate Signaling in Rice Plant *

He Long[1,2,3], Chen Xuan [1,2], Yang Jin[1,2], Zhang Tianye [1,2],
Li Juan [1,2], Li Jing [1], Zhong Kaili [2], Chen Jianping [1,2**],
Yang Jian [1,2**], Zhang Hengmu [1**]

(1. *Institute of Virology and Biotechnology*, *Zhejiang Academy of Agricultural Sciences*, *Hangzhou* 310021, *China*; 2. *Institute of Plant Virology*, *Ningbo University*, *Ningbo* 315211, *China*; 3. *College of Plant Protection*, *Nanjing Agricultural University*, *Nanjing* 210014, *China*)

Abstract: Rice black-streaked dwarf virus (RBSDV), a member of genus *Fijivirus*, family *Reoviridae*, causes severe losses to rice, maize and other cereal productions in East Asia. It contains ten dsRNA segments, most of which encode a single protein while the S5, S7 and S9 segments encode two proteins each (Zhang *et al.*, 2001). Its S5 segment is a bicistronic RNA and encodes two proteins (P5-1 and P5-2) (Yang *et al.*, 2014), in which P5-1 has been thought to be a component of viroplasm (Xie *et al.*, 2017). However, the role of P5-1 during viral infection largely remains unknown. SCF (for Skp1/Cullin1/F-box) complexes are key regulators of many cellular processes, including immune responses in plant (Cheng *et al.*, 2011), and sometimes become hijacked by viral pathogens (Lozano-Duran *et al.*, 2011). The underlying mechanism is poorly understood. In this study, RBSDV P5-1 protein were found to specifically interact *in vivo* and *in vitro* with rice OsCSN5A protein, a subunit of the COP9 signalosome (CSN) complex. Over-expression of P5-1 in rice cells did not inhibit the expression of jasmonate (JA) biosynthesis-associated genes but down-regulated the expression of JA responsive genes. Application of JA to the WT rice reduced its susceptibility to RBSDV infection. When JA was applied to the P5-1 transgenic plants, the RBSDV disease incidence was not changed compared with the mock plants. In P5-1-transgenic plants, the derubylation activity of the CSN complex on CUL1 can be specifically inhibited, leading to the alteration of SCF-mediated E3 ubiquitination. Knocking *OsCSN5A* or O*sCUL*1 down also significantly enhanced the susceptibility of rice plants to RBSDV. Based on above results, we proposed a working model, in which P5-1 inhibited the ubiquitination activity of SCF E3 ligases through interaction with OsCSN5A and then hindered the rubylation \ \ derubylation of CUL1, leading to repression of JA response pathway and thus promoting the viral infection on rice plants.

---

* Funding: The Key R&D Project of Zhejiang Province (2019C02018) and China (2016YFD0200800), and National Science and Technology Support Program (2012BAD19B03)

** Corresponding authors: Zhang Hengmu; E-mail: zhhengmu@ tsinghua. org. cn
Yang Jian, nather2008@ 163. com
Chen Jianping, jpchen2001@ 126. com

DNA fragment RPA190-pc participates in regulating metlaxyl resistance in *Phytophthora capsici*[*]

Wang Weiyan[**], Liu Dong, Zhuo Xin, Wang Yiye, Song Zhiqiang,
Zhang Huajian, Chen Fangxin, Pan Yuemin, Gao Zhimou[***]
(*College of Plant Protection*, *Anhui Agricultural University*, *Hefei* 230036)

Abstract: Metalaxyl is one of the main fungicides for the control of pepper blight caused by *Phytophthora capsici*. Due to long-term heavy use, the resistance of *P. capsici* to metalaxyl is becoming more and more serious. In order to reveal the molecular mechanism of *P. capsici* resistance to metalaxyl, the homologous alignment analysis of the protective domain RPOLA-N of *RPA*190 gene in *P. infestans* was conducted, and the candidate DNA fragment RPA190-pc, which may be involved in the metalaxyl resistance of *P. capsici*, was excavated. This gene was cloned and sequenced, and a conserved RPOLA-N domain was found to be contained in it. Two silencing converters and one over-expression convertor were obtained by using PEG mediated protoplast transformation method, and the expression of RPA190-pc in each convertor was analyzed by qRT-PCR. The results showed that the gene silencing efficiency of SD1-9 C-3 and SD1-9 C-4 was above 60%. The gene expression efficiency of the over-expressing transformant OESD1-9-1 was three times that of the control strain M_{tr}SD1-9. The resistance levels of these three transformants to metalaxyl were also determined. The results showed that the EC_{50} value of metalaxyl to SD1-9 C-3 was 505.096 5 μg/mL, and the EC_{50} value of M_t^r SD1-9 was decreased 24.23%, the EC_{50} value of SD1-9 C-4 was 67.035 4 μg/mL, the EC_{50} value of M_t^r SD1-9 was decreased by 89.94%; and the EC_{50} value of OESD1-9-1 was 15 573.023 6 μg/mL. The EC_{50}value for M_t^r SD1-9 was increased by 23.26 times. The above results indicate that the candidate DNA fragment RPA190-pc is involved in the regulation of *P. capsici* resistance to metalaxyl.

Key words: *Phytophthora capsici*; metalaxyl resistance; gene silencing; protoplast transformation

* Funding: The National Natural Science Foundation of China (Grant No. 31671977)

** First Author: Wang Weiyan, doctoral student, major in mycology and plant fungal diseases

*** Corresponding author: Gao Zhimou; E-mail: gaozhimou@126.com

甘蓝型油菜 *TLP-Kinase* 基因的鉴定及其对核盘菌抗性的功能研究

钟　雪*，左　蓉，刘　杰，李　祥，李　艳，童超波，刘胜毅，白泽涛**
（中国农业科学院油料作物研究所，农业农村部油料作物生物学与遗传育种重点实验室，武汉　430062）

摘　要： 油菜是我国最主要的油料作物，有很高的经济价值。由核盘菌引起的菌核病能够导致油菜减产，产油量降低，近年来已经成为我国油菜高产稳产的主要限制因子之一。寻找关键的抗病基因，研究其抗病分子机制，进而应用到抗病育种是防治菌核病的有效方法。类甜蛋白（thaumatin-like proteins，TLP）是一种病程相关蛋白，具有抗真菌、响应生物及非生物胁迫等功能。本研究通过对甘蓝型油菜中的 *TLP* 家族的鉴定及结构域分析，发现 *TLP-Kinase* 融合基因的存在，依据甘蓝型油菜接种核盘菌后不同组织的转录组数据筛选出一个候选基因 *BnTLP-K*，并克隆了该基因的全长 CDS 区域，利用农杆菌转化法转化拟南芥，获得了该基因的过表达材料；同时构建了 *BnTLP-K* 在拟南芥中的同源基因 *AT4G*18250 的干扰载体，获得了转基因拟南芥干扰株系，通过荧光定量 PCR 检测，转基因植株中的 *AT4G*18250 基因表达量显著降低。此外，成功对 *BnTLP-K* 进行了体外原核表达，目前正在纯化蛋白。后续研究将进一步分析已获得的转基因材料及纯化蛋白对核盘菌的抗性及具体的作用机制。本研究通过正向遗传学深入探索 *BnTLP-K* 基因在油菜菌核病抗性中的作用及其机制，将为油菜的抗病性遗传改良提供新的基因源和理论指导。

关键词： 甘蓝型油菜；菌核病；*TLP-Kinase* 融合基因；抗病分子机制

* 第一作者：钟雪，硕士研究生，主要从事油菜抗菌核病研究
** 通信作者：白泽涛；E-mail：baizetao_2005@163.com

甘蓝型油菜 BnTLP-PRP 基因对核盘菌抗性的功能研究

石美娟*，左　蓉，刘　杰，赵传纪，何贻洲，李　艳，
高　峰，李　祥，刘越英，程晓辉，刘胜毅，白泽涛**
（中国农业科学院油料作物研究所，农业农村部油料作物生物学与
遗传育种重点实验室，武汉　430062）

摘　要：油菜是我国重要的油料作物，由核盘菌引起的油菜菌核病严重影响油菜的产量与品质。为了抵御核盘菌的侵染，油菜通过病原相关分子模式触发的免疫反应 PTI 和效应因子触发的免疫反应 ETI 有效的激发体内的防御反应，表达防御相关蛋白。大量研究表明甜类蛋白 TLP 是一类抗病相关基因，能有效的抑制核盘菌生长。前期通过生物信息学分析发现油菜体内少数 TLP 的 C 端融合了抗菌肽 PRP，为了研究 TLP 对核盘菌的抗性及融合基因之间的关系，本研究首先克隆到油菜 TLP-PRP 全长基因和含有 TLP domain、PRP domain 的片段。通过农杆菌介导的转基因技术在拟南芥中过表达该基因。目前已成功的获得过表达和 RNAi 材料，后期将通过活体接菌实验验证 TLP-PRP 基因及不同结构域对核盘菌的抑菌效果。同时通过原核表达实验成功表达出 TLP-PRP 蛋白，下一步将纯化蛋白进行体外蛋白抑菌实验。该研究有助于甘蓝型油菜核盘菌抗性材料的筛选，为改良甘蓝型油菜抗菌核病育种提供理论基础和新的靶基因源。

关键词：甘蓝型油菜；核盘菌；TLP-PRP 融合基因；抑菌效果；基因功能

* 第一作者：石美娟，硕士研究生，主要从事油菜抗菌核病研究
** 通信作者：白泽涛；E-mail：baizetao_2005@163.com

非编码 RNA 在植物抗病过程中响应的研究进展*

杨 帆**，赵 丹，范海燕，朱晓峰，王媛媛，段玉玺，陈立杰***
（沈阳农业大学植物保护学院，沈阳 110866）

摘 要：非编码 RNA 是生物中部分不表达蛋白质的基因，包括微小 RNA（miRNA）、长链非编码 RNA（lncRNA）和环状 RNA（circRNA）等。近年来在植物中研究发现很多非编码 RNA 能够参与植物抵御外界病害胁迫。本文阐述了 miRNA、circRNA 和 lncRNA 的分子特征和在植物抗病中的作用进行了综述，以期从非编码 RNA 角度认识植物的抗病分子机理。

关键词：植物；非编码 RNA；miRNA；circRNA；lncRNA

Research advance of response of non-coding RNA in plant disease resistance*

Yang Fan**, Zhao Dan, Fan Haiyan, Zhu Xiaofeng,
Wang Yuanyuan, Duan Yuxi, Chen Lijie***
(*College of Plant Protection*, *Shenyang Agricultural University*, *Shenyang* 110866, *China*)

Abstract: Non-coding RNAs are genes that do not express proteins in organisms, including microRNAs, lncRNAs and circRNAs. In recent years, studies have found that many non-coding RNAs can participate in plants against external disease stress. This paper discusses the molecular characteristics of miRNA, circRNA and lncRNA and their role in plant disease resistance, in order to understand the molecular mechanism of disease resistance from the perspective of non-coding RNA.

Key words: plants; non-coding RNA; miRNA; circRNA; lncRNA

非编码 RNA 是当今国际生物学研究的热点之一。包括 miRNA、circRNA 和 lncRNA 可以通过调控 NBS-LRR 基因的表达、植物应激反应和一些与植物抗病相关的转录因子，从而影响植物生长发育，在植物响应生物胁迫方面起着重要的调控作用。

1 MiRNA 在植物抗病过程中的作用

1.1 植物 miRNA 的发现和功能研究

现阶段 miRNA 的功能研究比较深入，自从第一个 miRNA 基因（lin-4）在秀丽隐杆线虫中被发现和鉴定以来，已经 25 年的时间过去了，目前 miRNA 的研究比较广泛和深入，涉及了几乎所有的常见物种，截至 2019 年 3 月已经发现了38 589余种成熟 miRNA。miRNA 能够在许多物种

* 项目资助：“十三五”国家重点研发计划项目（2017YFD0201104）
** 第一作者：杨帆，在读博士，研究方向：生防菌诱导植物抗病性机理研究；E-mail：yangjingdong2333@163.com
*** 通信作者：陈立杰，博士生导师，研究方向：植物病理学和植物线虫学；E-mail：chenlijie0210@163.com

的生长发育、信号转导以及应对生物胁迫响应中起调节作用。

1.2　MicroRNA 在植物抗病过程中的响应研究

有研究表明 miR482b 与植物抗病有关，当番茄受到晚疫病菌侵染时会导致 miR482b 表达量的降低，而当在番茄体内进行 miR482b 的沉默时，会导致其靶基因 NBS-LRR 的大量表达，从而增强了植物抗性（Jiang *et al.*，2018）。Chen 等在转基因烟草中的研究表明 miR396a-5p 在烟草抵抗烟草疫霉菌侵染时产生了应激反应，miR396a-5p 的表达量呈下降趋势，过表达烟草 miR396a-5p 能够显著降低对烟草疫霉的抗性（L. Chen *et al.*，2015）。另有研究表明，miR319 / TCP4 作为系统性防御应答者和调节根结线虫（RKN）系统性防御反应的调节因子的作用是通过 JA 介导的（Zhao *et al.*，2015）。这些研究说明 miRNA 在植物响应生物胁迫中发挥重要作用。

2　circRNA 在植物抗病过程中的作用

2.1　植物 circRNA 的发现和发展

circRNA 是一种新型的、内源性的生物体内广泛存在的 ncRNA，能在转录和转录后水平调节基因的表达。1976 年 Sanger 等在高等植物中发现一种为单链环状闭合的特殊 RNA，并将这种特殊 RNA 称为环状 RNA，这是第一次环状 RNA 的概念被提出。如今在很多植物上已经挖掘到了一些 circRNA 的存在，比如在冷处理后的番茄、水稻和热胁迫拟南芥中分别确定了 854 个，2 354 个和 1 583个 circRNA 的存在，这种大规模 circRNA 的出现引起了研究者的广泛重视，除了存在相关的预测分析外，还存在更细致地对其功能的预测和后续的相关实验验证。

2.2　CircRNA 在植物抗病过程中的响应研究

多项研究表明，circRNA 响应生物胁迫。在猕猴桃中发现 circRNA 对病原体入侵具有特异性响应（Wang *et al.*，2017c）。在马铃薯中，429 个差异表达 circRNA 响应肉毒杆菌巴西亚种感染（Zhou *et al.*，2018b）。棉花感染黄萎病后共发现 280 个差异表达的 circRNA，这些差异表达的 circRNA 的来源基因主要富集在刺激反应（Xiang *et al.*，2018）。Ghorbani 等（2018）比较伊朗花叶病毒侵染玉米植株叶片与未侵染植株叶片之间共有的 circRNA，发现 160 个发生差异表达。Wang 等（2018b）发现番茄叶片感染黄化曲叶病毒病与对照之间分别有 32 个和 83 个 circRNA 特异表达。这些结果表明 circRNA 响应并可能参与调控生物胁迫，在植物发育中发挥重要作用，但是这需要过表达或者敲除试验等证据进一步证实。

3　lncRNA 在植物抗病过程中的作用

3.1　植物 lncRNA 的发现和发展

lncRNA 是指长度在 200nt 上的不具有编码蛋白的 RNA，无长的开放阅读框（Open Reading Frame，ORF），lncRNA 种类很多，并已被大量发现能够在表观遗传水平、转录水平和转录后水平调控基因表达，广泛参与了生物体生长发育、生理和抗病等过程，细胞内主要涉及基因转录、转录后调控、mRNA 降解和调节蛋白活性等生命过程。

3.2　LncRNA 在植物抗病过程中的响应研究

有研究发现，当 lncRNA23468 在番茄中过表达时，miR482b 及其靶基因 NBS-LRR 的表达分别出现显著的降低和升高，抗病性得以增强；而沉默 lncRNA23468 则导致 miR482b 的积累，NBS-LRR 减少，抗病性下降（Jiang *et al.*，2019）。另有研究表明，在 WRKY1 过表达的番茄叶中，lncRNA33732 的过表达导致 RBOH 基因表达的增加，并且 lncRNA33732 的沉默导致 RBOH 基因表达降低，这表明 lncRNA33732 可能影响 RBOH 基因的表达（Jun *et al.*，2018）。

4　问题与展望

近年来在植物中研究发现一些非编码 RNA 与寄主病害密切相关，如 miR482b、miR396a-5p、

miR319、lncRNA23468 和 lncRNA33732 等。已知这些非编码 RNA 是在生物体内普遍存在的内源性小分子 RNA，也是植物生长发育的重要调控因子，能够通过作用于其靶基因而在植物抵御生物胁迫中起着关键作用。但是与植物非编码 RNA 相关研究仍有一些科学问题有待解决，包括植物中决定作物复杂性状 miRNA 的作用机制、lncRNA 受生物胁迫调控的机制、circRNA 的功能验证及其分子机制仍有待验证。

参考文献

Chen L，Luan Y，Zhai J. 2015. Sp-miR396a-5p acts as a stress-responsive genes regulator by conferring tolerance to abiotic stresses and susceptibility to Phytophthora nicotianae infection in transgenic tobacco [J]. Plant Cell Reports，34 (12)：2013-2025.

Jiang N，Meng J，Cui J，*et al.* 2018. Function identification of miR482b，a negative regulator during tomato resistance to Phytophthora infestans [J]. Horticulture Research，5 (1)：9.

Lee R C，Feinbaum R L，Ambros V. 1993. The C. elegans heterochronic gene lin-4 encodes small RNAs with antisense complementarity to lin-14 [J]. Cell，75 (5)：843.

Memczak S，Jens M，Elefsinioti A，*et al.* 2013. Circular RNAs are a large class of animal RNAs with regulatory potency [J]. Nature，495 (7441)：333-338.

Wang Z P，Liu Y F，Li D W，*et al.* 2017c. Identification of circular RNAs in kiwifruit and their species-specific response to bacterial canker pathogen invasion [J]. Frontiers in Plant Science，8：413.

Xiang L X，Cai C W，Cheng J R，*et al.* 2018. Identification of circularRNAs and their targets in Gossypium under Verticillium wilt stress based on RNA-seq [J]. Peer J，6：e4500.

Xu S，Xiao S，Qiu C，*et al.* 2017. Transcriptome-wide identification and functional investigation of circular RNA in the teleost large yellow croaker (*Larimichthys crocea*) [J]. Mar Genomics，32：71.

Zhao W，Li Z，Fan J，*et al.* 2015. Identification of jasmonic acid-associated microRNAs and characterization of the regulatory roles of the miR319/TCP4 module under root-knot nematode stress in tomato [J]. Journal of Experimental Botany，66 (15)：4653-4667.

Zhou R，Zhu Y X，Zhao J，*et al.* 2018b. Transcriptome-wide identification and characterization of potato circular RNAs in response to pectobacterium carotovorum subspecies brasiliense infection [J]. International Journal of Molecular Sciences，19：71.

拟南芥抗病性增强突变体 *aggie5* 的基因图位克隆*

胡　滢[1**]，魏君君[1]，齐　婷[1]，单丽波[2]，何　平[3]，孙文献[1]，崔福浩[1***]
（1. 中国农业大学植物保护学院，农业部作物有害生物监测与绿色防控重点实验室，
北京　100193；2. *Department of Plant Pathology & Microbiology, and*
Institute for Plant Genomics & Biotechnology, Texas A&M University,
College Station, TX 77843, USA；3. *Department of Biochemistry &*
Biophysics, and Institute for Plant Genomics & Biotechnology,
Texas A&M University, College Station, TX 77843, USA）

摘　要：植物细胞表面的免疫受体能感知诸如细菌鞭毛蛋白、真菌几丁质等保守的微生物相关分子模式（Microbe-associated Molecular Patterns，MAMPs），进而激发植物 MAPK 磷酸化、活性氧爆发、免疫基因表达等立体免疫反应。我们通过对前期建立的 *pFRK1*：*LUC* 转基因拟南芥 EMS 突变体库筛选，获得了细菌鞭毛蛋白小肽 flg22 处理后 *pFRK1*：*LUC* 活性显著升高的突变体 *aggie5*。免疫表型检测发现，*aggie5* 受 flg22 处理后的活性氧爆发显著增强。与此一致的是，*aggie5* 对丁香假单胞细菌的抗性显著提高。图位克隆和基因组测序发现，引起 *pFRK1*：*LUC* 活性升高的突变基因与花的发育相关，这与突变体的早花表型一致。解析 Aggie5 调控植物免疫的分子机制，有利于深入了解植物先天免疫信号调控网络以及揭示免疫与开花之间的互作。

关键词：拟南芥；先天免疫；信号传导；抗病分子机制

* 基金项目：中国农业大学基本科研业务费（15057006）
** 第一作者：胡滢，硕士研究生，研究方向：植物与病原细菌互作；E-mail：1017697252@ qq. com
*** 通信作者：崔福浩，副教授，主要从事植物与病原细菌、真菌互作；E-mail：cuifuhao@ 163. com

ath-miRz 通过靶向 *AtTAR1* 调控拟南芥对寄生疫霉菌的抗性*

勾秀红[1,3**]，钟成承[1,3]，张培玲[2,3]，单卫星[2,3***]
（1. 西北农林科技大学植保学院，杨凌 712100；2. 西北农林科技大学农学院，杨凌 712100；
3. 西北农林科技大学旱区作物逆境生物学国家重点实验室，杨凌 712100）

摘 要：寄生疫霉菌（*Phytophthora parasitica*）寄主范围广泛，能够侵染多种植物，引起严重的农作物灾害从而造成巨大的经济损失。miRNAs 是一类非编码的小 RNA 分子，可通过序列特异的方式在转录水平或转录后水平调控基因表达，在植物生长发育和防卫反应中发挥着重要作用，因此研究植物内源 miRNA 在与寄生疫霉菌互作过程的作用对于对寄生疫霉菌的防控具有重要意义。本研究以拟南芥-寄生疫霉菌亲和互作体系为基础，通过拟南芥接种寄生疫霉菌的高通量测序数据分析，获得一个在寄生疫霉菌侵染后下调表达的拟南芥内源 miRNA，命名为 ath-miRz。ath-miRz 的拟南芥过表达转化植株（OEmiRz）接菌结果显示 ath-miRz 增强了植物对寄生疫霉菌的感病性，表明其负调控植物对寄生疫霉菌的防卫反应。笔者通过生物信息学分析得到一个 ath-miRz 的候选靶标基因，命名为 *AtTAR1*。本氏烟叶片的共表达实验表明 ath-miRz 能够调控靶标序列的表达。在拟南芥中沉默 *AtTAR1* 后，接菌结果显示对寄生疫霉菌的感病性增强，我们因此推测 ath-miRz 可能通过调控 *AtTAR1* 的表达负调控拟南芥对寄生疫霉菌的免疫反应。

关键词：miRNA；寄生疫霉菌；ath-miRz；*AtTAR1*

---

* 基金项目："作物抗病育种与遗传改良学科创新引智基地" 111 项目

** 第一作者：勾秀红，在读博士生，研究方向：植物病理学；E-mail：xhgou@ nwafu. edu. cn

*** 通信作者：单卫星，教授；E-mail：wxshan@ nwafu. edu. cn

一个分泌型 DAMP 分子 19C55 参与烟草对疫霉菌的抗性研究

文曲江[1,3]，孟玉玲[2,3]，单卫星[2,3]*
（1. 西北农林科技大学植物保护学院，杨凌 712100；2. 西北农林科技大学农学院，杨凌 712100；3. 西北农林科技大学旱区作物逆境生物学国家重点实验室，杨凌 712100）

摘 要：寄生疫霉菌（*Phytophthora parasitica*）引起的黑胫病威胁烟草等经济作物的可持续性生产。然而，我们对于烟草与寄生疫霉菌的互作机制，以及可用于抗病育种的相关基因知之甚少。我们通过根癌农杆菌（*Agrobacterium tumefaciens*）介导烟草瞬时表达实验，鉴定到一个能引起烟草细胞坏死的免疫相关基因 19*C*55。19*C*55 基因在寄生疫霉菌侵染 12 h 和 24 h 显著上调，亦可被 SA 诱导表达。亚细胞定位分析表明，19C55 主要在细胞质膜累积。烟草瞬时表达结果表明，19C55 能增强烟草对寄生疫霉菌的抗性，VIGS 实验显示 19*C*55 沉默可降低烟草对疫霉菌的抗性，这表明 19C55 参与了烟草对寄生疫霉菌的抗性。Western 结果显示 19C55 能在 5 min、10 min、15 min 时显著激活 MPK3 和 MPK6。定量分析发现 19*C*55 可以诱导 PTI 相关基因上调表达。此外，19C55 可增强 PTI（Pattern Triggered Immunity）过程中，如 flg22（*flagellin peptide* 22）诱导的胼胝质沉积。这些结果表明 19C55 可作为一个 DAMP 分子激活植物 PTI 防卫通路。

关键词：寄生疫霉菌；19C55；DAMPs；抗病型

* 通信作者：单卫星，教授；E-mail：wxshan@nwafu.edu.cn

拟南芥基因 *VQ28* 负调控植物对疫霉菌抗性的机制研究*

蓝星杰[1,3**]，曹　华[1,3]，单卫星[2,3***]

（1. 西北农林科技大学植物保护学院，杨凌　712100；2. 西北农林科技大学农学院，杨凌　712100；3. 西北农林科技大学旱区作物逆境生物学国家重点实验室，杨凌　712100）

摘　要：寄生疫霉菌（*Phytophthora parasitica*）和大豆疫霉菌（*P. sojae*）均为重要作物病原菌，能够分别与模式植物拟南芥亲和互作与非亲和互作。*VQ28* 属于拟南芥 VQ motif-containing proteins 家族，研究发现，*VQ28* 基因在感大豆疫霉菌拟南芥突变体 *esp*1（*enhanced susceptibility to Phytophthora*）中特异上调。进一步的基因表达分析表明，*VQ28* 受疫霉菌和 MeJA 诱导表达。拟南芥接菌结果表明，过表达 *VQ28* 可促进寄生疫霉菌侵染，并解除植物对大豆疫霉菌的非寄主抗性。烟草瞬时表达结果表明，*VQ28*-GFP 可促进寄生疫霉菌侵染。与此一致，敲除 *VQ28* 可抑制寄生疫霉菌侵染。亚细胞定位分析表明 *VQ28* 在细胞核和质膜上均有分布。此外，荧光素酶互补试验表明 *VQ28* 与植物防卫相关的转录因子 WRKY51 和 WRKY33 互作。综上所述，*VQ28* 作为一个感病因子，其可能通过与植物防卫相关的转录因子互作从而负调控植物对疫霉菌的抗性。

关键词：*VQ28*；寄生疫霉菌；感病因子；植物免疫

---

* 基金项目：“作物抗病育种与遗传改良学科创新引智基地”111 项目

** 第一作者：蓝星杰，在读博士生，研究方向为植物与病原卵菌互作；E-mail：xjlan@ nwafu. edu. cn

*** 通信作者：单卫星，教授；E-mail：wxshan@ nwafu. edu. cn

玉米 DNA 甲基化相关基因的克隆及抗病功能分析*

杭天露**，刘　琼，王　其，谢珊珊，丁　婷***，江海洋***
（安徽农业大学生命科学学院，作物抗逆育种与减灾国家地方联合工程实验室，合肥　230036）

摘　要：相关文献已报道拮抗内生细菌 DZSY21 可以在玉米叶片中稳定定殖，并且其互作系统对玉米小斑病具有良好的抗病性（Ting Ding *et al.*，2017）。前期研究中，通过甲基化敏感扩增多态性分析（MSAP）发现内生细菌 DZSY21 处理前后玉米全基因组 DNA 的甲基化水平下降了 2.4%，进一步结合玉米全基因组 DNA 甲基化测序（WGBS）结果，并利用 PFam 对显著差异性甲基化基因蛋白域进行功能预测，最终筛选出 4 个与 DNA 甲基化相关抗病基因，但其在玉米中的功能和机制尚未清楚。该试验选择其中一个与抗病相关的 DNA 甲基化基因 GRM8 开展研究。克隆测序结果表明，基因 GRM8 编码了 349 个氨基酸，包含一个 IBR 结构域，是一种新型的锌指蛋白，利用玉米原生质体中瞬时表达的亚细胞定位技术发现基因 GRM8 定位于细胞核和细胞膜上。实时定量逆转录 PCR（qRT-PCR）结果表明玉米植株在玉米弯孢菌和细菌性枯萎病菌诱导下，基因 GRM8 的表达量发生上调。将 GRM8 基因转入拟南芥开展功能分析，接种丁香假单胞菌（*Pseudomonas syringae*），黄单胞菌（*Xanthomonas oryzae*）后，过表达 GRM8 基因的转基因拟南芥相对于野生型拟南芥，其抗病性明显提高；该研究结果为明晰玉米 DNA 甲基化相关基因的抗病功能和调控网络提供了一定的理论基础。

关键词：玉米；甲基化；抗病基因

---

* 基金项目：国家重点研发计划子课题（2017YFD0201106-09）

** 第一作者：杭天露，硕士研究生，主要从植物病害防治研究；E-mail：hangtianlu1995@163.com

*** 通信作者：丁婷，博士，教授，主要从植物病害生物防治及生防微生物分子生物学研究；E-mail：dingting98@126.com
江海洋，博士，教授，主要从玉米抗逆及抗病分子生物学研究；E-mail：hyjiang@ahau.edu.cn

谷子*SiWRKY03*基因的分子特征与抗病反应中的表达研究*

宋振君[1,2**]，李志勇[1]，王永芳[1]，全建章[1]，白　辉[1***]，董志平[1***]
（1. 河北省农林科学院谷子研究所，国家谷子改良中心，河北省杂粮重点实验室，
石家庄　050035；2. 河北师范大学生命科学学院，石家庄　050024）

摘　要：为了解谷子中*SiWRKY03*的功能，本研究利用生物信息学软件分析了其生物学特征，采用Real-time PCR技术检测了*SiWRKY03*转录因子在谷子抗锈病过程中的表达丰度变化。结果表明，*SiWRKY03*基因的开放阅读框全长为1 137 bp，编码378个氨基酸，预测分子量为39.73 ku，理论等电点pI为5.91，含有一个WRKY保守结构域。该蛋白质二级结构的最大元件是无规则卷曲，最小元件为β-转角。进化分析表明，SiWRKY03与*Panicum miliaceum*（RLM93064.1）和*Panicum hallii*（PAN29607.1）的氨基酸序列同源性最高，为99%。在谷子响应锈菌胁迫反应的120h内，*SiWRKY03*基因在抗病反应的12h和36h上调表达，而感病反应过程中无显著性变化，推测*SiWRKY03*在谷子抗锈病反应中起正调控作用。上述试验结果为进一步研究*SiWRKY03*基因功能与抗病机制奠定理论基础。

关键词：谷子；谷子锈病；WRKY转录因子；荧光定量PCR；基因表达

---

* 基金项目：国家自然科学基金项目（31872880）；河北省农林科学院创新工程（2019-4-2-3）；国家现代农业产业技术体系（CARS-07-13.5-A8）；河北省优秀专家出国培训项目

** 第一作者：宋振君，硕士研究生，专业方向为谷子分子生物学；E-mail：szj1018@163.com
*** 通信作者：白辉，博士，副研究员，从事谷子抗病分子生物学研究；E-mail：baihui_mbb@126.com
董志平，硕士，研究员，从事谷子病虫害研究；E-mail：dzping001@163.com

水稻与稻瘟菌互作中感病相关因子的筛选与鉴定

徐海娇*，常清乐，范 军**

（中国农业大学植物保护学院，北京 100193）

摘 要：稻瘟病是一种世界性的水稻病害，严重影响了我国的水稻生产和粮食安全。在植物—病原互作过程中广泛存在着基因转录重编程的现象，这些重编程影响了植物对病原菌的抗感性。目前关于植物感病基因和感病途径的研究仍知之甚少，因此获得互作中的差异表达基因并对其进行直接的功能筛选是一个重要的功能解析方法。

本研究利用农杆菌介导的大麦瞬时表达系统，研究水稻与稻瘟菌互作过程中差异表达的基因对大麦感病性的影响。通过接种稻瘟菌 P131，分析该菌株在瞬时表达候选基因组织中的生长状况，以筛选能够促进发病等潜在的调控植物感病通路的关键因子。初步筛选共获得 22 个基因能显著促进发病使黄化增强，其中有 8 个基因直接表达即可引起黄化，接种 P131 后黄化增加的有 14 个。对 P131 接种后 2d 的部分样品进行生物量检测，引起 P131 生长量增加了 3.5~5.5 倍，它们分别为编码稻瘟菌的半乳糖苷酶基因，编码水稻的锌指结构蛋白、MYB 类蛋白、几丁质酶、剪接因子、WRKY 蛋白以及其他类蛋白的基因。后续研究将进一步分析这些基因在水稻与稻瘟菌互作过程中的功能和具体作用机制，该研究将为稻瘟病的感病机制研究提供新的依据。

关键词：稻瘟病；稻瘟菌；感病基因筛选；感病基因鉴定

* 第一作者：徐海娇，博士研究生，主要研究方向植物病原物致病机理

** 通信作者：范军，教授，主要研究方向植物病原物致病机理及植物数量抗病性的遗传和分子机理；E-mail：jfan@cau.edu.cn

苹果树腐烂病菌外泌蛋白 Vmhp-1 互作靶标的鉴定及其干扰免疫功能的研究*

高 晨**，许 铭，刘召阳，高宇琪，黄丽丽***，冯 浩***
（旱区作物逆境生物学国家重点实验室，西北农林科技大学植物保护学院，杨凌 712100）

摘 要：苹果树腐烂病是由黑腐皮壳属真菌 *Valsa mali* 引起的枝干病害。解析病菌与寄主的互作机理对病害防控新策略的研发具有重要意义。笔者实验室前期研究发现苹果树腐烂病菌 *Vmhp*-1 在侵染过程中上调表达，并通过酵母异源表达后证明 Vmhp-1 是一种外泌蛋白，更重要的是缺失后突变体致病力较野生型显著降低。在此基础上，本研究利用非寄主本氏烟瞬时表达系统发现 Vmhp-1 会抑制由 Bax 引起的细胞坏死。亚细胞定位分析发现 Vmhp-1 定位在烟草细胞膜上，而去掉信号肽 Vmhp-1 定位在细胞膜和细胞核。Vmhp-1 瞬时表达后，水杨酸（SA）通路（NbPR1 和 NbPR2）、茉莉酸（JA）通路（NbPR4 和 NbLOX）和乙烯（ETH）通路（ERF1）上的 5 个 maker 基因均下调表达，且 Vmhp-1 可以抑制 INF1 诱导的活性氧迸发和胼胝质的沉积。同时研究发现，Vmhp-1 瞬时表达后降低了烟草对核盘菌（*S. sclerotiorum*）的抗性水平。进而利用酵母双杂交系统（Y_2H）、双分子荧光（Bifc）和免疫共沉淀技术（Co IP）鉴定到 2 个与 Vmhp-1 互作的候选植物靶标。本研究结果为揭示外泌蛋白 Vmhp-1 调控寄主免疫反应的分子机理奠定了重要基础。

关键词：苹果树腐烂病菌；外泌蛋白；靶标鉴定；植物免疫

---

* 基金项目：国家自然科学基金项目（3197170375）
** 第一作者：高晨，西北农林科技大学植物保护学院硕士研究生；E-mail：1056206877@ qq. com
*** 通信作者：冯浩，教授；E-mail：xiaosong04005@ 163. com
黄丽丽，教授；E-mail：huanglili@ nwsuaf. edu. cn

野油菜黄单胞菌——十字花科植物之间“攻防战”的分子机理

曹雪强[*]，周　莲，Diab Abdelgader，何亚文[**]
（微生物代谢国家重点实验室，上海交通大学生命科学技术学院，上海　200240）

摘　要：黄单胞菌与其寄主之间的相互作用是一场复杂的“攻防战”：一方面，病原细菌产生的信号分子和代谢物可以影响植物的生长和发育；另一方面，植物源的次生代谢物也能影响黄单胞菌致病因子的产生。

菌黄素是黄单胞菌产生的一类芳香基多烯磷酯类黄素色，能保护细菌抵抗光伤害，协助细菌在植物表层和组织内生存，是黄单胞菌重要的致病因子。笔者课题组在野油菜黄单胞菌（*Xcc*）中深入阐述了菌黄素生物合成的分子机理：*xanB*2 编码新型双功能分支酸裂解酶，水解分支酸同时产生 3-羟基苯甲酸（3-HBA）和 4-羟基苯甲酸（4-HBA），其中 3-HBA 作为前体参与菌黄素的生物合成；XanA2 负责激活 3-HBA 腺苷酰化，通过两步反应加载至 *xan*C 编码的酰基载体蛋白上；菌黄素的芳香基多烯链依赖Ⅱ型聚酮合酶途径进行合成。笔者还发现，在 3-HBA 含量降低时，*Xcc* 可利用 4-HBA 合成菌黄素。*Xcc* 利用不同前体物合成菌黄素的特性，有利于其在寄主植物内合成菌黄素，从而辅助其进一步侵染。

辅酶 Q（CoQ）是有氧呼吸途径中电子传递链的关键组成部分。CoQ 的生物合成机制在大肠杆菌和酵母菌中研究较深入，而在植物病原细菌中鲜有报道。我们研究发现 XanB2 产生的 4-HBA 是 CoQ 的合成前体，生物信息学和遗传学分析表明 *ubi*A^{Xc}、*ubi*B^{Xc}、*ubi*E^{Xc}、*ubi*G^{Xc}、*ubi*H^{Xc}、*ubi*I^{Xc}、*ubi*J^{Xc}、*ubi*K^{Xc}与大肠杆菌中的同源蛋白具有相似功能。Coq7^{Xc}具有“Fe-Fe”单加氧酶活性，催化苯环 C-6 的羟基化，其催化机理类似于酵母菌同源蛋白 Coq7。敲除 CoQ 合成关键基因显著降低 *Xcc* 的致病性。黄单胞菌长期进化的 CoQ 生物合成基因多样性可能有助于其适应与植物互作过程中的复杂环境。

黄单胞菌 DSF 信号分子介导的群体感应与细菌致病性密切相关。为了研究黄单胞菌与植物互作过程中植物代谢产物对 DSF 群体感应的影响，笔者把寄主植物水解粗提物、植物来源的氨基酸、有机酸以及激素添加至 *Xcc* 的培养基，检测 DSF-家族信号分子的生物合成。笔者发现部分植物来源的物质能促进 DSF-家族信号分子的合成，其中增长最多的信号分子是 BDSF。另外，*Xcc* 侵染寄主植物后，在植物体内产生的信号分子也是以 BDSF 为主。黄单胞菌致病因子生物合成及其与植物互作分子机理的研究为开发选择性的黄单胞菌防控农药提供了潜在靶标。

关键词：黄单胞菌；菌黄素；辅酶 Q；DSF 信号分子介导

* 第一作者：曹雪强，博士后，研究方向：黄单胞菌菌黄素的生物合成；E-mail：xqcao2014@ sjtu. edu. cn
** 通信作者：何亚文，教授；E-mail：yawenhe@ sjtu. edu. cn

不同猕猴桃品种对溃疡病的抗性差异及其机理研究*

张 迪**，高小宁，韩 宁，赵志博，秦虎强，黄丽丽***

（西北农林科技大学植物保护学院，旱区作物逆境生物学国家重点实验室，杨凌 712100）

摘 要：目前由 *Pseudomonas syringae* pv. *actinidiae*（Psa）引起的细菌性溃疡病已成为猕猴桃生产中威胁最大的病害，在猕猴桃生产中造成了巨大的经济损失。为揭示其抗病机制，进而为利用抗病品种防治病害提供科学依据，本研究通过对离体枝条进行定量接种试验，分析 9 个常见猕猴桃品种的抗病性，测定接种前后不同时间点 6 个抗病相关基因表达量及防御酶中过氧化物酶（POD）、苯丙氨酸解氨酶（PAL）、超氧化物歧化酶（SOD）和过氧化氢酶（CAT）活性的变化。结果显示 9 个品种的抗病性差异显著，徐香（*Actinidia deliciosa* cv. *Xuxiang*）抗病性最强，红阳（*A. chinensis* cv. *Hongyang*）抗病性最弱。抗性相关基因 PR1、PR5、POD、PAL 在徐香中显症前显著上调表达；而在红阳中这些基因则是在显症后才显著上调表达，且表达量显著低于徐香。徐香的 POD 和 PAL 活性显著高于红阳，且显症前达到峰值，与基因表达趋势一致；而 CAT 和 SOD 活性及相关基因表达在抗、感品种中没有显著差异。综上所述，PR1、PR5 和 POD、PAL 在徐香抵御溃疡病侵染时发挥着重要作用。

关键词：猕猴桃；溃疡病；防御酶；基因表达

---

* 基金项目：陕西省重点研发计划项目（2017ZDCXL-NY-03-02，2018TSCXL-NY-01）；西北农林科技大学试验示范站（基地）科技成果推广项目（TGZX2018-23）

** 第一作者：张迪，硕士研究生；E-mail：zhangdi1278@163.com

*** 通信作者：黄丽丽，教授；E-mail：huanglili@nwsuaf.edu.cn

甘蔗赤腐病菌变异及甘蔗抗病性研究*

李 婕**，李文凤，张荣跃，王晓燕，单红丽，尹 炯，罗志明，仓晓燕，黄应昆***
（云南省农业科学院甘蔗研究所，云南省甘蔗遗传改良重点实验室，开远 661699）

摘 要：由镰孢炭疽菌（*Colletotrichum falcatum* Went.）引起的甘蔗赤腐病是造成甘蔗严重损失的重要真菌病害之一，通常被称为甘蔗的“癌症”，造成云南临沧、孟连、石屏多片蔗区甘蔗成片死亡，已由次要病害上升为主要病害。由于该病原菌频繁变异导致防控困难，目前对该病害药剂防控效果不理想，仍无有效、彻底的根治措施。因此，甘蔗赤腐病病菌变异研究及甘蔗抗病研究是实现病害持久可持续控制的必由之路。本文结合国内外最新研究，重点从甘蔗赤腐病病菌形态变异、致病性变异、分子变异分析了甘蔗赤腐病菌遗传变异，论述了甘蔗赤腐病快速分子检测技术及甘蔗抗病研究进展；并就深入开展甘蔗赤腐病研究进行展望，以期为我国甘蔗赤腐病的研究和防控提供理论依据。

关键词：甘蔗；赤腐病；变异；分子检测；抗病

* 基金项目：国家现代农业产业技术体系（糖料）建设专项资金资助（CARS-170303）；“云岭产业技术领军人才”培养项目“甘蔗有害生物防控”（2018LJRC56）；云南省现代农业产业技术体系建设专项资金资助

** 第一作者：李婕，硕士，研究实习员，主要从事甘蔗病害研究；E-mail：lijie0988@163.com

*** 通信作者：黄应昆，研究员，从事甘蔗病害防控研究；E-mail：huangyk64@163.com

甘蔗抗褐锈病基因定位亲本间多态性 SSR 标记筛选*

单红丽**，李文凤，黄应昆***，王晓燕，张荣跃，李　婕，尹　炯，仓晓燕
（云南省农业科学院甘蔗研究所，云南省甘蔗遗传改良重点实验室，开远　661699）

摘　要：由黑顶柄锈菌（*Puccina melanocephala* H. Sydow & P. Sydow）引起的褐锈病是一种重要的世界性甘蔗病害，发掘新的抗病基因，选育持久抗性品种是防治该病最经济有效的方法，而建立完整的、高密度的分子遗传图谱是发掘和定位甘蔗抗褐锈病基因的重要前提基础。为获得更多构建遗传连锁图谱可用的多态性 SSR 标记，本研究以 6 个高抗甘蔗褐锈病品种和 6 个高感甘蔗褐锈病品种为亲本，筛选在亲本间条带清晰、多态性明显、重复性较好的引物。结果表明，组合‘柳城 03-1137’×‘德蔗 93-88’、‘Mex 105’×‘粤糖 00-236’亲本间的多态性引物数最多，分别占 52.38%和 47.62%；其中，mSSCIR34 等 10 对引物在组合‘柳城 03-1137’×‘德蔗 93-88’中多态性最好，mSSCIR38 等 7 对引物在‘Mex 105’×‘粤糖 00-236’多态性最好。因此，mSSCIR34 等 10 对多态性引物用于‘柳城 03-1137’×‘德蔗 93-88’为亲本材料的作图群体，或 mSSCIR38 等 7 对多态性引物用于‘Mex 105’×‘粤糖 00-236’为亲本材料的作图群体，将更有利于构建分子遗传图谱，为抗褐锈病新基因定位和开发与其紧密连锁的分子标记奠定了良好基础。

关键词：甘蔗；褐锈病；抗病基因；亲本；SSR 标记

---

* 基金项目：国家自然科学基金项目（31660419）；国家现代农业产业技术体系（糖料）建设专项资金（CARS-170303）；云岭产业技术领军人才培养项目“甘蔗有害生物防控”（2018LJRC56）；云南省现代农业产业技术体系建设专项资金资助

** 第一作者：单红丽，硕士，助理研究员，主要从事甘蔗病害研究；E-mail：shhldlw@163.com

*** 通信作者：黄应昆，研究员，从事甘蔗病害防控研究；E-mail：huangyk64@163.com

甘蔗新品种及主栽品种对甘蔗梢腐病的自然抗性评价*

李文凤**，张荣跃，单红丽，王晓燕，李 婕，尹 炯，罗志明，仓晓燕，黄应昆***
（云南省农业科学院甘蔗研究所，云南省甘蔗遗传改良重点实验室，开远 661699）

摘 要：不同的甘蔗品种对甘蔗梢腐病的抗性不一，筛选和种植抗病品种是防治甘蔗梢腐病最为经济有效的措施。为明确近年国家及省甘蔗体系育成的新品种及云南蔗区各地主栽品种对甘蔗梢腐病的抗性，筛选抗梢腐病优良新品种供生产上推广应用。2015—2018 年对云南保山、开远、临沧、德宏 4 个区域化试验站示范新品种及云南蔗区各地主栽品种进行自然抗性调查评价。新品种自然抗性评价结合区域化试验进行，在 10 月底梢腐病发病稳定后，各处理小区采用 3 点（3 行）取样，每点（行）顺序连续调查 100 株，共 300 株，记录调查总株数及发病株数，计算发病株率；云南蔗区各地主栽品种自然抗性调查评价，在 10 月底梢腐病发病稳定后，选择代表性田块采用 3 点（3 行）取样，每点（行）顺序连续调查 100 株，共 300 株，记录调查总株数及发病株数，计算病株率。根据各品种病株率按 1~5 级划分进行甘蔗梢腐病自然抗性评价。田间自然发病调查结果表明，51 个新品种中 33 个表现中抗到高抗，18 个表现为感病到高感；20 个主栽品种中 10 个表现中抗到高抗，10 个表现为感病到高感。根据甘蔗新品种及主栽品种对梢腐病的自然抗性评价结果，推荐选种‘云蔗 03-194’、‘云蔗 05-51’、‘柳城 05-136’、‘柳城 07-500’、‘福农 38 号’、‘海蔗 22 号’、‘粤甘 46 号’、‘粤糖 40 号’、‘桂糖 30 号’、‘桂糖 44 号’等抗病优良新品种，淘汰‘粤糖 93-159’、‘粤糖 86-368’、‘新台糖 25 号’、‘新台糖 1 号’、‘盈育 91-59’、‘川糖 79-15’等感病主栽品种。区域内甘蔗种植品种要多样化，早中晚熟多品种搭配，可抑制梢腐病暴发流行。

关键词：甘蔗；新品种及主栽品种；梢腐病；自然抗性

* 基金项目：国家现代农业产业技术体系（糖料）建设专项资金（CARS-170303）；云岭产业技术领军人才培养项目“甘蔗有害生物防控”（2018LJRC56）；云南省现代农业产业技术体系建设专项资金资助

** 第一作者：李文凤，研究员，主要从事甘蔗病害研究；E-mail：ynlwf@ 163. com

*** 通信作者：黄应昆，研究员，从事甘蔗病害防控研究；E-mail：huangyk64@ 163. com

响应黄瓜绿斑驳花叶病毒侵染的黄瓜 miR159 靶基因鉴定和功能预测

李晓宇[1*]，梁超琼[1,2]，苗　朔[1]，罗来鑫[1]，李健强[1**]
（1. 中国农业大学植物保护学院，种子病害检验与防控北京市重点实验室，北京　100193；
2. 中国农业大学园艺学院，北京　100193）

摘　要：黄瓜绿斑驳花叶病毒（*Cucumber green mottle mosaic virus*，CGMMV）是帚状病毒科（Virgaviridae）烟草花叶病毒属（*Tobamovirus*）成员，近年来已在我国多地传播蔓延，严重影响葫芦科作物的产量及品质。由于病毒基因变异频率高、变异速度快，存在抗病品种的抗性易丧失等问题，因此病毒病较难防治。MicroRNA（miRNA）作为植物内源非编码的小 RNA 分子，在植物生长发育、抵御逆境胁迫及响应病原物侵染过程中发挥重要的调节作用，因此，鉴定受 miRNA 调控的靶基因并研究 miRNA 在植物与病原物互作过程中的作用具有重要意义。研究团队前期试验结果表明，在黄瓜与 CGMMV 互作过程中，miR159 通过指导剪切或翻译抑制发挥其对下游靶基因的负调控作用；此外，通过降解组测序鉴定了响应 CGMMV 侵染的黄瓜 miR159 的 3 个靶基因 XM004140579. 2、XM004140875. 2 和 XM011658282. 1，生物信息学分析结果表明其均为 MYB 类转录因子，可能在黄瓜抵御 CGMMV 侵染过程中发挥重要调控作用。本研究以中农 16 号黄瓜的 3 叶期幼苗为试验材料，将 CGMMV 接种 20d 后的黄瓜叶片样品采用 Trizol 法提取总 RNA，采用 5′-RLM-RACE 方法验证了 miR159 对上述 3 个靶基因的剪切作用并计算其剪切效率，结果表明：miR159 在其 5′端的第 10 位核苷酸处对其靶基因 XM004140579. 2、XM004140875. 2 和 XM011658282. 1 进行特异性剪切，切割频率分别为 23/25、37/38、25/39，据此认为 miR159 通过剪切作用负调控可能与抗病相关的靶基因。后续研究将通过 miR159 的过表达或抑制表达试验，揭示 miR159 在黄瓜-CGMMV 互作过程中的具体功能，为黄瓜抗绿斑驳花叶病毒的分子机制研究提供理论参考。

关键词：黄瓜绿斑驳花叶病毒；黄瓜；miR159；靶基因；抗病毒

* 第一作者：李晓宇，硕士研究生；E-mail：lixiaoyu1996@163. com
** 通信作者：李健强，教授，主要从事种子病理学研究；E-mail：lijq231@cau. edu. cn

水稻抗病蛋白 RGA5_ HMA 结构域突变体的重组表达、纯化及晶体生长

马梦琪，刘 洋，张 鑫，赵 鹤，彭友良，刘俊峰
（中国农业大学植物保护学院植物病理学系，北京 100193）

摘 要：由稻瘟菌（*Magnaporthe oryzae*）引起的稻瘟病是水稻中危害最大、最为普遍的真菌病害，在世界范围内的水稻每年因感染稻瘟病严重危害水稻的产量与质量。种植抗病品种是防治病害最为经济有效的措施。开发新型抗病基因将为指导水稻的持久、广谱抗病育种提供理论基础。

水稻抗病蛋白 RGA5 的 C 末端存在重金属结合结构域 HMA 直接参与识别稻瘟病菌的效应蛋白 AVR1-CO39 和 AVR-Pia。对该 HMA 结构域进行改造可以拓宽其识别范围，获得识别稻瘟病菌中与 AVR1-CO39、AVR-Pia 结构相似的其他效应蛋白的能力。本研究根据已经解析的 RGA5_HMA/AVR1-CO39 复合物的晶体结构，设计多个 RGA5_HMA 结构域的突变体。利用原核表达系统，将构建好的 RGA5_HMA 结构域突变体的重组表达载体转化到大肠杆菌表达菌株中，利用 IPTG 诱导表达筛选出最合适的蛋白表达体系。并通过亲和层析、凝胶过滤层析等技术对目的蛋白进行分离纯化，获得纯度较高且均一性较好的蛋白样品。利用座滴气相扩散法进行目的蛋白结晶条件的筛选，已经获得一个 RGA5_HMA 结构域突变体的晶体，这一结果将为解析突变体结构，设计和优化识别多种稻瘟病菌效应蛋白的免疫受体提供基础。

关键词：水稻；RGA5_HMA 结构域；原核表达；蛋白纯化

水稻 VIGS 技术体系的建立及其在水稻抗纹枯病中的应用*

赵 美**，万 俊，周而勋***，舒灿伟***
（华南农业大学植物病理学系，广东省微生物信号与作物病害重点实验室，广州 510642）

摘 要：水稻纹枯病是水稻生产上的重要真菌病害，其病原菌为立枯丝核菌（*Rhizoctonia solani* AG1-IA），该病菌寄主范围广、腐生性强，可侵染多达 260 种植物。由于缺少稳定的遗传转化体系，目前有关水稻纹枯病菌生长发育和致病相关基因的功能及其与水稻的互作机理研究较少，直接影响水稻高效稳定抗性品种的培育。因此，本研究利用水稻纹枯病菌基因组 10 847 个蛋白质序列，通过病原菌与寄主互作数据库和碳水化合物活性酶类数据库进行预测，综合分析选出 45 个基因作为候选靶标基因。然后建立水稻病毒诱导的基因沉默技术（virus-induced gene silencing，VIGS），构建 45 个候选基因的烟草脆裂病毒（*Tobacco rattle virus*，TRV）干扰载体。通过农杆菌注射的方法转化本氏烟草（*Nicotina benthamiana*）后用高致病力的水稻纹枯病菌菌株 GD118 进行接种，统计病情指数并检测真菌的生物量及靶标基因的转录水平。本研究最后发现转化了海藻糖磷酸脂酶基因 *Rstpp* 的本氏烟草的病情指数与对照相比显著减低。下一步将利用寄主诱导的基因沉默技术（host-induced gene silencing，HIGS）验证 *Rstpp* 基因的沉默在水稻抗纹枯病中的作用，并可望获得抗纹枯病的水稻新品种。本研究建立了水稻的 VIGS 技术体系，可以用于筛选水稻抗纹枯病菌的致病关键基因，为水稻抗性品种的培育奠定基础。

关键词：水稻纹枯病；立枯丝核菌；致病关键基因筛选；病毒诱导的基因沉默；寄主诱导的基因沉默

---

* 基金项目：国家自然科学基金青年科学基金项目（31801677）
** 第一作者：赵美，博士生，研究方向：植物病原真菌及真菌病害；E-mail：429605899@ qq. com
*** 通信作者：周而勋，教授，博导，研究方向：植物病原真菌及真菌病害；E-mail：exzhou@ scau. edu. cn
舒灿伟，副教授，硕导，研究方向：植物病原真菌及真菌病害；E-mail：shucanwei@ scau. edu. cn

水稻抗性相关蛋白 SIP4 在植物免疫中的功能研究*

高　涵**，方安菲，郑馨航，邱姗姗，李月娇，孙文献***
（中国农业大学植物保护学院，农业部作物有害生物监测与
绿色防控重点实验室，北京　100193）

摘　要：SCRE8 稻曲病菌中关键效应蛋白，对于稻曲菌的致病力起到重要作用。实验室前期工作通过酵母双杂交筛选到 SCRE8 与水稻 SIP4 蛋白互作，该互作在水稻原生质体中得到了验证。但是离体 Pull-down 实验显示 SCRE8 与 SIP4 相互作用并不是直接的。SIP4 属于水稻蛋白家族，其成员包括 SIP1-SIP8。有报道推测这一蛋白家族成员在植物抗病性中发挥作用。因此，本论文对该蛋白在水稻抗病性中的功能进行了研究。

实验结果表明，SIP4 能够抑制烟草由 BAX 诱导的细胞死亡。此外，也研究了这一基因家族在 PAMPs 诱导后的表达情况，发现在 flg22 诱导下 SIP2、SIP4 和 SIP6 表达发生变化，在几丁质诱导时 SIP6 表达发生变化。结果暗示这类蛋白参与水稻抗病。

其次，前期工作通过酵母双杂交实验初步认为 SIP1-SIP5 均与 SCRE8 互作。这里利用水稻原生质体瞬时表达体系与 Co-IP 技术验证了这些互作关系。

最后，对从日本获得的水稻 Tos17 插入的 sip4 突变体进行了鉴定，证实 SIP4 在突变体中表达被破坏。后续对该突变体进行了稻曲菌、稻瘟菌、细菌性条斑病菌（*Xanthomonas oryzae* pv. *oryzicola*，*Xoc*）和白叶枯病菌（*Xanthomonas oryzae* pv. *oryzae*，*Xoo*）接菌，并与野生型水稻日本晴品种的接菌结果进行比较。除 *Xoo* 接菌后突变体与野生型没有明显差异外，接种其他菌后，突变体均表现出比野生型更加感病。因此，确定 SIP4 参与水稻的抗病过程。此外，还发现部分家族成员存在自身互作，比如 SIP1、SIP3 和 SIP4；并且部分成员间也发生相互作用，如 SIP1 与 SIP4。这些互作在酵母和原生质体内均得到验证。这些结果暗示这些蛋白能够在体内形成多聚体发挥生物学功能。

研究结果为深入揭示 SIP4 蛋白在水稻抗病过程中发挥功能提供了重要基础。

关键词：植物免疫；SIP4；水稻；抗病性

* 基金项目：稻曲病菌致病关键效应蛋白的鉴定及其毒性功能的分子机理（21026050）

** 第一作者：高涵，博士研究生，研究方向：植物与病原菌互作；E-mail：8659869624@qq.com

*** 通信作者：孙文献，教授，主要从事植物与病原细菌、真菌互作；E-mail：08042@cau.edu.cn

基于晚疫病菌效应子识别策略挖掘马铃薯栽培种‘合作 88’潜在抗病基因[*]

栾宏瑛[**]，郑英转，王洪洋[***]
（云南师范大学马铃薯科学研究院，昆明 650500）

摘　要：四倍体马铃薯栽培种‘合作 88’是我国西南地区的主栽品种，具有鲜食品质、适合加工、抗晚疫病性等优良性状。为了深入挖掘和分析‘合作 88’的潜在抗病基因，挑选了晚疫病菌侵染马铃薯时早期上调表达的 68 个 RXLR 类效应子基因，将它们分别克隆到 PVX 病毒植物表达载体 pGR106 上，采用农杆菌牙签穿刺方法在‘合作 88’上进行效应子基因瞬时表达。据过敏反应（Hypersensitive response，HR）发生与否来推断马铃薯中是否存在潜在抗病基因。结果表明，5 个效应子基因，包括无毒基因 Avr2 家族成员 *PITG*_23008，无毒基因 Avr-blb2 家族成员 *PexRD*39，*PexRD*3，*PITG*_10232，*PITG*_07555，能够在‘合作 88’材料上诱导 HR，预示‘合作 88’中除含有已知抗病基因 *Rpi-R*2 和 *Rpi-blb*2 外，还应有其他潜在抗病基因。

关键词：马铃薯；‘合作 88’；RXLR 效应子；抗病基因

---

* 基金项目：国家自然科学基金（31800134）

** 第一作者：栾宏瑛，硕士研究生，主要从事马铃薯抗病分子遗传育种研究；E-mail：1276224361@ qq. com

*** 通信作者：王洪洋，博士，讲师，主要从事马铃薯与致病疫霉菌互作研究；E-mail：hongyang8318@ 126. com

大豆疫霉 RXLR 效应子抑制植物 PCD 的靶标筛选

靳雨婷，刘美彤，王群青
（山东省农业微生物重点实验室，山东农业大学植物保护学院，泰安 271018）

摘 要：大豆疫霉菌（*Phytophthora sojae*）引起的大豆根茎腐病是大豆生产上的毁灭性病害，每年在全球范围内造成的经济损失高达数十亿美元。疫霉菌在侵染过程中分泌大量的 RXLR 效应分子进入植物细胞抑制寄主防卫反应。解析 RXLR 效应分子的毒性机制有助于发现控制疫霉菌病害的新途径。前期研究发现抑制植物程序性细胞死亡（PCD）是大豆疫霉 RXLR 效应分子的重要毒性功能，但目前对抑制机制的了解还不够深入。前期实验已证明 RXLR 效应分子 Avh5 蛋白 C 端的 W-motif 是抑制植物 PCD 功能的主要蛋白域，并预测和验证了其中关键的氨基酸位点。研究发现，抑制植物的 PCD 反应是病原菌在侵染前期的主要策略，而 RXLR 效应分子家族普遍具有的 W-domain 被证明与抑制植物的 PCD 有关。本研究通过酵母双杂交筛选 W-domain 的植物寄主靶标，利用 mating 筛库法在鉴定到直接与其互作的两个 E3 泛素连接酶。经 GST pull-down 技术进一步验证两个蛋白直接互作。在植物中，泛素-26S 蛋白酶体系统（Ub/26S proteasome system，UPS）特别是 E3 泛素连接酶蛋白已经被报道与植物的免疫反应有关。SGT1 和 RARI 能够促进基础防卫反应中必须要 E3 泛素连接酶中的 CRLs 类蛋白特异性的聚集，而 SGT1 和 RAR1 是植物 R 基因介导的抗性信号传递中重要的基因，说明 E3 泛素连接酶能够与 SGT1 和 RAR1 相互作用调控植物的免疫反应。本研究将进一步解析疫霉菌 RXLR 效应分子 W-domain 参与抑制植物 PCD 的机制，分析 E3 泛素连接酶在抗疫病中的作用，为针对性地开发农药靶标提供理论基础。

关键词：大豆疫霉；程序性细胞死亡；E3 泛素连接酶；W-domain

一种大豆疫霉新型 PAMP 的鉴定和功能初步研究

胡玉瑶，韩　超，贾玉丽，王群青
（山东省农业微生物重点实验室，山东农业大学植物保护学院，泰安　271018）

摘　要：大豆疫霉引起的大豆根腐病是一种毁灭性病害。植物先天免疫系统识别病原菌的保守病原相关模式分子（pathogen/microbe-associated molecular patterns，PAMP）可触发寄主免疫反应（PAMP-triggered immunity，PTI）。利用病原菌的 PAMP 或激发子激活植物免疫系统，提高植物系统抗病性已成为病害防治的重要策略。研究发现，大豆疫霉编码胞外纤维素酶基因 *PsCBH*1 在病原卵菌和真菌中高度保守，并在侵染初期显著上调表达。通过毕赤酵母表达系统表达并纯化该蛋白，发现其能够在烟草、大豆和番茄上诱导植物细胞死亡，推测其为一种新型大豆疫霉菌 PAMP。检测发现 PsCBH1 蛋白能有效水解 β-1，4-葡聚和微晶纤维素 Avicel。利用定点突变等方法，将水解酶活性位点突变，发现 PsCBH1 的激发子活性和水解活性之间没有直接关系，说明 PsCBH1 激发植物免疫并不依赖其酶活性或寄主细胞壁的降解产物。研究结果有望揭示一类新型广谱激发子的作用机制，为大豆抗疫霉病资源筛选和设计防治新策略提供理论依据。

关键词：大豆疫霉；PAMP；激发子；胞外水解酶

Osa-miR159 对水稻稻瘟病的免疫调控机理研究

陈金凤，周士歆，樊　晶，李　燕，王文明
（四川农业大学水稻研究所，成都　611130）

摘　要：稻瘟病是水稻最为严重的病害之一，在水稻的各个部位及整个生长期均可发生，严重影响水稻产量，威胁粮食安全。近年来的研究表明，水稻 miRNA 可通过对其靶基因的抑制微调水稻对稻瘟病的抗性。在前期的研究中，笔者发现 Osa-miR159 在感病材料丽江新团黑谷（LTH）和抗病单基因系材料 IRBLkm-Ts 中的累积量均显著升高，提示 Osa-miR159 可能参与调控水稻的基础防御反应。人工模拟靶标 Target mimicry 可以与 miRNA 定向结合，从而抑制了 miRNA 对靶基因的抑制作用。因此，笔者通过对拟南芥内源性 IPS（INDUCED BY PHOSPHATE STARVATION 1）基因的定向改造构建了诱捕靶标材料 MIMIC159 和 Osa-miR159 过表达材料 OX159 的水稻转基因材料，对它们进行划伤滴菌和喷雾接菌，结果显示 OX159 材料对稻瘟病抗性减弱，而 MIM159 的抗性增强。并且，早期防御相关基因 *LOC_Os04g*10010 在喷菌 12h 后显著下调，说明 Osa-miR159 负调控水稻稻瘟病抗性。MYB 转录因子为植物转录因子中最大的家族之一，参与植物多种生长发育和抗病调控过程。*LOC_Os06g*40330 编码 MYB 家族转录因子，已被证明是 Osa-miR159 的靶基因之一，在 OX159 中其转录水平被显著抑制，而在 MIM159 中显著上调。另外，稻瘟菌强致病生理小种 RBS22 可诱导 *LOC_Os06g*40330 的转录，说明 miR59 可能通过抑制 *LOC_Os06g*40330 的表达来调控水稻对稻瘟病菌的免疫反应。研究 Osa-miR159 及其靶基因在水稻稻瘟病抗性中的作用，有利于挖掘新的抗性种质资源，为水稻的抗病育种提供基础理论依据。

关键词：水稻；稻瘟病；Osa-miR159；免疫调控

甘蔗新品种对甘蔗褐锈病的自然抗性评价*

李文凤**，王晓燕，单红丽，张荣跃，李　婕，尹　炯，罗志明，仓晓燕，黄应昆***
（云南省农业科学院甘蔗研究所，云南省甘蔗遗传改良重点实验室，开远　661699）

摘　要：甘蔗褐锈病发生与品种抗性密切相关，不同的甘蔗品种对黑顶柄锈菌的抗性不一，筛选和种植抗病品种是防治甘蔗褐锈病最为经济有效的措施。为明确近年国家及省甘蔗体系育成的 50 个新品种对黑顶柄锈菌的抗性，筛选抗褐锈病优良新品种供生产上推广应用。2015—2018 年通过在云南保山、开远、临沧、德宏 4 个区域化试验站，采用田间自然抗性调查与分子标记辅助鉴定抗性基因的方法，对中国近年选育的 50 个新品种及 2 个主栽品种进行自然抗性评价及抗黑顶柄锈菌基因 *Bru*1 的分子检测。根据叶片上病斑有无及占叶面积的比率按 1~9 级划分进行甘蔗褐锈病自然抗性评价。田间自然发病调查结果表明，50 个新品种中 13 个新品种表现高抗，16 个新品种表现抗病，3 个新品种表现中抗，18 个新品种表现为中感到高感；2 个主栽品种‘新台糖 16 号’和‘新台糖 22 号’均表现高抗。根据自然抗性评价及分子检测结果推荐选种‘云蔗 05-51’、‘云蔗 05-49’、‘柳城 05-136’、‘柳城 07-500’、‘福农 38 号’、‘福农 0335’、‘粤甘 34 号’、‘粤糖 40 号’、‘桂糖 30 号’、‘桂糖 32 号’等抗病优良新品种，早中晚熟多品种搭配，可抑制甘蔗褐锈病暴发流行。

关键词：甘蔗；新品种；褐锈病；自然抗性

---

* 基金项目：国家现代农业产业技术体系（糖料）建设专项资金（CARS-170303）；云岭产业技术领军人才培养项目“甘蔗有害生物防控”（2018LJRC56）；云南省现代农业产业技术体系建设专项资金资助

** 第一作者：李文凤，研究员，主要从事甘蔗病害研究；E-mail：ynlwf@ 163. com

*** 通信作者：黄应昆，研究员，从事甘蔗病害防控研究；E-mail：huangyk64@ 163. com

拟南芥广谱抗病蛋白 RPW8.2 通过多个亚细胞器协同调节其诱导的抗性和细胞死亡*

黄衍焱[1]，张凌荔[1]，马先锋[1,2,4]，赵继群[1]，He Ping[3]，Xiao Shunyuan[2]，王文明[1**]

（1. 四川农业大学水稻研究所，成都　611130；2. *Institute of Biosciences and Biotechnology Research*，*Department of Plant Science and Landscape Architecture*，*University of Maryland*，*College Park*，*MD* 20850，*USA*；3. *Department of Biochemistry & Biophysics*，*Texas A&M University*，*College Station*，*TX* 77843，*USA*；4. 湖南农业大学园艺园林学院，长沙　410128）

摘　要：*RPW*8.2 是从拟南芥生态型 MS-0 中克隆得到的一个对白粉病菌具有广谱抗性的基因，该蛋白的表达受白粉菌的诱导，特异的锚定于吸器外质膜（extra-haustorial membrane，EHM）上。其介导的抗性反应包括 SA 的积累、PR 基因的升高、H_2O_2 的积累、胼胝质的沉积和 HR 反应，但产生抗性的机制依然没有明确。前期研究发现 RPW8.2 含有 2 核定位信号和 3 个核输出信号，笔者通过融合外源核定位信号或核输出信号的方式，发现 RPW8.2 在细胞核中启动抗性，在细胞质中诱导细胞死亡。然后以 RPW8.2 中核质穿梭相关信号为节点，分别从 N 端或 C 端进行截短，发现 RPW8.2 C 端在细胞质内诱导细胞死亡且 N 端含有多个抑制原件。深入研究发现两个决定 RPW8.2 定位于吸器的元件是抑制细胞死亡的关键位点。另外，通过对 RPW8.2 C 端转基因材料的研究发现，叶绿体的降解是启动细胞死亡的关键。进一步研究证实衰老相关基因 *AtWRKY*53、*AtSAG*13、*AtRPS*17 和 *AtRBCS* 参与并调节了 RPW8.2 的抗性反应。本研究为进一步深入研究 RPW8.2 的功能奠定了基础。

关键词：RPW8.2；白粉菌；广谱抗性；核定位信号；核输出信号

---

* 基金项目：国家自然科学基金（31371931；31672090）

** 通信作者：王文明，研究员，主要从事植物广谱抗病分子机制研究；E-mail：j316wenmingwang@163.com

Fungal protein pmTFB3 activation of RPW8. 2-mediated cell death by competitive binding of RPW8. 2 repressor protein phosphatase type 2C during infection*

Zhao Jinghao**, Yong Zu, Dang Wenqiang, Hu Zijin, Wang Wenming***
(*Rice Research Institute*, *Sichuan Agriculture University*, *Chengdu* 611130, *China*)

Abstract: Powdery mildewdifferentiates a feeding structure named the haustorium to extract nutrients from plant hosts during infection. The atypical resistance (R) protein RPW8. 2 specifically targets to haustorium, where it activates salicylic acid (SA) pathway-dependent resistance to *Golovinomyces* spp. However, how RPW8. 2 activates defense remains largely uncharacterized. Here, we found that a powdery mildew protein RNA polymerase Ⅱ transcription factor-like protein B subunit 3 (pmTFB3) interacts with RPW8. 2 and the phytochrome-associated protein phosphatase type 2C (PAPP2C) in yeast and *in planta*. In this study, we found that powdery mildew produces haustoria, through which pmTFB3 was secreted and translocated into host cells. We further demonstrated that RPW8. 2 recognize pmTFB3 and protein redistribution from plasma membrane into nuclear. In addition, the protein competition assay shows RPW8. 2 repressor protein PAPP2C have more affinity to pmTFB3 than RPW8. 2. Followed serine / threonine phosphatase assay, we found that pmTFB3 has ability to activate S-factor PAPP2C to negative regulation host immune. Taken together, our data indicate that pmTFB3 is functionally connected with RPW8. 2-mediated defense, and PAPP2C was selected as a directly target by powdery mildew.

---

* Funding: This work was supported by the National Natural Science Foundation of China (grants 31371931 and 31672090 to W-MW)

** First author: Zhao Jinghao, PhD student, zhaojinghao402@gmail.com

*** Corresponding author: Wang Wenming; E-mail: j316wenmingwang@sicau.edu.cn

Osa-miR172a 通过 AP2 调控水稻稻瘟病抗性*

马晓春，王　贺，李续濮，鲁均华，樊　晶，李　燕，王文明**
（四川农业大学水稻研究所，成都　611130）

摘　要：植物的先天免疫在抵御病原菌入侵的过程中发挥着不可替代的作用，它会随着植物的生长周期而发生改变。在植物中，miRNA 调控着一系列的生物过程，包括生长、发育、应对各种生物和非生物胁迫等。越来越多的证据表明 miRNA 会参与植物对病原物的免疫过程。在笔者实验室前期的工作中，通过对接菌后的水稻叶片深度测序分析，得到一些对水稻稻瘟病侵染表现出不同反应的 OsmiRNA，其中包括 miR172a。对 miR172a 过表达株系 *OX172a* 接种稻瘟病菌，结果显示 miR172a 过表达株系增强了水稻对稻瘟病的抗病性，而模拟靶标转基因株系 *MIMIC172a* 增强了水稻对稻瘟病的感病性。随后构建 miR172a 靶基因 AP2 过表达和敲除的克隆，转化水稻，接种稻瘟病菌显示 AP2 过表达株系增强了水稻对稻瘟病的感病性，AP2 敲除株系增强了水稻对稻瘟病的抗病性。这一结果与 *OX172a* 和 *MIMIC172a* 接菌结果一致。上述结果表明 miR172a 通过抑制 AP2 的表达来增强对稻瘟病的抗性。

关键词：miR172a；水稻；稻瘟病

* 基金项目：国家自然科学基金（31430072；31471761）

** 通信作者：王文明，研究员，主要从事植物广谱抗病分子机制研究；E-mail：j316wenmingwang@163.com

Osa-miR1320 调控稻瘟病抗性及水稻的生长发育*

党文强，朱　勇，赵志学，刘信娴，黄衍焱，樊　晶，李　燕，王文明**
（四川农业大学水稻研究所，成都　611130）

摘　要：水稻稻瘟病是一种非常严重的病害，轻则导致水稻减产，重则颗粒无收。研究发现 MicroRNAs 不仅能够参与调控水稻的生长发育等过程，而且与免疫反应息息相关。通过对比分析感病材料 LTH 和抗病材料 IRBLkm-Ts 在接种稻瘟菌前后 MicroRNAs 表达量的变化，筛选响应稻瘟菌免疫反应的 MicroRNAs。本试验以其中一个候选 MicroRNAs Os miR1320 为研究目标，首先通过荧光定量 PCR，在接种稻瘟菌的 LTH 和 IRBLkm-Ts 材料中验证确认 miRNA1320 参与稻瘟菌诱导的免疫反应。然后分别构建 miRNA1320 过表达（OX1320）和 MIM1320 转基因材料。发现 OX1320 相比 WT 具有更强的稻瘟菌抗性，主要表现为病斑变小、附着胞及侵染菌丝形成较慢（即侵染进程较慢）、过氧化氢累积量升高。而 MIM1320 转基因材料则相反。另外，OX1320 转基因植株矮小，但对产量并无明显影响。这说明 Os miR1320 不仅正调控水稻稻瘟菌抗性，而且调控水稻生长发育。

关键词：水稻；MicroRNAs；稻瘟病；免疫反应

---

* 基金项目：国家自然科学基金（31430072；31471761）

** 通信作者：王文明，研究员，主要从事植物广谱抗病分子机制研究；E-mail：j316wenmingwang@163.com

Osa-miR1425 调控水稻稻瘟病抗性*

杨雪梅**，曹小龙，王　贺，陈金凤，樊　晶，李　燕，王文明***
（四川农业大学水稻研究所，成都　611130）

摘　要：近年来，越来越多的报道表明，miRNA 参与调控水稻稻瘟病抗性。笔者实验室前期通过对普感水稻材料丽江新团黑谷（LTH）和抗病单基因系材料 IRBLkm-Ts（含 Pikm 抗性位点）接种稻瘟病菌后不同时间点的小 RNA 高通量测序数据分析发现，miR1425 的累积量在抗病和感病材料间存在显著差异，但其在水稻稻瘟病抗性中的作用尚不明确。

笔者通过 RT-PCR，检测了 LTH 和 IRBLkm-Ts 接种稻瘟病菌后不同时间点 miR1425 的累积量来分析 miR1425 在水稻稻瘟病抗性中的作用；并且构建了 miR1425 的过表达材料 OX1425 及其诱捕靶标过表达材料 MIM1425（过表达与 miR1425 互补的序列从而解除 miR1425 对靶基因的抑制作用），对其进行稻瘟菌抗性分析。结果表明，在接菌后 24 h 后，miR1425 在感病和抗病材料中的累积量都逐渐增加，提示 miR1425 可能调节水稻基础免疫来调控水稻稻瘟病抗性。然后，对 OX1425 和 MIM1425 材料进行喷菌后发现，与对照相比，OX1425 对稻瘟病的抗性增强，而 MIM1425 相反，说明 miR1425 通过抑制靶基因的表达正向调控稻瘟病抗性。进一步笔者利用 CRISPR/Cas9 技术构建了靶基因 *LOC_ Os*08*g*01650 的敲除突变体后进行接菌发现，与对照相比，突变体对稻瘟病的抗性增强。上述结果表明，Osa-miR1425 可能通过 Rf 家族靶基因来正调控水稻对稻瘟病的抗性。

关键词：水稻；miR1425；稻瘟病菌；抗性

* 基金项目：国家自然科学基金（31430072；31471761）
** 第一作者：杨雪梅，硕士研究生，主要从事 miRNA 调控水稻稻瘟病抗性研究；E-mail：1429730601@ qq. com
*** 通信作者：王文明，研究员，主要从事植物广谱抗病分子机制研究；E-mail：j316wenmingwang@ 163. com

Osa-miR156 negative regulates rice immunity against blast disease via *SPL*14*

Zhang Lingli**, Zheng Yaping, Zhou Shixin, Wang He,
Wan Liangfang, Zhao Jiqun, Wang Wenming***

(*Rice Research Institute*, *Sichuan Agricultural University*, *Chengdu* 611130, *China*)

Abstract: Rice blast disease caused by *Magnaporthe oryzae* (*M. oryzae*) is one of the most destructive fungal diseases worldwide. MicroRNAs (miRNAs) play an important role to regulate plant growth and defense. Here, we show that miR156 acts as a negative regulator in rice immunity against *M. oryzae* by target SPL14. The accumulation of miR156 was increased in a susceptible accession Lijiangxin TuanHeigu (LTH) upon 12 and 24 hours post-inoculation (hpi), whereas decreased at 12hpi, but recovered at 24hpi in a resistant accession *IRBLkm-Ts*. Transgenic lines overexpressing Osa-miR156 (*OX*156) displayed higher susceptibility to M. oryzae, conversely, transgenic lines overexpressing target mimicry of Osa-miR156 (*MIM*156) to improve the expression of target genes showed highly resistance associating with increased expression of defense-related genes. Overexpression of Osa-miR156 in OX156 facilitate the infection process of *M. oryzae*, while silence of Osa-miR156 compromised fungus penetration and extention in MIM156.

MiR156 is a highly conserved miRNA family targeting SPL (Squamosa Promoter Binding Protein Like) genes. The accumulation of *OsSPL*14 was decreased in *OX*156 and increased in *MIM*156. Intriguing, the expression of *OsWRKY*45, which was identified as the target of SPL14, was also decreased in OX156 whereas increased in MIM156, indicating miR156 regulated blast disease resistance via post-transcriptional suppression on SPL14.

Key words: miRNA; OsSPL14; PTI, ETI; rice resistance; *Magnaporthe oryzae*

* Funding: This work was supported by the National Natural Science Foundation of China (No. 31430072, 31471761 and 31672090)

** First author: Zhang Lingli, PhD student; E-mail: zhanglnc@126.com

*** Corresponding author: Wang Wenming; E-mail: j316wenmingwang@sicau.edu.cn

Osa-microRNA167d facilitates infection of *Magnaporthe oryzae* by suppressing OsARF12*

Zhao Zhixue**, Feng Qin, Wang He, Zhu Yong, Li Yan, Wang Wenming***

(*Rice Research Institute*, *Sichuan Agricultural University*, *Chengdu* 611130, *China*)

Abstract: Rice blast disease caused by *Magnaporthe oryzae* is harmful for global rice industry, it endangers whole plant include root, leaf, spike and grain. But the commercial pesticide rapidly lost their function because of the diversity of rice blast fungal. it's significative to study the innate immunity of rice against rice blast. MicroRNAs are a class of endogenous noncoding mRNA which usually functioning by repressing target genes expression. lots of researches have verified that mircoRNAs take part in multiple-development and immunity process. But the role of microRNAs in rice blast is still not clear.

In our work, we reported that *osa-miR*167*d* negatively regulates rice blast disease by suppressing OsARF12: over-expressing *osa-miR*167*d* had a large lesion and supported fungal growth upon spraying or punching ways, further cytobiological experiments showed that over-expressing *osa-miR*167*d* mutants promote infection process of rice blast in leaf sheath upon GFP-tagged strain GZ8 infection. While over-expressing target mimicry which is capable of capturing *osa-miR*167*d* improved expression of *osa-miR*167*d*'s target genes, and the disease assay results showed that *MIM*167*d* had a less lesion and inhibited fungal growth. Meanwhile infection process of GZ8 in *MIM*167*d* mutants was repressed. To further analysis the molecular mechanism, we got transgenic plants of Os*ARF*12 and Os*ARF* 25 which are the target genes of *osa-miR*167*d*. Finally we found that *arf* 12 mutants increased susceptibility to *M. oryzae*, but *arf* 25 mutants didn't show any difference in rice blast resistance compared with WT.

* Funding: This work was supported by the National Natural Science Foundation of China (No. 31430072, 31471761 and 31672090)

** First author: Zhao Zhixue, Ph. D. student; E-mail: 1005795703@qq.com

*** Corresponding author: Wang Wenming; E-mail: j316wenmingwang@sicau.edu.cn

Osa-miR393a regulates rice blast resistance and affects growth and development*

Wang Liangfang, Zhou Shixin, Zhang Lingli, Wang He, Tong Ying, Wang Wenming**

(*Rice Research Institute*, *Sichuan Agricultural University*, *Chengdu* 611130, *China*)

Abstract: Rice blast is a fungal disease caused by the infection of *Magnaporthe oryzae*, which is one of the major diseases of rice and poses a serious threat to global food security. miRNA is a non-coding single-stranded RNA molecule composed of 20-24 nucleotides, which plays an important role in regulating plant growth and immune response. In the preliminary study, more than 30 candidate miRNAs were identified to involve in the immune response against rice blast. In this study, we discovered that miR393a is down-regulated in the susceptible material LTH and up-regulated in the disease-resistant material IR-BLkm - Ts, indicating miR393a may involve in the immunity against rice blast. Therefore, we constructed transgenic lines overexpressing miR393a (OX393a) and overexpressing mimicry of target gene for miR393a (MIM393a). Upon inoculation of *Magnaporthe oryzae* stain Guy 11, OX393a lines display enhanced resistant against rice blast, but MIM393a is more susceptible, suggesting Os-miR393a positively regulates rice blast resistance. In addition, the OX393a lines show early flowering and lower height compared with WT, indicating miR393a also involves in regulating rice growth and development. However, the mechanism of miR393a underling regulates the resistance, growth and development still more further study.

Key words: Os-miR393a; Rice blast resistance; Early flowering and lower height

* Funding: This work was supported by the National Natural Science Foundation of China (No. 31430072, 31471761 and 31672090)

** Corresponding author: Wang Wenming; E-mail: j316wenmingwang@ sicau. edu. cn

Osa-miR162 activates the resistance to *Magnaporthe oryzae* in rice*

Li Xupu, Ma Xiaochun, Wang He, Lu Junhua, Fang Jin, Li Yan, Wang Wenming**
(*Rice Research Institute*, *Sichuan Agricultural University*, *Chengdu* 611130, *China*)

Abstract: MicroRNAs (miRNAs) play important roles in rice defense response to *Magnaporthe oryzae*, the causative agent of rice blast disease. *Osa-miR162a* belongs to a conserved miRNA family targeting Dicer-like (DCL) genes, and DCL1 acts as a negative regulator in rice immunity. Here, we showed that miR162a play a positive role in rice immunity against *M. oryzae* by suppressing its target gene. The accumulation of miR162 was increased in resistant rice but reduced in susceptible rice upon *M. oryzae* infection. Transgenic rice lines overexpressing *Osa-miR162a* enhanced resistance to rice blast disease. By contrast a target mimicry to block miR162a showed highly susceptible to blast fungal. Taken together Our results indicate that miR162a positively regulates rice immunity against *M. oryzae* by repressing DCL1 gene.

Key words: miR162; resistance; rice blast

* Funding: This work was supported by the National Natural Science Foundation of China (No. 31430072, 31471761 and 31672090)
** Corresponding author: Wang Wenming; E-mail: j316wenmingwang@ sicau. edu. cn

大麦特异 TGA 转录因子 HvbZIP254 与 NPR1 蛋白互作介导植物抗病反应*

李欢鹏[1**]，吴娇娇[1]，赵淑清[1]，尚小凤[1]，刘大群[1,3]，刘　博[2***]，王逍冬[1***]
（1. 河北农业大学植物保护学院，保定　071000；2. 中国农业科学院植物保护研究所，北京　100193；3. 中国农业科学院研究生院，北京　100081）

摘　要：植物应对初生病原物侵染时，会在侵染区域外围产生对次生病原物的系统获得抗性（Systemic Acquired Resistance，SAR）。作为 SAR 过程中的关键调控因子，小麦 wNPR1 蛋白表现出非常保守的蛋白互作特性，其可与 *bZIP* 转录因子家族中的多个 TGA 蛋白互作，诱导 *PR* 基因的表达。然而，在麦类作物 SAR 过程中，*bZIP* 转录因子家族的调控网络尚不明确。本研究通过在全基因组水平分析大麦 *bZIP* 转录因子家族基因在 *NPR*1 基因介导的 SAR 类似反应中的表达模式，鉴定得到了一个大麦特异的差异表达基因 *HvbZIP*254，其编码蛋白在进化树中独立于其他已知 TGA 蛋白。利用农杆菌介导的基因瞬时表达技术，发现重组蛋白 HvbZIP254-GFP 定位于植物细胞核。利用酵母双杂交和双分子荧光互补技术，证明 HvbZIP10 与 wNPR1 存在蛋白互作，且互作定位同样为细胞核。进一步制备得到的小麦转基因材料 *Ubi*：*HvbZIP*254 表现出对稻瘟菌（*Magnaporthe oryzae*）菌株 P131 及小麦条锈菌（*Puccinia striiformis* f. sp. *tritici*）毒性小种 CYR32 的更高抗性水平。综上所述，本研究报道了一个大麦特异的 TGA 转录因子 HvbZIP254，其通过与 NPR1 蛋白互作介导植物抗病反应。小麦转基因材料 *Ubi*：*HvbZIP*254 未来有望作为创新性种质资源用于小麦抗病遗传改良。

关键词：大麦；TGA 转录因子；bZIP 转录因子；NPR1；蛋白互作；植物抗病

---

* 基金项目：国家自然科学基金青年基金（31701776）；河北省优秀青年科学基金（C2018204091）

** 第一作者：李欢鹏，硕士，研究方向为植物病理学；E-mail：809763271@ qq. com
*** 通信作者：王逍冬，博士，副教授，主要从事植物免疫学研究；E-mail：zhbwxd@ hebau. edu. cn
刘博，博士，主要从事植物与病原菌分子互作研究；E-mail：bliu@ ippcaas. cn

全基因组水平鉴定具有保守蛋白基序的小麦叶锈菌候选效应因子*

赵淑清[1**]，尚小凤[1**]，李欢鹏[1]，吴娇娇[1]，刘大群[1,2***]，王逍冬[1***]
（1. 河北农业大学植物保护学院，保定　071000；2. 中国农业科学院研究生院，北京　100081）

摘　要：锈菌的专性寄生依赖于其向寄主植物分泌的各种效应因子，抑制植物防卫反应。一般认为，锈菌通过特殊结构吸器，向吸器外间质分泌效应因子，而部分效应因子需要进一步转移至植物细胞内以抑制相应靶标。已有一些保守蛋白基序被认为与效应因子跨膜运输显著相关，例如，来自卵菌效应因子的 RxLR 和 CRN 保守基序、来自白粉菌效应因子的 Y/F/WxC 基序和来自赤霉菌效应因子的［SG］-P-C-［KR］-P 基序。本课题组前期研究发现，小麦锈菌中存在具有 RxLR 保守基序的毒性效应因子 PNPi，其可通过靶标植物胞内 NPR1 蛋白抑制植物抗病反应。然而，在小麦锈菌中，其他具有保守蛋白基序的候选效应因子仍亟待明确。本研究利用RNA-seq 技术，对感病小麦品种‘中国春’接种叶锈菌（*Puccinia triticina*）毒性小种 PHTT（P）后 4d、6d、8 d 叶片及锈菌夏孢子萌发样品进行了转录组测序。利用 *Pt* 1-1-BBBD Race 1 参考基因组进行转录组拼装，共注释了17 976个基因，包括2 284个新转录本。基因差异表达分析表明，小麦叶锈菌侵染过程中有3 149个基因显著上调表达（Log2Fold Change ＞ 1，q-value<0. 05），而有1 613个基因在锈菌孢子萌发阶段表达量较高（Log2Fold Change<-1，q-value<0. 05）。上调基因中，编码分泌蛋白、糖转运蛋白、氨基酸渗透酶和蛋白激酶的基因富集明显。利用 HMM 特征及蛋白同源分析，初步筛选得到了 12 个差异表达的 RxLR 类候选效应因子和 11 个差异表达的Y/F/WxC 类候选效应因子，未比对到具有 CRN 和［SG］-P-C-［KR］-P 蛋白基序的效应因子。利用荧光实时定量 qRT-PCR 技术，验证了部分候选效应因子的基因表达模式。利用酵母双杂交技术，初步探索了 RxLR 类候选效应因子的潜在植物靶标。利用农杆菌介导的基因瞬时表达技术，明确了上述候选效应因子在植物程序性细胞死亡中的功能。综上所述，本研究利用转录组学方法，从全基因组水平初步明确了小麦叶锈菌毒性小种 PHTT（P）侵染过程中的转录调控网络。转录组中鉴定得到的差异表达基因是了解叶锈菌毒性机制的宝贵资源，而具有保守基序的锈菌效应因子可作为进一步研究锈菌效应因子的跨膜转运及胞内靶标的良好模型。

关键词：保守蛋白基序；小麦叶锈菌；效应因子；RxLR、Y/F/WxC

*　基金项目：国家自然科学基金青年基金（31701776）；河北省优秀青年科学基金（C2018204091）
**　第一作者：赵淑清，硕士，研究方向为植物病理学；E-mail：1059037315@ qq. com
尚小凤，硕士，研究方向为资源利用与植物保护；E-mail：1787796738@ qq. com
***　通信作者：王逍冬，博士，副教授，主要从事植物免疫学研究；E-mail：zhbwxd@ hebau. edu. cn
刘大群，博士，教授，主要从事生防与分子植病研究；E-mail：liudaqun@ caas. cn

利用麦类作物系统获得抗性 SAR 关键转录调控因子 *HvWRKY*165 与 *HvWRKY*213 基因提高小麦抗病水平*

吴娇娇[1**]，李欢鹏[1**]，高　静[1]，尚小凤[1]，刘大群[1,3]，王晓杰[2***]，王逍冬[1***]

(1. 河北农业大学植物保护学院，保定　071000；2. 西北农林科技大学植物保护学院，杨凌　712100；3. 中国农业科学院研究生院，北京　100081)

摘　要：系统获得抗性（Systemic Acquired Resistance，SAR）是一种可诱导的植物抗病模式，具有对次生病原物的广谱抗性特征。在拟南芥中，受 *NPR*1 基因调控，病原菌直接侵染或 SA/BTH/INA 处理均可诱导产生 SAR 并转录激活 *PR* 基因。本课题组前期研究表明，麦类作物中，*NPR*1 基因在丁香假单胞菌（*Pseudomonas syringae* pv. *tomato*）DC3000 诱导的获得抗性（Acquired Resistance，AR）中参与调控 *PR* 基因的表达。然而，BTH 处理虽然可显著提高麦类作物对白粉病和叶锈病等多种病害的抗病水平，却与 *NPR*1 基因相关性较低。为了明确 *NPR*1 基因在麦类作物 BTH 诱导抗性中的功能，本研究对大麦转基因材料 *Ubi*：*wNPR*1 和 *Ubi*：*HvNPR*1-*Kd* 进行了 BTH 处理情况下的 RNA-seq 测序。结果表明，大多数 BTH 诱导基因（*BCI*）的表达均不受 *NPR*1 基因调控。而在大麦转基因材料 *Ubi*：*wNPR*1 中，少数的 *PR* 基因对 BTH 处理变得更加敏感。利用荧光实时定量 qRT-PCR 技术，我们验证了部分 *PR* 基因和 *BCI* 基因的表达模式。大量 *WRKY* 转录因子受 BTH 处理显著诱导表达，其中少数 *WRKY* 基因表达模式与 *NPR*1 的表达水平相关。综合课题组已有转录组文库，发现 *HvWRKY*165 与 *HvWRKY*213 基因在 BTH 诱导抗性和 *NPR*1 基因介导的 AR 反应中均显著上调表达，推测为麦类作物系统获得抗性关键转录调控因子。进一步制备得到的小麦转基因材料 *Ubi*：*HvWRKY*165 和 *Ubi*：*HvWRKY*213 表现出对稻瘟菌（*Magnaporthe oryzae*）菌株 P131 及小麦条锈菌（*Puccinia striiformis* f. sp. *tritici*）毒性小种 CYR32 的更高抗性水平。利用荧光实时定量 qRT-PCR 技术，初步明确了部分 *PR* 基因受 *HvWRKY*165 和 *HvWRKY*213 基因的调控情况。综上所述，本研究架构了麦类作物 BTH 诱导抗性的转录调控网络及其与 *NPR*1 基因的关系，从全基因组水平探索了该生物学过程中的关键 *WRKY* 转录因子。小麦转基因材料 *Ubi*：*HvWRKY*165 和 *Ubi*：*HvWRKY*213 未来有望作为创新性种质资源用于小麦抗病遗传改良。

关键词：麦类作物；系统获得抗性；*WRKY* 转录因子；小麦；抗病

---

* 基金项目：国家自然科学基金青年基金（31701776）；河北省优秀青年科学基金（C2018204091）；旱区作物逆境生物学国家重点实验室开放课题（CSBAAKF2018009）

** 第一作者：吴娇娇，硕士，研究方向为植物病理学；E-mail：2410071703@qq.com
李欢鹏，硕士，研究方向为植物病理学；E-mail：809763271@qq.com

*** 通信作者：王逍冬，博士，副教授，主要从事植物免疫学研究；E-mail：zhbwxd@hebau.edu.cn
王晓杰，博士，教授，主要从事植物免疫学研究；E-mail：wangxiaojie@nwsuaf.edu.cn

小麦 F-box 基因 *TaSKIP27-like* 响应生物及非生物逆境的表达分析[*]

孟钰玉[1**]，魏春茹[1]，范润侨[1]，于秀梅[1,2***]，刘大群[2***]

（1. 河北农业大学生命科学学院，河北省植物生理与分子病理学重点实验室，保定　071001；
2. 河北省农作物病虫害生物防治工程技术研究中心，保定　071001）

摘　要：F-box 蛋白是一类 N 端含有 F-box 结构域的蛋白，其在泛素介导的蛋白质水解过程中负责待降解底物的识别。该家族成员众多，广泛地参与了植物生长发育、自交不亲和、衰老及生物/非生物胁迫等多个生物学过程。笔者实验室前期对小麦 F-box 家族进行了系统地分类鉴定，发现 TaSKIP27-like N 端含有一个 F-box 结构域，C 端无明显保守结构域，属于 FBU 类型，dPCR 显示 *TaSKIP27-like* 在灌浆期叶片中高表达，并受热胁迫大幅上调表达。为探讨 F-box 基因 *TaSKIP27-like* 在小麦中是否受到生物及其他非生物逆境的诱导，本研究对小麦 *TaSKIP27-like* 进行了全长 CDS 扩增和表达模式分析。首先，以接种非亲和叶锈菌株 05-19-43②的小麦抗叶锈病近等基因系 Tc*Lr*15 为材料，通过 RT-PCR 扩增获得 1 个 495bp 的核苷酸序列。该序列包含完整的开放阅读框，可编码一条长度为 164aa 的多肽链，等电点为 9.6，进化树分析表明其与山羊草属（*Aegilops*）的 *AetSKIP27-like* 一致性最高。其次，以叶锈菌株 05-19-43②（非亲和菌株）和 05-19-137③（亲和组合）接种、外源激素（ABA、SA、MeJA）诱导、盐胁迫和干旱处理的 Tc*Lr*15 为材料，利用 qRT-PCR 技术对这个基因在各组合中的表达模式进行分析。*TaSKIP27-like* 基因在 12h 和 24h 的抗病组合中的表达量均高于感病组合，在抗病组合中，24h 的表达量达到最大，为 0h 的 2.8 倍；外源激素 SA 处理小麦叶片 24h 时表达量达到最大值，为对照组的 4.3 倍，ABA 处理 48h 时表达量达到最大值，为对照组的 4.9 倍，但该基因受 MeJA 的影响较小；盐胁迫处理 0.5h 后该基因的表达量即快速上升，在 2h 时达到最大，为 0h 的 18.1 倍；干旱处理 0.5h 后表达量达到最大值，为对照组的 2.7 倍。这些实验结果表明 *TaSKIP27-like* 基因对 SA、ABA、盐和干旱等胁迫都有响应。*TaSKIP27-like* 基因在不同生物及非生物逆境下的表达结果为深入解析其功能奠定了重要基础。

关键词：小麦；F-box；表达模式

---

* 基金项目：旱区作物逆境生物学国家重点实验室 2018 年开放课题（CSBAAKF2018008）；河北省高等学校科学技术研究项目（ZD2019086）；国家自然科学基金（31301649）

** 第一作者：孟钰玉，在读硕士研究生，从事植物抗病机理研究

*** 通信作者：于秀梅，副教授，从事植物抗病机理研究

刘大群，教授，从事分子植物病理学和植物病害生物防治研究

浙江省建德市草莓灰霉病发生流行动态监测与腐霉利抗性现状

郑 远[1]，王华弟[2]，戴德江[2]，沈 颖[2]，赵帅锋[3]，吴鉴艳[1]，张传清[1]
（1. 浙江农林大学农业与食品科学学院，临安 311300；2. 浙江省农产品质量监督检验测试中心，杭州 310020；3. 浙江省建德市植保站，建德 311600）

摘　要： 灰霉病是草莓上普遍发生的一种全球性真菌病害，为了探明灰霉病在浙江省草莓主产区建德市的发生规律，本文连续 34 年（1985—2018 年）监测该地区草莓灰霉病的发病规律，揭示 2 月下旬到草莓采收期为草莓灰霉病发生期，3—4 月为发病高峰期；统计 34 年间草莓灰霉病叶片、花朵、果实发病与危害损失关系的历史资料，采用相关回归的分析方法，建立了危害损失预测模型为 $Y=-0.769+0.042X_{11}+0.045X_{21}+0.178X_{31}$（$X_{11}$为叶片发病率，$X_{21}$为花朵发病率，$X_{31}$为果实发病率）。采用区分剂量法测定了采自 2015—2017 年的灰霉病菌对腐霉利的抗性，结果显示抗性频率高达 100%，分子机制研究表明抗性的产生均由 *OS*-1 基因发生碱基突变引起，单点突变型为第 365 位密码子由 ATC 突变为 AGC 或 AAC，双点突变型为第 369 位密码子由 CAG 突变为 CCG、第 373 位由 AAC 突变为 AGC。

关键词： 灰霉病；流行动态；区分剂量法；腐霉利；抗性机制

室内条件下不同马铃薯品种抗黄萎病的鉴定*

康立茹，贾瑞芳，张园园，张　键，张之为，周洪友，赵　君**
（内蒙古农业大学园艺与植物保护学院，呼和浩特　010018）

摘　要：筛选并培育马铃薯抗黄萎病品种是防控该病害非常有效的措施，进行抗性鉴定是培育抗病品种的重要基础。本研究在室内条件下利用伤根接种法对收集到的34份马铃薯品种进行了抗黄萎病鉴定，结果表明只有中薯21号是高抗黄萎病的品种，病情指数仅为5.0；表现为中抗水平的品种有11个，约占供试品种的32.4%，以‘中薯2号’、‘青薯9号’、‘中薯18号’等为代表；表现为中感的品种有16个，占供试品种的47.1%，主要包括‘大西洋’、‘红美’、‘克新1号’等品种；剩余的6个品种如‘康尼贝克’、‘兴佳’、‘中薯9号’等均表现为高感黄萎病，占供试品种的17.6%。本研究中未见有对黄萎病呈现免疫的品种。

关键词：马铃薯；黄萎病；抗性鉴定

* 基金项目：国家自然科学基金（31860495）
** 通信作者：赵君；E-mail：zhaojun@ imau. edu. cn

植物利用重金属抗病的探索*

郭 超**，胡净净，高文强，曹志艳，董金皋，周丽宏***
（河北农业大学生命科学学院，河北省植物生理与分子病理学重点实验室，保定 071000）

摘 要：近年来，由于污水灌溉、农药化肥的大量施用和使用超标物品等人类活动引起重金属含量超出其背景值，重金属污染不仅严重破坏生态系统，导致环境污染，还会使植物的根、茎、叶片等器官大量富集重金属，从而导致植物出现生长缓慢、植株矮小等现象，严重影响农作物的产量和质量。镉（Cd）并不是人体所需的必需元素，Cd 和其化合物均具有一定的毒性，当其通过食物链在人体中富集超过人体所耐受限度时，人体则可能出现慢性或急性的中毒反应，严重时甚至会导致死亡。有研究表明，当植食性昆虫虫食和机械损伤均会使在 Cd^{2+} 植物叶片中的含量增加，那么引起 Cd^{2+} 富集含量增加后，叶片中 Cd^{2+} 富集范围是否会在损伤部位增加？植物是否会利用重金属进行自身防御？

本研究利用针刺模拟昆虫虫食，对含有 Cd^{2+} 的拟南芥（*Arabidopsis thaliana*）叶片进行机械损伤、对拟南芥和大白菜进行灰葡萄孢菌接种发病后，利用双硫腙与 Cd^{2+} 高度螯合的作用，用双硫腙对损伤叶片及接种灰葡萄孢菌叶片进行染色。

本研究发现：损伤 0~24h 后，损伤处 Cd^{2+} 的富集范围无明显变化；48h 后损伤处 Cd^{2+} 的富集范围明显增加；在 120h，Cd^{2+} 的富集范围与对照相比无明显差异。本实验探究了拟南芥在受到损伤时，Cd^{2+} 在叶片中富集规律的变化情况，模拟虫食后，重金属富集在损伤处，为植物会利用重金属抗病提供了数据支持。

关键词：重金属；机械损伤；拟南芥；灰葡萄孢菌

---

* 基金项目：河北农业大学科研启动项目（ZD201610）

** 第一作者：郭超，研究生，生物工程；E-mail：Guochaoos@ 163. com
*** 通信作者：周丽宏，副教授，主要从事植物物理研究；E-mail：lihongzhou12@ 126. com

小豆抗锈病的组织学及生理生化机制的初步研究*

徐　菁，孙伟娜，殷丽华，徐晓丹，柯希望**，左豫虎**
（黑龙江八一农垦大学农学院，国家杂粮工程技术研究中心，大庆　163319）

摘　要： 由豇豆单胞锈菌（*Uromyces vignae*）引起的小豆锈病是小豆（*Vigna angularis*）生产上危害最为严重的病害之一，严重影响小豆的产量和品质。然而，小豆—豇豆单胞锈菌互作机理不明、寄主抗病机制不清，是制约小豆锈病可持续控制的重要因素。因此，明确小豆锈病菌侵染过程，揭示小豆抗锈病的生理及分子机理，是实现小豆锈病绿色防控的重要基础。

本研究采用荧光显微观察了锈菌在感病品种宝清红、抗病品种 VaHR136 及免疫品种 1-D-3 上的侵染过程，结果表明，感病品种接种后 12 h 夏孢子萌发并在气孔上方形成附着胞，接种后 24 h 产生气孔下囊，接种后 48 h 形成吸器，接种后 5 d 胞间菌丝大面积扩展，叶片出现花斑，至接种后 8 d 可产生大量夏孢子堆。而免疫及抗病品种中，病菌附着胞形成率显著下降，气孔下囊畸形率升高，吸器周围寄主细胞过敏性坏死及胞壁沉积物的形成有效阻止了病菌的扩展与繁殖。此外，感病品种中苯丙氨酸解氨酶（PAL）活性于接种后 24 h 开始升高，随后下降，且 PAL 活性整体处于较低水平，而免疫及抗病品种中 PAL 活性于接种后 12 h 即迅速升高，并在整个侵染过程中保持较高水平。相比感病品种，抗性品种中 H_2O_2 含量在侵染后期显著高于感病品种。

上述结果表明，PAL 对抗菌物质代谢及细胞壁结构的调控，以及 H_2O_2 对下游防卫反应的激活在小豆抗锈病过程中发挥了重要作用。本研究结果将为深入揭示小豆抗锈病机理奠定基础，为小豆锈病的可持续控制提供理论依据。

关键词： 小豆；豇豆单胞锈菌；侵染过程；抗病机理

---

* 基金项目：国家自然科学基金（31501629）；黑龙江省优势特色学科建设项目；黑龙江省青年创新人才项目（UNPYSCT-2016201，UNPYSCT-2017113）；黑龙江八一农垦大学三纵三横团队计划（TDJH201801）；黑龙江省农垦总局科技计划（HNK125B-08-08A）

** 通信作者：柯希望，讲师，主要从事植物病理学方向的研究；E-mail：kexylh@163.com
左豫虎，教授，主要从事植物病理学方向的研究；E-mail：zuoyhu@163.com

小豆 EG45 基因的亚细胞定位及其在抗病中的功能分析*

孙伟娜，徐　菁，殷丽华，徐晓丹，柯希望**，左豫虎**
（黑龙江八一农垦大学农学院，国家杂粮工程技术研究中心，大庆　163319）

摘　要：由豇豆单胞锈菌（*Uromyces vignae*）引起的小豆锈病在我国小豆种植区内普遍发生，且危害严重。笔者课题组前期鉴定获得了多个抗锈病小豆品种，在此基础上应用 RNA-seq 技术分析了抗病品种应答锈菌侵染的基因表达情况，发现小豆中一个植物利钠肽基因家族成员 *VaEG*45 于接种后显著上调，推测该基因在小豆抗锈病中发挥重要作用。

为明确 *VaEG*45 在植物抗病中的作用，本研究利用 qRT-PCR 分析了该基因在抗、感不同品种中应答锈菌侵染的表达情况，结果表明，感病品种中，*VaEG*45 在病菌侵染的不同阶段表达水平无显著变化，但在抗病品种中，*VaEG*45 于接种后 5 d 和 8 d 显著上调表达，表明该基因在抗锈菌侵染过程中起重要作用。同时，以基因组序列为参考克隆了 *VaEG*45 基因，该基因无内含子，编码区全长 396 bp，编码 131 个氨基酸，进一步利用 pCambia1302-GFP 构建 *VaEG*45 瞬时表达载体（pCambia1302-EG45-GFP），并采用半叶法将表达载体注射烟草叶片，以空载体为对照，注射后 2 d 观察基因的亚细胞定位发现，*VaEG*45 定位于细胞膜上。同时，分析了 *VaEG*45 的瞬时表达对烟草-灰葡萄孢菌（*Botrytis cinerea*）互作的影响。结果表明，瞬时表达 *VaEG*45 的烟草叶片表现出明显的灰霉病抗性，与对照相比，病斑直径下降了 39. 05%。

上述结果表明 *VaEG*45 显著提高了烟草对灰霉病的抗性，说明该基因在植物抵抗病原真菌侵染过程中发挥正调控作用，有关 *VaEG*45 参与植物抗病机理方面的研究工作目前仍在进行中。

关键词：小豆；小豆锈病；瞬时表达；抗病基因

* 基金项目：国家自然科学基金（31501629）；黑龙江省优势特色学科建设项目；黑龙江省青年创新人才项目（UNPYSCT-2016201，UNPYSCT-2017113）；黑龙江八一农垦大学三纵三横团队计划（TDJH201801）；黑龙江省农垦总局科技计划（HNK125B-08-08A）

** 通信作者：柯希望，讲师，主要从事植物病理学方向的研究；E-mail：kexylh@ 163. com
左豫虎，教授，主要从事植物病理学方向的研究；E-mail：zuoyhu@ 163. com

*GhMYB*43 调控棉花对黄萎病菌抗性的机制解析

佚　名

摘　要：植物在自然环境中生存，随时都要面临病原菌和植食性昆虫的侵害。但是植物不会坐以待毙，在长期的进化与选择过程中，植物像人一样建立了属于自己的防御系统，去感知和响应生物与非生物胁迫类型，激活植物体内响应的应答路径，对生物胁迫与非生物胁迫做出响应的应答来保护植物自身不受侵害。本课题组通过前期研究工作，克隆了 MYB 转录因子 *ChMYB*43 基因全长，并表明其受黄萎病菌诱导表达。本研究以 *GhMYB*43 为研究对象，探讨其在响应棉花黄萎病菌入侵后的功能及其抗病分子机制，取得以下主要结果：

（1）利用棉花和拟南芥酵母转录因子库，筛选得到的可结合在 GhLac1 启动子上的转录因子，其中一个为 D6-4，从棉花 jin668 中克隆其全长进行序列比对后，发现与拟南芥中 AtMYB43 最为同源，因此该基因命名为 *GhMYB*43。在接种棉花黄萎病菌 V991 后，*GhMYB*43 受黄萎病菌诱导先下调后上调表达。

（2）利用病毒介导的基因沉默（VIGS）技术抑制 *GhMYB*43 表达后，植株在接种黄萎病菌 V991 后表现为更加抗病。

（3）在拟南芥中异源表达 *GhMYB*43，转基因拟南芥表现为耐盐和甘露醇。因而推测 *GhMYB*43 在调控棉花非生物逆境中发挥了重要的作用。

关键词：棉花；黄萎病；次生代谢；MYB；非生物逆境

ε-聚赖氨酸对烟草赤星病的抑菌效果和相关基因表达的影响*

刘　鹤**，陈建光**，夏子豪，安梦楠***，吴元华***
（沈阳农业大学植物保护学院，沈阳　110866）

摘　要：ε-聚赖氨酸（ε-PL）是由微生物代谢产生的一种水溶性、可生物降解且具广谱抗微生物活性的多肽类物质，其主要用于食品防腐剂上，但在农业上对植物病原真菌的抑制作用和机制尚未见报道。本文以烟草赤星病菌作为研究对象，利用从细黄链霉菌辽宁致病变种中分离纯化获得的具有自主知识产权的 ε-PL（发明专利受理号：201910330737.2）进行平板抑菌试验、离体叶接种试验、孢子萌发抑制试验和实时荧光定量 PCR（Real-Time qPCR）试验，研究了 ε-PL 对烟草赤星病的防治作用及机理。结果显示 ε-PL 在烟草赤星病菌培养基周围可产生清晰明显的抑菌圈，且最小抑菌浓度（MIC 值）为 5μg/mL；离体叶接种试验表明，浓度大于 10μg/mL 的 ε-PL 即对烟草离体叶上病斑扩展有明显抑制作用；ε-PL 对病菌菌丝生长的抑制试验表明，200 μg/mL ε-PL 处理 3 d 时对烟草赤星病菌菌丝生长的抑制率达 92%，处理 7d 的抑制率也高达 87%左右，表明 ε-PL 稳定性好，抑制作用强，可以长时间控制菌丝生长；ε-PL 对病菌孢子萌发和芽管形态影响显微观测表明，当 25 μg/mL 的 ε-PL 处理 24h 后，烟草赤星病菌孢子萌发产生的芽管会出现皱缩、畸形等变化；当浓度达到 200 μg/mL 时，对孢子萌发的抑制率接近 98%；ε-PL 对烟草赤星病菌菌丝萌发及生长相关基因如 *Alternaria alternata* 环腺苷酸依赖性蛋白激酶催化亚基（AAPK1）、环腺苷酸依赖性蛋白激酶 A 基因（PKA）、甘油醛-3-磷酸脱氢酶基因（GAPDH）和 5.8s 核糖体 rRNA 代表基因进行 Real-Time qPCR 检测。结果显示前 3 个基因在 ε-PL处理后发生显著下调表达，从分子水平揭示了 ε-PL 对真菌抑制的作用机制。以上研究为 ε-PL在农业领域上的应用进一步提供理论依据，并为其将来投入商品化的生产奠定科学基础。

关键词：ε-聚赖氨酸；烟草赤星病；抗真菌机制

---

* 项目基金：国家重点研发项目（2017YFD0201104）；国家重点研发项目（2017YFE04900）

** 第一作者：刘鹤，硕士研究生，从事植物病理学研究；E-mail：2363598334@ qq. com
陈建光，博士研究生，从事植物病理学研究；E-mail：chenjianguang2010@ 126. com
*** 通信作者：安梦楠，讲师，从事植物病理学研究；E-mail：anmengnan1984@ 163. com
吴元华，教授，从事植物病理学研究；E-mail：wuyh09@ syau. edu. cn

Comparative transcriptome profiling of mRNA and lncRNA related to tea leaves infected by *Phoma segeticola* var. *camelliae*[*]

Li Dongxue[**], Wang Xue, Yin Qiaoxiu, Ren Yafeng, Bao Xingtao, Dharmasena Dissanayake-saman-pradeep, Wu Xian, Song Baoan, Chen Zhuo[***]

(*State Key Laboratory Breeding Base of Green Pesticide and Agricultural Bioengineering*, *Guizhou University*, *Guiyang* 550025, *China*)

Abstract: Non-coding RNAs (ncRNAs) such as long non-coding lncRNA have been reported to play important roles in the regulation of many biological processes. The rapid development of omics sequencing technology has promoted the identification of thousands of lncRNAs in plant species, but the role of lncRNAs in plant-pathogen interactions remains largely unexplored.

We identified significant differentially expressed genes (DEGs) andlncRNAs (DELs), and then construct lncRNA-mRNA networks using comparative transcriptome analysis of the leaves of tea plant (*Camellia sinensis*) infected and uninfected by *Phoma segeticola* var. *camelliae*, coupled with bioinformatics. A total of 1615 DEGs and 561 DELs (including 146 up-regulated lncRNAs and 415 down-regulated lncRNAs, $P<0.05$ and $|\log_{2foldchange}| \geq 1$) were identified between tea leaves infected and uninfected by *P. segeticola* var. *camelliae*. The network of co-localization was constructed for the analysis of the interaction of DEGs and DELs. These DEGs may be cis-regulated by DELs were enriched in 10 terms containing with NADP binding and sucrose alpha-glucosidase activity, *et al* ($P<0.05$). By the enrichment analysis of KEGG, 4 DELs were most enriched in the pathway of biosynthesis of amino acids ($P<0.05$). Integrated the analysis of the data of DEGs and DELs, 6 pairs of significant co-expression pairs were screened, with 5 positively correlated pairs (COR $\geq$ 0.7) and 1 negatively correlated pair (COR$\leq$ $-$0.7). As the differentially expressed trends of lncRNAs, lncRNA118096 (MSTRG. 70289. 1) and lncRNA20211 (MSTRG. 12101. 1) may play crucial roles in plant defense mechanism. In summary, the present study extends the lncRNA database of the tea leaves and these lncRNAs may provide reference on the study of plant-pathogen interactions in the future.

Key words: tea leaves; *Phoma segeticola* var. *camelliae*; lncRNA; co-expressing

---

* Funding: This work was supported by National Key Research Development Program of China (2017YFD0200308) and its Post-subsidy project (2018-5262), and the Major Science and Technology Projects in Guizhou Province (No. 2012-6012)

** First author: Li Dongxue; E-mail: gydxli@ aliyun. com

*** Corresponding author: Chen Zhuo; E-mail: gychenzhuo@ aliyun. com

水杨酸诱导水稻叶片的磷酸化蛋白质组学分析*

孙冉冉[1,2]，聂燕芳[1,3]，张 健[1,2]，王振中[1,2]，李云锋[1,2**]

（1. 华南农业大学广东省微生物信号与作物病害重点实验室，广州 510642；2. 华南农业大学农学院，广州 510642；3. 华南农业大学材料与能源学院，广州 510642）

摘 要：水杨酸（Salicylic acid，SA）是一种重要的信号分子，能够诱导水稻的抗病性。蛋白质磷酸化是一种重要的蛋白质翻译后修饰，在植物抗性信号转导途径中起着重要的作用。开展 SA 诱导的水稻磷酸化蛋白质组变化，对于全面了解 SA 诱导水稻的抗病性机制具有重要的意义。

以抗稻瘟病近等基因系水稻 CO39（不含已知抗稻瘟病基因）及 C101LAC（含 *Pi*-1 抗稻瘟病基因）为材料，用 SA 喷雾接种水稻，于接种后的 12 h 和 24 h 取样。经叶片总蛋白质的提取、磷酸化蛋白质的富集、双向电泳（2-DE）和凝胶染色，获得了不同时间段的磷酸化蛋白质 Pro-Q Diamond 特异性染色 2-DE 图谱和硝酸银染色 2-DE 图谱。用 PDQuest 8.0 软件进行图像分析，共获得了 47 个差异表达的磷酸化蛋白质。

采用 MALDI-TOF/TOF 质谱技术对差异表达的磷酸化蛋白质进行了分析，成功鉴定了其中的 40 个磷酸化蛋白质点；主要参与光合作用、防卫反应、抗氧化作用、蛋白质合成与降解、氨基酸代谢和能量代谢等功能。

关键词：水杨酸；水稻；磷酸化蛋白质组；双向电泳

---

* 基金项目：国家自然科学基金（31671968）；广州市科技计划项目（201804010119）；广东省科技计划项目（2016A020210099）

** 通信作者：李云锋；E-mail：yunfengli@ scau. edu. cn

外力触碰表皮毛诱导乙烯合成酶基因表达[*]

胡净净[**]，郭　超，高文强，董金皋，周丽宏[***]
（河北农业大学生命科学学院，河北省植物生理与分子病理学重点实验室，保定　071000）

摘　要： 表皮毛是植物表皮细胞分化而来的毛状附属物，是植物与外界环境直接接触的屏障，具有减轻外界病原生物侵害、紫外线损伤、机械损伤的作用。拟南芥的表皮毛底部形态类似奶嘴结构，外力触碰时易发生屈曲形变和应力集中，由此可能会引发不同的防御反应。在植物的生长发育过程中，乙烯作为一种应激激素，参与植物的生物应激和非生物应激反应。有研究报道，ACC 合成酶（ACS）在乙烯的生物合成途径中发挥重要作用，由一个多基因家族编码，那么，外力触碰拟南芥表皮毛是否会引起其表达变化呢？

本研究利用 GUS 组织化学染色检测外力触碰 *AtACS5pro*：*GUS* 和 *AtACS7pro*：*GUS* 拟南芥植株表皮毛后基因表达变化情况，结果显示触碰表皮毛，诱导 ACC 合成酶在两种植株表皮毛基部及周围细胞表达。研究结果为进一步探索植物表皮毛抵抗外界生物及非生物刺激的应激反应提供了理论支撑。

关键词： 表皮毛；乙烯合成酶；拟南芥；外力触碰

---

* 基金项目：河北农业大学科研启动项目（ZD201610）
** 第一作者：胡净净，研究生，植物学；E-mail：hujinghc@163.com
*** 通信作者：周丽宏，副教授，主要从事植物物理研究；E-mail：lihongzhou12@126.com

第七部分
病害防治

Antifungal activity of HSAF against *Colletotrichum fructicola* and its possible mechanisms of action[*]

Li Chaohui[**], Tang Bao, Sun Weibo, Zhao Yancun, Liu Fengquan[***]

(*Institute of Plant Protection, Jiangsu Key Laboratory for Food Quality and Safety-State Key Laboratory Cultivation Base of Ministry of Science and Technology, Jiangsu Academy of Agricultural Sciences, Nanjing* 210014, *China*)

Abstract: Anthracnose caused by *Colletotrichum fructicola* is one of the most important diseases in pear fruit, result in huge economic losses. Pear anthracnose is primarily an orchard disease but also may lead to significant post-harvest losses during the storage period. HSAF is an antifungal natural product isolated from the biocontrol agent *Lysobacter enzymogenes* and is regarded as a potential biological pesticide with novel mode of action. HSAF was previously demonstrated to show antifungal activity against a number of plant pathogenic fungi, but its efficacy against *Colletotrichum* species remained unclear. In current study, the substantially antifungal activities of HSAF on mycelial growth, conidia germination and apperssorium formation of *C. fructicola* were observed in a dose-dependent manner. We found that HSAF was fungicidal toward *Colletotrichum fructicola* with half maximal effective concentration (EC_{50}) of 0.69 μg/mL when assayed mycelia growth on PDA. Meanwhile, hyphae grown on HSAF added plates displayed defects such as clustered branching, curling, swelling and depolarized growth. In the presence of HSAF (4 μg/mL), conidia germination were significanly delayed and the germ tubes growth were inhibited, however appressorium could still formed after 24h incubation. Indeed, many HSAF-treated conidia featured small surface protuberances that could reflect failed attempts at forming a germ tube. HSAF at the concentration of 8 μg/mL could totally block the conidia germination of *C. fructicola*. In conclusion, our results validate HSAF as a potential drug to treat anthracnose caused by *Colletotrichum fructicola*.

Key words: Biocontrol; Antifungal activity; *Colletotrichum fructicola*; Anthracnose; HSAF

* 基金项目：江苏省重点研发计划（BE2018389）；国家梨产业技术体系（CARS-28-16）

** 第一作者：李朝辉，助理研究员，主要从事植物病理学研究；E-mail：chaohuili@yeah.net

*** 通信作者：刘凤权，研究员，博士生导师，主要从事植物病理学研究；E-mail：fqliu20011@sina.com

Occurrence of fludioxonil resistance in *Botrytis cinerea* from greenhouse tomato in China[*]

Zhou Feng [1,2**], Song Yulu[1], Li Shuai[1], Gao Yuqing[1],
Ruan Kang[1], Fan Yuchuang[1], Guo Hanxue[1], Li Chengwei[2***]

(1. *College of Resources & Environmental Science*, *Henan Institute of Science and Technology*, *Xinxiang*, 453003, *China*; 2. *Henan Engineering Research Center of Crop Genome Editing*, *Xinxiang*, 453003, *China*)

Abstract: Gray mold caused by the fungus *Botrytis cinerea* is one of the most important diseases worldwide. Fludioxonil has high activity against *B. cinerea*, and it has been reported that resistant mutants of *B. cinerea* emerged in some regions due to its multi-frequency and unscientific uses. Fludioxonil at 5 μg/mL was used as a discriminatory dose to detect resistance in candidate isolates, and 50% effective concentration values were determined for all fludioxonil-resistant isolates and some sensitive isolates. One high level field-fludioxonil-resistant isolate was detected. Compared with the sensitive isolates of *B. cinerea*, the resistant isolate was more sensitive to osmotic pressure, glycerol content significantly higher and less pathogenic on detached leaves of tomato and cucumber. Sequence analysis indicated that two amino acid sites mutations at the (Histidine kinases, Adenylyl cyclases, Methy binding proteins, Phosphatases; HAMP) HAMP domain of (Histidine Kinase, HK) HK (BcOs1) in field fludioxonil-resistant isolate, and the HAMP domain of HK is necessary to recognizing the stimulation from the compounds, the two mutations in this domain may be related to the fludioxonil-resistance in *B. cinerea*. These results will reveal the molecular mechanism of the function of HAMP domain of Bcos1 in fludioxonil-resistance in *B. cinerea*, and these studies have important theoretical and practical significance for the development of new fungicides and fludioxonil-resistance management strategies.

Key words: Fludioxonil; Resistance mechanism; *Botrytis cinerea*; Bcos1; Biological characteristics

---

* Funding: The Key Scientific and Technological Research Projects in Henan Province (No. 192102110056) and Foundation for High-level Talents in Henan Institute of Science and Technology (No. 2018022)

** First author: Zhou Feng; E-mail: zfhist@ 163. com

*** Corresponding author: Li Chengwei; E-mail: lichengweiwau@ hotmail. com

Current status of fungicides use for *Trichosanthes kirilowii* in anqing city and toxicity test of fungicides to *Colletootrichum orbiculare**

Li Ping**, Liu Dong, Bi Zhangyou, Wu Chengfang
(*Department of Horticulture and Landscape, Anqing Vocational and Technical College, Anqing* 246003, *China*)

Abstract: As a minor crop, the industry of *Trichosanthes kirilowii* has developed on a large scale and consequently its disease and pests occurred seriously. There is potential safety hazard that can't be ignored since the blind use of pesticides. In order to screen high efficiency and low toxicity fungicide, the effects of 8 fungicides on *Colletootrichum orbiculare* were tested with the mycelium growth rate in laboratory. The results showed that tebuconazole, difenoconazole, prochloraz, zhongshengmycin and diethyl methyl-cyprozoate had significant inhibition activities on mycelium growth, with the EC_{50} values ranged from 0. 102 6 to 0. 821 9 μg/mL, indicating the pathogen was sensitive to the above fungicides. The inhibitory effect of bio-fungicide zhongshengmycin was stronger than that of chemical fungicides such as tebuconazole and trifloxystrobin - tebuconazole, with the EC_{50} values 0. 775 4, 0. 821 9 and 1. 576 3 μg/mL, respectively. In field application, bio-fungicides were selected and used alternately with common fungicides to slow the generation of chemical fungicide resistance and improve the quality of trichosanthes.

Key words: *Trichosanthes kirilowii*; *Colletootrichum orbiculare*; fungicides; toxicity

* 基金项目：安徽省小宗特色作物（瓜蒌）农药使用情况调查及主要有害生物防治药剂筛选课题；安徽省高校优秀青年人才支持计划重点项目（gxyqZD2016516，gxyqZD2018120）

** 第一作者：李萍，博士，副教授，主要从事植物真菌病害防治研究；E-mail：liping05515156@163. com

Isolation and identification of antifungal metabolites from the biocontrol strain *Pseudomonas chlororaphis* TC3*

Sun Weibo**, Jiang Tianping, Li Chaohui, Zhao Yancun, Liu Fengquan***

(*Institute of Plant Protection, Jiangsu Key Laboratory for Food Quality and Safety-State Key Laboratory Cultivation Base of Ministry of Science and Technology, Jiangsu Academy of Agricultural Sciences, Nanjing* 210014, *China*)

Abstract: *Pseudomonas* spp. is an important bacterium that plays a biological control function around the rhizosphere soil of plants. It has a wide variety that widely distributed in nature. The main target of this kind of bacterium in biological control is phytopathogenic fungi. *Pseudomonas* can produce various types of secondary metabolites to exert their biocontrol functions, and different types of *Pseudomonas* produce different metabolites. Pear is the third largest fruit tree species in China, and the occurrence of pests and diseases seriously affects the quality and yield of pears. The researchers are committed to searching and developing anti-fungal metabolites from natural source of biocontrol bacterium. We isolated a biocontrol bacteria named TC3 from the rhizosphere soil of pear trees in Tongchuan, Shaanxi Province. In the plate antagonistic experiment, the bacteria showed excellent inhibitory activities against various pear pathogenic fungi. The strain was identified as *Pseudomonas chlororaphis* based on a 16S rDNA BLAST and the strain's physiological and biochemical characterization. Specific primers were used to detect the antibiotic-producing gene of the strain, and the strain was cultured on a large scale of NB medium. The synthesis genes of 2, 4-diacetylphloroglucinol, phenazine-1-carboxylic acid and pyrrolnitrin were found in strain TC3, and the antifungal activities of the monomer compounds were determined. The results showed that the secondary metabolites produced by *Pseudomonas chlororaphis* TC3 had good antifungal activities. This study can provide basis for the development and utilization of *Pseudomonas* biocontrol strains, as well as the discovery of new active ingredients, and the development of new biological pesticides.

Key words: Biocontrol; *Pseudomonas*; Antifungal; Structure identification

---

* Funding: This work was supported by grants from the Jiangsu Provincial Key Technology Support Program (BE2018389), the Earmarked Fund for China Agriculture Research System (CARS-28-16), National Key Research and Development Program (2018YFD0201400)

** First author: Sun Weibo, Assistant Researcher; E-mail: sunweibo.1001@163.com

*** Corresponding author: Liu Fengquan, Researcher; E-mail: fqliu20011@sina.com

Sensitivity of *Curvularia coicis* to pyraclostrobin and its control efficacy against *Coix* leaf blight in South Fujian Province*

Dai Yuli**, Gan Lin, Ruan Hongchun, Shi Niuniu,
Du Yixin, Chen Furu, Yang Xiujuan***

(*Fujian Key Laboratory for Monitoring and Integrated Management of Crop Pests, Institute of Plant Protection, Fujian Academy of Agricultural Sciences, Fuzhou, Fujian* 350013, *China*)

Abstract: Seventy-two *Curvularia coicis* isolates from 4 regions in South Fujian Province were tested for their sensitivity to pyraclostrobin using mycelial growth method. And the efficacy of pyraclostrobin for the control of *Coix* leaf blight (CLB) was evaluated under greenhouse conditions. The results indicated that the tested isolates were sensitive to pyraclostrobin. And the EC_{50} values of these isolates to pyraclostrobin ranged from 1.11 μg/mL to 5.06 μg/mL, with the mean value of (3.44±0.98) μg/mL. The frequency curve of pyraclostrobin was continuous and unimodal, and following the normal distribution. Hence, the mean EC_{50} value of (3.44 ± 0.98) μg/mL can be recognized as the baseline sensitivity of *C. coicis* to pyraclostrobin. Results of pot experiments showed that the spray of 25% pyraclostrobin EC at 250 μg/mL exhibited high efficacy for the control of CLB, with the control efficacy ranging from 78.2% to 79.7%. This efficacy was as good as that resulted from the spray of 50% iprodione SC at 500 μg/mL, and significantly higher than the efficacy of reference fungicide (80% mancozeb WP at 800 μg/mL). This study will guide the reasonable selection of fungicides for effective management of CLB in the future.

Key words: *Curvularia coicis*; pyraclostrobin; baseline sensitivity; *Coix* leaf blight; control efficacy

---

* 基金项目：福建省属公益类科研院所专项（2017R1025-3；2018R1025-1）；国家重点研发计划（2018YFD0200706）；福建省农业科学院青年科技英才百人计划项目（YC2016-4）；福建省农业科学院植物保护创新团队（STIT2017-1-8）

** 第一作者：代玉立，助理研究员，博士，研究方向：真菌学及植物真菌病害；E-mail：dai841225@126.com

*** 通信作者：杨秀娟，研究员，研究方向：植物病理学；E-mail：yxjzb@126.com

Biocontrol and Its Mechanism of *Bacillus amyloliquefaciens* and *Bacillus subtilis* on soybean Phytophthora blight*

Li Kunyuan[1**], Hu Jiulong[1**], Liu Dong[1,2**],
Wang Weiyan[1], Liu Xiao[1], Li Dandan[1], Gao Zhimou[1***]
(1. *College of plant protection*, *Anhui Agricultural University*, *Hefei* 230036, *China*;
2. *Department of Horticulture and Landscape*, *Anqing Vocational and Technical College*, *Anqing* 246003, *China*)

Abstract: With the improper application of fungicides, *Phytophthora sojae* begins to develop resistance to fungicides, and biological control is one of the potential ways to control it. We screened two strains of *Bacillus* (*Bacillus amyloliquefaciens* JDF3 and *Bacillus subtilis* RSS-1), which had an efficient inhibitory effect on *P. sojae*. They could inhibit mycelial growth, the germination of cysts, and the swimming of motile zoospores. To elucidate the response of *P. sojae* under the stress of *B. amyloliquefaciens* and *B. subtilis*, and the molecular mechanism of biological control, comparative transcriptome analysis was applied. Transcriptome analysis revealed that the expression gene of *P. sojae* showed significant changes, and a total of 1 616 differentially expressed genes (DEGs) were detected. They participated in two major types of regulation, namely, "specificity" regulation and "common" regulation. They might inhibit the growth of *P. sojae* mainly by inhibiting the activity of ribosome. A pot experiment indicated that *B. amyloliquefaciens* and *B. subtilis* enhanced the resistance of soybean to *P. sojae*, and their control effects were 70.7% and 65.5%, respectively. In addition, *B. amyloliquefaciens* fermentation broth could induce an active oxygen burst, NO production, callose deposition, and lignification. *B. subtilis* could also stimulate the system to develop the resistance of soybean by lignification and phytoalexin.

Key words: *Phytophthora sojae*; *Bacillus amyloliquefaciens*; *B. subtilis*; Biocontrol

---

* 基金项目：国家自然科学基金面上项目（31671977）；安徽省高校优秀青年人才项目（gxyqZD2018120）；安徽省高校优秀青年人才项目（gxyqZD016516）

** 第一作者：李坤缘，硕士生；胡九龙，硕士生；刘冬，博士生，研究方向均为真菌学及植物真菌病害
Li Kunyuan, Hu Jiulong and Liu Dong contributed equally to the paper

*** 通信作者：高智谋，教授，主要研究方向为真菌学及植物真菌病害；E-mail：gaozhimou@ 126. com

菘蓝抗菌肽 IiR-AMP1 的抑菌作用机制*

吴 佳**，董五辈***

（华中农业大学植物科技学院，武汉 430070）

摘 要：抗菌肽（antimicrobial peptide，AMP）是普遍存在于生物体内，能够抵御外来病原物侵染的一类小分子多肽，因具有分子量小、水溶性好、稳定性高以及不易产生抗药性等诸多优点，被国内外许多研究者作为一种新型抗生素开发应用。大多数 AMP 能够干扰病原菌与生命活动相关的生物学过程而具有广谱的抑菌活性。研究表明抗菌肽的抑菌机制多种多样，其中阳离子抗菌肽能够与细菌细胞壁上的磷壁酸或细胞膜上的脂多糖层相结合并破坏细胞壁或细胞膜结构的完整性，产生穿孔现象，进而致使细菌无法完成生命活动必要的生物学过程而死亡。

IiR-AMP1 是我们从中草药菘蓝中分离得到的一个新的抗菌肽，对多种病原真、细菌都有强烈的抑菌作用，对线虫的取食性行为也有一定的影响。且该抗菌肽具有很好的热稳定性、化学试剂稳定性，在不同 pH 环境中，仍然维持较高的抑菌活性。本研究构建了该基因的枯草芽孢杆菌表达载体，成功表达了 IiR-AMP1 抗菌肽，并将含基因菌株命名为 IiR1，笔者对该基因的抑菌机制作了初步探究。凝胶阻滞实验结果表明该抗菌肽无法直接破坏核酸结构；SEM 研究结果表明，IiR1 菌株在摇培后出现了大量的扭曲、变形或者破碎的菌体，而对照菌株的菌体则是表面光滑的正常杆状形态，由此可见 IiR-AMP1 对寄主细胞壁产生破坏；流氏细胞分析发现：该菌株的 PI 相对染色率高达 88%，而对照菌株的 PI 相对染色率为 1.75%；进行激光共聚焦显微镜观察，IiR1 菌株出现了红色荧光而对照菌株没有观察到红色荧光；荧光探针检测结果表明：该菌株细胞质膜内部流动性显著降低，细胞膜两侧电势差显著增加；胞外环境检测到的钾离子浓度、核酸浓度均显著增加，进一步说明细胞膜受到破坏，形成了孔洞，导致胞内物质外流。总之，本研究分离得到了一个全新的来源于中草药植物的抗菌基因，其抑菌机制是破坏细胞壁或细胞膜完整性，使细胞发生穿孔、皱缩、变形，最终胞内物质外流，细胞因无法维持正常的生命活动而死亡。本研究对发现的新抗菌基因的抑菌机制进行了初步探究，为病原菌-寄主互作理论提供了一定的支撑，新抗菌肽在控制植物病虫害的生物防治中，具有很大的潜力。

关键词：菘蓝；抗菌肽；Li-AMP1；LiR1；抑菌机制

---

* 基金项目：玉米抗病转基因育种新材料获得及新基因克隆（课题编号：2016ZX08003-001-006）

** 第一作者：吴佳，2016 级博士研究生，分子植物病理学

*** 通信作者：董五辈；E-mail：dwb@mail.hzau.edu.cn

川渝地区水稻纹枯病菌对噻呋酰胺及水稻稻曲病菌对戊唑醇敏感性基线的建立[*]

傅宇航[**]，彭复蓉，余 洋，杨宇衡，方安菲，毕朝位[***]
（西南大学植物保护学院，重庆 400715）

摘 要：水稻纹枯病和水稻稻曲病均是水稻上的重要病害，对川渝地区水稻生产造成重大损失，目前化学防治是防治这两种病害的主要措施。其中琥珀酸脱氢酶合成抑制剂（噻呋酰胺）、麦角甾醇合成抑制剂（戊唑醇）等新型杀菌剂具有广谱、高效、选择性高、作用靶标独特等优点，被广泛应用于水稻纹枯病和水稻稻曲病的化学防治中。但是由于这两类杀菌剂作用位点单一，长期且大规模的使用易导致抗药性产生。因此，本研究测定了川渝地区水稻纹枯病菌（*Rhizoctonia solani* Kühn）对噻呋酰胺、水稻稻曲病菌［*Ustilaginoidea virens*（Cooke）Tak］对戊唑醇的敏感性，研究结果如下：

（1）采用菌丝生长速率法测定。2018 年分离自川渝地区 100 株水稻纹枯病菌对噻呋酰胺的敏感性，结果表明其 EC_{50}值在 0.0077～0.1053μg/mL，最不敏感菌株的 EC_{50}是最敏感的 13.7 倍，平均 EC_{50}值为 0.0384±0.0207μg/mL，敏感性频率分布呈连续性单峰曲线，接近正态分布，没有出现敏感性下降的抗性群体。因此确定以 0.0384±0.0207μg/mL 作为川渝地区水稻纹枯病菌对噻呋酰胺的敏感性基线。

（2）采用菌丝生长速率法测定。2018 年分离自川渝地区 100 株水稻稻曲病菌对戊唑醇的敏感性，结果表明其 EC_{50}值在 0.0178～0.1282μg/mL，最不敏感菌株的 EC_{50}是最敏感的 7.2 倍，平均 EC_{50}值为 0.0567±0.0222μg/mL，敏感性频率分布呈连续性单峰曲线，接近正态分布，没有出现敏感性下降的抗性群体。因此确定以 0.0567±0.0222μg/mL 作为川渝地区水稻稻曲病菌对戊唑醇的敏感性基线。

以上研究结果为监测水稻纹枯病菌对噻呋酰胺及水稻稻曲病菌对戊唑醇的田间抗性奠定了基础。

关键词：水稻稻曲病菌；水稻纹枯病菌；戊唑醇；噻呋酰胺；敏感性基线

---

* 基金项目：国家重点研发计划“华南及西南水稻化肥农药减施技术集成研究与示范”（2018YFD0200300）；中央高校基本科研业务费专项资金（XDJK2017B026）

** 第一作者：傅宇航，硕士研究生，主要从事植物真菌病害方面的研究；E-mail：1047223522@ qq. com

*** 通信作者：毕朝位，副教授，主要从事植物真菌病害及病原菌抗药性研究；E-mail：chwbi@ swu. edu. cn

桃褐腐病拮抗放线菌的筛选鉴定及抑菌机理研究*

陈美均**，李珊珊，董国菊，马冠华，陈国康***
（西南大学植物保护学院，重庆　400715）

摘　要： 桃褐腐病是严重为害核果类水果生产的世界性病害，我国褐腐病菌的主要优势种是美澳型核果链核盘菌（*Monilinia fructicola*）。目前，关于放线菌防治植物病害的报道很多，但是针对桃褐腐病拮抗放线菌的研究鲜见报道。因此本研究从非农田生态系统（如森林、竹林、草地等）采集了15份根际土壤样品，通过稀释平板涂布法于高氏一号培养基中对土壤中的放线菌进行了分离纯化，共获得410株放线菌。以桃褐腐病病菌作为目的菌，采用对峙培养法，对所得菌株进行了初筛和复筛，得到27株具有较好生防潜力的放线菌菌株，对菌丝生长的抑制率范围为61.11%~82.32%。在后续的抑菌机理试验中，拮抗菌的无菌发酵滤液对桃褐腐菌丝生长的抑制作用测定，只有XDS1-5和HCD1-10两株菌株有明显抑制作用，抑制率分别为80.29%、70.64%。采用双皿对扣法测定拮抗菌体外挥发性成分对褐腐菌的影响试验中，其挥发性产物对褐腐菌菌丝的生长无明显抑制作用，但均在不同程度上抑制了褐腐病菌分生孢子的产生，对照CK的孢子量为4.3×10^6个孢子/mL，其中有ZBY1-8，CC1-18，LCD2-1等14株放线菌菌株几乎完全抑制了桃褐腐病菌孢子的产生，显示出了较好的生防潜力。而XDS1-5和HCD1-10两株菌株的孢子量分别为1.67×10^4个孢子/mL、1.5×10^4个孢子/mL，对褐腐病菌孢子的产生有一定的抑制作用。后续笔者将进一步对拮抗菌挥发性抑菌代谢产物进行鉴定和分析。对这27株放线菌菌株用16S rDNA的细菌通用引物27F/1492R对分离物进行PCR扩增及测序，测定结果获得的序列在NCBI上比对，初步鉴定为链霉菌属（*Streptomyces* sp.）。

关键词： 桃褐腐病菌；放线菌；抑菌机理；挥发性代谢产物

---

* 基金项目：国家重点研发计划子课题：优良放线菌筛选和生防特性评价（2017YFD0201108）；植被结构优化及其配套措施控制桃、梨重要病害技术（2018YFD0201402-6）

** 第一作者：陈美均，硕士研究生，主要从事植物病害与生物防治研究；E-mail：792292421@qq.com

*** 通信作者：陈国康，副教授，主要研究方向为芸薹属作物根肿病、植物线虫及生物防治；E-mail：chenguokang@swu.edu.cn

生物杀菌剂 B1619 水分散粒剂加工工艺的研究*

陈志谊[1**]，刘永锋[1]，刘邮洲[1]，乔俊卿[1]，陆　凡[2]，邱　光[2]
（1. 江苏省农业科学院植物保护研究所，南京　210014；
2. 江苏省苏科农化有限责任公司，南京　210014）

摘　要：近年来，由土传病害（枯萎病、青枯病、根腐病、立枯病、黄萎病和根结线虫等）引起的设施番茄连作障碍频繁暴发，严重威胁设施番茄生产的可持续发展。目前生产上防治设施蔬菜土传病害主要采取“夏季高温闷棚+化学农药处理土壤”的措施，但防治效果不理想，土传病害发生依然严重。同时大剂量化学农药处理土壤易造成设施生态环境污染，增加番茄中有毒化学物质的残留，对人类健康带来严重危害。因此，必须突破目前设施蔬菜土传病害以化学农药为主的防治对策，开拓以生物农药为核心的无公害防治设施蔬菜土传病害的技术体系。

1.2 亿活芽孢/g 解淀粉芽孢杆菌 B1619 水分散粒剂（简称：生物杀菌剂 B1619）是由江苏省农科院植保所植病生防研究室研发、具有自主知识产权的技术产品（ZL201210208366.9、ZL 2016101170003.2），并独家转让给江苏省苏科农化有限责任公司，获得农药正式登记证（PD20171746）。经过试验示范结果表明，该产品能够有效防控设施番茄枯萎病；无毒无致病性，对人畜安全；能够保护设施农业生态环境，降低化学农药使用量；并对设施蔬菜生长有一定的促生作用。

（1）生物杀菌剂 B1619 水分散粒剂加工工艺流程。将助剂和填料混合，初步粉碎后再进行气流粉碎，加入生防菌 B1619 发酵液进行捏合，造粒，然后放入电热恒温干燥箱中烘干，筛分得到产品，最后利用平板计数法测定水分散粒剂中 B1619 的含菌量（活芽孢/g）。工艺流程见如下图示。

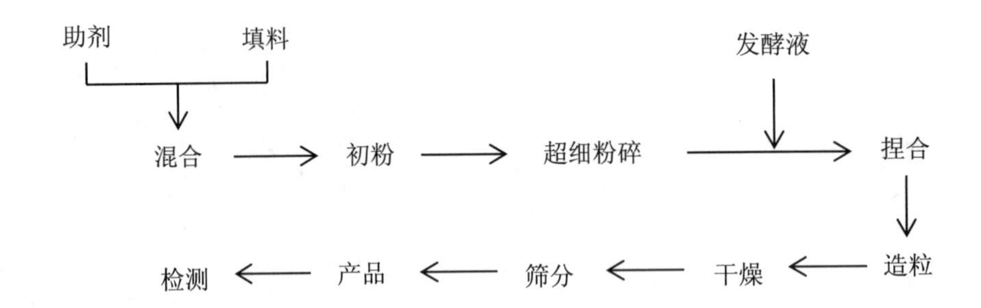

（2）生物杀菌剂 B1619 水分散粒剂主要助剂配比和颗粒制备条件。通过对多种的助剂和填料对生防菌 B1619 安全性的筛选，比较了各种助剂和填料及其组合的理化性能，科学配伍 B1619 水分散粒剂的产业化生产配方。通过单因子水平筛选和正交试验，对 B1619 水分散粒剂主要助剂配比和颗粒制备条件进行了优化。3 个主要助剂的最佳配比组合为：硅酸镁铝 1.0%，萘磺酸盐甲醛 3.0%，硫酸铵 4.5%，优化后颗粒含菌量提高了 152.0%；颗粒制备条件 3 个主要因子的

* 基金项目：国家重点研发计划（2017YFD0201101）

** 第一作者及通信作者：陈志谊，研究员，主要从事植物病害生物防治技术及机制研究；E-mail：chzy@jaas.ac.cn

最佳组合为：粒径1.5mm、烘干温度65℃、烘干时间20min，优化后颗粒含菌量提高了160.0%。测定了生物杀菌剂B1619水分散粒剂产品的性能，结果表明，该产品含菌量大于1.2×10^8CFU/g，悬浮率为90%，润湿时间为15s，崩解时间30s，水分含量小于5%，热贮稳定性合格；所得样品的各项技术指标均符合产品标准要求。

关键词：生物杀菌剂B1619；水分散粒剂；加工工艺

花生果腐病病原鉴定及化学药剂的室内筛选*

于　静**，马　骏，许曼琳，张　霞，郭志青，吴菊香，董炜博，迟玉成***
（山东省花生研究所，青岛　266100）

摘　要：花生是我国主要的经济作物，在其整个生长过程中会受各种病原微生物的侵染产生病害导致产量和品质受损，造成经济损失。花生果腐病是花生荚果上的一种重要病害，一旦发生会造成荚果腐烂落果，重病田可致绝收。近年来，笔者对全国各大花生产区的花生果腐病进行调查和样品采集，并从新鲜发病组织分离鉴定该病的病原，认为群结腐霉菌是引起花生烂果的主要致病菌。为了更加有效防治该病害，寻找防治花生果腐病的有效药剂，本研究对 8 种化学药剂进行室内筛选，实验结果表明，96.8% 咯菌腈和 95%噻呋酰胺对引起花生果腐病的病原菌群结腐霉有较好的抑菌效果，EC_{50}分别 3.162×10^{-5}mg/L 和 1.995×10^{-5}mg/L，较低浓度的药剂就可以达到很好的抑制效果。96.8%咯菌腈和 95%噻呋酰胺是防治花生果腐病有应用潜力的药剂品种。

关键词：花生；果腐病；群结腐霉；病原鉴定；化学防治

---

* 基金项目：国家自然基金（31801711）；山东省自然科学基金（ZR2018LC015）；国家重点研发计划项目（2018YFD0201007）；青岛市民生科技计划（17-3-3-70-nsh）；山东省现代农业产业技术体系花生创新团队建设项目（SDAIT-04-07）；山东省农业科学院农业科技创新工程（CXGC2018E21；06210214442019，2-18-43）

** 第一作者：于静，博士，助理研究员；E-mail：iamyujing2008@126.com

*** 通信作者：迟玉成，博士，研究员；E-mail：87626681@163.com

基于多拷贝序列扩增的寄生疫霉菌检测体系构建*

王瑢笙[1,2]**，单卫星[2,3]***

（1. 沈阳农业大学植保学院，沈阳 110161；2. 西北农林科技大学农学院，杨凌 712100；
3. 旱区作物逆境生物学国家重点实验室，杨凌 712100）

摘 要：寄生疫霉菌（*Phytophthora parasitica*）是分布广泛的土传植物病原菌。其寄主范围广泛，可侵染超过250个属的植物；特别是烟草黑胫病等作物疫霉病，发病迅速，难以防治，因此建立快速、精准、灵敏、高效、经济且便于操作的病原菌检测技术，是实现寄生疫霉菌田间早期诊断与防控的关键。本研究通过疫霉属比较基因组学分析，鉴定到寄生疫霉菌基因组中一个特异的多拷贝序列PpM34。笔者以该序列为靶标设计了SYBR检测引物，构建了寄生疫霉菌SYBR快速检测体系。一系列qPCR实验结果表明，该检测体系高度特异，检测下限为4 fg/μL。综上，本研究建立了一个高度特异、灵敏的寄生疫霉菌快速检测体系。

关键词：寄生疫霉菌；病原菌检测；比较基因组学；多拷贝序列

---

* 基金项目："作物抗病育种与遗传改良学科创新引智基地"111项目

** 第一作者：王瑢笙，在读博士生；E-mail：814518413@qq.com

*** 通信作者：单卫星，教授；E-mail：wxshan@nwafu.edu.cn

两种烟草根际拮抗菌调控青枯病的效果及抑菌机理研究*

张欣悦**，罗翠琴，陈小洁，王 其，王 璐，丁 婷***
（安徽农业大学植物保护学院，合肥 230036）

摘 要：为进一步丰富烟草青枯病的生防资源，该研究对安徽省黄山及宣城烟区的健康烟株根际土壤进行烟草根际细菌的分离纯化，共得到烟草根际细菌 178 株。利用抑菌圈法，筛选出两株抗青枯病菌的根际细菌 GZYCT-4 和 GZYCT-9，培养 48 h 的抑菌圈直径分别为 16. 64 mm 和 18. 90 mm，上述 2 株拮抗根际细菌均能有效地防控烟草青枯病，降低发病程度。系统发育分析结果表明 GZYCT-4 和 GZYCT-9 均属于芽孢杆菌。采用 PCR 扩增技术初步明确菌株 GZYCT-4 和 GZYCT-9 均具有合成脂肽类物质的能力，且其脂肽类粗提物对烟草青枯病菌的抑菌圈直径分别达到 16. 24mm 和 20. 71mm。通过趋化性试验探究烟草根系分泌物及有机酸对拮抗菌 GZYCT-4 和 GZYCT-9 在烟草根部定殖的影响，结果显示 GZYCT-4 对‘红花大金元’根系分泌物有较强的趋化反应，且‘红花大金元’根系分泌物产生的草酸对 GZYCT-4 吸引作用较大，而 GZYCT-9 则对 K326 根系分泌物有较好的趋化反应，K326 根系分泌物产生的柠檬酸对 GZYCT-9 有较好的吸引作用。本研究结果为进一步开发利用优良的生防菌株提供理论依据。

关键词：根际细菌；烟草青枯病菌；脂肽；趋化性

* 基金项目：贵州省烟草公司科技项目“烟草内生菌和土壤拮抗菌对黔南烟草土传病害的调控研究”（201619）；国家级大学生创新训练项目（201810364047）

** 第一作者：张欣悦，硕士研究生，主要从植物病害防治研究；E-mail：1162550824@ qq. com

*** 通信作者：丁婷，博士，教授，主要从植物病害生物防治及生防微生物分子生物学研究；E-mail：dingting98@ 126. com

细菌素 SyrM 杀菌机理的初步研究*

李俊州**，尹平仪，苑轲轲，李盈曦，彭友良，范　军***
（中国农业大学植物保护学院，北京　100193）

摘　要：细菌素是一类可以杀死近缘种属细菌的蛋白质或多肽，具有潜在的生防应用前景，明确细菌素的杀菌机理对细菌病害的生物防治具有重要意义。笔者实验室前期从丁香假单胞菌番茄致病变种 *Pseudomonas syringae* pv. *tomato* DC3000 菌株中鉴定了大肠菌素 M 类细菌素 SyrM，发现该细菌素可以介导细菌间竞争。在此基础上，笔者近期通过转座子突变和高通量细菌生长测定体系，从 12 000个克隆中筛选到 16 株对细菌素不敏感的突变体菌株。分别对这 16 株突变体进行回补，均能恢复对细菌素敏感的表型。序列相似性分析表明，这些突变基因与蛋白的组装、修饰以及物质转运途径等相关。这暗示 SyrM 的作用方式是复杂的，本研究将为后续解析其杀菌机理奠定基础。

关键词：丁香假单胞菌；细菌素；SyrM；杀菌机理

* 基金项目：国家重点研发计划项目子课题（2017YFD0201106-01C）

** 第一作者：李俊州，博士研究生，研究方向：植物病原物致病机理；E-mail：lijunzhou2010@163.com

*** 通信作者：范军，教授，研究方向：植物病原物致病机理及植物数量抗病性的遗传和分子机理；E-mail：jfan@cau.edu.cn

辣椒疫霉抗甲霜灵基因的分子标记研究*

卓　新**，王艺烨，刘　冬，姜庆雨，潘月敏，张华建，陈方新，高智谋***
（安徽农业大学植物保护学院，合肥　230036）

摘　要：辣椒疫霉（*Phytophthora capsici*）是一种重要的植物病原卵菌，寄主范围较广，是农业生产上重要性有害生物之一。目前生产上主要是使用甲霜灵及其复配剂对其进行防治，但由于甲霜灵作用位点单一，长期使用后，田间很容易产生抗药性菌株，而且抗药问题日益严重，加之关于辣椒疫霉对甲霜灵抗性的分子机理尚不清楚，给抗药性治理带来困难。因此，找到与抗甲霜灵相关的基因尤为重要。本研究通过 SSR 和 ISSR 两种分子标记来寻找与抗甲霜灵相关的基因，旨在为抗药性基因的定位和克隆提供必要的实验基础，取得的主要结果如下：

（1）辣椒疫霉抗甲霜灵突变菌株的筛选。用甲霜灵对供试的辣椒疫霉敏感菌株诱变后获得了 2 株抗甲霜灵突变菌株 SD1-9 和 SH1-7。实验结果发现抗性突变菌株的生长速率要比野生敏感型菌株的生长速率慢；同时发现抗性突变菌株和野生敏感型菌株的菌丝和孢子囊在形态上无明显差异。菌丝都是白色，无隔，偶有瘤状或节状膨大；也都是呈近直角形分枝，分枝处有缢缩。突变菌株和野生菌株的孢子囊的形状也都不规则，有的是近球形的，有的是近长椭圆形或者倒梨形的，乳突都很明显。通过测定抗性突变菌株的抗性水平，结果表明，抗性突变菌株 SH1-7 的 EC_{50} 为 10.025 7 μg/mL，抗性水平为 16.097 8 倍；抗性突变菌株 SD1-9 的 EC_{50} 值为 26.062 9μg/mL，抗性水平为 58.581 5倍。

（2）辣椒疫霉抗甲霜灵基因的标记。利用 SSR 分子标记发现在辣椒疫霉抗性菌株 LN3、LN4、LN5、JX1、JX2、JX3 和 JX4 中扩增出差异条带，进一步通过 ISSR 分子标记同样也在上述 7 个菌株中发现差异条带。将引物 UBC873 和 UBC864 中的差异条带回收测序后，在 NCBI 中进行 Blastx 发现差异序列 1 与恶疫霉（*Phytophthora cactorum*）的假定蛋白 PC110_g 19195 相似度最高，为 64.58%；差异序列 2 与恶疫霉（*P. cactorum*）的假定蛋白 PC110_g 17981 相似度最高，为 81.02%；差异序列 3 在 NCBI 以及 JGI 比对后发现无论是在核酸水平还是蛋白水平，比对结果都没有发现与之相类似的基因或是蛋白。设计特异性引物对差异序列 1 验证时发现 LN3、LN4 和 LN5 中扩增的条带要明显比敏感菌株亮，通过 RT-PCR 验证发现差异序列 1 在辣椒疫霉抗性菌株 LN3、LN4 和 LN5 中相关表达量要高于敏感菌株 NM1。差异序列 2 进行特异性引物验证的时候在敏感与抗性菌株中均扩增出了条带，但通过 RT-PCR 验证发现差异序列 2 在辣椒疫霉抗性菌株 LN3、LN4 和 LN5 中相关表达量要低于敏感菌株 NM1。差异序列 3 特异性引物验证时只在抗性菌株 LN3、LN4、LN5、JX1、JX2、JX3 和 JX4 中扩增出了目的条带，通过 RT-PCR 验证该序列在抗性菌株 LN3、LN4、LN5、JX1、JX2、JX3 和 JX4 中的表达量确实要高于敏感菌株 NM1。

（3）SSR 和 ISSR 两种分子标记的比较。从 103 对 SSR 引物中筛选出的 13 对引物，对 22 个辣椒疫霉菌株进行扩增，结果显示 7 个不同地区的 Shannon's 多样性指数从大到小顺序为：江西>上海>辽宁>江苏>内蒙古>安徽>山东。从 47 个 ISSR 引物中筛选出了 20 个引物，扩增结果显示 7

* 基金项目：国家自然科学基金面上项目（31671977）

** 第一作者：卓新，硕士生，研究方向为真菌学及植物真菌病害

*** 通信作者：高智谋，教授，主要研究方向为真菌学及植物真菌病害；E-mail：gaozhimou@126.com

个地区的 Shannon's 多样性指数从大到小顺序为：山东>江西>上海>辽宁>安徽>江苏>内蒙古。两种分子标记聚类结果都显示在阈值为 0. 66 时，供试菌株与地理来源存在着一定的相关性；在阈值为 0. 66 时，辣椒疫霉对甲霜灵的敏感性与相应的聚类分组之间存在着一定的相关性。对 SSR 和 ISSR 两种分子标记得到的遗传相似系数矩阵进行相关性分析，相关性系数为 0. 434，说明在 0. 05 的显著水平上呈弱相关。

本研究结果为接下来辣椒疫霉抗甲霜灵基因的精细定位和克隆以及辣椒疫霉抗甲霜灵的分子机制研究打下了必要的实验基础，同时也为辣椒疫霉甲霜灵抗药性的监测和治理提供理论依据。

关键词：辣椒疫霉；甲霜灵；抗药性；分子标记

辣椒疫霉对甲霜灵的抗性监测及诱变研究*

王艺烨**，卓　新，潘广学，陈方新，张华建，潘月敏，高智谋***
（安徽农业大学植物保护学院，合肥　230036）

摘　要：辣椒疫霉（*Phytophthora capsici*）是引致辣椒疫病的重要病原卵菌，而甲霜灵是防治疫病的主要杀菌剂之一，但随着甲霜灵的大量使用，疫霉菌对甲霜灵的抗药性成为主要问题。本文对来自不同地区的26株辣椒疫霉菌株进行抗性测定，并选取部分敏感菌株进行抗性突变菌株的诱导，并对抗性突变菌株及其敏感菌株的生物学特性和致病力进行比较研究，为这些地区的辣椒疫霉对甲霜灵抗药性风险评价和综合防治提供理论依据，并为后期利用分子生物学技术进行抗药性基因的定位与克隆，以便最终揭示该菌对甲霜灵抗药性的遗传本质提供必要的试验基础和理论支持。研究结果如下：

（1）不同地区辣椒疫霉菌株对甲霜灵的敏感性。采用菌丝生长速率法测定了26株供试辣椒疫霉对甲霜灵的敏感性，结果表明，有12株为敏感菌株，14株为中间菌株。8个省（区、市）中，辽宁省的5株辣椒疫霉菌株均为中间菌株，说明辽宁省辣椒疫霉对甲霜灵产生了抗药性。除甘肃和内蒙古各一株敏感菌株外，其他省（市）辣椒疫霉菌株的敏感性在敏感和中间范围内均有分布。

（2）辣椒疫霉抗甲霜灵突变株的诱导筛选。以12个敏感菌株为供试菌株，采用药剂驯化、菌丝块紫外光照射、游动孢子紫外光照射3种方法进行抗甲霜灵诱变。结果，通过药剂驯化的方式诱导突变菌株获得了6株对甲霜灵敏感性较低的辣椒疫霉菌株，且EC_{50}值最高的SD1-9，其EC_{50}值为26.062 9μg/mL；经游动孢子紫外照射法诱导获得2株突变菌株，而菌丝块紫外照射诱导没有突变菌株产生。

（3）辣椒疫霉抗甲霜灵突变株的生物学性状。选取的4株辣椒疫霉突变株及其对应的亲本敏感菌株其菌落均呈白色近圆形，在显微镜下观察其菌丝形态突变菌株及其对应的亲本敏感菌株无明显差异。突变菌株与其对应的亲本菌敏感菌株之间菌丝平均生长速率并无明显差异。通过在倒置荧光显微镜下观察各菌株孢子囊形态，结果表明敏感菌株以及其诱导出的抗性菌株在孢子囊形态方面并无明显差异。各敏感菌株及其对应的突变菌株间，其孢子囊长、宽、长宽比、柄长以及乳突高度的平均值差异不大。

（4）辣椒疫霉抗甲霜灵突变株对辣椒植株及不同果实上的致病力。采用下胚轴伤口接种法测定了辣椒疫霉敏感菌株与突变菌株在新苏椒5号及青丰羊角王这2个辣椒品种的植株的致病力，结果表明，接种植株发病症状均为植株茎秆明显萎蔫，且通过对比各植株病斑大小发现，供试菌株对不同辣椒植株品种的致病力有差异，且敏感菌株和突变菌株对于同一种植株致病力的差异并不显著。

供试菌株也能侵染青色圆椒、荷兰黄瓜、番茄、茄子以及红色尖椒。通过病斑大小数据进行分析，表明供试菌株在不同果实上的致病力是有差异的，且不同菌株对同一果实的致病力存在一定的差异；敏感菌株及其相对应的突变菌株其在同一种果实上致病力无明显差异。

关键词：辣椒疫霉；甲霜灵；抗性监测；诱变

---

* 基金项目：国家自然科学基金面上项目（31671977）

** 第一作者：王艺烨，硕士生，研究方向为真菌学及植物真菌病害

*** 通信作者：高智谋，教授，主要研究方向为真菌学及植物真菌病害；E-mail：gaozhimou@126.com

健身栽培措施对油菜菌核病的控制效应*

蒋冰心[1**]，柯章祥[2]，钱志恒[2]，李坤缘[1]，屈　阳[1]，潘月敏[1]，高智谋[1***]
（1. 安徽农业大学植物保护学院，合肥　230036；
2. 安徽省肥东县农技推广中心，肥东　231600）

油菜是我国在种植结构调整中大力倡导种植的主要油料作物，油菜菌核病（*Sclerotinia sclerotiorum*）是严重影响油菜优质安全生产的重大生物灾害，特别是近年来，由于气候变化、主栽油菜品种感病以及轻简化栽培等诸多因素的影响，我国油菜菌核病在包括安徽在内的长江中下游地区危害逐年加重，连年暴发成灾，严重制约了油菜生产。目前，由于抗病品种匮乏，对于油菜菌核病的防治主要采用药剂防治的措施，而农药的长期大量使用又导致“3R”问题日益突出，因此，迫切需要寻求替代化学农药的安全有效的防治措施。笔者开展了栽培措施对油菜菌核病的控制效应试验研究，旨在考察考察深沟高垄或清沟排水、中耕培土、摘除老黄叶、花期叶面喷施B、K肥等措施对油菜菌核病的防治效果及综合控病增产效果。

1　试验方法

1.1　试验处理和田间设计

1.1.1　单一措施对油菜菌核病的防治效果试验

（1）深沟高垄或清沟排水：设处理区、对照区，小区面积150m²。处理区种植整地时深沟高垄，3月中旬至成熟期清沟排水；对照区按常规管理。处理区和对照区对比排列，3次重复。

（2）中耕培土：设处理区、对照区，小区面积150m²。处理区3月中旬进行中耕培土；对照区不进行中耕培土。处理区和对照区对比排列，3次重复。

（3）摘除老黄叶：设处理区、对照区，小区面积150m²。处理区摘3月中旬至封行前除老黄叶，带出田外集中销毁；对照区不进行此项操作。处理区和对照区对比排列，3次重复。

（4）花期叶面喷施B、K肥：设处理区、对照区，小区面积150m²。处理区初花期采用卫士手动喷雾器喷施磷酸二氢钾+速效液体硼肥，每亩（1亩≈667m²）用水2 kg。对照区采用卫士手动喷雾器喷施等量清水。处理区和对照区对比排列，3次重复。

1.1.2　复合措施的综合控病增产效果试验示范

处理区：面积5亩，实施深沟高垄或清沟排水、中耕培土、摘除老黄叶、花期叶面喷施B、K肥等措施：①深沟高垄或清沟排水：种植整地时深沟高垄，3月中旬至成熟期清沟排水；②中耕培土：3月中旬进行；③摘除老黄叶：3月中旬至封行前进行；④施肥：磷酸二氢钾+速效液体硼肥（浓度同上），在初花期喷施，每亩用水2 kg。

对照区：面积1亩，不实施上述措施。

1.2　病情调查与产量测定

于油菜角果发育期（5月上旬），每小区按棋盘式取样调查10点共200株，按分级标准记载

---

* 基金项目：国家重点研发计划项目（2018YFD0200902-2）
** 第一作者：蒋冰心，硕士，实验师，研究方向为真菌学及植物真菌病害
*** 通信作者：高智谋，教授，主要研究方向为真菌学及植物真菌病害；E-mail：gaozhimou@126.com

病株严重度，计算茎病株率和病情指数。

于油菜黄熟期，每小区按棋盘式取 10 点，每点随机取 10 株，共 100 株，按株测定主茎株高、主茎粗、分枝数、第一分枝高度、结荚数和角粒数，计算理论产量。

2 结果与分析

2.1 试验实施情况

试验地点位于肥东县石塘镇阚东村。2018 年 10 月至 2019 年 5 月底，由肥东县昊锐生态农业种植专业合作社按照试验方案实施：试验田前茬为水稻田，2018 年 10 月 24 日进行土壤封闭处理，每亩用药为 72%异丙甲草胺乳油 150 mL。2018 年 10 月 21 日撒播，品种为‘荣华油 6 号’，连片种植 300 亩左右，播种量 0. 3 kg，亩用基肥‘3 个 15’复合肥 35 kg、尿素 10 kg，2019 年 3 月 12—13 日进行了清沟排水、中耕培土；3 月 17—18 日进行了摘除老黄叶、花期叶面喷施 B、K 肥。5 月 13 日进行病情调查，5 月 20 日取样考种，测定小区油菜籽理论产量。

2.2 单一栽培措施的控病增产效果

据 2019 年 5 月 13 日取样调查，深沟高垄、中耕培土、花期叶面喷施 B、K 肥及摘除老黄叶对菌核病的防治效果分别为：7. 88%、52. 49%、20. 57%、47. 33%，增产效果分别为 3. 21%、18. 38%、16. 37%、13. 81%（表 1）。对菌核病的防治效果以中耕培土最佳，其次为摘除老黄叶。增产效果也以中耕培土最佳，其次为花期叶面喷施 B、K 肥。

2.3 复合栽培措施的控病增产效果

据 2019 年 5 月 13 日（角果发育期）病情调查，健身栽培技术示范区菌核病平均茎病株率 4. 67%，病指 1. 59；对照区平均茎病株率 7. 33%，病指 3. 67，示范区平均防治效果为 54. 77%。据 5 月 20 日取样测定，健身栽培技术示范区理论产量 195. 80 kg/亩，对照区 153. 97 kg/亩，示范区增产 21. 36%（表 2）。

3 小结

（1）在试验考察的 4 种单一栽培措施中，对菌核病的防治效果以中耕培土最佳，其次为摘除老黄叶。增产效果也以中耕培土最佳，其次为花期叶面喷施 B、K 肥。

（2）健身栽培技术示范区（实施深沟高垄或清沟排水、中耕培土、摘除老黄叶、花期叶面喷施 B、K 肥）菌核病平均防治效果为 54. 77%，增产 21. 36%。

参考文献

高学文，陈孝仁 . 2018. 农业植物病理学［M］. 北京：中国农业出版社 .

全国农业技术推广服务中心 . NY/T 2038—2011 油菜菌核病测报技术规范［S］.

表 1 单一栽培措施对油菜菌核病的防控及增产效果（肥东，2019）

处理	面积（m^2）	茎病株率（%）	茎病情指数	防效（%）	主茎株高（cm）	主茎粗（cm）	分枝数	第一分枝高度（cm）	结荚数	角粒数	理论产量（kg/亩）	增产（%）
深沟高垄（清沟排水）	150	4.00	1.87	7.88	133.0	0.9	5.0	53.5	126.9	23	164.99	3.21
对照（常规起沟）	150	4.33	2.03		130.7	0.9	4.8	55.7	123.1	21	159.70	
中耕培土	150	3.67	1.24	52.49	133.1	1.2	6.9	37.4	214.3	18.8	188.49	18.38
对照（不中耕培土）	150	5.33	2.61		132.1	1.1	4.9	50.7	148.3	17.2	153.84	
花期叶面喷施 B、K 肥	150	7.00	3.36	20.57	134.5	1.1	5.4	45.0	160.4	22.6	184.50	16.37
对照（花期叶面不喷施 B、K 肥）	150	8.33	4.23		137.5	1.0	4.7	56.5	138.3	18.5	154.30	
摘除老黄叶	150	3.33	1.75	47.33	145.3	1.2	5.6	54.3	180.1	20.5	190.70	13.81
对照（不摘除老黄叶）	150	5.67	3.33		138.5	1.1	4.5	60.9	148.9	19.0	164.37	

表 2 复合健身栽培措施对油菜菌核病的防控及增产效果（肥东，2019）

处理	面积（m^2）	茎病株率（%）	茎病情指数	防效（%）	主茎株高（cm）	主茎粗（cm）	分枝数	第一分枝高度（cm）	结荚数	角粒数	理论产量（kg/亩）	增产（%）
复合健身栽培措施	3334	4.67	1.59	54.77	143.0	1.1	5.9	53.3	177.4	22.6	195.80	21.36
对照（常规）	667	7.33	3.67		141.3	1.1	4.9	55.3	158.7	16.08	153.97	

香蕉枯萎病根系土壤病原及拮抗微生物相关功能基因监测分析*

邓 涛**，翟子翔，杨腊英，周 游，汪 军，黄俊生***

（中国热带农业科学院环境与植物保护研究所，农业农村部热带作物有害生物综合治理重点实验室，海南省热带农业有害生物监测与控制重点实验室，海口 571101）

摘 要：重茬障碍正严重影响我国的农业生产，包括因连作导致土壤营养物质不均衡等原因引起的生理性病害以及因病原菌发生严重而导致的病理性病害。根系土壤微生物菌群失衡，即有害的病原微生物在土壤中的大量增殖，而有益微生物在土壤中的数量大幅降低，是病理性重茬病害发生的主要原因。香蕉枯萎病作为一种毁灭性的土传香蕉病害，主要由尖孢镰刀菌古巴专化型（*Fusarium oxysporum* f. sp. *cubense*）的 1 号和 4 号小种侵染引起的病理性重茬病害。目前，主要的生物防治手段是向病区土壤中施加含有对 *Foc*-4 具有拮抗能力的微生物制剂，包括芽孢杆菌（*Bacillus* sp.）、木霉（*Trichoderma* sp.）、放线菌（*Actinomyces* sp.）等，其中芽孢杆菌与木霉使用最为广泛。当前对根系土壤中的有益及有害微生物的数量监测研究，多局限于涂板计数与 16S、ITS 测序。前者操作烦琐结果置信度低，而后者靶标性不足。本实验通过荧光定量 PCR 扩增香蕉根系土壤中 *Foc*-4 的相关致病基因与芽孢杆菌、木霉的相关拮抗基因，通过对有害及有益微生物功能性基因丰度的定量监测，联系植株的发病状况，从土壤有害及有益微生物相关功能基因的角度，探究香蕉枯萎病这一病理性重茬病害的发生规律。本研究为香蕉枯萎病的田间预防治提供技术支持，并探讨了对土壤中有害及有益微生物功能性基因丰度的定量监测这一方法，在土传病害监测及克服重茬障碍中应用的可行性。

关键词：重茬；香蕉枯萎病；土传病害；有益微生物；功能基因

---

* 基金项目：国家重点研发计划项目（No. 2017YFD020060206）；海南自然科学基金创新研究团队项目（No. 2017CXTD016）；中国热带农业科学院基本科研业务费专项资金（No. 1630042018013）；中国热带农业科学院环境与植物保护研究所自主选题项目（No. hzsjy2017004）；广西创新驱动发展专项资金项目（No. 桂科 AA18118011-2）

** 第一作者：邓涛，硕士研究生，研究方向：土壤微生物；E-mail：2811776989@ qq. com

*** 通信作者：黄俊生，研究员，研究方向：分子植物病理学及生物防治；E-mail：H888111@ 126. com

外源施用dsRNA防治小西葫芦黄花叶病毒*

解昆仑**，古勤生***
（中国农业科学院郑州果树研究所，郑州　450000）

摘　要：小西葫芦黄花叶病毒（*Zucchini yellow mosaic virus*，ZYMV）是马铃薯Y病毒属的主要成员之一，能够侵染葫芦科作物，并可以通过种传、蚜虫和机械方式等途径进行传播。由于广泛的传播方式，它一旦发生就很容易扩散开来，会在田间造成严重的为害。自RNAi现象发现以来，人们就尝试将其应用到病毒病的防治中，自2003年Tenllado等利用体外喷施dsRNA的方法来防治病毒病以来，一种利用外源dsRNA防治病毒病的新方法得以出现。本研究利用含有双T7启动子的L4440载体以及RNAⅢ酶缺陷型菌株HT115构建了能够稳定表达和保存dsRNA的原核表达体系，针对ZYMV的3′UTR、6K2、HC-Pro、P3、NIb等5个片段选择了200~350bp的保守区域构建了dsRNA的原核表达体系，利用外源喷施的方法对ZYMV进行防治。经检测发现不同的dsRNA片段的防效是有差异的，在接种病毒后21 d发现其中防效最好的是HC-Pro片段，防效能达到95%，而防效相对较差的是6K2片段，其防效也能达到80%。本实验利用喷施的方法筛选出了防效较好的dsRNA片段，对后续该方法的实际应用具有很好的指导意义。

关键词：西瓜；小西葫芦黄花叶病毒；外源喷施；dsRNA

---

* 基金项目：国家西甜瓜产业体系（CARS-26-13）；中国农业科学院科技创新工程项目（CAAS-ASTIP-2018-ZFRI）；中央级公益性科研院所基本科研业务费专项（1610192019503）

** 第一作者：解昆仑，硕士研究生；E-mail：18595820659@163.com
*** 通信作者：古勤生，研究员，主要从事瓜类病害及其防控研究；E-mail：guqinsheng@caas.cn

2 株拮抗内生真菌对成熟期葡萄灰霉病的防治*

任苗苗**，王忠兴[2]，顾沛雯[1]***

（1. 宁夏大学农学院，银川 750021；2. 宁夏大学葡萄酒学院，银川 750021）

摘 要：为明确拮抗内生真菌 $XKZKDF_{27}$和 NQ8GⅡ4对成熟期葡萄灰霉病的防治作用。进行了皿内平板对峙，成熟期离体葡萄果粒和成熟期田间防效试验。在皿内平板对峙试验中拮抗内生真菌 $XKZKDF_{27}$和NQ8GⅡ4菌丝和发酵滤液都能够很好地抑制葡萄灰霉病菌菌丝的生长，其抑制率均在 60%左右。成熟期离体葡萄果粒和田间药效试验表明，拮抗内生真菌 $XKZKDF_{27}$和 NQ8GⅡ4发酵滤液对成熟期葡萄果实也具有较好的预防保护作用，其成熟期田间平均防效分别为 70.29%和 67.32%，显著高于市售生防农药的防治效果，与化学药剂腐霉利防效相当，不仅没有药害现象发生且还能促进生长。本试验证明 2 株拮抗内生真菌 $XKZKDF_{27}$和 NQ8GⅡ4 对成熟期葡萄灰霉病具有很好的预防作用，值得在葡萄生产上推广使用。

关键词：拮抗内生真菌；葡萄灰霉病；生防作用；葡萄成熟期

---

* 基金项目：宁夏“十三五”重大科技项目——酿酒葡萄安全生产关键技术研究（2016BZ06）

** 第一作者：任苗苗，硕士研究生，研究方向资源利用与植物保护；E-mail：1721788107@ qq. com

*** 通信作者：顾沛雯，教授，主要从事植物病理学与生物防治方面的研究；E-mail：gupeiwen2013@ 126. com

基于 GIS 的葡萄霜霉病菌田间越冬量与分布研究*

杜　娟**，李文学，顾沛雯***
（宁夏大学农学院，银川　750021）

摘　要：构建葡萄霜霉病菌（*Plasmopara viticola*）的实时荧光定量 PCR（real-time PCR）检测体系，量化葡萄霜霉病菌的田间越冬量，运用地理信息系统（GIS）和地统计学方法分析贺兰山东麓酒庄葡萄园葡萄霜霉病菌田间越冬量的空间结构，并采用普通 Kriging 插值法模拟葡萄霜霉病菌田间越冬量的空间分布，为贺兰山东麓葡萄霜霉病的早期预警提供理论依据。结果表明：研究设计的 real-time PCR 引物特异性高，以携带目的基因片段的重组质粒为标准品，引物灵敏度为 1×10^3 copies/μL，构建的 real-time PCR 循环阈值（Ct）与质粒浓度的标准线性曲线为 $y = 41.419 - 3.931x$，相关系数 $r^2 = 0.9989$，扩增效率为 80%，线性范围达 7 个数量级，在 $1\times10^3 \sim 1\times10^9$ copies/μL 呈现良好的线性关系；对葡萄园土壤中葡萄霜霉病菌进行 real-time PCR 检测，田间病原菌越冬量 DNA 为 307.76～3 230.56 copies/μL。构建的定量检测葡萄霜霉病菌越冬的 real-time PCR 体系灵敏度优于常规 PCR，结合 GIS 和地统计学的方法研究田间葡萄霜霉菌越冬水平，为葡萄霜霉病的田间越冬水平提供了量化工具，同时也为葡萄霜霉病的早期预警和科学防控提供了科学的理论依据。

关键词：葡萄霜霉病菌；越冬量；地理信息系统（GIS）；实时荧光定量 PCR

---

* 基金项目：宁夏回族自治区“十三五”重大科技项目——酿酒葡萄安全生产关键技术研究（2016BZ06）
** 第一作者：杜娟，硕士研究生，生物防治与菌物资源利用；1291524369@qq.com
*** 通信作者：顾沛雯，教授，从事植物病理学与生物防治方面研究；gupeiwen2019@nxu.edu.cn

云南盈江澳洲坚果病害发生情况调查及防治建议*

李庆磊**，韩长志***
（西南林业大学生物多样性保护学院，云南省森林灾害预警与控制重点实验室，昆明 650224）

摘 要：澳洲坚果 *Macadamia integrifolia* 是山龙眼科澳洲坚果属多年生常绿果树，主要分布在澳大利亚、巴西、中国等热带和亚热带地区。我国自 1910 年开始引种该树，主要的栽培区域为云南、广西等地。云南省盈江县澳洲坚果引种始于 1985 年，目前投产面积和鲜果产量是全国之首，种植面积13 746. 67hm^2，投产面积2 666. 67hm^2，产量4 097t，农业产值6 145万元。近些年，随着盈江县澳洲坚果种植面积的逐年扩大，其病害的发生情况日益严重，然而，学术界对其病害的研究报道较少，对于澳洲坚果上的病害尚缺乏较为系统性的研究，极大地阻碍着该产业的健康、有序和快速发展。因此，为了明确澳洲坚果上主要病害的发生情况，本研究通过对位于盈江县太平镇、旧城镇、支那乡等地的澳洲坚果种植地区开展实地调研，明确澳洲坚果上常见病害有炭疽病、速衰病、慢衰病、枝枯病、叶斑病以及根腐病等，上述病害严重制约着该植物的健康成长，威胁着澳洲坚果产业的发展，同时，通过对病害标本进行室内病原菌分离、鉴定以及生物学特性研究，明确了造成澳洲坚果病害的病原菌及其特性；此外，通过对健康根际土壤中生防菌筛选、室内对峙培养等研究，获得用于防治澳洲坚果炭疽病、根腐病等病菌的生防菌株，为今后进一步实现云南省盈江县澳洲坚果种植过程中病害防治提供重要的理论基础。

关键词：澳洲坚果；病害；发生情况；防治措施

---

* 基金项目：云南省贫困地区、民族地区和革命老区人才支持计划科技人员专项计划

** 第一作者：李庆磊，硕士生；E-mail：2454675663@ qq. com

*** 通信作者：韩长志，博士，副教授，研究方向：经济林木病害生物防治与真菌分子生物学；E-mail：hanchangzhi2010@ 163. com

蓝莓根腐病研究进展*

蔡旺芸**，祝友朋，韩长志***
（西南林业大学，云南省森林灾害预警与控制重点实验室，昆明 650224）

摘 要：蓝莓也称蓝浆果，是越橘类的一种灌木，果实为小浆果，具有很高的经济价值和营养价值，以及较好的开发利用价值和市场前景。近几年，蓝莓产业发展迅速。根腐病作为蓝莓生产上重要的真菌病害之一，主要为害根部，根部受到感染后，植株有很明显的表现，树势整体呈现衰弱之势，如同营养缺乏的症状，叶片褪绿变黄，侧根先坏死腐烂，后扩散到整个根系，导致植株生长缓慢，最后植株彻底死亡。对蓝莓的生长产生了严重的影响，严重时导致蓝莓绝产，极大地危害着云南蓝莓产业的健康、有序、快速发展。目前，学术界对蓝莓根腐病病原菌的报道不尽相同，有棒孢拟盘多毛孢（*Pestalotiopsis clavispora*）、冬青丽赤壳菌（*Calonectria ilicicola*）、科氏丽赤壳（*Calonectria colhounill*）、腐皮镰刀菌（*Fusarium solani*）、层出镰刀菌（*Fusarium proliferatum*）、尖孢镰刀菌（*Fusarium oxysporum*）、木贼镰孢菌（*Fusarium equiseti*）、疫霉属（*Phytophthora* sp.）、腐霉（*Pythium sterilum*）、蜜环菌（*Armillariella mellea*）、寄生帚梗柱孢菌（*Cylindrocladium parasiticum*）、丝核菌属（*Rhizoctonia* spp.）等12种。本文对这12种病原菌导致的根腐病进行致病性、分布范围等进行分析，为我国蓝莓根腐病的研究提供借鉴和参考，为我国蓝莓根腐病的防治提供一定的参考依据。

关键词：蓝莓根腐病；病原菌；研究进展

---

* 基金项目：国家级大学生创新创业训练计划项目（项目编号：201710677013）；云南省大学生创新创业训练计划项目（项目编号：S201710677013）

** 第一作者：蔡旺芸，本科生；E-mail：770034492@qq.com

*** 通信作者：韩长志，博士，副教授，研究方向：经济林木病害生物防治与真菌分子生物学；E-mail：hanchangzhi2010@163.com

香港米埔红树林植物根际土壤细菌的抗植物致病菌活性成分研究*

王开玲[1,2**]，李飞腾[3]，马　瑞[3]，王　钰[4]，徐　颖[4]，郝凌云[4***]
（1. 大理大学药物研究所，大理　671000；2. 大理大学药学与化学学院，大理　671000；3. 大理大学公共卫生学院，大理　671000；4. 深圳大学生命与海洋科学学院，深圳　518060）

摘　要：随着人们对与农业杀菌剂应用所产生的相关环境污染问题的日益关注，研发用于植物病害管理的环境友好型生物杀菌剂迫在眉睫。本研究在已采集的香港米埔红树林自然保护区的老鼠簕植物根际土壤样品中分离鉴定得到 113 株细菌。对这些细菌发酵液的乙酸乙酯粗浸膏进行体外抗植物病原菌（细菌：丁香假单胞菌番茄致病变种 *Pseudomonas syringae* pv. *tomato*（*Pst*）DC3000 和胡萝卜软腐果胶杆菌 *Pectobacterium carotovorum*（*Pec*）；真菌：番茄早疫病菌 *Alternaria solani* HLJ65，稻瘟病菌 *Magnaporthe oryzae* P131，玉米小斑病菌 *Cochliobolus heterostrophus* Ch16 和炭疽病菌 *Colletotrichum fructicola* cf1）活性筛选发现，26%的菌株在浓度为 50μg/mL 对多种农业上重要的植物致病菌具有抗菌作用。其中，部分菌株发酵液的乙酸乙酯粗浸膏在具有抑制茶树炭疽病菌和稻瘟病菌生长的同时，还能够显著抑制水果和水稻中炭疽病和稻瘟病的发展。此外，10 株具有较好的抗菌活性菌株，在其发酵液的乙酸乙酯粗浸膏的活性浓度 25μg/mL 下，对小球藻的正常生长表现出低毒甚至无毒特性。本研究结果充分证明了来自香港米埔红树林自然保护区的红树林植物根际微生物具有较好的环境友好型“绿色农药”开发潜力，并为低毒高效的天然产物农业抗菌剂的研发提供菌株基础。

关键词：红树林；根际土壤；细菌；抗菌剂；植物病原菌

---

* 基金项目：国家自然科学基金青年基金项目（41706155）；广东省自然科学基金自由探索项目（2017A030313227）

** 第一作者：王开玲，博士，特聘副研究员，主要从事微生物药用资源开发与利用；E-mail：kailingw@ 163. com
*** 通信作者：郝凌云，博士，特聘副研究员，主要从事植物病理学研究；E-mail：haolingyun@ szu. edu. cn

橡胶炭疽病菌生防菌分离鉴定及生防作用研究*

翟纯鑫[1**]，梁艳琼[2]，谭施北[2]，吴伟怀[2]，习金根[2]，李 锐[2]，
郑金龙[2]，黄 兴[2]，陆 英[2]，贺春萍[2***]，易克贤[2***]
（1. 南京农业大学植物保护学院，南京 210095；2. 中国热带农业科学院环境与植物保护研究所，农业农村部热带农林有害生物入侵检测与控制重点实验室，海南省热带农业有害生物检测监控重点实验室，海口 571101）

摘 要： 橡胶树炭疽病是橡胶树上重要的叶部病害之一，主要由胶孢炭疽菌（*Colletotrichum gloeosporioides*）和尖孢炭疽菌（*Colletotrichum acutatum*）侵染造成的。化学药剂防治是防治橡胶炭疽病最有效的方法之一。长期使用化学药剂，易导致病原菌产生抗药性，污染环境及造成农药残留等问题，因此亟待寻求一种更持久、更安全的防治措施。生物防治尤其是高效生防菌的筛选利用已成为近年来植物病害防治研究的热点，被认为是比较安全和有效的防控措施。采用平板对峙培养法、抑菌圈法和菌丝生长速率法综合筛选出对橡胶树炭疽病具有生防作用的菌株，结果表明：从假花生、柱花草及不间作的根际土壤中筛选获得 56 株生防菌。从假花生根际土壤筛选到 15 株生防菌，其中 Jhs1-10、Jhs1-14、Jhs2-7 等 6 株生防菌抑制效果较好。从柱花草根际土壤筛选到 17 株生防菌，其中 Zhc1-1、Zhc1-9 等 13 株生防菌抑制效果较好，抑菌率均达到 75%以上。从不间作土壤筛选到 24 株生防菌，其中 BJZ1-2、BJZ1-3 等 19 株生防菌抑制效果较好，抑菌圈直径达到 20. 360±0. 123mm 以上。通过形态学观察、16SrDNA 序列以及 gyrB 基因进行分类鉴定，生防菌大部分属于芽孢杆菌属的解淀粉芽孢杆菌（*Bacillus amyloliquefaciens*）、枯草芽孢杆菌（*B. subtilis*）、短芽孢杆菌（*Brevibacillus brevis*）和地衣芽孢杆菌（*B. lincheniformis* ）。通过平板对峙培养发现这些生防菌对西瓜枯萎病菌（*Fusarium oxysporum*）、橡胶红根病菌（*Ganoderma pseudoferreum*）、橡胶褐根病菌（*Phellinus noxius*）、柱花草炭疽病菌（*C. gloeosporioides*）等病原菌均有不同程度的抑制。

关键词： 橡胶炭疽病菌；生防菌；分离鉴定；生防作用

* 基金项目：国家天然橡胶产业技术体系建设项目（No. nycytx-34-GW2-4-3；CARS-34-GW8）；海南省科协青年科技英才学术创新计划项目（QCXM201714）

** 第一作者：翟纯鑫，在读研究生，研究方向：植物保护；E-mail：zcx490405@ 163. com
*** 通信作者：贺春萍，硕士，研究员，研究方向：植物病理；E-mail：hechunppp@ 163. com
易克贤，博士，研究员，研究方向：分子抗性育种；E-mail：yikexian@ 126. com

枯草芽孢杆菌 Czk1 与杀菌剂协同防治橡胶根病*

谢 立[1**]，董文敏[3]，梁艳琼[2]，翟纯鑫[3]，吴伟怀[2]，李 锐[2]，贺春萍[2***]，易克贤[2]
（1. 海南大学林学院，海口 570228；2. 中国热带农业科学院环境与植物保护研究所，
农业农村部热带农林有害生物入侵检测与控制重点实验室，海南省热带农业有害生物检测监控
重点实验室，海口 571101；3. 南京农业大学植物保护学院，南京 210095）

摘 要：根病是橡胶树上三大毁灭性病害之一，防治十分困难。该病害是限制我国橡胶产量提高的关键生物因子。本研究以枯草芽孢杆菌 Czk1 为生防菌，测定 7 种杀菌剂对橡胶红根病菌和褐根病菌的室内毒力，筛选毒力最强且能与 Czk1 具很好相容性的化学杀菌剂；将筛选获得的杀菌剂与 Czk1 进行复配，探讨复配剂对橡胶红根病菌和褐根病菌的室内毒力效果，旨在构建由生防菌与化学药剂组成的菌-药复剂，为田间防治提供理论基础。实验结果表明：‘根康’可以在较低的浓度下抑制红根病菌和褐根病菌的生长，也能与 Czk1 很好的相容。将‘根康’（EC_{50} = 0.05 μg/mL）和生防菌 Czk1（$EC_{50}=1\times10^{8}$ μg/mL）混配，当 V（Czk1）：V（根康）= 7∶3 时，对褐根病菌抑制的增效作用最高，增效比为 1.09；将‘根康’（EC_{50} = 0.5μg/mL）和 Czk1（$EC_{50}=6\times10^{7}$ μg/mL）混配，当 V（Czk1）：V（根康）= 7∶3 时，对红根病菌抑制的增效作用也最高，增效比为 1.03。菌药复配剂防效显著优于单剂‘根康’和生防菌 Czk1 的防效，且‘根康’使用量只有单剂使用量的 1/3，表明二者复配有明显的增效作用，且大大降低了化学药剂的使用量。分别在 5d、10d、15d 观测发现，菌药复配剂的抑菌效果持续保持。

关键词：橡胶红根病；橡胶褐根病；枯草芽孢杆菌；菌药复配；协同防治

---

* 基金项目：国家重点研发计划项目（No. 2018YFD0201100）；国家天然橡胶产业技术体系建设专项资金资助项目（No. CARS-33-GW-BC1）

** 第一作者：谢立，硕士研究生，研究方向：林业；E-mail：13178981326@163.com

*** 通信作者：贺春萍，硕士，研究员，研究方向：植物病理学；E-mail：hechunppp@163.com

西沙群岛土壤生防细菌分离鉴定及抑菌作用*

王　义[1,2**]，李　敏[2]，赵　超[1]，高兆银[2]，李春霞[2]，周子骞[2,3]，洪小雨[2]，胡美姣[2***]
（1. 海南大学生命科学与药学院，海口　570228；2. 中国热带农业科学院环境与植物保护研究所，海口　571101；3. 华中农业大学植物科学技术学院，武汉　430070）

摘　要： 西沙群岛是中国南海四大群岛之一，由永乐群岛和宣德群岛组成，属热带海洋性季风气候。西沙群岛常年高温高湿，且各岛礁相对独立，人为干预少，热带盐生植物资源丰富，潜在的植物生防微生物资源巨大。2018—2019 年，本实验从 12 个岛礁（宣德群岛 6 个岛礁，永乐群岛 6 个岛礁）共收集 36 份土样，从中分离获得 400 株细菌分离物。以草海桐叶斑病菌（*Alternaria longipes*）为靶标，通过 PDA 平板对峙法，从中筛选出抑菌效果明显的菌株 21 株，通过形态学及 16S rDNA 序列测定，鉴定为链霉菌（*Streptomyces* spp.）和芽孢杆菌（*Bacillus* spp.），分别是 14 株和 7 株。以鉴定的 21 株菌为生防菌株，以 4 种西沙原植物主要病原真菌 *A. longipes*、*Colletotrichum gloeosporioides*、*Lasiodiplodia theobromae* 和 *Phomopsis liquidambari* 为靶标菌株，进行 PDA 平板对峙培养试验。结果表明，3 株链霉菌（YXD1-6、GZSD2-1 和 NZSZ-1）对 4 种病原菌的菌丝生长抑制率均大于 60%，其中 YXD1-6 对 *A. longipes* 的抑菌效果最好，达到 72.15%。3 株芽孢杆菌（YXDZ-1、JQDZ-3 和 GQDZ-1）对 4 种病原菌的菌丝生长抑制率大于 50%，其中 GQDZ-1 对 *C. gloeosporioides* 的抑菌效果最好，达到 67.46%。本研究为热带岛礁植物病害生防资源的挖掘提供了数据资料，为热带岛礁植物生物防治提供参考依据。

关键词： 西沙群岛；生防细菌；分离鉴定；抑菌作用

---

* 基金项目：农业农村部财政专项项目（NFZX2018）

** 第一作者：王义，在读研究生，微生物专业；E-mail：wang_yi_w@ sina. com

*** 通信作者：胡美姣，博士，研究员，从事热带果蔬采后保鲜与病害防治研究；E-mail：humeijiao320@ 163. com

天然橡胶航空植保新技术与“一带一路”热带植保科技合作

黄贵修
（中国热带农业科学院国际合作处，海口 571101）

摘　要：天然橡胶是重要的战略物资和工业原料，因其具有合成橡胶不可替代的优良特性，在航空、航海、医疗和重型汽车制造业等领域广为应用。而病虫草害问题一直是国内天然橡胶产业发展最为严重的生物性限制因素之一，本文主要介绍结合抗性利用、兼治多种叶部病害药剂研制、病虫害疫情监测、飞防增效助剂研发及无人直升机施药的天然橡胶航空植保新技术；同时，对“一带一路”热带国家农业病虫草害及其生防资源联合调查与开发评价、抗病虫种质引进与创制、联合实验室及综合试验站共建、人力资源合作开发等热带植保科技交流合作进行概述及展望。

关键词：天然橡胶；航空植保；新技术；科技合作

戊唑醇抑制苹果树腐烂病菌的细胞学研究*

高　双**，田润泽，刘召阳，冯　浩，黄丽丽***
（西北农林科技大学植物病理学系，旱区作物逆境生物学国家重点实验室，杨凌　712100）

摘　要：为了明确苹果树腐烂病防控常用药剂戊唑醇的杀菌机理，本研究利用显微技术观察了戊唑醇对病菌孢子的萌发、菌丝形态及细胞结构的影响。结果发现，戊唑醇能够抑制孢子萌发，但不影响孢子的吸胀，主要是抑制芽管的伸长，使芽管畸形、增粗、分枝增多等，进而无法正常侵入寄主。受杀菌剂处理的苹果树腐烂病菌菌丝形态和细胞结构均发生明显变化，形态变化主要为菌丝顶端肿胀，分枝增多，菌丝增粗，细胞壁破裂，原生质外渗等；结构的变化主要为细胞壁不规则增厚，线粒体增多、膜增厚或不规则缢缩，细胞核增多、核仁弥散，细胞隔膜增多、不规则增厚，细胞液泡化严重，形成空腔，原生质外渗，细胞最终坏死等。同时，在已坏死的菌丝内可发现子菌丝，且子菌丝异常，主要表现为细胞壁不规则增厚，线粒体数量增多、细胞质坏死等。本研究明确了戊唑醇对苹果树腐烂病菌的细胞学作用机理，为该药剂的田间使用提供了重要理论依据。

关键词：苹果树腐烂病菌；戊唑醇；组织细胞学；超微结构

* 基金项目：国家重点研发计划（2016YED0201100）；西北农林科技大学试验示范站科技成果推广项目（TGZX2018-37）
** 第一作者：高双，硕士研究生，研究方向为植物病理学
*** 通信作者：黄丽丽，教授，研究方向为植物病害综合治理

复合高效配方药剂对甘蔗梢腐病防控效果评价*

李文凤**，张荣跃，王晓燕，单红丽，李　婕，尹　炯，罗志明，
仓晓燕，黄应昆***
（云南省农业科学院甘蔗研究所，云南省甘蔗遗传改良重点实验室，开远　661699）

摘　要：近年云南蔗区多雨高湿加上大面积种植的主栽品种‘粤糖 93-159’‘新台糖 25 号’新台糖 1 号‘和’川糖 79-15‘等高感甘蔗梢腐病，导致甘蔗梢腐病在云南临沧、玉溪、西双版纳、普洱、红河等主产蔗区大面积暴发危害成灾，减产减糖严重，甘蔗生产受到严峻灾害威胁。为筛选防控甘蔗梢腐病的复合高效配方药剂及精准施药技术，选用 50%多菌灵 WP、75%百菌清 WP、25%嘧菌酯 EC、25%吡唑醚菌酯 SC、30%苯甲嘧菌酯 SC 进行人工叶面喷施田间药效试验和生产示范验证。试验结果及综合评价分析显示，（50%多菌灵 WP 1 500 g+75%百菌清 WP 1 500 g+磷酸二氢钾 2 400 g+农用增效助剂 300 mL/hm^2）、（25%吡唑醚菌脂 SC 600 mL+磷酸二氢钾 2 400 g+农用增效助剂 300 mL/hm^2）等 2 种药剂配方处理对甘蔗梢腐病均具有良好的防治效果，2 种药剂配方处理的病株率均在 8.96%以下，其防效均达 90.59%以上，显著高于对照药剂配方处理（75%百菌清 WP 1 500 g+磷酸二氢钾 2 400 g+农用增效助剂 300 mL/hm^2）和（50%多菌灵 WP 1 500 g+磷酸二氢钾 2 400 g+农用增效助剂 300 mL/hm^2）的防效 57.4%和 67.37%。2 种药剂配方处理防控效果显著、稳定，推荐为防控甘蔗梢腐病最佳药剂配方，可在 7—8 月发病初期，按 2 种药剂配方每公顷用水 900 kg，采用电动背负式喷雾器人工叶面喷施、7~10 d 喷 1 次，连喷 2 次，可有效控制甘蔗梢腐病暴发流行。

关键词：复合高效配方药剂；甘蔗梢腐病；防效评价

---

* 基金项目：国家现代农业产业技术体系（糖料）建设专项资金（CARS-170303）；云岭产业技术领军人才培养项目“甘蔗有害生物防控”（2018LJRC56）；云南省现代农业产业技术体系建设专项资金资助

** 第一作者：李文凤，研究员，主要从事甘蔗病害研究；E-mail：ynlwf@ 163. com
*** 通信作者：黄应昆，研究员，从事甘蔗病害防控研究；E-mail：huangyk64@ 163. com

药用植物肿节风内生真菌的鉴定及抗菌活性研究*

宋利沙[1**]，蒋 妮[1,2***]，蓝祖栽[1]，张占江[1]
（1. 广西壮族自治区药用植物园，南宁 530023；2. 广西大学农学院，南宁 530004）

摘 要：研究肿节风内生真菌的抗菌活性，并筛选出具有较强的抗菌活性菌株。采用平板对峙法筛选具有抗菌活性的内生真菌；结合形态特征和ITS序列对分离到的内生真菌进行鉴定。从肿节风的不同组织中分离到111株内生真菌，选择不同形态类型48株进行分子鉴定，分属于9个目20个属26个种。筛选出抗菌活性强的菌株RJ-1和JJ-1，对15种病原真菌和病原细菌大肠杆菌具有较强抑制作用。这两菌株分别鉴定为橡胶生拟茎点霉 *Phomopsis heveicola* 和新壳梭孢菌 *Neofusicoccum parvum*。肿节风存在丰富的内生真菌物种，RJ-1和JJ-1可作为抗菌活性菌株进行深入研究。

关键词：肿节风；内生真菌；抗菌活性；鉴定

---

* 基金项目：广西药用植物园青年基金（桂药基201801）；广西科技基地人才专项（桂科AD16380013）
** 第一作者：宋利沙，博士，助理研究员，从事中药材病虫害防治；E-mail：lishasong@126.com
*** 通信作者：蒋妮，硕士，副研究员，从事中药材病虫害防治；E-mail：jiangni292@126.com

猕猴桃溃疡病菌生物型鉴定及药剂筛选*

鄢明峰[1**]，邹曼飞[1]，强　遥[1]，张凯东[1]，李帮明[2]，涂贵庆[2]，蒋军喜[1***]
（1. 江西农业大学农学院，南昌　330045；2. 江西省奉新农业局，奉新　330700）

摘　要：猕猴桃细菌性溃疡病（简称猕猴桃溃疡病）是当前为害世界猕猴桃产业发展最为严重的病害之一。自 2008 年前后起，猕猴桃溃疡病在江西省奉新县陆续发生，并造成近年病情不断加重，直至出现因溃疡病为害而发生毁园现象。本文在前期将奉新猕猴桃溃疡病菌鉴定为丁香假单胞菌猕猴桃致病变种（*P. syringae* pv. *actinidiae*）的基础上，进一步对其生物型进行了鉴定，并对其防治药剂进行了室内筛选，现将研究结果报道如下。

（1）以猕猴桃溃疡病菌的 6 个管家基因 *acnB*、*cts*、*gapA*、*gyrB*、*pfk* 及 *rpoD* 作为靶基因，进行多位点序列分析（MLSA），结果表明，奉新县猕猴桃溃疡病菌均为生物型 3（biovar 3）。随后，利用猕猴桃溃疡病菌生物型的特异性引物进一步验证了该结果的可靠性。

（2）在药剂初筛的基础上，采用牛津杯法测定了 13 种杀菌剂对猕猴桃溃疡病菌的室内毒力，结果表明：不同的杀菌剂对溃疡病菌株的抑菌效果相差很大。0.3%四霉素和 20%溴硝醇的抑菌作用最佳，EC_{50}均小于 0.1mg/L，其 EC_{50}分别为 0.002mg/L 和 0.068mg/L，其次，80%乙蒜素和 1.5%噻霉酮也具有较强的毒力，其 EC_{50}分别为 7.104mg/L 和 19.974mg/L；其他 9 种杀菌剂 3%中生菌素、2%春雷霉素、72%农用链霉素、80%代森锰锌、15%络氨铜、50%氯溴异氰尿酸、47%春雷·王铜、27.12%碱式硫酸铜、46.1%氢氧化铜的抑菌作用则较差。

关键词：猕猴桃溃疡病菌；生物型鉴定；杀菌剂室内筛选

* 基金项目：江西省科技计划项目（20181ACF60017）；国家自然科学基金（31460452）
** 第一作者：鄢明峰，硕士生，主要从事分子植物病理学研究；E-mail：136043695@qq.com
*** 通信作者：蒋军喜，教授，主要从事植物病害综合治理研究；E-mail：jxau2011@126.com

奉新猕猴桃3种真菌病害病原鉴定及室内药剂筛选*

周　英[1**]，邹曼飞[1]，张凯东[1]，强　遥[1]，赵尚高[2]，李帮明[2]，蒋军喜[1***]
（1. 江西农业大学农学院，南昌　330045；2. 江西省奉新农业局，奉新　330700）

摘　要：猕猴桃是江西省奉新县的重要和特色果树，发展猕猴桃生产对促进该县经济发展，带动农民增收致富具有重要作用。然而，随着该县猕猴桃种植规模的不断扩大，病害发生种类不断增多，为害程度不断加重。本文针对近年来奉新县猕猴桃生长期出现的黑斑病、白纹羽病及黑霉病3种真菌病害进行病原鉴定、生物学特性试验及室内药剂筛选，目的是查明这些病害的发生原因，并为其发生规律和防治研究奠定基础，现将研究结果报道如下：

（1）通过病菌分离、病菌培养性状、形态学观察、致病性测定及分子生物学手段，对江西奉新县猕猴桃生长期出现的3种真菌病害的病原进行鉴定，确定了奉新县猕猴桃黑斑病的病原为链格孢（*Alternaria alternata*），猕猴桃白纹羽病病原为褐座坚壳菌（*Rosellinia necatrix*），猕猴桃黑霉病病原为猕猴桃假尾孢（*Pseudocercospora actinidiae*）。

（2）在26种杀菌剂对*A. alternata*的室内毒力测定试验中，10%申嗪霉素、50%咪鲜胺锰盐、25%已唑醇、75%肟菌・戊唑醇、25%丙环唑、10%苯醚甲环唑、75%戊唑・嘧菌酯和40%氟硅唑这8种杀菌剂对*A. alternata*具有强毒力，EC_{50}值介于0. 1821～0. 7633 μg/mL；在26种杀菌剂对*R. necatrix*的室内毒力测定试验中，10%申嗪霉素、22. 5%啶氧菌酯、50%咪鲜胺锰盐、45%敌磺钠、50%多菌灵、25%丙环唑、25%嘧菌酯、70%甲基硫菌灵、10%多抗霉素、40%五氯硝基苯、30%恶霉灵、40%氟硅唑和0. 3%四霉素这12种杀菌剂对*R. necatrix*具有强毒力，EC_{50}值介于0. 0001～0. 6861 μg/mL。

关键词：猕猴桃黑斑病；猕猴桃白纹羽病；猕猴桃黑霉病；病原鉴定；杀菌剂室内筛选

---

* 基金项目：江西省科技计划项目（20181ACF60017）；国家自然科学基金（31460452）

** 第一作者：周英，硕士生，主要从事植物病害综合治理研究；E-mail：1373583449@ qq. com
*** 通信作者：蒋军喜，教授，主要从事植物病害综合治理研究；E-mail：jxau2011@ 126. com

肿节风炭疽病拮抗真菌筛选及作用机理研究*

宋利沙[1**]，蒋　妮[1,2***]，蓝祖栽[1]，张占江[1]
（1. 广西壮族自治区药用植物园，南宁　530023；2. 广西大学农学院，南宁　530004）

摘　要：筛选肿节风炭疽病（*Colletotrichum dematiu*）拮抗真菌，为新型生物制剂的开发提供基础材料。采用平板稀释法，从健康肿节风根系土壤中分离、纯化真菌，并采用平板对峙法筛选肿节风炭疽病拮抗真菌。平板对峙试验结果表明，菌株 JT-8、JT-3 和 DT-5 对肿节风炭疽病菌具有较强的拮抗作用，拮抗作用最强的菌株是 JT-8，抑制率达 91.57%。抗菌谱测定结果表明，JT-8、JT-3、DT-5 菌株对供试的 14 种病原真菌均有明显的拮抗作用，拮抗作用最强的菌株同样是 JT-8，平均抑制率达 89.88%，该菌株可产生几丁质酶、纤维素酶和蛋白酶，可降解病原真菌细胞壁中的纤维素和蛋白质，抑制病原真菌菌丝生长，病原真菌菌丝表现为扭曲、断裂，分支缠绕，菌丝颜色加深等。通过形态学和 ITS 测序鉴定，发现 JT-8 菌株为淡紫紫孢菌（*Purpureocillium lilacinum*）。JT-8 菌株能产生多种抗菌活性物质，且具有广谱抑菌活性，具有较好的生防开发潜能。

关键词：肿节风；炭疽病；拮抗真菌；抗菌活性；生物防治

---

* 基金项目：广西药用植物园青年基金（桂药基 201801）；广西科技基地人才专项（桂科 AD16380013）

** 第一作者：宋利沙，博士，助理研究员，从事中药材病虫害防治；E-mail：lishasong@126.com

*** 通信作者：蒋妮，硕士，副研究员，从事中药材病虫害防治；E-mail：jiangni292@126.com

贝莱斯芽孢杆菌 HN-2 调控抑菌物质合成相关基因的研究*

韦丹丹**，许沛冬，缪卫国，刘文波***，靳鹏飞***
（海南大学植物保护学院，热带农林生物灾害绿色防控教育部重点实验室，海口　570228）

摘　要：芽孢杆菌（*Bacillus*）是常见的生防菌株，产生多种抑菌物质且大多具有很好的稳定性，对人畜无害，在农业中被广泛运用。关于芽孢杆菌的很多研究已进入分子水平，随着转座子突变技术的大量运用，该技术转而运用于研究生防菌的生防机制已成为热点。HN-2 为本实验室分离鉴定的贝莱斯芽孢杆菌（*Bacillus velezensis*），室内平板实验显示对杧果炭疽病菌（*Colletotrichum gloeosporioides* Penz）和水稻白叶枯病菌（*Xanthomonas oryzae* pv. *oryzae*, Xoo）等多种植物病原真菌细菌均有明显的抑菌效果。为明确调控该菌株生防能力的基因的功能，本研究构建贝莱斯芽孢杆菌 HN-2 的突变体文库，然后通过定点敲除贝莱斯芽孢杆菌 HN-2 中与抑菌相关基因，研究其代谢机理。本研究已成功建立贝莱斯芽孢杆菌 HN-2 的转化体系，将含有转座子TnYLB-1 的穿梭载体 pMarA 转化到 HN-2 菌株中，再高温诱导 TnYLB-1 转座子随机插入基因组序列中，通过抗性平板筛选以及 PCR 验证，得到突变体 1 830个；血平板筛选得到溶血圈有显著变化的突变体 16 个，利用 Southern 杂交技术确定 13 个突变体中的转座子已成功地单拷贝、随机插入在 HN-2 菌株的基因组中。通过反向 PCR 克隆转座子插入位点的侧翼序列，克隆鉴定被插入基因的功能，转而定点敲除 HN-2 菌株的该段基因，观察 HN-2 菌株抑菌能力的变化，从而验证被插入基因的功能。本研究对研究贝莱斯芽孢杆菌 HN-2 活性基因功能、揭示其生防作用机理以及合成生物学提供理论依据。

关键词：贝莱斯芽孢杆菌；突变体库；抑菌机理；合成生物学

* 基金项目：海南大学科研启动经费（No. KYQD（ZR）1842）；海南大学青年教师基金（hdkyxj201708）；海南省产学研项目（No. CXY20140038）

** 第一作者：韦丹丹，硕士研究生，植物保护专业；E-mail：18014693035@163.com

*** 通信作者：靳鹏飞，讲师，主要从事微生物相关研究；E-mail：jinpengfei@hainanu.edu.cn
刘文波，高级实验师，主要从事分子植物病理相关研究；E-mail：sauch@163.com

贝莱斯芽孢杆菌 HN-2 抑菌活性的初探*

谭 峥**，韦丹丹，刘文波，缪卫国，靳鹏飞***

（海南大学植物保护学院，热带农林生物灾害绿色防控教育部重点实验室，海口 570228）

摘 要：芽孢杆菌（*Bacillus*）是一类可产生对植物病原真菌和细菌具有拮抗作用的活性物质的一类革兰氏阳性杆状细菌，其被广泛用于生物防治领域中。前期研究发现贝莱斯芽孢杆菌 HN-2 对多种植物病原真菌和细菌具有较好的抑菌活性。因此在前期的基础上对该菌株进行进一步研究，通过正丁醇萃取的方法，利用滤纸片扩散法对贝莱斯芽孢杆菌 HN-2 的发酵液中的活性成分进行提取并与几种常用农药（叶枯唑、春雷霉素、杆菌肽）进行活性比较。同时测定了正丁醇提取物耐受强酸强碱、高温以及强紫外线等环境下的稳定性。经过研究发现：贝莱斯芽孢杆菌 HN-2 正丁醇提取物、杆菌肽、春雷霉素和叶枯唑对稻黄单胞菌水稻致病变种（*Xanthomonas oryzae* pv. *oryzae*，*Xoo*）抑菌圈直径分别为 29.79 ± 0.18 mm、27.71 ± 0.64 mm、18.71 ± 0.10 mm和 16.35 ± 0.45 mm，结果表明在相同浓度下，生防菌 HN-2 正丁醇提取物具有很好的抑菌效果。稳定性试验结果表明，贝莱斯芽孢杆菌 HN-2 正丁醇提取物的稳定性良好，在强酸性环境下活性不会受到影响，在强碱性条件下亦可保持 68.52%的抑菌活性，紫外线照射 3 h 和 121 ℃下处理 30 min 后的正丁醇提取物仍具有较高的抑菌活性。综上所述，生防菌 HN-2 的正丁醇提取物具有较好的抑菌活性，是一种高效稳定的生物源农药，具有继续深入精细化分离纯化的研究意义，并在未来的绿色农药的发展中具有较好的发展潜力。

关键词：贝莱斯芽孢杆菌；抑菌活性；农药；稳定性

---

* 基金项目：芽孢杆菌活性成分与代谢机理研究（KYQD（ZR）1842）；海南自然科学基金创新研究团队项目（2016CXTD002）；海南省重点研发计划项目（ZDYF2016208）

** 第一作者：谭峥，本科生，从事生防农药研究；E-mail：18789096089@163.com

*** 通信作者：靳鹏飞，讲师，主要从事微生物相关研究；E-mail：jinpengfei@hainanu.edu.cn

贝莱斯芽孢杆菌 HN-2 的分离鉴定及抑菌活性研究*

王　雨**，谭　峥，韦丹丹，刘文波，缪卫国，靳鹏飞***
（海南大学植物保护学院，热带农林生物灾害绿色防控教育部重点实验室，海口　570228）

摘　要：HN-2 菌株为本实验室分离得到的一株生防细菌，为探究生防细菌 HN-2 的分类学地位以及明确 HN-2 发酵液中主要抑菌活性化合物的种类。通过采用形态学观察结合现代分子生物学的手段对生防菌 HN-2 进行分类鉴定，同时运用各种提取方式提取生防菌 HN-2 发酵上清液中的活性物质，以杧果炭疽菌（*Colletotrichum gloeosporioides* Penz）作为靶标进行活性测试、利用飞行时间质谱（MALDI-TOF-MS）对有活性的正丁醇提取物进行组分分析，同时采用聚合酶链式反应（PCR）对 HN-2 菌株中合成脂肽类物质的合成与调控基因进行检测、比对、构建系统发育树，并且初步测试 HN-2 菌株正丁醇提取物在杧果上的保护作用。结果表明 HN-2 菌株为贝莱斯芽孢杆菌（*Bacillus velezensis*），其发酵上清液的正丁醇萃取得到的粗提物活性最好，抑菌圈大小为 20.93 ± 0.3 mm，半最大效应浓度（EC_{50}）为 70.62 μg/mL；通过飞行时间质谱数据结合基因检测的结果分析发现正丁醇提取物中主要活性成分为脂肽类物质，其中含有表面活性素（surfactin）、伊枯草菌素（iturins）和泛革素（fengycin）等；通过显微观察发现正丁醇粗提物可以明显抑制杧果炭疽病菌的生长，HN-2 正丁醇提取物处理后的杧果 15d 内未出现杧果炭疽病病状，对杧果果实具有较好的保护作用。贝莱斯芽孢杆菌 HN-2 的主要活性物质为脂肽类物质，其对植物病原真菌有较好的防治效果，具有进一步深入研究和开发应用的潜力。

关键词：贝莱斯芽孢杆菌；脂肽类化合物；抑菌活性；杧果炭疽菌；EC_{50}

* 基金项目：芽孢杆菌活性成分与代谢机理研究（KYQD（ZR）1842）；海南自然科学基金创新研究团队项目（2016CXTD002）；海南省重点研发计划项目（ZDYF2016208）

** 第一作者：王雨，硕士研究生，从事生防农药研究；E-mail：15383468897@163.com
*** 通信作者：靳鹏飞，讲师，主要从事微生物相关研究；E-mail：jinpengfei@hainanu.edu.cn

番茄立枯病生防细菌的筛选及其生防机制初探*

李凤芳**，许萌杏，袁高庆，林 纬，吴小刚，黎起秦***

（广西大学农学院，南宁 530004）

摘 要：番茄立枯病（Tomato damping-off）是番茄苗期重要的土传病害之一，目前防治该病害的方法主要是通过使用化学药剂来加强苗期管理，但长期使用化学药剂会产生防效下降以及污染环境问题。本研究在筛选出对番茄立枯病菌（*Rhizoctonia solani*）具有较强抑制作用的生防细菌的基础上，初步分析备选菌株的生防机制，为探讨利用生防菌防治番茄立枯病奠定基础。结果表明，采用土壤稀释分离法，从广西以及北京市的 7 不同乡镇采集到 105 份土样中，共分离到 2 929个细菌菌株；通过平板对峙法，获得对番茄立枯病菌有较强抑制作用（抑菌圈直径 0. 5~3. 5cm）的细菌 103 株；根据抑菌圈直径大小，对其中抑菌作用较强的 48 个菌株进行了盆栽生测试验，菌株 B11-4 对番茄立枯病的室内防治效果最好，达 75. 9%。根据生理生化鉴定，结合 16S rDNA 鉴定结果，将菌株 B11-4 鉴定为绿针假单胞菌（*Pseudomonas chlororaphis*）。通过光学显微镜与扫描电镜观察，菌株 B11-4 处理的病原菌菌丝表现出严重变形，或缢缩或膨大。利用抗利福平的诱变菌株 B11-4^{Ri}（对番茄立枯病菌的抑制作用与野生菌株相同）检测该菌株在番茄根部的定殖能力，将初始浓度为 10^8CFU/mL 的菌株 B11-4^{Ri}淋到番茄植株的根部，45d 后，番茄植株的根围土中含菌株 B11-4 的量为 4. 75×10^3CFU/g 土，淋根 60d 后，含菌量依然达 3. 1×10^3CFU/g 土。用结晶紫染色的方法对该菌株产生物膜（biofilm）能力测定，其与对照菌株荧光假单胞菌 2P24 的生物膜产生量处于相同水平。采用 PCR 方法对菌株 B11-4 抗生素吩嗪-1-羧酸（phenazine-1-carboxylic acid，PCA）2，4-二乙酰基间苯三酚（2，4-diacetylphloroglucinol，2，4-DAPG）、藤黄绿脓菌素（pyoluteorin，PLT）和吡咯菌素（pyrrolnitrin，PRN）的基因进行检测，结果表明菌株 B11-4 含有编码产生 PCA 和 PRN 基因，不含有编码产生 2，4-DAPG 和 PLT 的基因，推测菌株 B11-4 可能产生 PCA 和 PRN，可能不产生 2，4-DAPG 和 PLT。采用 Salkowski 比色法测定该菌株产生促生物质生长素（auxin，IAA）情况，结果表明该菌株可产生 IAA。采用平板检测的方法检测该菌株铁载体（siderophores）、蛋白酶（protease）产生情况，结果显示该菌株可产生铁载体、可形成较为明显的蛋白水解圈。

综上所述，菌株 B11-4 对番茄立枯病有较好的室内防治效果，该菌株主要通过抑制病原菌菌丝生长、稳定定殖于植物根围、产生抗生素、促生物质、蛋白酶等物质来发挥其生防作用。

关键词：番茄立枯病；生防细菌；绿针假单胞菌；生防机制

* 基金项目：广西科技重大专项（编号：桂科 AA17204041-3）

** 第一作者：李凤芳，在读硕士生，研究方向为植物病害及其防治研究；E-mail：1530046698@ qq. com

*** 通信作者：黎起秦，博士，教授，主要从事植物病害及其防治研究；E-mail：qqli5806@ gxu. edu. cn

11种药剂对猕猴桃细菌性溃疡病病原菌的室内药效测定*

潘　慧**，邓　蕾，陈美艳，李　黎***，钟彩虹***

（1. 中国科学院植物种质创新与特色农业重点实验室，中国科学院武汉植物园，武汉　430074；2. 中国科学院种子创新研究院，武汉　430074；3. 中国科学院猕猴桃产业技术工程实验室，武汉　430074）

摘　要：为明确猕猴桃细菌性溃疡病菌（*Pseudomonas syringae* pv. *actinidiae*）对不同杀菌剂的敏感性，采用紫外分光光度计法，在室内测定11种杀菌剂对该病菌的药效。试验结果表明：参试药剂对猕猴桃溃疡病菌均有不同程度的抑制作用。其中，80%乙蒜素4 000倍液乳油抑菌率最高，为91.58%，其次为0.15%四霉素梧宁霉素600倍液水剂和2%春雷霉素500倍液水剂，抑菌率分别为83.51%和81.95%，前三者与其他药剂抑菌效果差异显著；而80%波尔多液400倍液可湿性粉剂、20%叶枯唑600倍液可湿性粉剂和30%碱式硫酸铜300倍液悬浮剂等5种参试药剂对猕猴桃溃疡病菌的抑菌效果较差，与其他药剂的抑菌效果差异显著，其中80%波尔多液400倍液可湿性粉剂抑菌率最低，仅为2.31%。由此，乙蒜素、四霉素、梧宁霉素、春雷霉素可进一步在田间进行药效验证，以确定最佳药剂。

关键词：猕猴桃细菌性溃疡病；杀菌剂；药效

---

* 基金项目：国家自然科学基金青年科学基金项目（31701974）；湖北省自然科学基金面上项目（2017CFB443）；湖北省技术创新专项（重大项目）（2016ABA109）；湖北省农业科技创新行动项目“特色水果生态高效栽培与采后处理”项目；武汉市科技局前资助科技计划（2018020401011307）

** 第一作者：潘慧，硕士，工程师，主要从事猕猴桃病害鉴定及防治等研究；E-mail：panhui@ wbgcas. cn

*** 通信作者：李黎，副研究员，主要从事猕猴桃病害研究；E-mail：lili@ wbgcas. cn

钟彩虹，研究员，主要从事猕猴桃资源挖掘、育种、病害等相关研究；E-mail：zhongch1969@ 163. com

双苯菌胺与不同杀菌剂的交互抗药性分析*

程星凯**，王梓桐，梁　莉，王治文，孙铭优，刘鹏飞***
（中国农业大学植物病理学系，北京　100193）

摘　要：立枯丝核菌（*Rhizoctonia solani*）是自然界中普遍存在的一种土传病原真菌，腐生能力强，寄主范围广，能够侵染棉花、水稻、小麦、玉米、马铃薯、大豆以及多种十字花科蔬菜，从而引起农作物的大量减产，造成巨大的经济损失。我国创制性二苯胺类化合物——双苯菌胺杀菌谱广，抑菌活性高，对植物病原真菌和卵菌均具有良好的抑制活性，在植物病害防治中具有广阔的市场前景。但目前病原菌对双苯菌胺的抗性机制尚不明确，有关此方面的研究也相对缺乏。鉴于此，本研究开展了立枯丝核菌对双苯菌胺的抗性机制研究，以便为进一步制定科学的病害管理方案，避免和延缓田间抗药性的发生和发展提供参考。

通过室内药剂驯化从立枯丝核菌亲本菌株 X19 共筛选到 20 株抗双苯菌胺突变体，其中 5 株突变体的抗性倍数大于 50 倍。经过十代无药平板培养后，检测发现 3 株菌（X19-02、X19-18 和 X19-20）具有稳定的抗性。随后结合双苯菌胺与生产上常用杀菌剂的交互抗药性进行了测定。结果表明，双苯菌胺与氟啶胺的相关系数 ρ 为 0.864，与三苯基氯化锡相关系数 ρ 为 0.830，具有较强的相关性。与氰霜唑、2，4-二硝基苯酚、咯菌腈的相关系数 ρ 分别为 0.714、0.713 和 0.671，具有中等程度的相关性。立枯丝核菌抗性突变体同时对解偶联剂、ATP 合酶抑制剂、QiI 抑制剂和渗透信号传递抑制剂等多种作用机制杀菌剂产生了抗性，其原因可能是在药剂选择压力下，病原菌对药剂外排或代谢能力增强，从而引发多药抗性，需要进一步研究以明确立枯丝核菌对双苯菌胺的抗性机制。

关键词：立枯丝核菌；双苯菌胺；交互抗性

---

* 基金项目：国家自然科学基金（U160310041，31772192）

** 第一作者：程星凯，博士研究生，植物病理学；E-mail：xingkai210@126.com

*** 通信作者：刘鹏飞，副教授，博士，主要从事植物病理学研究；E-mail：pengfeiliu@cau.edu.cn

双苯菌胺对立枯丝核菌的作用机制研究*

梁　莉**，代　探，孙铭优，程星凯，王治文，刘鹏飞***
（中国农业大学植物病理学系，北京　100193）

摘　要：由立枯丝核菌引起的棉花立枯病是棉花苗期的重要病害，发病严重时可造成重大经济损失。双苯菌胺是沈阳化工研究院有限公司自主研发的新型杀菌剂，具有高效、低毒、广谱的特点，离体下对立枯丝核菌显示出高抑菌活性。实验室前期研究表明该药剂作用于辣椒疫霉呼吸过程，但其作用机制尚不清楚。本研究旨在明确双苯菌胺对立枯丝核菌的作用机制，可为其他二芳胺类药剂的作用机制解析提供参考。本研究采用菌丝生长速率法测定了双苯菌胺和其他6类不同作用机制能量合成抑制剂对立枯丝核菌野生型菌株的 EC_{50}，这些抑制剂包括NADH氧化还原酶抑制剂、QoIs、QiIs、SDHIs、ATP合酶抑制剂和解偶联剂。然后以此为依据，利用代谢组学方法进行了基于药剂作用下立枯丝核菌代谢指纹的GC-MS分析，对双苯菌胺和其他6类共13种呼吸作用抑制剂进行了层次聚类。结果显示，电子传递链上的复合物Ⅰ、Ⅱ和Ⅲ抑制剂分别聚类且整体聚为一个分支，在呼吸作用中主要影响氧化过程；而ATP合酶抑制剂和解偶联剂聚为另一个分支，在呼吸作用中主要影响磷酸化过程。双苯菌胺与磷酸化过程抑制剂聚为一类，推测其具有ATP合酶抑制或解偶联活性。进一步研究了双苯菌胺对立枯丝核菌ATP含量、呼吸速率和线粒体膜电位的影响。结果显示，双苯菌胺和氟啶胺分别在0.01~0.5 μg/mL和0.2~5 μg/mL浓度范围内，显著抑制立枯丝核菌体ATP合成，促进呼吸速率，并降低线粒体膜电位，且这些影响均随着浓度提高而增强，表现出了典型的解偶联剂的特征。研究可为其他创制性杀菌剂的作用机制解析提供参考，为双苯菌胺的市场开发和田间科学使用提供依据。

关键词：立枯丝核菌；双苯菌胺；代谢组学；解偶联；ATP合酶抑制活性

---

* 基金项目：国家自然科学基金（U160310041）

** 第一作者：梁莉，硕士研究生，植物病理学；E-mail：17600853226@163.com

*** 通信作者：刘鹏飞，副教授，博士，从事植物病理学研究；E-mail：pengfeiliu@cau.edu.cn

桃果实腐烂病原菌鉴定及其对 4 种杀菌剂的敏感性测定*

薛昭霖**，张　灿，张博瑞，刘西莉***
（中国农业大学植物病理学系，北京　100193）

摘　要：桃树是重要的核果类果树，在我国分布范围广，栽种面积大，其中福建省是我国桃主产区之一。2018 年，在福建省进行桃树病害调查时发现其果实出现腐烂现象，可造成 20%~30%的经济损失。为了明确该病原菌的种类，采用组织分离纯化的方法获得分离物，结合形态学观察、致病力测定和 rDNA-ITS、翻译延长因子 TEF 等序列对比的方法对分离物进行鉴定。结果表明，引起桃果实腐烂的病原菌包括腐皮镰刀菌 *Fusarium solani*、禾谷镰刀菌 *Fusarium graminearum*、九州镰孢菌 *Fusarium kyushuense*、层出镰刀菌 *Fusarium proliferatum*。

采用菌丝生长速率法，测定了上述病原菌对生产中常用的 4 种不同作用机制杀菌剂的敏感性，包括多菌灵、咯菌腈、戊唑醇和氰烯菌酯，以期为该类病害防控提供参考。结果表明，不同病原菌对 4 种药剂的敏感性存在差异：咯菌腈对 4 种供试病原菌的抑制活性最好，其 EC_{50} 介于 0.003~0.1 μg/mL；多菌灵的 EC_{50} 均小于 1.0 μg/mL；氰烯菌酯和戊唑醇对腐皮镰刀菌的 EC_{50} 均大于 50 μg/mL，氰烯菌酯对其他 3 种供试镰刀菌的 EC_{50} 值均小于 0.3 μg/mL，戊唑醇对其他 3 种供试镰刀菌的 EC_{50} 值介于 0.1~3 μg/mL。研究结果为进一步开展复配或交替使用上述药剂防治该地区的桃果实腐烂病提供了指导。

关键词：镰刀菌；杀菌剂；敏感性

---

* 基金项目：重点研发计划（2016YFD0201305-6）

** 第一作者：薛昭霖，在读博士研究生；E-mail：xuezhaolin1215@163.com

*** 通信作者：刘西莉，教授，主要从事植物病原卵菌与杀菌剂互作研究；E-mail：seedling@cau.edu.cn

新型呼吸抑制剂唑嘧菌胺的生物学活性与抗性机制初探*

高续恒**，李成成，苗建强，刘西莉***
（西北农林科技大学植物保护学院，杨凌 712100）

摘 要： 唑嘧菌胺是由巴斯夫公司开发的一种新型卵菌抑制剂，主要用于防治马铃薯、葡萄、黄瓜和辣椒等作物上的疫霉病与霜霉病。唑嘧菌胺属三唑并嘧啶类杀菌剂，可以与 bc1 复合体（复合物Ⅲ）上 Qo-标桩菌素位点（Qo-distal）结合，阻断呼吸电子传递链的电子传递，故杀菌剂抗性行动委员会 FRAC 将其归为 QoSI 类杀菌剂。最新研究结果表明，唑嘧菌胺是一种双作用位点杀菌剂，不仅可以与复合物Ⅲ上的 Qo-distal 位点结合，还可以与 Qi-位点结合。分子对接结果显示，唑嘧菌胺与细胞色素 bc1 复合体的结合方式与传统 QoI 与 QiI 类杀菌剂不同，故认为该药剂与嘧菌酯、氰霜唑、烯肟菌酯等常用卵菌抑制剂均不存在交互抗性。

本研究以辣椒疫霉为研究对象，测定了唑嘧菌胺对其不同发育阶段的影响，结果表明，唑嘧菌胺可以使辣椒疫霉游动孢子迅速休止，并且对孢子囊形成、游动孢子释放、游动孢子萌发等具有良好抑制活性，不同发育阶段的 EC_{50} 值均小于 0.1 μg/mL；进一步测定了唑嘧菌胺对 6 种疫霉和 18 种腐霉的菌丝生长的抑制作用。结果显示，唑嘧菌胺对荔枝霜疫霉 *Peronophythora litchii*、大豆疫霉 *Phytophthora sojae*、贵阳腐霉 *Pythium guiyangense*、寡雄腐霉 *Py. oligandrum*、瓜果腐霉 *Py. aphanidermatum*、宽雄腐霉 *Py. dissotocum*、簇囊腐霉 *Py. torulosum* 等 7 种卵菌表现出有较好的抑菌活性，EC_{50} 均小于 5 μg/mL。

以荔枝霜疫霉为靶标菌进行唑嘧菌胺抗性机制研究。采用室内药剂驯化菌饼的方法对 5 株亲本进行抗药性诱导，共计获得 10 株抗性突变体，抗性倍数均在 500 倍以上。进一步扩增比对了亲本菌株和突变体之间的 CYTB 序列，分析发现抗药性突变体 CYTB 上的 33 位丝氨酸突变为亮氨酸（S33L），228 位天冬氨酸突变为天冬酰胺（D228N）。S33L 为主要突变类型，D228N 为次要突变类型，2 种点突变是否是导致荔枝霜疫霉对唑嘧菌胺产生抗性的主要原因，还有待于进一步通过转化分析验证。

关键词： 唑嘧菌胺；辣椒疫霉；荔枝霜疫霉；生物学活性；抗性机制

---

* 基金项目：科技部重点研发计划（编号：2016YFD0201305-6）
** 第一作者：高续恒，在读博士研究生；E-mail：xuhenggao@163.com
*** 通信作者：刘西莉，教授，主要从事植物病原菌与杀菌剂互作研究；E-mail：seedling@cau.edu.cn

稻瘟病菌对唑菌酯的抗性风险评估和抗性机制研究*

赵国森**，邓　琳，刘思博，郭圣洁，张思玥，苗建强，刘西莉***
（西北农林科技大学植物保护学院，杨凌　712100）

摘　要：水稻稻瘟病是一种世界性水稻病害，与纹枯病、白叶枯病、稻曲病并称为水稻四大病害。一般造成水稻减产 10%~20%，严重时可减产 40%~50%，有时甚至绝收，化学防治仍是稻瘟病防治的重要手段。唑菌酯（pyraoxystrobin）是由沈阳化工研究院创制并开发的高效、广谱的 QoI 类杀菌剂，目前关于稻瘟病菌对该药剂的抗性风险及抗性分子机制尚未见研究报道。

本研究采用菌丝生长速率法测定了 109 株采自我国主要水稻产区的水稻稻瘟病菌对唑菌酯的敏感性，其 EC_{50}值分布于 0. 001 5~ 0. 022 0 μg/mL，平均 EC_{50}值为 0. 009 4 μg/mL，敏感性分布呈单侧峰曲线，未出现抗药性亚群体，可将该曲线作为稻病瘟菌对唑菌酯的敏感性基线。

通过室内药剂驯化的方法，从 7 株敏感菌株中筛选获得了 15 株抗药性突变体，突变频率为 1.25×10^{-3}，抗性水平在 1 112. 98~5 415. 46倍，抗药性性状能稳定遗传。与亲本菌株相比，抗性突变体的菌丝生长速率、孢子产生能力和孢子萌发率与亲本无显著性差异。交互抗药性结果表明，唑菌酯与嘧菌酯存在正交互抗药性，与田间生产中防治稻瘟病常用药剂多菌灵、稻瘟灵、咪鲜胺和百菌清无交互抗药性。根据药剂驯化获得抗药性突变体的难易程度，突变体的抗性倍数及其生存适合度，交互抗药性结果以及同类药剂田间使用情况等，综合分析表明，稻瘟病菌对唑菌酯可能存在中到高等抗性风险。进一步克隆了抗药突变体及其亲本的 *cytb* 基因，CYTB 氨基酸序列比对结果表明，15 株高抗突变体 143 位甘氨酸均突变为丝氨酸（G143S），该点突变已被报道能够引起多种植物病原菌对 QoI 类杀菌剂的抗性，因此我们推测 G143S 点突变导致了稻瘟病菌对唑菌酯的抗性。

本研究结果将为唑菌酯在稻瘟病防治过程中的科学合理使用以及制定有效延缓其抗性发生发展的治理策略提供重要依据。

关键词：稻瘟病菌；唑菌酯；风险评估；抗性机制

* 基金项目：大学生创业训练计划项目（201810712102）
** 第一作者：赵国森，在读本科生；E-mail：1184552004@ qq. com
*** 通信作者：刘西莉，教授，主要从事杀菌剂药理学及病原物抗药性研究；E-mail：seedling@ nwsafu. edu. cn

氟吡菌酰胺对黄瓜和草莓白粉病菌的抑菌活性研究*

张博瑞**，孟德豪，黄中乔，刘西莉***
（中国农业大学植物保护学院，北京　100193）

摘　要：黄瓜白粉病是黄瓜种植中常见病害，苗期至成株期均可发生，主要侵害叶片，发病初期在叶面或背面，幼茎上产生白色近圆形的小粉斑，重病时整个叶片布满白粉，叶片变黄、干枯，植株死亡，对黄瓜生产威胁极大，生产上迫切需要高效杀菌剂用于该病害的防治。草莓白粉病亦是草莓种植中常见病害，苗期至成株期均可发生，主要侵害植株叶片、花、果、果梗和叶柄，严重者花蕾不能开放，影响果实膨大，降低产量和品质。目前生产中白粉病菌已经对多种常用杀菌剂产生了普遍的抗药性，因此，筛选安全、高效的新药剂用于该病害的防治具有重要的现实意义。

本研究采用 10^5 个孢子/mL 的白粉病菌孢子悬浮液接种离体叶片（盘）法分别测定了供试药剂氟吡菌酰胺及对照药剂戊唑醇和醚菌酯对黄瓜白粉病菌和草莓白粉病菌的抑菌活性。研究结果表明，氟吡菌酰胺对黄瓜白粉病菌具有优异的抑制活性，其有效抑制中浓度为 0.225mg/L，优于生产上常规药剂戊唑醇的有效抑制中浓度 1.961mg/L；氟吡菌酰胺对草莓白粉病菌的有效抑制中浓度为 4.16mg/L，与生产上常规新型药剂醚菌酯的防效相当。

由于氟吡菌酰胺，戊唑醇和醚菌酯属于不同作用机制杀菌剂，相关研究表明，病原菌对该 3 类药剂的抗性机制不同，白粉病菌对氟吡菌酰胺与戊唑醇和醚菌酯之间无交互抗药性。因此，生产中可以通过交替或复配使用上述不同作用机制的杀菌剂来有效防控白粉病的为害，并有助于延缓抗药性的发生和发展。

关键词：白粉病菌；氟吡菌酰胺；抑菌作用

* 基金项目：科技部重点研发计划（编号：2016YFD0201305-6）
** 第一作者：张博瑞，在读硕士研究生；E-mail：zhangbr96@163.com
*** 通信作者：刘西莉，教授，主要从事植物病原菌与杀菌剂互作研究；E-mail：seedling@nwafu.edu.cn

解淀粉芽孢杆菌 Jt84 干悬浮剂加工工艺及其贮存稳定性研究*

张荣胜**，王法国，齐中强，杜　艳，刘永锋***
（江苏省农业科学院植物保护研究所，南京　210014）

摘　要：生防芽孢杆菌防治植物病害主要通过竞争营养与空间、产生具有抑菌和溶菌活性代谢产物、诱导植株产生系统抗病性等多方面协同作用，同时生防芽孢杆菌发挥作用的前提是在植株上有效定殖。解淀粉芽孢杆菌 Jt84 菌株对水稻真菌病害具有良好的防治效果。本研究通过喷雾干燥技术，开发一种生防菌新剂型杀菌剂。采用 Plackett-Burman 试验及 Box-Behnken 试验，对解淀粉芽孢杆菌 Jt84 干悬浮剂喷雾干燥工艺进行了优化。结合生产实际确定干悬浮剂制备条件分别为：进风温度 118.9 ℃，进样速率 25.7 r/min（约为 680 mL/h），压力 0.2 MPa，出风温度 80 ℃，喷头口径 0.7 mm，可获得活菌量为 6.8×10^9 CFU/g 解淀粉芽孢杆菌 Jt84 干悬浮剂产品，经喷雾干燥试验验证该理论预测值与实际值无显著差异。该干悬浮剂产品室温存放 90d 时，菌含量为 4.5×10^9CFU/g；360d 时，菌含数不低于 4.0×10^9CFU/g。干悬浮剂在加工后 0~90 d 内菌含量明显减少，下降了约 33.8%；在 90~360d 内，干悬浮剂中菌含量缓慢减少，菌含量维持相对稳定状态，随着存贮时间的进一步增加，540d 时 Jt84 干悬浮剂中菌含量下降到 2.4×10^9 CFU/g。

关键词：解淀粉芽孢杆菌；加工工艺；菌含量；贮存稳定性

---

* 基金项目：国家重点研发计划（2016YFD0300706）；江苏省自主创新资金项目 CX17（3021）
** 第一作者：张荣胜，博士，副研究员，主要从事水稻病害生物防治及其应用技术研究；E-mail：r_szhang@163.com
*** 通信作者：刘永锋，研究员，主要从事植物病害致病机制及其防控技术研究；E-mail：liuyf@jaas.ac.cn

sacB 介导的绿针假单胞菌 YL-1 遗传操作方法*

周亚秋**，张婷婷，乔俊卿，刘永锋，刘邮洲***
（江苏省农业科学院植物保护研究所，南京　210014）

摘　要：绿针假单胞菌（*Pseudomonas chlororaphis* YL-1）是从大豆根围分离获得的、对多种病原细菌和病原真菌有较强抑制作用的生防细菌。为了探明绿针假单胞菌 YL-1 生防相关基因的功能，本研究以嗜铁素转录调控因子 *Pvds* 编码基因为对象，建立了一套基于负选择标记基因 *Sacb* 的绿针假单胞菌无标记基因敲除技术，构建重组质粒 PEX18-*Pvds*，通过改良的细菌接合转移技术将重组质粒导入野生型菌株 YL-1 中，利用同源重组技术获得缺失突变株 Δ*Pvds*。生防相关性状研究结果表明，与野生型菌株 YL-1 相比，突变株 Δ*Pvds* 泳动能力和生长能力未发生改变，但是群集运动能力显著下降。同时突变株 Δ*Pvds* 合成嗜铁素的能力也显著下降，*Pvds* 基因互补后突变株能恢复合成嗜铁素的功能。本文结果表明，已成功建立了适用于YL-1 的基因定向敲除技术和功能基因互补体系，为深入研究 YL-1 的生防机制奠定了重要基础。

关键词：绿针假单胞菌；YL-1；*sacB* 介导；基因敲除；基因互补

---

* 基金项目：国家自然科学基金（31672076）；江苏省重点研发计划（BE2018359）；苏州市科技计划项目（No. SNG2018095）

** 第一作者：周亚秋，硕士，主要从事植物病害生物防治的研究；E-mail：369736603@ qq. com
*** 通信作者：刘邮洲，研究员，主要从事植物病害生物防治的研究；E-mail：shitouren88888@ 163. com

植物病虫害自动识别技术浅析*

常　月**，马占鸿***

（中国农业大学植物病理学系，北京　100193）

摘　要：传统的病虫害识别工作依赖于植物保护学专家的人为识别，随着计算机技术的发展，人工智能技术逐渐被应用到病虫害识别上，人们可通过计算机模型实现病虫害的机器识别。

虽然基于人为设计特征的病虫害识别方法取得了较好的识别效果，但特征的选取依赖于人为的筛选，所出现的参数数量有限，识别的准确率受专业知识的影响很大；此外，对于背景复杂的图像来说，由于缺乏对图像数据库中图像背景的整体分析，基于病症或病状或害虫特点所选取的特征很难实现目标与复杂背景的分离。相比较之下，深度学习网络具有强大的学习能力，在训练过程中无需人工提取特征，它会从某一病虫害的大量图像数据中自主提取图像的全局特征，可以选出成千上万的参数。

在病虫害自动识别技术的发展过程中，图像数据库是最为基础的支撑，也是最有收集难度的工作。例如：除了要考虑图像数量和清晰度外，还要考虑不同时间、地点及环境条件下同一种病原物引发的病症或病状的差别，这就对图像采集和构建数据库提出了更多的要求。识别模型面对的是大众用户，用户上传的可能是任意环境条件下、任意发病时间内的不同发病程度的病虫害图像，如何保证在多种变化中提取到典型的特征成为了衡量数据库质量的关键。数据库的构建工作离不开植保专业人士的参与，它需要构建者对病虫害的发病特点、传播规律、全国分布情况等内容的熟知，在构建数据库时挑选合适的照片，使提取的特征有代表性，这样识别模型才能在大范围的推广中仍可以保持较高的识别准确率。

从事植物病虫害研究的学者需要探究出一套系统的病虫害图像数据库构建方法，为此项研究提供宝贵的经验，助力人工智能技术在植保行业中的应用，进而促进农业大数据的发展。

关键词：病虫害自动识别；人工智能；深度学习；图像数据库

* 基金项目：国家重点研发计划项目（2016YFD0201302）

** 第一作者：常月，硕士研究生，主要从事植物病害流行学研究；E-mail：239654967@qq.com

*** 通信作者：马占鸿，教授、博士生导师，主要从事植物病害流行学和宏观植物病理学研究；E-mail：mazh@cau.edu.cn

疫霉菌对氟噻唑吡乙酮的抗性分子机制*

刘小飞**，李桂香，李成成，苗建强***，刘西莉
（西北农林科技大学植物保护学院，杨凌 712100）

摘　要：氟噻唑吡乙酮（通用名称 oxathiapiprolin）是美国杜邦公司 2007 年研发出的结构新颖的杀菌剂，其化学结构中包含了吡唑环、噻唑环和异噁唑啉环，是迄今为止生物活性最高的一类杀菌剂。国内外前期研究中，通过紫外诱导和药剂驯化的方法分别获得了辣椒疫霉和烟草疫霉对氟噻唑吡乙酮的系列抗性突变体，进一步研究发现靶标蛋白 ORP1 存在 20 种点突变类型（L733W、S768I/F/K/Y、G770A/I/P/V/L、ΔG818ΔF819、N837I/F/Y、G839W、P861H、L863W/F、I877F/Y）。笔者课题组前期通过 CRISPR/Cas9 定点编辑的方法明确了 ORP1 上 G770V 和 G839W 位点的突变与疫霉菌对氟噻唑吡乙酮的抗性相关，而目前报道的以上其他突变类型与病原菌抗药性的相关性，以及其引起的抗药性水平及对疫霉菌生物学表型的影响尚属未知。

本研究以野生型大豆疫霉 P6497 的 *PsORP*1 为编辑靶标，分别构建 PsORP1 蛋白 ORD 结构域上除 L863F 之外的 17 种点突变的 CRISPR/Cas9 定点编辑同源替换载体。通过遗传转化的方法，获得系列 *PsORP*1 定点突变阳性转化子。敏感性测定结果显示，8 种点突变类型（L733W、S768Y/F、N837F/Y、P861H、L863W、I817Y）均能导致大豆疫霉对氟噻唑吡乙酮的高水平抗性，抗性倍数大于 2000 倍，其余点突变类型引起的抗性倍数均小于 200 倍，推测不同突变类型所引起的药剂与靶蛋白之间结合能力的改变不同。进一步测定了不同突变类型转化子与野生型菌株在菌丝生长速率、孢子囊产量、游动孢子产量、休止孢萌发和卵孢子产量方面的差异。结果表明，不同类型转化子在菌丝生长速率、孢子囊产量、游动孢子产量方面与野生型菌株相当或低于亲本，在休止孢萌发及卵孢子产量方面与亲本无显著差异，表明不同类型的大豆疫霉抗性突变体具有与亲本相当或较低的田间生存适合度。

综上所述，疫霉菌 ORP1 上 L733W、S768Y/F、N837F/Y、P861H、L863W、I817Y 的单位点突变可引起病原菌对氟噻唑吡乙酮的高水平抗性，田间生产中须针对这些潜在的突变位点进行抗性分子检测和早期预警。

关键词：疫霉；氟噻唑吡乙酮；突变位点；抗性机制

* 基金项目：西北农林科技大学科研启动经费；陕西省自然科学基础研究计划（2019JQ-301）
** 第一作者：刘小飞，在读硕士研究生；E-mail：517071890@qq.com
*** 通信作者：苗建强，讲师，主要从事杀菌剂药理学及病原物抗药性研究；E-mail：mjq2018@nwafu.edu.cn

杧果细菌性黑斑病生防菌的筛选及防治效果研究

喻群芳，漆艳香，张辉强，张　贺，蒲金基*

（中国热带农业科学院环境与植物保护研究所；农业农村部热带作物有害生物综合治理重点实验室；海南省热带作物病虫害生物防治工程技术研究中心，海口　571101）

摘　要：杧果细菌性黑斑病是杧果生产中为害最严重的病害之一，该病害由柑橘黄单胞菌杧果致病变种（*Xanthomonas citri* pv. *mangiferaeindicae*）引起，主要为害杧果叶片、枝条、花芽和果实。杧果细菌性黑斑病在我国广东、广西、海南、云南、福建、四川等杧果主产区普遍发生。一般造成产量损失达 15%~30%，严重时高达 50%以上。对该病的防治主要是化学农药防治，而化学防治会增加杧果质量安全和环境污染风险，且容易使病原菌产生耐药性等问题，因此寻找安全高效的防治方法对于提高杧果的产量和质量有重要意义。

本研究从病害的生物防治的角度出发，对荧光假单胞菌、蜡质芽孢杆菌、甲基营养型芽孢杆菌和绿针假单胞菌等 4 种生防菌对杧果细菌性黑斑病的抑菌能力和田间防效进行测定，4 种生防菌的抑菌圈直径分别为 26. 51 mm、24. 12 mm、23. 78 mm 和 20. 32mm。田间防效测定结果表明，4 种生防菌均能有效防治杧果细菌性黑斑病，其中荧光假单胞菌的效果最好，平均防效达到 64. 04%，其次是甲基营养型芽孢杆菌、蜡质芽孢杆菌和绿针假单胞菌，平均防效为 62. 84%、60. 80%和 60. 13%。

关键词：杧果细菌性黑斑病；生物防治；防治效果

* 通信作者：蒲金基

基于氧化海藻酸钠合成银纳米颗粒的抗植物病原真菌活性研究*

向顺雨[1,3]**，施　焕[1]，刘昌云[1]，廖舒悦[1]，董国菊[1]，汪代斌[4]，陈海涛[4]，
张　帅[4]，黄　进[2,3]，孙现超[1,3]***
（1. 西南大学植物保护学院，重庆　400615，2. 西南大学化学化工学院，重庆　400615，
3. 软物质材料化学与功能制造重庆市重点实验室，重庆　400615；4. 中国烟草公司重庆市公司烟草科学研究所，重庆　400716）

摘　要：化学农药的大量使用所带来的负面问题迫使科研工作者研究与开发更为安全高效的新型农药。研究发现，银纳米颗粒具有良好的抗菌性能，但是，传统合成银纳米颗粒方法中由于要使用有毒的化学试剂，合成过程中生成有毒副产物，合成出的银纳米颗粒稳定性分散性较差，粒径分布不均一等问题限制了其在生物科学研究中的使用。因此，探索绿色合成稳定性强，分散性好，粒径分布均一的银纳米颗粒的方法颇为重要。在这里笔者提出了利用醛基化海藻酸钠一步合成银纳米颗粒（A-SNPs）的方法，该方法可以稳定合成5~30nm的银纳米颗粒，将其冷冻干燥成粉末后依然能稳定分散在水中。此外，对比试验证实了A-SNPs比无表面活性剂的纯银纳米颗粒（n-NPs）有更高的稳定性与分散性，同时其对于植物病原真菌有更强的抑制作用。进一步研究发现，A-SNPs可以通过抑制柑橘炭疽病菌丝正常生理形态，破坏细胞膜通透性，抑制蛋白质的合成，破坏DNA结构与抑制DNA复制等来影响真菌的正常生长发育。同时，离体试验也发现了低浓度的A-SNPs能够有效抑制病菌侵染柑橘果实，较高浓度的A-SNPs能够完全抑制病菌侵染。最后，水稻和本氏烟种子萌发实验也证实了A-SNPs对植物无明显毒性，不会影响植物的正常生长发育。该A-SNPs具有良好的生物相容性与高效的抗植物病原微生物的能力，因此该合成银纳米颗粒的方法能够被用于植物保护研究领域中。

关键词：银纳米颗粒；醛基化海藻酸钠；绿色合成；抗真菌；抗菌机制

---

* 基金项目：国家自然科学基金（31670148，31870147）；重庆市社会事业与民生保障创新专项（cstc2016shmszx0368）；中国烟草公司重庆市公司科技项目（NY20180401070010，NY20180401070001，NY20180401070008）

** 第一作者：向顺雨，硕士研究生，从事植物病理学研究

*** 通信作者：孙现超，博士，研究员，博士生导师，主要从事植物病毒学及植物病害防控研究；E-mail：sunxianchao@163.com

烟草赤星病拮抗菌的筛选鉴定及拮抗机理初步研究*

谢中玉[1**]，李　斌[2]，李　晗[1]，汪代斌[3]，陈海涛[3]，
徐　宸[3]，张　帅[3]，陈德鑫[4]，孙现超[1***]
（1. 西南大学植物保护学院，重庆　400716；2. 中国烟草总公司四川省公司，
成都　610000；3. 中国烟草公司重庆市公司烟草科学研究所，重庆　400716；4. 中国
农业科学院烟草研究所，青岛　266101）

摘　要：从四川、贵州和山东的 8 个县市共采集健康烟株的根际土壤 40 份，用平板稀释法及平板划线分离法从土壤中分离、纯化出 224 株细菌菌株。通过平板对峙培养，筛选到 15 株抑制烟草赤星病菌（*Alternaria alternata*）活性较强的菌株，其抑菌率均能达到 40% 以上，其中 LZ88 抑菌活性最强，抑菌率达 72.22%。对菌株发酵滤液进行抑菌活性测定，筛选出 4 株对烟草赤星病菌有抑制作用的菌株。对 4 株菌株进行了生理生化鉴定及 16S rDNA、gyrA 基因序列分析，鉴定 4 株菌株均为芽孢杆菌属，分别为莫海威芽孢杆菌、解淀粉芽孢杆菌和枯草芽孢杆菌，其中菌株 WL58 和 AQ60 为解淀粉芽孢杆菌。菌株发酵液处理烟叶后测定其抗病相关酶活性，发现其可以显著提高 PPO 和 POD 酶的活性。同时为了更好地了解拮抗菌株所产抗菌物质的理化性质，利用硫酸铵沉淀法粗提蛋白，测定了粗提蛋白的抑菌活性及理化性质的研究，发现拮抗菌株产生的抗菌物质有较好的稳定性。温室盆栽防效表明菌株对烟草赤星病有一定的防效。通过本实验主要为生物防治提供更多的菌种资源，为生物防治技术的发展奠定良好的基础。

关键词：拮抗菌；烟草赤星病；拮抗机理；盆栽防效

---

* 基金项目：中国烟草公司四川省公司科技项目（SCYC201703）；中国烟草公司重庆市公司科技项目（NY20180401070010，NY20180401070001，NY20180401070008）

** 第一作者：谢中玉，硕士研究生，从事植物病理学研究

*** 通信作者：孙现超，博士，研究员，博士生导师，主要从事植物病毒学及植物病害控制研究；E-mail：sunxianchao@163.com

纳米氧化锌和纳米二氧化硅抗 TMV 活性及机制初步研究*

蔡 璘[1**]，刘昌云[1]，刘朝龙[1]，贾环宇[1]，丰 慧[1]，李 斌[2]，汪代斌[3]，陈海涛[3]，徐 宸[3]，陈德鑫[4]，孙现超[1***]

（1. 西南大学植物保护学院，重庆 400716；2. 中国烟草总公司四川省公司，成都 610000；3. 中国烟草公司重庆市公司烟草科学研究所，重庆 400716；4. 中国农业科学院烟草研究所，青岛 266101）

摘 要：烟草花叶病毒（*Tobacco mosaic virus*，TMV）是为害最严重的植物病毒之一。近年来，利用纳米技术在控制植物病原体侵染方面显示出较好应用前景。纳米氧化锌（ZnONPs）和纳米二氧化硅（SiO_2NPs）具有无毒、对人体细胞具有较好生物相容性和容易获得等优点，且能作为植物生长营养元素补充剂，在生产创新农药和提高作物产量方面具有显著发展潜力。本文合成了性质稳定的 ZnONPs 和 SiO_2NPs，并分析二者对 TMV 及其侵染、植物生长和抗氧化系统、抗病相关基因表达的影响，探究了植物对纳米材料的吸收和运输情况，从而综合分析 ZnONPs 和 SiO2NPs 的抗病毒活性及机制。结果表明，100μg/mL 的 ZnONPs 和 SiO_2NPs 与 TMV 体外混合孵育 2h，透射电子显微镜（TEM）观察到 TMV 粒子发生大量聚集和断裂；此混合液摩擦接种到本氏烟，接种后 2d 内 TMV 在接种叶片上的侵染和积累明显低于对照处理，但是接种后 7d TMV 的侵染量与清水对照无显著差异；而预先采用 100μg/mL ZnONPs 和 SiO_2NPs 喷施叶片，次日接种 TMV 的处理方式对接种后 7d 的 TMV 侵染积累仍有显著抑制作用。进一步的植物抗性相关酶和基因分析表明，两种纳米材料都能显著提高 POD 和 CAT 酶活性，并引起叶片活性氧积累；水杨酸（SA）信号通路相关基因 *PR*1 和 *PR*2 也显著高于对照；对应 SA 激素含量显示 ZnONPs 处理后显著升高，而 SiO_2NPs 处理则没有显著变化。TEM 超薄切片观察到经 100μg/mL ZnONPs 和 SiO_2NPs 处理的本氏烟植株能通过叶片吸收纳米材料，电感耦合等离子体质谱仪进一步证实这两种纳米材料能至上而下运输至植株根部；且处理后的植株干重和鲜重都显著高于清水处理。本研究明确了 ZnONPs 和 SiO_2NPs 对 TMV 的直接钝化作用，并且能诱导植物免疫抗性，研究结果为深入探索 ZnONPs 和 SiO_2NPs 诱导植物免疫抗性机理奠定基础，为利用纳米技术进行病害防治提供重要参考。

关键词：TMV；ZnONPs；SiO_2NPs；植物免疫

---

* 基金项目：国家自然科学基金（31670148，31870147）；中国烟草公司四川省公司科技项目（SCYC201703）；中国烟草公司重庆市公司科技项目（NY20180401070010，NY20180401070001，NY20180401070008）

** 第一作者：蔡璘，博士研究生，从事植物病理学研究

*** 通信作者：孙现超，博士，研究员，博士生导师，主要从事植物病毒学及植物病害防控研究；E-mail：sunxianchao@163.com

纤维素纳米晶（CNC）表面阳离子化抗辣椒疫霉活性研究*

向顺雨[1,3]**，廖舒悦[1]，施　焕[1]，袁梦婷[1]，樊光进[1]，
马冠华[1]，黄　进[2,3]，孙现超[1,3]***
（1. 西南大学植物保护学院，重庆　400615，2. 西南大学化学化工学院，
重庆　400615，3. 软物质材料化学与功能制造重庆市重点实验室，重庆　400615）

摘　要：纤维素纳米晶（CNC）作为一种理想的高分子纳米载体，已广泛应用于各个研究领域，然而，在植物保护和农业等研究领域中却鲜有报道。本研究将十六烷基三甲基溴化铵（CTAB）通过简单的静电作用吸附在纤维素纳米晶（CNC）表面而形成一种新型生物基纳米抗真菌材料（CNC@ CTAB）。用傅里叶变换红外仪、Zeta 电位仪、元素分析仪、X 射线衍射仪、AFM 和 TEM 对 CNC@ CTAB 进行了表征，证实 CTAB 成功富集在 CNC 表面。抗菌试验证实了 CNC@ CTAB 对辣椒疫霉具有较高的拮抗活性，且抗菌效果远高于纯 CTAB。对辣椒疫霉菌丝细胞膜通透性进行测定发现 CNC@ CTAB 相比于纯 CTAB 对细胞膜破坏性更强。进一步研究发现，CNC@ CTAB 表面 CTAB 包覆率越高，其抗菌活性越强。最终辣椒叶片离体试验证明了 CNC@ CTAB 具有较强的防治病害的能力。由于该材料具有合成方法简单，材料绿色环保，且具高效的抗菌活性，为合成新型绿色高效纳米抗菌材料提供了良好思路。

关键词：纤维素纳米晶；十六烷基三甲基溴化铵；辣椒疫霉。

---

* 基金项目：国家自然科学基金（31670148，31870147）；重庆市社会事业与民生保障创新专项（cstc2016shmszx0368）；中国烟草公司重庆市公司科技项目（NY20180401070010，NY20180401070001，NY20180401070008）

** 第一作者：向顺雨，硕士研究生，从事植物病理学研究

*** 通信作者：孙现超，博士，研究员，博士生导师，主要从事植物病毒学及植物病害防控研究；E-mail：sunxianchao@163. com

重庆涪陵烟区烟草棒孢霉叶斑病病原鉴定*

李　晗[1**]，吴　杰[2]，冉　茂[2]，汪代斌[2]，陈海涛[2]，
徐　宸[2]，李　斌[2]，陈德鑫[4]，孙现超[1***]
（1. 西南大学植物保护学院，重庆　400716；2. 中国烟草公司重庆市公司烟草科学研究所，重庆　400716；3. 中国烟草总公司四川省公司，成都　610000；4. 中国农业科学院烟草研究所，青岛　266101）

摘　要：自2016年烟草棒孢霉在重庆涪陵烟区首次报道以来，烟草棒孢霉叶斑病的发生有逐年扩展的趋势。为明确重庆烟区烟草棒孢霉叶斑病的病原，对田间采集病叶进行病原物分离获得三种病原菌。依据柯赫氏法则，将病原菌分别接种于活体‘云烟87’叶片上测定其致病性，只有一种真菌能成功侵染，并在叶片上呈现密集褐色斑点，分离得到与接种物一致的真菌。该菌在PDA培养基上菌落为深褐色，培养基表面大多形成毡毛状的菌丝层，中心菌丝凹陷外圈突出，边缘菌丝呈现红色，长时间培养菌落有红色分泌物。显微镜下观察菌丝分枝，有隔膜。分生孢子梗单支，单生或丛生，排列稀疏，大小为81~218（140±33）μm×4~9（6.7±1.0）μm。分生孢子常单生，或2~6个串生于顶端，分生孢子直立或稍弯，圆柱形或倒棍棒形，半透明至浅橄榄色。利用CTAB法提取致病菌的DNA，选取真菌转录间隔区通用引物ITS1/ITS4对该致病菌进行PCR扩增，获得测序结果在GenBank进行比对，确定病原为多主棒孢菌［*Corynespora cassiicola*（Berk. & Curt.）Wei］，又叫山扁豆棒孢，属半知菌亚门（deuteromycotina），丝孢目（hyphomycetales），棒孢属（*Corynespora* Güssow）。通过对病原物形态和分子鉴定，明确了重庆烟草棒孢霉叶斑病的病原物，为后续烟草棒孢霉叶斑病的绿色防控提供理论依据。

关键词：重庆；烟草棒孢霉病；病原；鉴定

* 基金项目：中国烟草公司四川省公司科技项目（SCYC201703）；中国烟草公司重庆市公司科技项目（NY20180401070010，NY20180401070001，NY20180401070008）

** 第一作者：谢中玉，硕士研究生，从事植物病理学研究

*** 通信作者：孙现超，博士，研究员，博士生导师，主要从事植物病毒学及植物病害控制研究；E-mail：sunxianchao@163.com

双重功能载药缓释水凝胶诱导植物抗烟草花叶病毒活性研究*

向顺雨[1**]，吕 星[1]，刘昌云[1]，施 焕[1]，李欣雨[1]，薛 杨[1]，汪代斌[2]，孙现超[1***]
（1. 西南大学植物保护学院，重庆 400615；2. 重庆市烟草科学研究所，重庆 400615）

摘 要：如何不断提高植物对系统性病害的抗性一直是研究的热点。本文报道了一种长效缓释诱导植物抗性与促进植物生长的双功能载药缓释水凝胶。其核心是把氨基寡糖通过静电作用在海藻酸钠-香菇多糖载药水凝胶（AL-hydrogel）表面形成一层氨基寡糖薄膜，合成出海藻酸钠-香菇多糖-氨基寡糖水凝胶（ALA-hydrogel）。本试验采用了红外光谱、zeta电位测试、扫描电镜和元素分析等方法去验证了氨基寡糖薄膜的形成。药物释放对比试验证实了ALA-hydrogel的缓释香菇多糖的能力更强，缓释速率更稳定，且缓释时间相比于AL-hydrogel至少延长了2倍。将ALA-hydrogel和AL-hydrogel埋入土壤中对比诱导植物抗TMV的能力。结果表明，ALA-hydrogel能够更显著的持续诱导植物抗性，能够更有效的抑制TMV侵染本氏烟与抑制TMV在本氏烟植物体中移动。同时，氨基寡糖薄膜的形成还可以促进AL-hydrogel释放钙离子，最终促进本氏烟（*N. benthamiana*）的生长。本研究突破了传统植物病毒病药剂的使用方式，通过根部施用载药水凝胶实现缓释诱抗药剂和释放营养元素来持续诱导植物抗TMV抗性，拥有潜在的农业生产运用的价值。

关键词：双重功能水凝胶；香菇多糖；氨基寡糖；TMV

---

* 基金项目：国家自然科学基金（31670148，31870147）；中国烟草公司重庆市公司科技项目（NY20180401070010，NY20180401070001，NY20180401070008）

** 第一作者：向顺雨，硕士研究生，从事植物病理学研究

*** 通信作者：孙现超，博士，研究员，博士生导师，主要从事植物病毒学及植物病害防控研究；E-mail：sunxianchao@163. com

菌株 S17-377 的鉴定及其对水稻纹枯病防治作用机制*

李雪婷**，郑树仁，聂倩文，刘　璐，郑通文，万伟杰，姜明明，孙正祥***，周　燚
（长江大学农学院，荆州　434025）

摘　要：从我国 9 个地区的 45 份水稻根际土样中分离筛选出 1 株对水稻纹枯病菌（*Rhizoctonia solani*）具有强抑菌作用的生防细菌 S17-377，经形态特征、16S rDNA 和 *recA* 序列鉴定为吡咯伯克霍尔德氏菌（*Burkholderia pyrrocinia*），在 GenBank 中登录号分别为 MK601672 和 MK603132。从诱导抗病作用、定殖动态和拮抗作用等方面研究了菌株 S17-377 的生防作用机制。结果显示，S17-377 发酵液对水稻植株具有较强的抗病诱导作用，处理后水稻叶鞘的 PAL 和 POD 活性呈上升趋势，PPO 活性先上升后下降。此外，S17-377 还可激活水稻防御基因（*NH*1、PR1a、PR10、*PAL/ZB*8、*LOX*、AOS_2）的表达。生防菌 S17-377 可产生蛋白酶、几丁质酶和嗜铁素，但不能产生纤维素酶。通过 GFP 标记获得标记菌株 S17-377GFP，转化后菌株 S17-377GFP 和原始菌株 S17-377 形态一致、生长差异不明显、遗传稳定性达 75%以上，且不影响其对水稻纹枯病菌的抑菌作用。通过荧光显微镜观察发现，菌株 S17-377GFP 可定殖于水稻叶表皮、叶肉细胞和维管束中。采用盐酸沉淀、硫酸铵沉淀及有机溶剂萃取法粗提生防菌株的活性成分，发现正丁醇粗提物、蛋白类粗提物和脂肽类粗提物（浓度为 50 μg/mL）对水稻纹枯病菌的的抑菌率分别为 95.23%、94.43% 和 75.11%，其中正丁醇粗提物对纹枯病菌的 EC_{50} 为 8.95μg/mL。通过 GC-MS 检测正丁醇提取物，发现粗提物中含有 38 种化合物，其中匹配度高达 90%以上的化合物有 7 种：9-甲基-十九烷、甘氨酸-L-脯氨酸、正十七烷、邻苯二甲酸二丁酯、棕榈酸、邻苯二甲酸二（2-乙基己）酯、烟碱。该研究表明，菌株 S17-377 是一株对水稻纹枯病具有良好防治作用的生防菌株，是水稻纹枯病绿色防控的有效菌种资源。

关键词：水稻纹枯病；生防细菌；防治机制；防效

---

* 基金项目：湖北省教育厅重点项目（D20181304）

** 第一作者：李雪婷，硕士，植物病害生物防治；E-mail：1258903674@qq.com

*** 通信作者：孙正祥，博士，副教授，主要从事植物病害生物防治研究；E-mail：sunzhengxiang9904@126.com

酸性电解水在蔬菜病害防治上的研究进展*

刘　琪[1**]，张　鑫[1]，韩成贵[1]，杨普云[2]，王　颖[1***]
（1. 中国农业大学植物病理学系，农业生物技术国家重点实验室，
北京　100193；2. 全国农业技术推广服务中心，北京　100125）

摘　要：电解水（Electrolyzed Water）又叫做电生功能水，由稀盐溶液在含隔膜的电解槽中电解产生，电解槽阳极产生酸性电解水（Acidic Electrolyzed Water，AEW），阴极产生碱性电解水（Basic Electrolyzed Water，BEW）。其中酸性电解水的 pH 值较低，氧化还原电位高，有效氯浓度高，这些特性决定了其在杀菌消毒功能上的特效，应用在植物生产上可以广谱地防治植物病害，部分替代化学农药的使用，绿色环保，切实“双减”。

酸性电解水在植物病害防治上效果相当明显，对国内各地典型的代表植物的典型病害都有相当的防治效果。在蔬菜中，利用强酸性电解水处理马铃薯晚疫病时，有明显的病害防治效果，并可部分代替农药的使用，还有可能对马铃薯起到增产的效果（郑虚等，2011）。酸性电解水处理茄子时对灰霉病的防效达到 68.71%（聂小凤等，2009），对茄子黄萎病也有较好的防效（李信，2018）。处理番茄时对叶霉病的防效达 64.17%（郑磊等，2009）。而处理莴苣时对灰霉病的防效可达 68.7%，用酸性电解水稀释正常用药量 2/3 的克得灵的防效可达 88.25%，显著高于单一用药（胡宇舟等，2018）。酸性电解水对黄瓜白粉病有防治效果（刘先辉等，2018），pH 值为 2 时对白粉病防治效果最好，高达 72.12%，有效防治期长达喷施后 3 d（罗欣希等，2015），且先喷酸性电解水后喷碱性电解水防效更好，高达 75.1%（魏肖鹏等，2015；魏肖鹏等，2016）；酸性电解水还对黄瓜霜霉病有效，效果与化学农药相当，且用于稀释化学农药可明显减少农药的使用（孙雄军等，2018）。

自电解水引进中国已有 30 年的历史，这 30 年内国内农业发生着翻天覆地的变化，而绿色、环保、经济、可持续一直是发展的主题，电解水在农业上的应用和这些主题更是无比切合。而就目前的形式而言，电解水的作用机制等相关研究一直是世界难题。因此，电解水的发展应科研与应用齐头并进，在最短的时间攻克一切难题，更快更好的造福全人类。

关键词：酸性电解水；蔬菜病害；绿色农业

---

* 基金项目：国家重点研发计划（2016YFD0300710-1）
** 第一作者：刘琪，硕士研究生，主要植物病害绿色防控的研究；E-mail：liuqi0725@cau.edu.cn
*** 通信作者：王颖，副教授，主要从事植物病毒与寄主互作研究；E-mail：yingwang@cau.edu.cn

甜瓜枯萎病菌对三种杀菌剂的敏感性研究*

吴思颖**，张小芳，华晖晖，吴学宏***
（中国农业大学植物保护学院，北京　100193）

摘　要：甜瓜在世界范围内广泛栽培，作为高档果品深受人们喜爱。由尖孢镰刀菌甜瓜专化型引起的甜瓜枯萎病，是甜瓜上的重要真菌病害，在甜瓜的整个生育期都可以发生。目前，使用化学杀菌剂仍是生产上防治这种病害的重要手段。麦角甾醇生物合成抑制剂如戊唑醇、甲氧基丙烯酸酯类如嘧菌酯、苯并咪唑类如多菌灵、芳杂环异恶唑类如噁霉灵、苯并吡咯类如咯菌腈等对尖孢镰刀菌有较好的抑制效果。有研究证明，尖孢镰刀菌对咪鲜胺和咯菌腈的敏感性尤其显著。本研究从全国19个城市35个地点采集甜瓜枯萎病病害样品，分离鉴定得到135株尖孢镰刀菌甜瓜专化型。采用菌丝生长速率法测定其对咪鲜胺、戊唑醇和苯醚甲环唑的敏感性。结果表明，135株尖孢镰刀菌甜瓜专化型对三种杀菌剂均有一定的敏感性，并且随着杀菌剂浓度的升高菌株的敏感性增加。咪鲜胺对供试菌株的抑制效果最为显著，平均抑制中浓度（EC_{50}）为0.042 5 μg/mL，戊唑醇次之，平均EC_{50}为0.434 1 μg/mL，苯醚甲环唑的平均EC_{50}最高，为0.963 3 μg/mL。尖孢镰刀菌甜瓜专化型对三种杀菌剂的交互抗性分析表明，苯醚甲环唑与咪鲜胺呈现中度正交互抗性（咪鲜胺与苯醚甲环唑的相关系数为0.539），戊唑醇与苯醚甲环唑或咪鲜胺存在较弱的正交互抗性（咪鲜胺与戊唑醇的相关系数为0.157，苯醚甲环唑与戊唑醇的相关系数为0.374）。

关键词：甜瓜枯萎病；尖孢镰刀菌甜瓜专化型；杀菌剂；敏感性；交互抗性

* 基金项目：国家公益性行业（农业）科研专项（编号：201503110-14）

** 第一作者：吴思颖，硕士研究生，主要从事甜瓜枯萎病菌对杀菌剂的敏感性及其真菌病毒的研究；E-mail：799650234@qq.com

*** 通信作者：吴学宏，教授，主要从事植物病原真菌种类鉴定及其遗传多样性研究；E-mail：wuxuehong@cau.edu.cn

西瓜枯萎病菌对 3 种杀菌剂的敏感性测定*

韩 涛**，陈垦西，梁佳媛，吴学宏***

（中国农业大学植物保护学院，北京 100193）

摘 要：西瓜是世界重要的经济作物之一，由尖孢镰刀菌西瓜专化型（*Fusarium oxysporum* f. sp. *niveum*）引起的西瓜枯萎病对西瓜的产量和品质均造成严重影响。防治西瓜枯萎病的方法有多种，如抗病育种、农业防治（包括嫁接等）、生物防治、物理防治、化学防治等，其中施用化学杀菌剂在控制西瓜枯萎病对西瓜的产量和品种所造成的影响方面仍起到了重要的作用。本研究从来自我国 17 个省、市或自治区的西瓜枯萎病病害样品上分离得到尖孢镰刀菌西瓜专化型 177 株。采用菌丝生长速率法测定了其对苯醚甲环唑、戊唑醇和咪鲜胺的敏感性。三种杀菌剂对供试菌株均有明显的抑制作用，对供试菌株的平均 EC_{50} 值均小于 5 μg/mL，且浓度越高抑制效果越明显，但供试菌株对这 3 种杀菌剂的敏感性存在明显差异。苯醚甲环唑、戊唑醇和咪鲜胺对尖孢镰刀菌西瓜专化型的平均 EC_{50} 值分别为 1.099 1 μg/mL、0.456 7 μg/mL 和 0.049 2 μg/mL。以 3 种杀菌剂对供试菌株的 EC_{50} 的 5%修正均值作为尖孢镰刀菌西瓜专化型对 3 种杀菌剂的敏感基线，分别为 0.662 7 μg/mL、0.293 7 μg/mL 和 0.026 6 μg/mL。但是苯醚甲环唑对五株尖孢镰刀菌西瓜专化型的 EC_{50} 值大于 5 μg/mL，表现出敏感性有所降低，其 EC_{50} 值介于 7.494 7~11.341 1 μg/mL，这 5 株菌是否对苯醚甲环唑产生抗性还有待进一步研究。

关键词：西瓜枯萎病；尖孢镰刀菌西瓜专化型；杀菌剂；敏感基线

---

* 基金项目：国家公益性行业（农业）科研专项（编号：201503110-14）

** 第一作者：韩涛，硕士研究生，主要从事西瓜枯萎病菌对杀菌剂的敏感性及其真菌病毒的研究；E-mail：1131140436@qq.com

*** 通信作者：吴学宏，教授，主要从事植物病原真菌种类鉴定及其遗传多样性研究；E-mail：wuxuehong@cau.edu.cn

引起甜瓜叶部病害的链格孢菌对3种杀菌剂的敏感性*

刘 泉**，李郁婷，陈垦西，吴学宏***
（中国农业大学植物保护学院，北京 100193）

摘 要：甜瓜是一种重要的园艺作物。由链格孢菌（*Alternaria* spp.）引起的甜瓜叶部病害是甜瓜生产上重要的真菌病害之一，可使甜瓜含糖量降低，产量下降，严重影响甜瓜生产的经济效益。目前，在缺乏优良抗病品种的情况下，化学防治仍是控制该病害的主要手段。本研究于2016—2017年，从全国13个省、直辖市和自治区的甜瓜产区采集叶部病害样品，利用组织分离法和单孢纯化法得到链格孢菌，通过形态学观察和分子生物学方法明确引起甜瓜叶部病害的病原菌种类为交链格孢（*A. alternata*）和细极链格孢（*A. tenuissima*）。选取201株链格孢菌测定其对啶酰菌胺、苯醚甲环唑和咯菌腈的敏感性；结果表明，啶酰菌胺、苯醚甲环唑和咯菌腈对链格孢菌均有一定的抑制作用，且随着杀菌剂浓度的升高，抑制作用增强；3种杀菌剂对供试菌株的平均EC_{50}值分别为1.632 8 μg/mL、0.461 7 μg/mL和0.149 1 μg/mL，在$P<0.05$水平上呈显著性差异。其中啶酰菌胺对*A. tenuissima*和*A. alternata*的平均EC_{50}值分别为1.656 7 μg/mL和1.572 5 μg/mL，无显著性差异（$P>0.05$）；苯醚甲环唑对*A. tenuissima*和*A. alternata*的平均EC_{50}值分别为0.405 8 μg/mL和0.603 0 μg/mL，差异显著（$P<0.05$）；咯菌腈对*A. tenuissima*和*A. alternata*的平均EC_{50}值分别为0.112 0 μg/mL和0.149 1 μg/mL，差异显著（$P<0.05$）。三种杀菌剂中咯菌腈的毒力最强，苯醚甲环唑次之，啶酰菌胺略差。本研究结果可为生产中合理使用杀菌剂提供一定的理论依据。

关键词：甜瓜叶部病害；链格孢菌；杀菌剂；敏感性

* 基金项目：北京市西甜瓜产业创新团队（BAIC10-2019）
** 第一作者：刘泉，硕士研究生，主要从事引起甜瓜叶部病害的链格孢菌对杀菌剂的敏感性及其真菌病毒的研究；E-mail：2235262089@qq.com
*** 通信作者：吴学宏，教授，主要从事植物病原真菌种类鉴定及其遗传多样性研究；E-mail：wuxuehong@cau.edu.cn

防治瓜类细菌性果斑病的小分子化合物筛选研究*

芦 钰[1]**，Loic Deblais[2,3]，Gireesh Rajashekara[2]，
Sally A. Miller[3]，张海军[1]，吴 萍[1]，徐秀兰[1]***

（1. 北京市农林科学院蔬菜研究中心，农业部华北地区园艺作物生物学与种质创制重点实验室，农业部都市农业（北方）重点实验室，北京 100097；2. *Food and Animal Health Research Program*，*The Ohio State University*，*Ohio Agricultural Research and Development Center*，*Wooster*，*OH*，*USA*，44691；3. *Department of Plant Pathology*，*The Ohio State University*，*Ohio Agricultural Research and Development Center*，*Wooster*，*OH*，*USA*，44691）

摘 要：西瓜嗜酸菌（*Acidovorax citrulli*）是引起瓜类细菌性果斑病的重要种传细菌，对瓜类作物生产造成巨大经济损失，种子处理是预防和控制果斑病的关键环节。本研究利用高通量小分子化合物筛选技术，对 4 个小分子化合物库共计4 952个小分子化合物进行靶标菌的生长抑制初步筛选。初次筛选小分子化合物浓度为 100 μmol/L，以 *A. citrulli* 野生型菌株 Xu 3-14 为靶标，共筛选出具有致死和完全抑制效果的小分子化合物 127 个（选中率为 2.5%）。二次筛选发现部分化合物在对番茄细菌性溃疡病菌（*Clavibacter michiganensis* subsp. *michiganensis*）和十字花科黑腐病菌（*Xanthomonas campestris* pv. *campestris*）保持抑制作用的同时，对植物有益菌如荧光假单胞菌（*Pseudomonas fluorescens*）和枯草芽孢杆菌（*Bacillus subtilis*）的生长无显著影响。50%以上的化合物对拟南芥和西瓜没有表现植物毒性。综合分析二次筛选中小分子化合物的特异性、敏感性和植物毒性的评估结果，以及小分子化合物对 *A. citrulli* 种子到种苗传带抑制率的实验室测试结果，选出了 9 个化合物作为种子处理药剂对人工接菌的西瓜、甜瓜种子进行田间测试。最终选出 5 个显著抑制 *A. citrulli* 的小分子化合物，防病效果最高可达到95%。本研究筛选出的 5 种小分子化合物为 *A. citrulli* 特异性杀菌剂的开发提供了参考。

关键词：*Acidovorax citrulli*；瓜类细菌性果斑病；小分子化合物；高通量筛选

---

* 基金项目：北京市农林科学院科技创新能力专项（KJCX20180203）；北京市农林科学院院专项（yzx002）

** 第一作者：芦钰，在读硕士研究生；E-mail：luyucau@ 126. com

*** 通信作者：徐秀兰，副研究员；E-mail：xuxiulan@ nercv. org

12 种杀菌剂对禾谷镰刀菌（*Fusarium graminearum*）的毒力测定*

周　锋[1,2**]，范玉闯[1]，宋雨露[1]，李　帅[1]，高育青[1]，袁　康[1]，郭含雪[1]，闫红飞[3***]

（1. 河南科技学院资源与环境学院，新乡　453003；2. 河南省粮食作物基因组编辑工程技术研究中心，新乡　453003；3. 河北农业大学植物保护学院，保定　071001）

摘　要：由禾谷镰刀菌（*Fusarium graminearum*）侵染小麦引起的小麦赤霉病是当前小麦生产上的一种重要真菌病害。因尚未选育出能够有效抵抗禾谷镰刀菌的抗病品种，当前对其主要以化学药剂防治为主，但由于多频次及不科学的使用，导致当前生产上已有病原菌对部分杀菌剂产生抗药性的报道。为了进一步明确中部小麦主产区的禾谷镰刀菌对常用杀菌剂的敏感性，本研究采用菌丝生长速率法测定了 2018 年从中部小麦主产区分离到的 9 株禾谷镰刀菌（CM2，YX2-1-4，2xZ-4，30-30，HBXT2-2，2-1-5，CM1，SQ1-2 和 YN1-3-2）对 12 种常用杀菌剂的敏感性。结果表明：咯菌腈、啶酰菌胺、戊唑醇及吡唑醚菌酯对上述 9 株禾谷镰刀菌均表现出较强的抑制作用；其中，咯菌腈的 EC_{50}值分布范围为 0.012 6~0.071 6μg/mL，啶酰菌胺的 EC_{50}值分布范围为 0.077 2~0.417 1μg/mL，戊唑醇的 EC_{50}值分布范围为 0.083 4~0.740 2μg/mL，吡唑醚菌酯的 EC_{50}值分布范围为 0.251 6~0.573 1μg/mL；腐霉利、氟吡菌酰胺、苯醚甲环唑、菌核净、氟啶胺及嘧菌环胺次之；但其对多菌灵已经产生了较高水平的抗性。本研究为指导中部小麦主产区对小麦赤霉病的化学药剂防治提供参考，并为后期进一步开展禾谷镰刀菌对上述杀菌剂的抗药性机理研究提供实验材料。

关键词：禾谷镰刀菌；杀菌剂；毒力测定

---

* 基金项目：河南省科技攻关计划项目（192102110056）；校高层次人才科研项目（2018022）

** 第一作者：周锋，讲师，主要从事病原菌抗药性分子机制方面的研究；E-mail：zfhist@163.com

*** 通信作者：闫红飞，副教授，主要从事分子植物病理学方面的研究；E-mail：hongfeiyan2006@163.com

我国小麦白粉病防控中的品种抗性布局研究*

刘美玲，龚双军，曾凡松，史文琦，向礼波，薛敏峰，喻大昭，杨立军**
（湖北省农业科学院植保土肥所，农业部华中作物有害生物综合治理重点实验室，
农作物重大病虫草害防控湖北省重点实验室，武汉　430064）

摘　要：针对我国小麦白粉病菌越夏、不同生态麦区病菌毒性结构以及传播和流行规律，在白粉病越夏区与流行区之间、流行区与流行区之间合理布局种植不同的抗性品种可有效降低白粉病区域之间的传播流行。本研究在前期病圃鉴定筛选了有一定种植面积且抗性相对较好的 30 份品种，用采自我国小麦白粉病西南越夏区（云贵、川）、西北越夏区（陕甘）以及春季流行区长江中下游麦区（鄂、皖苏）、黄淮海麦区（豫鲁）、华北麦区（冀）等 121 个代表性菌株组成的 7 个菌群分别接种鉴定。结果表明，来自西南麦区的‘云麦 53’、‘贵农 775’，‘内麦 8 号’、‘内麦 9 号’、‘国豪麦 5 号’、‘国豪麦 6 号’等 6 个品种对所测定的 7 个菌群均抗性优异，可在西南白粉病越夏区种植以压低白粉病原基数、减少初侵染源；来自长江中游麦区的‘鄂麦 19’、‘襄麦 25’和长江下游的‘宁麦 13’和‘安农 1124’对云贵川和陕甘麦区菌群的抗性频率高于对当地菌群的抗性频率，在长江中下流麦区种植这 4 个品种可有效阻断和降低西南和陕甘麦区传入长江中下游麦区菌源的为害；来自于黄淮麦区的‘偃展 4110’、‘新麦 208’、‘豫麦 18-99’和‘矮抗 58’对云贵川、陕甘、湖北和皖苏麦区菌群抗性频率均高于黄淮海麦区的菌群，在黄淮海麦区种植这些品种可减少从西南云贵川、西北陕甘以及长江中下游麦区向黄淮海麦区菌源的传播；来自陕甘麦区的‘兰天 17 号’、‘兰天 22 号’和‘兰天 30’对长江中下游和黄淮海麦区菌群抗性频率要明显高于陕甘麦区菌群的抗性频率，表明在陕甘麦区种植这些品种可阻止长江中下游和黄淮海麦区的菌源向陕甘麦区传播。本研究结果为利用抗性品种合理布局防控小麦白粉病提供了依据。

关键词：小麦白粉病；品种布局；抗性分化；防控

* 基金项目：公益性行业科研专项（201303016）；湖北省农业科技创新项目（2016-620-000-001--15）；国家小麦产业技术体系（CARS-03-04B）；湖北省农科院领军人才项目（L2018013）

** 通信作者：杨立军，研究员；E-mail：yanglijun1993@163.com

18 种杀菌剂对咖啡炭疽病菌的毒力测定*

吴伟怀[1**]，余易兰[2]，朱孟烽[1]，梁艳琼[1]，陆　英[1]，
郑金龙[1]，黄　兴[1]，李　锐[1]，贺春萍[1]，易克贤[1***]
（1. 中国热带农业科学院环境与植物保护研究所，农业农村部热带作物有害生物综合治理重点实验室，海南省热带农业有害生物监测与控制重点实验室，海口　571101；
2. 海南大学热带农林学院，海口　570228）

摘　要：咖啡炭疽病是我国咖啡上常见病害，在海南、云南等地均有发生。本研究通过菌丝体生长速率法测定 18 种杀菌剂对咖啡炭疽病菌（*Colletotrichum gloeosporioid* Penz.）的毒力，以了解咖啡炭疽病菌对杀菌剂的敏感性。根据 EC_{50}值、EC_{95}值的实验结果可知，18 种供试药剂对咖啡炭疽病菌的毒力存在明显差异。其中咪鲜胺锰盐（50%可湿性粉剂）、戊唑醇（43%悬浮剂）、嘧菌酯（30%悬浮剂）等 3 种药剂其对应 EC_{50}值均小于 1μg/mL，介于 0.029~0.791 μg/mL；3 种药剂其对应 EC_{95}值分别为 2.886μg/mL、84.412μg/mL 和 30.499μg/mL，均小于 100 μg/mL。因此咪鲜胺锰盐、戊唑醇、嘧菌酯等 3 种药剂对咖啡炭疽病菌菌丝体生长的抑制作用最好。就炭疽病菌菌株对 18 种供试药剂敏感性而言，毒力回归方程中咪鲜胺锰盐的 b 值最大，其次是嘧菌酯的。由此表明，咖啡炭疽病菌对咪鲜胺锰盐和嘧菌酯的剂量反应变化比其他几种杀菌剂要敏感。

关键词：咖啡树；炭疽病菌；杀菌剂；毒力

---

* 基金项目：国家重点研发项目“特色经济作物化肥农药减施技术集成研究与示范”（2018YFD0201100）
** 第一作者：吴伟怀，副研究员，研究方向：植物病理；E-mail：weihuaiwu2002@163.com
*** 通信作者：易克贤，博士，研究员，研究方向：热带作物真菌病害及其抗性育种；E-mail：yikexian@126.com

大分子季铵盐对水稻纹枯病菌的抑菌特性研究*

钟伟强[1**]，古广武[2]，林雅铃[2***]，张安强[1***]
（1. 华南理工大学材料科学与工程学院，广州 510641；
2. 华南农业大学材料与能源学院，广州 510642）

摘　要：由立枯丝核菌（*R. solani*）引起的水稻纹枯病（rice sheath blight）是我国水稻三大病害之一。近年来，由于矮秆品种的推广和栽培水平的提高，水稻纹枯病的发生呈加重趋势。目前，水稻纹枯病的防治主要以使用井冈霉素为主，但从多年的使用情况来看，对水稻纹枯病的防效仅 50%~60%，且持效期短，需要频繁的田间施药，由此带来大量的劳力用工成本及抗药性问题。因此，研究开发新的防治药剂对于水稻的安全生产具有重要意义。菌丝与菌核是立枯丝核菌侵染水稻的两种形式，尤其是菌核作为立枯丝核菌（*R. solani*）在无性繁殖中的一种特殊形态，因其特殊的内外层结构而具有超强的生存能力。大分子季铵盐作为一种阳离子抗菌剂，不仅具有广谱抗菌性，能对立枯丝核菌进行有效抑制，还具有结构可控的优点，可通过调节亲疏水基团增强其吸附特性和渗透特性，从而对立枯丝核菌菌核形成有效的吸附及渗透，达到抑制菌丝与菌核萌发、阻断立枯丝核菌侵染循环的效果。本研究首先合成了一系列分子量不同的聚丙烯酸酯类季铵盐（PDMAEMA-BC），探讨分子量对立枯丝核菌抑菌活性的影响。结果表明，PDMAEMA-BC 的抗菌活性优于相应的单体（即：丙烯酸酯季铵盐，DMAEMA-BC），且在较小的聚合度下即可获得最佳的抗菌活性；在此基础上，进一步合成了一系列具有两亲性的聚丙烯酸酯季铵盐-聚硅氧烷两嵌段聚合物（PDMS-*b*-QPDMAEMA）。当聚丙烯酸酯季铵盐链段和聚硅氧烷链段的分子量都为 5000 的时候，两亲性大分子季铵盐获得了较优的吸附与渗透能力，对立枯丝核菌菌核表现出了良好的吸附、渗透以及抑制萌发的能力，在 400 μg/mL 的浓度下对立枯丝核菌菌核的萌发抑制率达到了 85%。上述研究结果为进一步研发可根治水稻纹枯病的大分子药物及其防治方法提供了有益的研究思路和良好的理论基础。

关键词：大分子季铵盐；两亲性；水稻纹枯病菌；菌核

---

* 基金项目：广州科技计划项目（201704020084，201803020015）；国家自然科学基金（31772202）

** 第一作者：钟伟强，博士研究生，研究方向为大分子季铵盐的合成及其在植物病害化学防治中的应用；E-mail：1152239907@qq.com

*** 通信作者：林雅铃，副教授，研究方向为大分子季铵盐的合成及其在植物病害化学防治中的应用；E-mail：linyaling@scau.edu.cn

张安强，教授，研究方向为大分子季铵盐的合成及其在植物病害化学防治中的应用；E-mail：aqzhang@scut.edu.cn

两亲性大分子季铵盐对香蕉枯萎病菌抑制作用*

常瑶瑶[1**]，钟伟强[2]，古广武[1]，张安强[2***]，林雅铃[1 ***]

（1. 华南农业大学材料与能源学院，广州 510642；
2. 华南理工大学材料科学与工程学院，广州 510641

摘 要：香蕉枯萎病是一种严重的土传病害，由尖孢镰刀菌古巴专化型（*Fumrium oxysporum* f. sp. *cubense*，Foc）侵染引起，其中4号生理小种（Foc4）几乎能侵染所有的香蕉，最具毁灭性。现有的防治方法多有不足，针对Foc4病原菌孢子能够长期在土壤中生存的特性，本课题组合成了一类高效低毒、具有两亲性的大分子季铵盐（PDMS-*b*-QPDMAEMA），研究了大分子季铵盐对Foc4的抑制作用和抑菌机理，大分子季铵盐在土壤中的吸附与迁移性能及其在土壤中的抑菌效果。结果表明：与亲水性的季铵盐均聚物（QPDMAEMA）相比，疏水嵌段的加入更易促使两亲性大分子季铵盐吸附到细胞表面，增强其抗菌性能；与小分子季铵盐（苯扎氯铵，BC）相比，PDMS-*b*-QPDMAEMA不仅可破坏Foc4细胞壁和细胞膜的完整性，导致细胞内容物渗漏，还会使细胞膜发生脂质过氧化，影响线粒体和基因组DNA的正常功能，导致Foc4死亡；PDMS-*b*-QPDMAEMA在土壤中易吸附、难淋溶；大分子季铵盐在土壤中对Foc4的抑菌性能受多种因素的影响，在适宜条件下对新增的Foc4病原菌孢子仍可保持60 d的抑菌能力。两亲性大分子季铵盐对Foc4的特殊抑菌机理及其对土壤中Foc4孢子的长效抑制特性，使其有望在香蕉枯萎病防治方面获得应用。

关键词：两亲性大分子季铵盐；香蕉枯萎病菌；抑菌机理；吸附淋溶

---

* 基金项目：广州科技计划项目（201704020084和201803020015）；国家自然科学基金（31772202）

** 第一作者：常瑶瑶，硕士研究生；E-mail：1129859132@ qq. com

*** 通信作者：张安强，教授，研究方向为大分子季铵盐的合成及其在植物病害化学防治中的应用；E-mail：aqzhang@ scut. edu. cn

林雅铃，副教授，研究方向为大分子季铵盐的合成及其在植物病害化学防治中的应用；E-mail：linyaling@ scau. edu. cn

生物防治在杧果病害中的研究与展望*

叶　子**，苏初连，夏　杨，刘晓妹，蒲金基，张　贺***

（中国热带农业科学院环境与植物保护研究所，农业农村部热带作物有害生物综合治理重点实验室，海南省热带农业有害生物监测与控制重点实验室，海口　571101）

摘　要：杧果因其果形美观、营养丰富、风味独特而被誉为“热带水果之王”。杧果生长过程中容易受到多种病害的影响，其中侵染性病害 88 种，病原物达 113 种之多。为了保障杧果的食用安全和生态环境安全，近年来许多科研工作者和生产者着重对杧果病害的生物防治方面进行了研究。生物防治技术在杧果病害方面的研究已日益成熟，涉及炭疽病等多个重要病害的生物防治，并有越来越多的研究证明了生物防治的可行性。杧果病害的生物防治技术主要包括植物提取物、生防微生物（细菌、酵母菌、放线菌等及其代谢产物）和诱抗剂的筛选、鉴定、应用等方面。随着生物防治不断的深入研究以及部分生防菌的批量生产与应用，生物防治必将成为杧果病害防治中的重要防治方法之一。

关键词：杧果病害；生物防治；生防菌

---

* 基金项目：国家重点研发专项（2017YFD0202100）；海南省重大科技计划项目（ZDKJ2017003）；中国热带农业科学院基本科研业务费专项资金（1630042017019）；农业农村部农业技术试验示范与服务支持项目“滇桂黔石漠化地区特色作物产业发展关键技术集成示范”

** 第一作者：叶子，硕士研究生，研究方向：植物保护；E-mail：675863094@ qq. com

*** 通信作者：张贺，硕士，助理研究员，研究方向：热带果树病理学；E-mail：atzzhef@ 163. com

河南省小麦赤霉病研究概况与绿色防控关键技术*

于思勤[1]，马忠华[2]，张　猛[3]
（1. 河南省植保植检站，郑州　450002；2. 浙江大学生物技术研究所，
杭州　310058；3. 河南农业大学植物保护学院，郑州　450002）

小麦赤霉病是典型的气候型病害，小麦扬花期遇连阴雨、结露、多雾等气候条件及田间湿度大有利于该病发生，赤霉病不仅造成小麦减产，而且病麦中含有镰刀菌代谢产生的致呕毒素等，严重影响小麦品质。

1　河南省小麦赤霉病研究概况

1.1　小麦赤霉病发生趋势

小麦赤霉病在20世纪70年代以前仅在河南省零星发生，80年代后发生频率增加，1989—1999年，发生面积达70万 hm^2 以上的中度流行年份2年，流行频率为18.2%。进入21世纪以来，赤霉病逐渐成为河南省小麦常发性病害，2000—2019年，中度以上流行年份12年，流行频率为60.0%，其中2012年、2016年、2018年大流行。豫南地区小麦赤霉病发生较重。

1.2　病残体带菌率及相关气象因子

小麦赤霉病菌主要依靠其有性子囊菌玉蜀黍赤霉菌越冬，成为第二年主要侵染来源。近3年研究结果证明，河南省麦田病残体平均带菌率在10%以上，带菌量完全能够满足小麦赤霉病大发生的需要。小麦抽穗扬花期遇到连阴雨天气条件是引起赤霉病流行的关键因素。

1.3　致病菌种类及其抗药性

2016—2018年，从河南省有关县（市）提供的病残体及病穗上分离到1 708个菌株，抗药性检测结果表明，固始县、淮滨县、息县、光山县、平桥区、唐河县、平舆县等地小麦赤霉病菌对多菌灵产生了不同程度的抗性，其中豫南麦区的固始县、淮滨县、息县、平桥区等地赤霉病病菌对多菌灵产生抗性菌株的平均比例达10.99%。

1.4　主栽小麦品种赤霉病自然发病调查

2018年河南省小麦田赤霉病病残体带菌率高、小麦扬花期间雨水偏多，田间湿度大，有利于病原菌的侵染和扩展蔓延，导致赤霉病偏重发生。2019年5月中旬调查43个主栽品种自然发病情况，结果表明，‘郑麦9023’、‘郑麦0943’、‘西农529’、‘西农9718’、‘西农979’、‘扬麦13’、‘扬麦15’、‘新麦21’、‘农大1108’等品种发病相对较轻。

1.5　不同杀菌剂对小麦赤霉病的防治效果

2016—2019年的药效试验结果表明，在小麦齐穗至扬花期，使用0.3%四霉素水剂、25%氰烯菌酯、40%丙硫唑、20%叶菌唑悬浮剂、40%丙唑·戊唑醇悬浮剂均能有效防治小麦赤霉病。

2　小麦赤霉病绿色防控关键技术

做好小麦赤霉病防控，必须坚持“预防为主、综合防治”植保方针，落实“政府主导、属地责任、联防联控”工作机制，大力推行小麦全生育期赤霉病绿色防控关键技术。

* 基金项目：河南省小麦产业技术体系（S2015-01-G08）；国家公益性行业专项（201303030）

2.1 选用抗病品种

在豫中南小麦赤霉病常发区，选择种植发病相对较轻的品种。

2.2 农业防治

推广小麦与花生、大豆、蔬菜等作物轮作，压低菌源基数；前茬秸秆要充分粉碎，深翻掩埋，减少土表裸露的病残体数量；适期适量播种，科学管理肥水，防止小麦群体过大和通风透光不良，促进小麦健壮生长，提高抗逆能力。

2.3 生物防治

在绿色和有机小麦生产基地，选用四霉素、井冈·蜡芽菌在扬花初期和末期喷施。

2.4 科学用药

选用戊唑醇、苯醚甲环唑等药剂处理种子，预防赤霉病造成的苗枯；小麦返青拔节期，使用内吸性杀菌剂预防赤霉病造成的茎基腐；小麦抽穗以后，根据赤霉病菌子囊壳发育进度、扬花期的气候条件等因素综合分析，在扬花初期和末期及时喷施氰烯菌酯、戊唑醇、戊唑·咪鲜胺、丙硫唑等与多菌灵不同类型的化学药剂，重点喷施小麦穗部，减轻穗腐和秆腐发生程度，减少小麦产量损失，降低毒素污染风险。

几种非化学药剂对韭菜灰霉病菌的室内毒力测定*

刘　梅**，赵亚林，李亚萌，黄金宝，周　莹，张　玮
（北京市农林科学院植物保护环境保护研究所，北京　100097）

摘　要：韭菜灰霉病是北方地区露地、保护地韭菜生产中一种常见的真菌病害，一般在每年的12月初发病，属于低温高湿型病害，通常引起叶片白斑、干尖以及湿腐等症状，随着韭菜的生长发育，病情逐渐加重，并可延续到翌年的2—3月，造成产量损失，严重时可减产30%以上，韭菜的食用性和商品性也受到极大影响。经对北京地区保护地韭菜灰霉病菌进行鉴定，发现主要为葱鳞葡萄孢霉（*Botrytis squamosal*），并对其生物学特性进行了研究。

生产中使用化学药剂仍是防治韭菜灰霉病的常用手段，目前登记的防治韭菜灰霉病的药剂主要是腐霉利，另外嘧霉胺、百菌清、多菌灵等药剂也被用于灰霉病防治。由于化学药剂的频繁超量使用以及灰霉病菌极易产生变异等因素，导致病原菌抗药性问题加重，田间防效不稳定，腐霉利等农残超标问题频现。为减少化学药剂的使用，延缓病原菌抗药性产生，本研究选择了枯草芽孢杆菌、寡雄腐霉、丁子香酚等11种非化学药剂进行了韭菜灰霉病菌的室内毒力测定，以明确这些药剂的抑菌活性，为筛选出有效的防治韭菜灰霉病菌的药剂以及减少化学农药的使用提供依据。采用生长速率法进行测定，抑菌效果最好的为多抗霉素和枯草芽孢杆菌，EC_{50}值为0. 004 06 μg/mL 、0. 007 592 μg/mL，其次是寡雄腐霉、蛇床子素，EC_{50}值为0. 273 69 μg/mL、0. 752 74 μg/mL，柠檬烯、苦参碱、香菇多糖的抑菌效果次之，EC_{50}值分别为1. 504 71 μg/mL、2. 391 77 μg/mL和5. 935 48 μg/mL，阿泰灵、大黄素甲醚和哈茨木霉的抑菌效果较差，EC_{50}值分别为88. 808 μg/mL、116. 852 μg/mL和202. 446 μg/mL。试验结果表明，多抗霉素、枯草芽孢杆菌以及寡雄腐霉、蛇床子素等在室内对韭菜灰霉病菌具有良好的抑菌效果，可在田间进行示范验证，筛选出几种高效安全的非化学药剂以实现化学药剂的替代，促进韭菜的品质提升。

关键词：韭菜灰霉病菌；非化学药剂；毒力测定

* 基金项目：国家重点研发计划（2016YED0201000）
** 第一作者：刘梅，助理研究员，从事植物病害综合防治；E-mail：liumeidmw@ 163. com

不同浓度诱抗剂诱导香蕉抗褐缘灰斑病的田间效果*

漆艳香，谢艺贤，丁兆建，曾凡云，彭 军，张 欣
（中国热带农业科学院环境与植物保护研究所，海口 571101；农业农村部热带作物有害生物综合治理重点实验室，海口 571101；海南省热带作物病虫害生物防治工程技术研究中心，海口 571101）

摘 要：球腔菌属（*Mycosphaerella*）的多种病原菌可引起香蕉叶斑病（又称香蕉褐缘灰斑病或香蕉尾孢菌叶斑病），是香蕉生产上常发性重要病害之一。该病在世界香蕉产区普遍发生，叶片受害率一般约 30%，严重时高达 80%，可造成产量损失高达三成以上，严重影响香蕉产量与品质。因此，对香蕉褐缘灰斑病的防控是香蕉生产中需要解决的重要问题。目前生产上尚无高抗或免疫褐缘灰斑病的香蕉品种，防治上仍以化学防治为主，但长期使用同类杀菌剂，已使病原菌对一些杀菌剂表现出抗药性，即使加大药剂用量，防效也不甚理想，还面临农药残留和环境污染等问题。近年来，出于对环境及食品安全的考虑，人们正在有意识的减少使用杀菌剂，而其替代品之一的诱抗剂逐渐成为研究热点。

在参考前人研究成果和文献报道的基础上，本试验以 25%丙环唑乳油 1 000倍液与清水分别作为药剂对照和空白对照，采用田间小区试验，比较了 5%氨基寡糖素水剂、20%异噻菌胺悬浮剂、禾之壮有机水溶肥 3 种诱抗剂不同浓度处理对香蕉褐缘灰斑病的诱导抗病作用。结果表明，供试诱抗剂各浓度处理的病情指数显著低于清水对照，对香蕉褐缘灰斑病均有较好的诱导效果，其中 5%氨基寡糖素水剂以浓度为 600 倍液时诱导效果最好，第 3 次药后 7 d 达 69. 19%；20%异噻菌胺悬浮剂以浓度为4 000倍液时诱导效果最好，第 3 次药后 7 d 达 69. 49%；禾之壮有机水溶肥以浓度为 600 倍液时诱导效果最好，第 3 次药后 7 d 达 66. 91%。在各自的最佳处理浓度下，供试诱抗剂各处理 4 次调查平均诱导效果相当，分别为 51. 52%，50. 85%和 50. 65%，且在第 3 次药后 28 d 时，诱导效果下降率仍均略低于对照药剂，持效期与之相当，28 d 以上。

关键词：香蕉褐缘灰斑病；诱抗剂；诱导抗病性

* 基金项目：国家重点研发计划（2017YFD0202100）；现代农业产业技术体系建设专项（CARS-31-07）；农业农村部南亚办专项（15182130108235271 2）

室内条件下锦苗标靶抑制列当寄生向日葵的研究*

柳慧卿**，石胜华，王　娜，张　键，赵　君***
（内蒙古农业大学园艺与植物保护学院，呼和浩特　010018）

摘　要：本研究以锦苗标靶为研究对象在不同的施用条件下以及列当寄生的不同阶段研究了其对列当寄生的抑制效果。结果表明，在瘤结形成前进行叶片喷雾处理，1∶1 000与1∶2 000稀释液喷雾15d后瘤结褐变率达100%；瘤结形成后进行叶片喷雾，15d后，三个不同的稀释倍数褐变率介于83.21%～100%。在瘤结形成前进行根系浇水处理，5d后，三种处理的稀释液瘤结褐变率均达100%；列当瘤结形成后浇水处理，只是在第10d，1∶000和1∶2 000倍液处理的瘤结的褐变率达到100%。整体看来，不论是哪种施用方法，在列当瘤结形成前施用的抑制效果会优于瘤结形成后以及幼茎形成后；且相同的稀释倍数下，不论在列当寄生的哪个阶段，浇水处理对列当寄生的抑制效果优于叶片喷雾处理。

关键词：锦苗标靶；向日葵列当；施用方法；瘤结变褐率

---

* 基金项目：特色油料产业技术体系CARS-14；内蒙古自治区科技计划项目（向日葵列当抗性种质资源创新及关键防控技术研究与集成示范）

** 第一作者：柳慧卿，硕士研究生，主要从事向日葵列当研究；E-mail：1611601920@qq.com
*** 通信作者：赵君，教授，主要从事向日葵病理研究；E-mail：zhaojun02@hotmail.com

pH 值对 4 株生防芽孢杆菌在西瓜根际定殖的影响*

李　丹**，李　妍，任争光，尚巧霞，赵晓燕***
（农业农村部华北都市农业重点实验室，北京农学院生物与资源环境学院，北京　102206）

摘　要：微生物在不同 pH 值条件下的生理生化状态不同，每种微生物生长都有其一定的 pH 值范围和最适 pH 值条件，在最适的 pH 值条件下，微生物的酶活性最高，生长率也最高，pH 值的突然改变会影响微生物的膜运输蛋白、破坏质膜以及抑制酶活性。在植物生产中，土壤的 pH 值不仅对土壤微生物的生长有很大的影响，也会影响微生物在植物根系的定殖。生防菌在植物根际的成功定殖是其能充分发挥作用的先决条件，为了解 4 株生防芽孢杆菌在西瓜根际的定殖能力与 pH 值条件，本实验通过显微镜观察法和细菌平板计数法测定了不同 pH 值（5.5~8.5）对 4 株生防芽孢杆菌的趋化作用、生物膜形成以及在西瓜根部定殖的影响。实验结果发现不同 pH 值对 4 株生防菌趋化性和生物膜的影响不同。其中菌株 TR2 和菌株 DJ1 在中性的环境下表现出较强的趋化性，CE 菌株在弱酸到弱碱的环境下表现出较强的趋化性，而 ZT4-2 菌株在弱酸性环境下表现出较强的趋化性。菌株 TR2 在弱碱性的条件下生物膜的形成能力最强，菌株 CE 、DJ1 、ZT4-2 在不同的 pH 值条件下生物膜的形成能力差异不明显。从在西瓜根部的定殖情况看，TR2、CE 、DJ1 在 pH 值为 7.5 时，根部定殖细菌量最多，ZT4-2 株生防菌在 pH 值为 7.0 时，根部定殖细菌量最多，说明 pH 值为 7.0~7.5 时，4 株生防菌在西瓜根系的定殖能力最强，但是菌株之间的最适 pH 值还是略有差异。西瓜生长的最适 pH 值为 6~7，本实验研究的 4 株生防菌在 pH 值为6~8 的范围内具有较明显的趋化作用和生物膜形成能力，与西瓜的生长 pH 值范围一致，所以可以在西瓜根部很好的定殖和繁殖。本实验结果为以后生防菌株在田间植株根际的定殖行为的研究提供了依据，为生防菌的田间应用奠定了基础。

关键词：pH 值；生防芽孢杆菌；定殖

* 基金项目：北京市教育委员会科研计划项目（KM201910020011）
** 第一作者：李丹，硕士研究生，研究方向：植物保护；E-mail：289773003@qq.com
*** 通信作者：赵晓燕，副教授，研究方向：植物真菌病害综合防治；E-mail：zhaoxy777@163.com

生防细菌 B8 抗病毒活性物质的分析及其作用机理研究*

厉彦芳**，谢菁菁，王春阳，赵秀香***
（沈阳农业大学植物保护学院，沈阳 110866）

摘 要：侧孢短芽孢杆菌（*Brevibacillus laterosporus*）具有繁殖快、营养简单、孢子形成能力强、抗逆性强、易在土壤中定殖等特点，是一类广泛应用的生防菌。目前，国内外已报道其具有杀线虫、抗菌、生产赖氨酸、溶磷、降解有机物以及抗肿瘤的作用。这类细菌具有明显的杀线虫作用，但对其抗病毒活性及其作用机理研究较少。侧孢短芽孢杆菌 B8 是一株对多种植物病原真菌具有拮抗作用的生防菌，研究发现该菌株对烟草花叶病毒（*Tobacco mosaic virus*，TMV）具有很强的体外钝化作用。本研究从侧孢短芽孢杆菌 B8 的发酵液中分离纯化出一种蛋白类的抗病毒活性物质，根据该蛋白的质谱结果克隆得到 *blb8* 基因，并在大肠杆菌 BL21（DE3）中成功表达及纯化了相对分子质量约为 12.8 ku 的融合重组蛋白。

笔者研究了重组蛋白 BLB8 对 TMV 的体外钝化作用及诱导烟草产生系统抗病性。将重组蛋白 BLB8 与 TMV 粒子进行体外钝化处理，使用透射电镜观察 TMV 粒子和半叶法接种心叶烟，结果表明重组蛋白 BLB8 能够破坏 TMV 粒子的外壳结构，使其发生断裂；当浓度为 200μg/mL 时，对心叶烟的枯斑抑制率达到 80%以上。通过半定量检测 HSR203J 基因在注射重组蛋白 BLB8 的烟草叶片中的表达情况，证明了 BLB8 能够引起烟草叶片产生典型的 HR 反应。利用实时荧光定量 PCR（RT-qPCR）技术检测重组蛋白 BLB8 对 TMV 积累量的影响及对烟草防御相关基因的诱导作用，结果表明使用重组蛋白 BLB8 处理的烟草 TMV 的积累量较对照低，烟草防御相关基因 *PR*5、*PR*1*a*、*PAL*、*NPR*1、*COI*1 和 *PDF*1.2 均出现不同程度地上调表达。表明 BLB8 可激发烟草防卫反应，诱导烟草产生系统抗性，提高烟草对病毒的抗性。

关键词：侧孢短芽孢杆菌；原核表达；体外钝化；系统抗病性

---

* 基金项目：国家重点研发计划（2017YFD0201104）
** 第一作者：厉彦芳，硕士研究生，研究方向为植物病害生物防治；E-mail：947072893@qq.com
*** 通信作者：赵秀香，博士，副教授，主要从事植物病害生物防治及植物病毒病害研究；E-mail：zhaoxx0772@163.com

微生物菌剂 DP-11 对苦瓜根际微生物群落结构的影响*

习慧君**，刘　闯，万鑫茹，文才艺，赵玉华，赵　莹***
（河南农业大学植物保护学院，郑州　450002）

摘　要：田间试验结果显示，未经 DP-11 处理的苦瓜植株发病率为 100%，经 DP-11 处理的苦瓜植株发病率为 4.5%，发病轻微。由此可见，微生物菌剂 DP-11 可显著防控苦瓜枯萎病。为研究微生物菌剂 DP-11 在防治苦瓜枯萎病过程中植株根际土壤微生物群落结构的变化情况，本试验利用基于 ITS 序列和 16S rRNA 基因的扩增子测序技术分析苦瓜根际土壤中真菌和细菌群落的差异。

ITS 测序结果显示，未经 DP-11 处理的土壤样品中平均获得63 900条序列，聚类分析得到了404 个操作分类单元（Operational taxonomic unit，OTU），经 DP-11 处理的土壤平均获得53 670条序列，聚类分析得到356 个操作分类单元（OTU）；其中 307 个 OTU 共同存在于两组处理中。经 OTU 注释分析后发现在属的分类水平中，未经 DP-11 处理的土壤中优势菌为假霉样真菌属（*Pseudallescheria*）和被孢霉属（*Mortierella*），所占比例达 27.1%和 21.3%，而在经 DP-11 处理的土壤中含量为 7.06%和 16.36%；经 DP-11 处理的土壤中的优势菌为热带头梗霉（*Cephaliophora*），所占比例为 15.71%，而在未经 DP-11 处理的土壤中含量仅为 0.81%；此外，致病菌镰孢菌属（*Fusarium*）在经 DP-11 处理后的土壤中的相对丰富度比未经 DP-11 处理土壤中的降低了 2.08%。16S rRNA 测序结果显示，共得到了169 152条有效序列，聚类分析结果显示未经 DP-11 处理土壤中得到1 916个 OTU，经 DP-11 处理土壤中得到1 900个 OTU，其中两组处理共同存在1 872个 OTU。物种注释分析结果发现其两组处理差距不明显，其优势菌均为杆菌属（*Bacillus*）和酸杆菌属（*Acidobacteria*）。扩增子分析发现经 DP-11 处理后的土壤中真菌群落种类和含量明显发生变化，且可以降低病原真菌的含量，但是对细菌的群落影响不显著。研究结果明确了微生物菌剂对苦瓜根际微生物的影响，为深入探究微生物菌剂的作用机理和生防应用奠定了良好的基础。

关键词：微生物菌剂；DP-11；根际土壤；微生物群落

---

* 基金项目：公益性行业（农业）科研专项资助项目（201503110）

** 第一作者：习慧君，在读硕士，植物病理学专业；E-mail：xihuijun1994@126.com

*** 通信作者：赵莹，博士，讲师，研究方向：植物病害生物防治；E-mail：nying2009@126.com

硼元素抑制黄瓜绿斑驳花叶病毒侵染引起西瓜倒瓤机制研究

毕馨月*，安梦楠，夏子豪**，吴元华**
（沈阳农业大学植物保护学院，沈阳　110866）

摘　要：西瓜（*Citrullus lanatus* Thunb.）被认为是世界上最受欢迎的、有营养的水果作物之一，在世界各地广泛种植，但是其产量受到病原物、虫害和非生物胁迫的严重威胁。黄瓜绿斑驳花叶病毒（*Cucumber green mottle mosaic virus*，CGMMV）是烟草花叶病毒属（*Tobamovirus*）成员，是我国重要检疫性有害生物，可以侵染多种葫芦科经济作物，尤其侵染西瓜，引起果肉出现暗红色水渍状病变，并伴有酸化和腐败，又称为"西瓜血瓤病"。本试验为了研究 CGMMV 引起西瓜倒瓤和硼元素（B）调控的分子机制，对处理后西瓜果实样本进行 RNA-seq 及差异分析。西瓜响应 CGMMV 侵染的 DEGs 在 GO 数据库中主要涉及生物学过程（biological process）有代谢过程（metabolic process）、细胞过程（cellular process）和个体组织过程（single-organism process）；涉及的细胞组分（cellular component）主要为细胞（cell）、细胞组分（cell part）、细胞器（organelle）和生物膜（membrane）等；涉及的分子功能（molecular function）主要有催化活性（catalytic activity）和结合（binding）等；KEGG 分析表明，DEGs 主要参与的代谢通路有植物激素信号转导（Plant hormone signal transduction）、淀粉和蔗糖的代谢（Starch and sucrose metabolism）、碳代谢（Carbon metabolism）、光合作用（Photosynthesis）、光合作用-天线蛋白（Photosynthesis-antenna proteins）等；西瓜响应 B 抑制 CGMMV 侵染的 DEGs 主要参与的代谢通路，除了植物激素信号转导、淀粉和蔗糖的代谢、碳代谢，还主要参与苯丙氨酸代谢（Phenylalanine metabolism）、苯丙素的生物合成（Phenylpropanoid biosynthesis）、氨基糖和核苷酸糖代谢（Amino sugar and nucleotide sugar metabolism）、氨基酸的生物合成（Biosynthesis of amino acids）、植物-病原互作（Plant-pathogen interaction）等。这些途径与西瓜果实细胞中细胞壁的合成并保持稳态，碳水化合物的运输与代谢，维持细胞膜的稳态与功能密切相关，对西瓜倒瓤有明显抑制作用。本研究通过对西瓜果实响应 CGMMV 侵染，以及喷施硼元素调控 CGMMV 侵染的转录组测序与分析，揭示 CGMMV 侵染引起西瓜倒瓤的分子机制，明确硼元素和倒瓤症状形成的关系，为 CGMMV 的绿色防控和采用基因调控创制持久抗病的西瓜提供理论依据。

关键词：西瓜（*Citrullus lanatus* Thunb.）；黄瓜绿斑驳花叶病毒（CGMMV）；硼元素（B）；倒瓤机制

* 第一作者：毕馨月，博士研究生，主要从事植物病毒学研究；E-mail：mj_bxy@163.com
** 通信作者：夏子豪，讲师，主要从事植物病毒学方向研究；E-mail：zihao8337@syau.edu.cn
吴元华，教授，博士生导师，主要从事植物病毒学和生物农药方向研究；E-mail：wuyh7799@163.com

山西省藜麦主要病害发生种类研究*

蔚天春**，姜晓东，贺建元，郭雪梅，郝晓娟，王建明***，李新凤***
（山西农业大学农学院，晋中 030801）

摘　要：藜麦（*Chenopodium quinoa*）是全谷全营养完全蛋白碱性食物，其单体作物即可满足人类所需全部营养。山西省自 2008 年于静乐县首次实现藜麦规模化种植以来，由于种植面积不断扩大，重茬种植等原因，目前我省各藜麦种植区，病虫害发生日趋严重，成为影响藜麦生产的主要限制因子。为促进山西省藜麦产业的健康发展，明确山西省藜麦病害发生情况已势在必行。

本研究通过对山西省各地区发生普遍而严重的藜麦病害进行调查，采集具有典型病害症状的罹病植株，分离纯化病原菌，经柯赫氏法则验证后，结合形态学与分子生物学方法，确定病原菌的分类地位。结果表明，山西省藜麦病害主要有：由 *Phoma apiicola* 引起的藜麦秆枯病、由 *phoma* 引起的藜麦黑杆病、由 *Colletotrichum* 引起的藜麦炭疽病、由 *Peronospora variabilis* 引起的藜麦霜霉病、及由 *Cercospora chenopodii* 侵染引起的藜麦叶斑病。藜麦秆枯病典型症状：主要为害藜麦茎秆，且多从茎秆分枝处开始发病，被侵染部位开始颜色变浅，后变为白色，并在其上产生同心轮纹状排列的小黑点，严重时，茎秆倒伏枯死。藜麦黑杆病：主要为害茎秆，病斑为不规则梭形，初期灰白色，后颜色逐渐变深，随着病斑扩大，病部中央变为黑色，边缘灰白色，严重时病斑连片，整个茎秆变为亮黑色，后期植株枯死。藜麦炭疽病：既为害茎秆，也为害叶片。通常为害嫩茎，茎秆变黑，并密生褐色小颗粒，使染病茎秆变得粗糙。为害叶片时，叶片上产生近圆形褐色病斑。藜麦霜霉病：发病初期，叶片出现淡黄色不规则病斑。严重时，病斑呈粉红色，叶背面出现粉红色霉层，叶片变黄脱落。藜麦叶斑病典型症状为：发病初期叶片出现圆形或近圆形淡黄色病斑，病斑中央浅褐色，外缘深褐色，边缘有黄色晕圈，后期病斑中央组织脱落呈穿孔状。这些研究结果为进一步研究山西省藜麦病害发生流行规律及病害综合防治奠定了理论基础。

关键词：藜麦；病害种类；分离；鉴定

---

* 基金项目：山西省重点研发项目（201803D221012-2）；山西省应用基础研究项目（201703D221006-3）

** 第一作者：蔚天春，硕士研究生，植物病理学专业；E-mail：757534368@qq.com

*** 通信作者：李新凤，副教授，主要从事植物真菌病害、植物病理生理及分子植物病理研究；E-mail：lxf1309@163.com
王建明，教授，主要从事植物病害综合防治，植物病理生理及分子植物病理研究；E-mail：jm.w@163.com

密克罗尼西亚联邦农业病虫草害调查初报*

唐庆华[1**]，黄贵修[2]，覃伟权[1***]，范海阔[1]，弓淑芳[1]，刘国道[3]，杨虎彪[4]
（1. 中国热带农业科学院椰子研究所，文昌　571339；2. 中国热带农业科学院国际合作处，海口　571101；3. 中国热带农业科学院院机关；4. 中国热带农业科学院热带作物品种资源研究所，海口　571101）

摘　要：密克罗尼西亚联邦（以下简称密国）是西太平洋中的一个岛国，属于典型的热带海洋性气候，其农作物主要有椰子、槟榔、香蕉、胡椒等。其中，椰子被称为“生命之树”（Tree of Life），经济地位非常重要，然而其椰子产业尚处于原始阶段，面临椰子病虫害严重、椰子产量低、产品开发程度低等问题。应密国总统彼得·克里斯琴（Peter M. Christian）对椰子产业发展技术支持的请求，在商务部的资助下中国热带农业科学院组织相关专家12人于2018年6月11日至7月5日，对该国雅浦、丘克、科斯雷和波纳佩4个州的农业官员、农民等进行了为期25d的栽培技术、病虫害防控技术培训与田间实习。在此期间，专家们对密国的农业病虫草害进行了较系统的考察。调查发现，密国椰子上共发现病害6种、害虫7种；槟榔病害3种；香蕉病害3种；杧果病害5种；番木瓜病害3种、害虫1种；柑橘病害4种、害虫1种；番石榴病害和害虫各1种；诺丽病害1种；蔬菜（辣椒、黄瓜、苦瓜、芋头）病虫害9种；鸡蛋花病害及害虫各1种；美人蕉病害1种；木薯病害5种、害虫及节肢动物5种；咖啡、胡椒病害共5种；甘蔗害虫2种；杂草10种。其中，为害严重的病虫草害有椰子泻血病、辣椒病毒病及青枯病，马里亚纳椰甲及深蓝椰甲、椰红蚧，盾叶鱼黄草。疑似病害椰子致死性黄化病、槟榔黄化病、柑橘黄龙病有待进一步进行病原鉴定、确认。此外，有20余种未知病原的病害17种、未知分类地位的害虫11种。

关键词：密克罗尼西亚联邦；农业；病虫草害；系统调查

---

* 基金项目：农业农村部“一带一路”热带项目：“椰子菌草种植示范园”（BARTP-06）；地方合作项目：“琼中县槟榔重要病虫害绿色防控技术研究、示范及推广”

** 第一作者：唐庆华，博士，助理研究员，研究方向为热带棕榈植物病害综合防控技术；E-mail：tchuna129@163.com
*** 通信作者：覃伟权，研究员，研究方向为热带棕榈植物病害综合防控技术；E-mail：QWQ268@163.com

油棕病害研究进展及存在的主要问题*

唐庆华**，曹红星，覃伟权***

（中国热带农业科学院椰子研究所，海南省热带油料作物生物学重点实验室，
农业部热带油料作物科学观测试验站，文昌　571339）

摘　要：油棕是生产效率最高的油料作物，被誉为“世界油王”。近年来，随着人们对油脂需求的不断提高，油棕产业发展迅速。但是，随着油棕种植面积的扩大，已经出现了一些严重甚至毁灭性的病害。本文对国内外油棕主要病害的发生、分布及防治情况等进行了系统介绍以期为我国从事油棕病害研究的科研人员提供参考依据。

关键词：油棕；病害；综合防治

Current status and Key Problemsin in diseases of oil palm (Continued)

Tang Qinghua, Cao Hongxing, Qin Weiquan***

(*Coconut Research Institute, Chinese Academy of Tropical Agricultural Sciences, Hainan Key Laboratory for Biology of Tropical Oil Crops, Scientific Observing and Experimental Station of Tropical Oil Plants of Minitary of Agriculture, Wenchang* 571339, *China*)

Abstract: Oil palm is the most productive oil - bearing crop, which is praised as Oil King. Recently, with the increasingly demand for edible oil worldwide, the oil palm industry has developed rapidly. However, as the area under it had expanded, there have been outbreaks of serious even devastating disease. A systemically introduction of main oil palm diseases, and their managements worldwide were gave in this review and our aim is to provide references for the researches who work on diseases of oil palm.

Key words: oil palm; disease; integrated management

油棕（*Elaeis guineensis* Jacg.）为棕榈科常绿高大乔木，是世界上生产效率最高的油料作物[1]，被誉为“世界油王”，现主要分布于东南亚、非洲和南美洲的广大热带国家和地区。目前，油棕已发展成为最具发展潜力的战略资源作物和世界第一大油料作物[2-5]，因而备受关注。随着油棕种植面积的迅速扩大世界上已出现了一些严重甚至毁灭性的病害[2-3]。据报道全球

* 基金项目：农业农村部农业技术试验示范与服务支持项目：油棕品种区域适应性试种（16RZNJ-5）

** 第一作者：唐庆华，博士，副研究员，主要从事热带油料和棕榈作物病害综合防治技术研究；E-mail：tchuna129@163.com

*** 通信作者：覃伟权，硕士，研究员，主要从事热带油料和棕榈作物植物保护综合防治研究；E-mail：QWQ268@163.com

共有32种油棕病害，其中有9种重要病害，19种次要病害以及4种主要营养失调症[2-4,6]，并且一些新病害不断地被发现、报道[7-9]。其中，毁灭性病害有枯萎病[10-11]、茎基干腐病[2]、茎基腐病[12-13]以及致死性黄化病等[14-15]。目前，我国已把油棕纳入战略规划[4]。但是，我国的油棕病害研究尚很薄弱，缺乏系统研究。为了促进我国油棕科研工作者了解油棕病害，本文对全世界范围内最近20年来油棕病害的发生、为害及防治情况进行了较为系统的总结，本综述旨在为我国油棕产业的顺利健康发展和油棕病害综合防治研究提供有益的帮助。

1　苗疫病

1.1　为害与症状

油棕苗疫病（Blast disease）在气候条件适宜和管理措施不力的苗圃均可发生。在非洲的科特迪瓦、尼日利亚、马达加斯加、南非、坦桑尼亚、喀麦隆以及亚洲的越南、老挝、柬埔寨、新加坡等国均有发生，造成损失严重[2,6]。该病在马来西亚、印度尼西亚、巴西和哥伦比亚也有发生，我国海南亦有报道[16]。发病初期，叶片失去正常光泽，萎蔫，呈黄绿色或鲜黄色，在叶尖形成红棕色或紫色斑块，坏死部分由顶端向下蔓延，几天内叶片枯死，深褐色，似火烧状。老叶先死，心部箭叶多坏死，幼苗几天内死亡。根部症状为初生根尖先受害，根部皮层腐烂，在病根的中柱或下表皮上可看到深褐色或黑色小菌核。苗疫病是苗期最重要的根病，感病油棕苗大多死亡，少数残存植株畸形、矮小，长势较弱，不适于移栽至大田。

1.2　病原

该病由薄片丝核菌 *Rhizoctonia lamellifera* 和华丽腐霉菌 *Pythiurn splenderns* 两种真菌复合侵染所致。腐霉菌为主要初侵略源，它能穿透薄壁细胞扩展。丝核菌在皮层组织破坏中起着重要作用，并能穿过先侵入的腐霉菌进行侵染。

1.3　防治

①在雨季将育秧盘中发育良好的幼苗适时种植于苗圃，避开幼苗感病期与疫病流行期。②在旱季要特别注意进行灌溉，确保整个苗期育苗袋中含有足够的水分并利用遮阳网遮荫。③避免偏施氮肥，必要时施用石灰以降低发病率。④清除病和苗圃附近的杂草。⑤可用福美双加敌菌酮淋灌苗圃土壤。

2　茎基干腐病

2.1　为害与症状

茎基干腐病（Dry basal rot）仅在西非和扎伊尔流行。在尼日利亚，3~8龄油棕发病严重，死亡率约为70%，该病曾摧毁该国一处油棕园；随后在尼日利亚、喀麦隆和加纳的几个地区小面积发生[2]。该病初期导致的油棕死亡很普遍，但后期发病油棕大多能恢复健康，迄今再未见严重暴发的报道。马来西亚的沙巴和马来亚也曾有疑似该病的报道过[6]。叶部症状通常在旱季末期显现，随后果穗和花序大量腐烂。该病内部症状表现为病株基部干腐，病健交界处维管束坏死。外部症状为下层叶片折断而上层叶片依然保持挺立，整株形成一个完整的环，这是刚发病油棕的特征。随后，上层叶片和箭叶发病，植株死亡。通常情况下，一些发病油棕可以恢复，但需要数年才能结果。

2.2　病原

病原为子囊菌门的奇异长喙壳菌 *Ceratocystis paradoxa*，其无性阶段属半知菌类的奇异根串珠霉 *Thielaviopsis paradoxa*。*C. paradoxa* 是一种土壤习居菌，广泛分布于非洲和亚洲的热带地区，可侵染多种作物。

2.3 防治

①加强栽培管理：合理种植；避免偏施氮肥，宜增施钾肥；清除病叶并集中烧毁。②化学防治：发病初期可选用克菌丹、王铜、波尔多液、甲基托布津、代森锰锌和异菌脲等药剂，每隔7~14 d喷施叶片1次，连续喷施2~3次可有效防治该病。发病严重时，先把病叶清除干净，然后再喷施上述药剂。③选育、种植抗（耐）病品种。

3 油棕枯萎病

3.1 为害与症状

油棕枯萎病（Fusarium wilt）又称维管枯萎病（Vascular wilt），该病最早报道于刚果，现在主要分布于科特迪瓦、扎伊尔、尼日利亚、加纳、喀麦隆、科特迪瓦、贝宁湾、圣多美和刚果，南美洲的巴西和厄瓜多尔局部也有暴发[10-11,17-22]。该病也是我国的检疫性病害之一。在非洲，枯萎病是最具毁灭性的病害，可造成油棕减产25%~50%甚至更高[22]。该病在成年油棕上有两种症状类型。“急性”症状表现为叶片干枯并迅速死亡，但仍保持挺立直至被风吹断。这种症状发病迅速，植株2~3个月内死亡。“慢性”症状表现为植株矮小，可存活数月甚至数年之久。外层叶片首先干枯并垂挂于树干上，随后逐渐蔓延到新叶。同时，新叶长出后仍保持挺立，但一般是萎黄色，且比正常叶片小很多，整个树冠扁平，树干顶部逐渐缩小。该病苗期也可发生。幼苗感病后，植株矮小，外轮叶片干枯。油棕的中层叶片和嫩叶首先变黄，随后长出的叶逐渐变短，病株外貌呈平顶状，外轮叶片干枯，箭叶变得细小、紧缩成束，淡黄色至象牙白色。该病典型的内部症状是维管束变褐，在成龄期和苗期均可发生。

3.2 病原

油棕尖镰孢菌（*Fusarium oxysporum* Schlechtend. f. sp. *elaeidis* Toovey），为半知菌类、镰刀菌属真菌。该菌为土壤习居菌，主要为害非洲油棕种质材料，其菌丝体和厚垣孢子可以在土壤中长期存活。

3.3 防治

该病防治非常困难。防治措施包括以下几点。①检疫措施：在进行油棕种子贸易时进行严格的中间检疫和入境后隔离检疫；使用杀菌剂对油棕种子进行喷雾处理。②加强田间管理：增施钾肥，重新定植油棕时须离旧树桩超过2 m；田间间种葛属植物，创造不利于病害发生的环境。③选育抗病品种。

4 油棕茎基腐病

4.1 为害与症状

油棕茎基腐病（Basal stem rot）又称灵芝菌茎干腐烂病（Ganoderma trunk rot），是一种致死性病害[12-13,23-31]。该病于1915年在西非的刚果首先报道，马来西亚和印度尼西亚亦有严重发生。此外，非洲的安哥拉、喀麦隆、坦桑尼亚、扎伊尔、加纳和尼日利亚、大洋洲的巴布亚新几内亚、中美洲的洪都拉斯、南美洲的哥伦比亚以及亚洲的泰国等均有报道。该病最初仅危害25龄以上的油棕，直到20世纪60年代逐渐加重并开始危害10~15龄油棕，也有研究发现定植12~24个月的油棕小苗也可被侵染[29]。在东南亚，该病曾造成马来西亚滨海地区油棕园平均减产50%，13龄以上的油棕园减产80%[23]；2003年的一项调查显示，该地区13龄以上的油棕园发病率仍达30%。在印度尼西亚苏门答腊岛北部，该病曾造成40%~50%的油棕发病，每公顷鲜果减产0.16 t。油棕从苗期至成株期均可受害。发病初期，植株表现为轻微萎蔫，生长缓慢，外轮叶片变黄，随后下部叶片逐渐黄化、下垂，果实和雄花停止发育，箭叶枯萎；发病严重时，除最嫩的叶片和箭叶仍保持挺立外，其他叶片全部断折并向下悬挂，在靠近树干处干枯的叶片呈斗篷状。活着的叶片也变成淡黄色，最后整株枯死，茎干内部组织腐烂。内部症状为在树干中央存在明显的圆形至不规则形坏死区。根部症状为根系腐烂变色，皮层组织碎裂，中柱显露，邻近中柱的组织变褐；湿度大时，在根系外长出白色的菌丝垫。低龄油棕表现症状6~24个月后整株黄化枯死，成龄油棕2~3年内死亡，在紧贴地面的茎基部长出纽扣状担子果，随后迅速长大呈托架状，其颜色、形状、大小各异。

4.2 病原

病原菌为狭长孢灵芝 *Ganoderma boninense* Pat.，属担子菌门、灵芝属真菌。该病可通过土壤中的病残体来传播，也可依靠担孢子通过气流传播；油棕苗期发病主要是菌丝体侵染根部引起，而6龄以上油棕发病主要是由于病菌分生孢子侵染引起。此外，病原菌还可能通过大伪瓢虫、锯白蚁等昆虫介体进行传播[26]。

4.3 防治

①农业防治：清除田间病残体并焚毁，深翻、耙地；培育健康种苗，使用无病土壤育苗；改良土壤，增施钙肥；种植豆科植物，改善园区生态环境。②化学防治：施用放线菌酮、己唑醇、三唑酮、三唑醇、萎锈灵、多菌灵、戊菌唑、十三吗啉等杀菌剂，可通过土壤淋透和茎干注射等方法施用。一般而言，茎干注射法比土壤淋透法效果较好，茎干注射萎锈灵和五氯硝基苯混合物可取得较好的防治效果。③生物防治：拮抗菌有木霉菌、放线菌、芽孢杆菌、哈茨木霉、绿粘帚霉等以及洋葱伯克霍尔德菌、绿脓杆菌、粘质沙雷氏等油棕内生菌。丛枝菌根真菌也有提高抗病性和增产的作用。④培养抗（耐）病品种。

5 猝萎病

5.1 为害与症状

猝萎病（Sudden wither）又称为凋萎病，是油棕上最严重的病害之一[2]。该病在哥伦比亚、厄瓜多尔和秘鲁已引起严重损失，如在哥伦比亚造成的损失超过90%。该病在委内瑞拉、苏里南、特里尼达岛、巴西和美国等均有分布。在苏里南，该病被称为“Hartrot”病，在巴西的Bahia也有一种类似的病害。该病症状为发育中的果穗全部突然腐烂，叶柄顶部红色褪去，老叶片快速干枯。下层叶片末端和中部先出现红棕色条纹，随后变为淡绿色、黄色、红褐色和灰色等。发病植株2~3周内死亡，根系快速腐烂、干枯。在哥伦比亚，有一种症状类似的被称为“Marchintez progresiva”的病害，但扩展较慢。该病从1龄油棕开始侵染，典型症状是开始时箭叶不发病；根部皮层腐烂，天气潮湿时皮质分解，但在干旱季节坏死，从中柱上脱落。腐烂组织可扩展至主干以及下层根系。

5.2 病原

病原为原生动物 *Phytomonas staheli*，属原生动物门、植生滴虫属。传播媒介有 *Myndus crudus*（*Haplaxius pallidus*）和 *Lincus lethifer*，可能还包括根潜蝇（*Sagalassa valida*）。

5.3 防治

①农业防治：加强田间管理，合理施肥、适时灌溉或排水，及时除草；铲除病株并烧毁。②化学防治：联合使用杀虫剂和除草剂控制传播媒介；③选育种植抗病品种。

6 高茎腐病

6.1 为害与症状

高茎腐病（Upper stem rot）是一种致死性的茎干腐烂病，该病只在泥炭土和内陆山谷中的油棕上发生严重[32]。该病在马来西亚、印度尼西亚和印度也有发生，但多为零星发生，影响不大[2]。通常情况下，下部叶片先变黄，该症状逐渐延伸到中部叶片，然后是箭叶。病菌孢子通过叶片基部进入茎部外周组织。棕色腐烂组织从叶片基部缓慢向内发展，茎干常在大风作用下崩溃。腐烂组织沿茎干向上、向下扩展，最终侵入油棕冠部导致植株死亡。病菌存在两种子实体（正常的和倒置的），灰褐色，边缘棕色，但仅仅在病菌侵染油棕叶片基部时才引起大量腐烂。该病仅侵染茎干，不能扩展至根部。

6.2 病原

病原为木层孔菌 *Phellinus*（Fomes）*noxium*，为担子菌门、木层孔菌属真菌；病原也可能为担子菌门、灵芝属 *Ganodeerma* 真菌[29]。

6.3 防治

①定期进行病害调查，早发现早防治。②切除茎干上的病斑，用煤焦油涂抹切面；铲除并烧毁折断的病株。③施用含钾肥料。

7 红环病

7.1 为害与症状

红环病（Red ring disease）又称红环斑病，该病仅存在于中南美洲，在委内瑞拉、苏里南、巴西、哥伦比亚和哥斯达黎加都有发生[33]。在哥斯达黎加，该病危害非常严重。在巴西，该病已造成一定损失。在未采取保护措施的地区，红环病可迅速扩展。冠层中部矮化，新展开的叶片紧紧地聚在一起，叶片皱缩、扭曲，有时紧贴着穗轴，叶片上有胶状物流出。随后冠层叶片缓慢变黄、干枯，穗轴浅棕色伴有黄色斑点。中间一两片叶呈古铜色，2~5 个月后叶片全部慢慢变成黄色或古铜色，但依然保持挺立。果穗腐烂，花序不能结果。

典型的内部症状是在茎干外缘 7~8 cm 处形成 1~2 cm 宽的棕色圆环。该症状在油棕基部最明显，该症状可向上延伸至冠层的叶柄和叶轴，观察横切面可以看到坏死区或坏死斑。但是，这种感染不能侵入茎尖组织或周围非常嫩的叶。

7.2 病原

病原为椰子红环线虫 *Rhadinaphelenchus*（原 *Bursaphelenchus* 或 *Aphelenchus*）*cocophilus*。该虫仅在油棕组织中繁殖，其传播介体昆虫为棕榈象甲 *Rhynchophorus palmarum*；在哥伦比亚，主要传播媒介为 *Metamasius hemipterus*。在巴西，在野生的 *Oenocarpus distichus* 油棕上同时存在 *R. cocophilus* 和 *R. palmarum*，这表明两者可能联合形成侵染源。

7.3 防治

①加强田间管理，在发病油棕园中采取定期的卫生措施，防止油棕受伤，定期对工具进行消毒同时对修剪过的叶片和果穗表面进行药剂处理，铲除发病油棕并烧毁。②信息素诱捕昆虫媒介 *R. palmarum*[33]。③采取生物防治措施控制 *R. palmarum*。

8 环斑病毒病（叶片斑驳病）

8.1 为害与症状

环斑病毒病（ringspot）又称叶片斑驳病（Leaf mottle 或 Mancha anular）、芽叶斑驳病（Bud Leaf mottle）。该病曾被称为环斑病（ring spot）和致死性黄化病（Fatal yellowing disease），但这两种名称已用于其他病害，本文用“环斑病毒病”以区分“环斑病”。环斑病毒病在厄瓜多尔和秘鲁常引起油棕死亡[34-35]。该病发病率为 3%~40%，但 18~24 个月的油棕发病率可达 95%。在哥伦比亚，1985 年发现该病，至 1988 年 2.5 龄的油棕受害面积约为 4 500 hm^2，发病率达 2%~45%。目前，该病在中南美洲主要油棕种植国厄瓜多尔、秘鲁、哥伦比亚、巴西、委内瑞拉、哥斯达黎加、洪都拉斯、墨西哥、尼加拉瓜等引起了广泛关注。在厄瓜多尔和秘鲁，现已有 10 万 hm^2 的油棕处于潜在病害风险之中。该病多发生在 1~4 龄的油棕上，5 龄以上的油棕上未发现该病。症状为箭叶和新生叶及叶轴上出现环状和椭圆形的斑点，随后斑点变黄；叶片黄化症状最终扩展到下部叶片，箭叶坏死；在初始症状出现的 3 个月内，下部叶片最终变褐、死亡，生长出来的果串也会坏死；第三、第四级根坏死，随后扩展到主根和次根的中心部位；靠近根部位的茎基部维管束变紫，随着症状进一步沿茎部向上扩展，呈现紫色的环状坏死斑块。一些植株在表现症状后 3~4 个月死亡；而一些油棕虽然表现叶部症状，但在几年内依然能继续生长并结果。该病发病严重时可扩散至整个油棕园。

8.2　病原

该病是由非洲油棕环斑病毒（*African oil palm ringspot virus*，AOPRV）引起的。AOPRV 为新鉴定的一个种，数曲线病毒科（Flexiviridae）[36]，但目前尚未明确到属[37]。

8.3　防治

目前，尚无有效防治该病的措施。推荐的措施有：加强检疫，包括禁止从病害发生地进口油棕种质材料以及强化入境检验。此外，在油棕园内间种豆科植物有利于抑制该病的发生。

9　致死性黄化病

9.1　为害与症状

油棕致死性黄化病（Fatal yellowing）是根据其典型症状命名的，该病又称为致死性芽腐病（lethal bud rot 或 pudrición de cogollo，amarelecimento fatal）[14-15]。该病已给中美洲和南美洲（如巴西、厄瓜多尔和苏里南）油棕造成了严重损失，一些种植园已经被完全摧毁，其他发病油棕园也损失严重，许多发病严重的油棕园也长期处于失管状态。该病典型症状是箭叶干腐或湿腐，在雨季幼嫩叶片黄化，但黄化症状在旱季消失。该病有“急性”和“慢性”2 种症状，典型的“急性”症状表现为当腐烂蔓延到生长点后可导致油棕植株大量死亡。该病发生时 4~6 片幼嫩箭叶粘在一起、不能展开，形状像一个“指挥棒”。此症状出现后 10~30 d 内箭叶开始腐烂并向下扩展，1~9 周内箭叶崩溃，当腐烂侵染生长点后植株死亡。“慢性”症状表现为尽管箭叶腐烂可向生长点扩展，但染病油棕通常能恢复。油棕致死性黄化病症状很多，但通常不包括典型的“小叶”症状，该病在不同国家或地区间症状和严重度差异很大，并不总是引起油棕死亡。该病可分为 2 个阶段：第一阶段可持续到 12 龄油棕，病害呈线性增长；第二阶段病害扩展加速，呈指数增加。

9.2　病因

该病病因广泛，包括卵菌、真菌、细菌、类病毒昆虫等[2,38-45]。卵菌为棕榈疫霉 *Phytophthora palmivora*[2,15]。真菌有串珠镰刀菌 *F. moniliforme*、奇异根串珠霉菌 *Thielaviopsis paradoxa*、尖孢镰刀菌 *Fusarium oxysporum*、球二孢菌 *Botryodiplodia* sp.、茄镰刀菌 *F. solani* 和 *Sclerophoma* sp.。该病病原也可能为细菌和类病毒，但尚需进一步的实验证据。此外，有报道认为该病害则与植原体有关[46-47]。一些昆虫（如 *Alurnus humerlis*）也可能与该病有关，但尚未确认这些昆虫和病害之间的确切联系。有报道认为该病是非病原生物引起的，而一些除草剂也可引起类似的症状。此外，该病还可能与季风有关[38,41]。总之，该病病因非常复杂，学界尚有争议[15]，黄化症状仅是一种常见的胁迫应激反应。

9.3　防治

该病防治困难，一些混合施用杀真菌剂和杀虫剂的防治措施都没有取得成功。目前推荐的措施有：①切除腐烂箭叶。②改善排水系统、深耕（改善土壤的通气性）、清除根部覆盖物，施肥时要特别注意降低（钙+镁）/钾比率，对超过 20 龄的发病油棕园松土后重新定植健康油棕。③种植抗（耐）病杂交种。

10 结语

在我国，对油棕病害研究基础薄弱，尚未开展系统研究。目前，已报道的油棕病害有果腐病[6]、枯萎病[7]、苗疫病[16]、拟茎点霉叶斑病[8]、小孢拟盘多毛孢叶斑病等[9]，笔者自 2000 年即开展油棕病害研究，通过近 6 年来对海南、云南等地的调查发现大田低龄油棕叶部病害较严重，对其生长发育影响较大，需要引起足够重视；但随着树龄增加，油棕对病害抗性增强，病害影响较小（本课题组未发表资料）。油棕是世界上生产效率最高的油料，也是最具发展潜力的战略资源作物，目前已引起广泛关注。但是，我国油棕种植业发展曲折缓慢，在经历了几度引种、试种失败的尝试后，直到最近 5 年才有较大发展。令人欣慰的是，目前我国已把油棕纳入战略规划[4]。但是，需要格外关注的是尽管我国尚未发现严重的油棕病害，然而非洲、美洲以及东南亚一些主要油棕种植国已暴发了油棕枯萎病、茎基腐病等毁灭性病害以及一些严重威胁油棕种植业的虫害。目前，我国已开始从马来西亚、印度尼西亚、巴西、缅甸等国引种油棕种苗，预计未来十年我国油棕引种、试种工作将非常频繁[48]，一些重要病虫害（如油棕枯萎病）随种子种苗等传入我国的风险显著增大。迄今，已有检疫性病害油棕枯萎病从非洲成功扩散到南美洲的案例[2]，这也给我们敲响了警钟，检疫部门以及引种单位需要格外重视和警惕。因此，在引进国外油棕优良品种时要采取严格的检疫措施；同时，针对油棕重大病虫害的预警、疫情监测或调查体系需要尽早建立，系统的油棕病害种类、危害、分布及生防资源调查等工作也亟待开展。最后，在油棕病害治理必须树立“预防为主，综合防治”的观念，有计划地采取防治措施。

参考文献

[1] Singh R, Ong A M, Low E T L, *et al.* Oil palm genome sequence reveals divergence of interfertile species in old and new worlds [J]. Nature, 2013, 500 (7462): 335-339.

[2] Corley R H V, Tinker P B. TheOil Palm (Fourth Edition) [M]. Oxford: Blackwell Science Ltd, 2003.

[3] Goh K J, Chiu S B, Paramanathan S. Agronomic Principles & Practices of Oil Palm Cultivation [M]. Malaysia: Agricultural Crop Trust, 2011.

[4] Lei X T, Cao H X. OilPalm (in Chinese) [M]. Beijing: China Agriculture Press, 2013.

[5] Cao J H, Lin W F, Zhang Y S. Strategic Researches on The Development of Oil Palm Industry in China (in Chinese) [J]. China Tropical Agriculture, 2009 (4): 15-18.

[6] Qin W Q, Zhu H. Identification and Management of Pests, Diseases and Rats in Palms (in Chinese) [M]. Beijing: China Agriculture Press, 2011: 49-68.

[7] Wei A M, Huang M J, Wang J. First report of pathogen identificaion and pathogenicity assay of wilt disease in *Neodypsis decaryi* and *Elaeis guineensis* (in Chinese) [J]. Guangdong Landscape Architecture, 2005, 30 (4): 33-34.

[8] Niu X Q, Tang Q H, Yu F Y, *et al.* Identification and biological characteristics of pathogen of leaf spot in oil palm (*Elaeis guineensis* Jacq) (in Chinese) [J]. Acta Agriculturae Jiangxi, 2011, 23 (11): 103-105.

[9] ShenH F, Zheng L, Zhang J X, *et al.* First report of *Pestalotiopsis microspora* causing leaf spot of oil palm (*Elaeis guineensis*) in China [J]. Plant Disease, 2014, 98 (10): 1429.

[10] Flood J. A review of *Fusarium* wilt of oil palm caused by *Fusarium oxysporum* f. sp. *elaeidis* [J]. Phytopathology, 2006, 96 (6): 660-662.

[11] Yu F Y, Zhu H, Qin W Q, *et al.* Reviews on basal stem rot of oil palm (in Chinese) [J]. China Tropical Agriculture, 2009, (2): 46-48.

[12] Paterson R R M. *Ganoderma* disease of oil palm: a white rot perspective necessary for integrated control [J]. Crop Protection, 2007, 26 (9): 1369-1376.

[13] Zhu H, Yu F Y, Lv L B, *et al.* Advances of research on basal stem rot of oil palm (in Chinese) [J]. Chi-

nese Agricultural Science Bulletin, 2008, 24 (9): 465-469.

[14] Beuther E, Wiese U, Lukacs N, *et al.* Fatal yellowing of oil palms: search for viroids and double-stranded RNA [J]. Journal of Phytopathology, 1992, 136 (4): 297-311.

[15] Torres G A, Sarria G A, Martinez G, *et al.* Bud rot caused by *Phytophthora palmivora*: a destructive emerging disease of oil palm [J]. Phytopathology, 2016, 106 (4): 320-329.

[16] Chen X X, Zeng H C, Ho H L, *et al.* Identification of *Pythium splendens*, a new record in Hainan and its pathogenicity test to oil palm (in Chinese) [J]. Journal of Yunnan Agricultural University, 2008, 23 (3): 222-225.

[17] Flood J, Mepsied R, Velez A, *et al.* Comparison of virulence of isolates of *Fusarium oxysporum* f. sp. *elaeidis* from Africa and South America [J]. Plant Pathology, 1993, 42 (2): 168-171.

[18] Flood J, Meosted R, Cooper R M. Population dynamics of *Fusariun* species on oil palm seeds following chemical and heat treatment [J]. Plant Pathology, 1994, 43 (1): 177-182.

[19] Mepsied K, Flood J, Cooper R M. *Fusarium* wilt of oil palm I. Possible causes of stunting [J]. Physiological and Molecular Plant Pathology, 1995, 46 (5): 361-372.

[20] Tengoua F F, Bakoume C. Basalstem rot and vascular wilt, two treats for the oil palm sector in Cameroon [J]. The Planter, 2005, 81 (947): 97-105.

[21] Cooper R M. *Fusarium* wilt of oil palm: a continuing threat to South East Asian plantations [J]. The Planter, 2011, 87 (1023): 409-418.

[22] Ntsefong G N, Ngando Ebongue G F, Paul K, *et al.* Control approaches against vascular wilt disease of *Elaeis guineensis* Jacq. caused by *Fusarium oxysporum* f. sp. *elaeidis* [J]. Journal of Biology and Life Science, 2012, 3 (1): 160-173.

[23] Lim T K, Chung G F, Ko W H. Basal stem rot of oil palm caused by *Ganoderma boninense* [J]. Plant Pathology Bulletin, 1992, 1 (3): 147-152.

[24] Ariffin D, Idris A S, Gurmit S. Status of *Ganoderma* in oil palm [M]. Flood J, Bridge P D, Holdernness M. *Ganoderma* disease of perennial crop. Wallingford: CABI Publishing, 2000: 49-68.

[25] Rao V, Lim C C, Chia C C, *et al.* Studies on *Ganoderma* spread and control [J]. The Planter, 2003, 79 (927): 367-383.

[26] Chung GF. Management of *Ganoderma* diseases in oil palm to minimize spreading in the fields [J]. The Planter, 2005, 81 (957): 765-773.

[27] Rees R W, Flood J, Hasan Y, *et al.* Effects of inoculums potential, shading and soil temperature on root infection of oil palm seedlings by the basal stem rot pathogen [J]. Plant Pathology, 2007, 56 (5): 862-870.

[28] Rahamath Bivi M, Siti Noor Farhana M, Khairulmazmi A, *et al.* Control of *Ganoderma boninense*: a causal agent of basal stem rot disease in oil palm with endophyte bacteria in vitro [J]. International Journal of Agriculture & Biology, 2010, 12 (6): 833-839.

[29] Chung G F. Management of *Ganoderma* diseases in oil palm plantations [J]. The Planter, 2011, 87 (1022): 325-339.

[30] Wong L C, Bong C F J, Idris A S. *Ganoderma* species associated with basal stem rot disease of oil palm [J]. American Journal of Applied Sciences, 2012, 9 (6): 879-885.

[31] Ho C L, Tan Y C, Yeoh K A, *et al.* De novo transcriptome analyses of host-fungal interactions in oil palm (*Elaeis guineensis* Jacq.) [J]. BMC Genomics, 2016, 17 (66): 1-19.

[32] Hasan Y, Foster H L, Flood J. Investigations on the causes of upper stem rot (USR) on standing mature oil palms [J]. Mycopathologia, 2005, 159 (1): 109-112.

[33] Oehlschlager A C, Chinchilla C, Castillo G, *et al.* Control of red ring disease by mass trapping of *Rhynchophorus palmarum* (Coleoptera: Curculionidae) [J]. The Florida Entomologist, 2002, 85 (3): 507-513.

[34] Morales F J, Lozano I, Sedano R, *et al.* Partial Characterization of a potyvirus infecting African oil palm in south America [J]. Journal of Phytopathology, 2002a, 150 (4-5): 297-301.

[35] Morales F J, Lozano I, Velasco A C, *et al.* Detection of a fovea-like virus in African oil palms affected by a lethal ' ringspot ' disease in south America [J]. Journal of Phytopathology, 2002b, 150 (11-12): 611-615.

[36] Lozano I, Morales F J, Martinez A K, *et al.* Molecular characterization and detection of African oil palm ringspot virus [J]. Journal of Phytopathology, 2009, 158 (3): 167-172.

[37] King A M Q, Adams M J, Carstens E B, *et al.* Virus taxonomy: ninth report of the international committee on taxonomy of viruses [M]. Netherlands: Academic Press, 2012.

[38] Swinburne T R. Fatal yellows, bud rot and spear rot of African oil palm-a comparison of the symptoms of these diseases in Brazil, Ecuador and Colombia [J]. The Planter, 1993, 69 (802): 15-23.

[39] Van de Lande H L. Studies on the epidemiology of spear rot in oil palm (*Elaeis guineensis* Jacq.) in Suriname [D]. The Netherlands: Wageningen Agricultural University, 1993a.

[40] Van de Lande H L. Spatio-temporal analysis of spear rot and ' marchitez sorpresiva ' in African oil palm in Surinam [J]. Netherlands Journal of Plant Pathology, 1993b, 99 (Supplement 3): 129-138.

[41] Van de Lande H L, Zadoks J C. Spatial patterns of spear rot in oil palm plantations in Surinam [J]. Plant Pathology, 1999, 48 (2): 189-201.

[42] Suwandi, Akino S, Kondo N. Common spear rot of oil palm in Indonesia [J]. Plant Disease, 2012, 96 (4): 537-543.

[43] Bergamin Filho A, Amorim L, Laranjeira F F, *et al.* Análise temporal do amarelecimento fatal do dendezeiro como ferramenta para elucidar sua etiologia [J]. Fitopatologia Brasileira, 1998, 23 (3): 391-396.

[44] Torres G A, Sarria G A, Varon F, *et al.* First report of bud rot caused by *Phytophthora palmivora* on African oil palm in Colombia [J]. Plant Disease, 2010, 94 (9): 1163-1163.

[45] Benítez É, García C. The history of research on oil palm bud rot (*Elaeis guineensis* Jacq.) in Colombia [J]. Agronomía Colombiana, 2014, 32 (3): 390-398.

[46] Kochu B M, Ramachandran N K. Distribution of spear rot disease of oil palm (*Elaeis guineensis* Jacq.) and its possible association with MLO disease of palm in Kerala, India [J]. The Planter, 1993, 69 (803): 59-66.

[47] Alvarez E, Mejía J F, Contaldo N, *et al.* ' *Candidatus* Phytoplasma asteris ' strains associated with oil palm lethal wilt in Colombia [J]. Plant Disease, 2014, 98 (3): 311-318.

[48] Tang Q H, Lei X T, Qin W Q. Integrated pest management and ecological security of oil palm in China (in Chinese) [J]. Chinese Agricultural Science Bulletin, 2014, 98 (3): 311-318.

油棕病害研究进展及存在的主要问题（续）*

唐庆华**，曹红星，覃伟权***

（中国热带农业科学院椰子研究所，海南省热带油料作物生物学重点实验室，
农业部热带油料作物科学观测试验站，文昌　571339）

摘　要：近年来，随着油棕种植面积的扩大，世界上已经出现了一些严重甚至毁灭性的病害。目前，我国已把油棕列入战略规划。但是，我国对油棕病害研究尚很薄弱。为了便于科研人员更好地开展病害防治工作，本文系统介绍了国内外主要油棕病害的发生、分布及防治的研究进展，然后分析了油棕病害研究中存在的问题，最后阐述了我国油棕病害综合防治的意义。本综述旨在为我国油棕产业的持续、顺利和健康发展提供有益的帮助。

关键词：油棕；病害；病原；综合防治；检疫

Current status and Key Problemsin in diseases of oil palm（Continued）

Tang Qinghua，Cao Hongxing，Qin Weiquan *

（*Coconut Research Institute*，*Chinese Academy of Tropical Agricultural Sciences*，
Hainan Key Laboratory for Biology of Tropical Oil Crops，*Scientific Observing*
and Experimental Station of Tropical Oil Plants of Minitary of Agriculture，
China，*Wenchang*，571339）

Abstract：Recently，with the area under oil palm had expanded in the world，there have been out-breaks of serious even devastating disease. At present，China has set a strategic planning to develop the oil palm industry. However，its research on oil palm disease is still very weak. In order to facilitate the researchers to study on the diseases more effectively，a systematical introduction of occurrences，distributions，and managements of important oil palm diseases were made in this review，and then，issues regard to problems in the studies of oil palm diseases were analyzed，finally the importance of integrated pest management in China was elaborated. The aim of this review is to provide some constructive insights for the sustainable，successful and healthy development the oil palm industry in China.

Key words：oil palm；disease；pathogen；integrated management；quarantine

油棕（*Elaeis guineensis* Jacg.）为全球生产效率最高的产油植物[1]，被誉为“世界油王”。迄

＊ 基金项目：农业农村部农业技术试验示范与服务支持项目（农垦）：油棕品种区域适应性试种（16RZNJ-5）

＊＊ 第一作者：唐庆华，副研究员，博士，主要从事热带油料和棕榈作物病害综合防治技术研究；E-mail：tchuna129@163. com

＊＊＊ 通信作者：覃伟权，硕士，研究员，主要从事热带油料和棕榈作物植物保护综合防治研究；E-mail：QWQ268@ 163. com

今，油棕已发展成为世界第一大油料作物和最具发展潜力的战略资源作物[2-4]。自从第二次世界大战后，随着全球人口急剧增加、经济发展、生活水平的提高，人类对油脂和石化能源需求越来越多。由于油棕产油能力极强以及作为生物能源的潜能巨大，因而备受关注。目前，东南亚、非洲和南美洲的许多热带国家和地区纷纷大力发展油棕产业。但是，随着种植面积的迅速扩大世界上已出现了多种重大甚至是毁灭性病害[5-6]。据报道，油棕病害共有 32 种[3,5]。其中，毁灭性病害有枯萎病[7-10]、茎基腐病[10-17]、致死性黄化病[18-19]等。目前，我国已制定了发展油棕产业的战略规划，但是我国对油棕病害研究尚很薄弱。为了促进我国油棕科研工作者对油棕病害研究现状的认识，本文对世界范围内最近 20 年来油棕病害的发生、危害及防治、存在的问题以及病害综合防治的意义进行了较为详细的介绍和阐述，本综述旨在为我国油棕科研人员提供参考依据，确保我国油棕产业的顺利健康发展。

1 箭叶腐烂-小叶病

1.1 为害与症状

箭叶腐烂-小叶病（Spear rot-little leaf disease）曾被称为芽腐-小叶病（Bud rot-little leaf）。该病曾在刚果南部地区严重发生，死亡率超过 30%。该病在西非其他地区和亚洲很少发生，且多仅限于一些特定的油棕材料，很少导致植株死亡[5]。该病最明显的症状是未展开的箭叶基部褐色湿腐。发病较轻时只为害叶片，造成相邻箭叶的小叶相继腐烂。穗轴发病后箭叶塌陷、倒伏；如果腐烂继续向下扩展，则可能引起芽腐症状，但只有到达生长点时油棕才会死亡。随后抽出的第一片复叶的羽轴基部畸形，之后长出特别短小、皱缩的“小叶”，但后续长出的叶片较长，叶片较少异常。

1.2 病因

引起“小叶”症状的病因很多。常见的病因是缺硼。在刚果，一种类似于 *Erwinia lathyri* 的细菌也可引起小叶症状。该病也可能是遗传性、生理性和季节性原因所致。

1.3 防治

该病并不严重，无需采取防治措施，但要保证充足的水分和肥料供应。

2 拟盘多毛孢叶斑病

2.1 为害与症状

拟盘多毛孢叶斑病（Pestalotiopsis leaf spot）又称叶枯病（Leaf wither）或灰枯病（Grey leaf blight）。该病在哥伦比亚、厄瓜多尔和洪都拉斯的部分地区发生严重，造成大量落叶，对产量影响很大。该病可导致成龄油棕园鲜果产量从 18~20 t/hm^2 降至 12~15t/hm^2，低龄油棕园产量从 11 t/hm^2 减少至 7~8 t/hm^2[5]。在拉丁美洲，该病发病也很严重。发病初期叶片上出现带有黄色晕圈的褐色小斑点。这些斑点很快扩展并汇集成褐色坏死斑，随后病斑变灰、易碎。病斑褐色和灰色带交界处有一条明显的分界线，在灰色部分可以发现许多小黑点，这是病原菌产生的分生孢子。

2.2 病原

病原为拟盘多毛孢菌 *Pestalotiopsis* sp.。常见的拟盘多毛孢有两种，但长蠕孢属和弯孢属真菌等也能侵染叶片。此外，*Leptopharsa gibbicarina* 和 *Peleopoda arcanella* 也有助于拟盘多毛孢菌侵染。

2.3 防治

①农业防治：合理施肥，注意多施含镁的肥料并保持钾和镁的平衡。②化学防治：喷施残杀威、杀螟硫磷、磷胺等农药；树干注射或在根部施用吸收内吸性杀菌剂。③生物防治：用白僵

菌、拟青霉菌防治虫媒。

3　黄斑病

3.1　为害与症状

黄斑病（Patch yellows）仅限于非洲，尽管分布广泛，但该病为零星发生且只侵染一小部分油棕。在马来西亚，有一种被称为“顶枯病”（wither tip）的类似病害，其病原为尖孢镰刀菌 *Fusarium oxysporum* 和腐皮镰刀菌 *F. solani*。该病发生在未展开的箭叶上。叶片展开后病斑在羽叶上对称出现，病斑圆形或椭圆形，周围有淡黄色的晕圈，有时中心呈棕色。随后斑块中心干枯、脱落，形成穿孔；黄色斑块继续扩展、颜色变暗，可以看到橙色小斑点。

3.2　病原

病原为尖孢镰刀菌 *Fusarium oxysporum*，属半知菌类、镰刀菌属。

3.3　防治

可通过育种筛选剔除易感油棕品系，但这种病害还没有严重到值得采取该措施的程度。

4　尾孢菌叶斑病

4.1　为害与症状

尾孢菌叶斑病（Cercospora Leaf spot）（又称雀斑病 freckle）在非洲发生普遍，但在亚洲或美洲尚未见报道。有报道该病可导致油棕产量降低超过 10%，但也有报道称该病对产量影响不大[20]。幼苗最幼嫩的叶片发病后形成半透明的小斑点，周围有黄绿色晕圈；斑点逐渐扩大，呈黑褐色。随后这些病斑汇集到一起，叶片干枯，呈灰褐色。

4.2　病原

病原为油棕尾孢菌 *Cercospora elaeidis*，属半知菌类、尾孢属真菌，该菌在田间可以存活数年。

4.3　防治

①加强苗圃管理，去除老叶和严重感染的叶片。②合理施肥，多施用钾肥并减少氮肥施用量。③化学防治，喷施代森锰锌和苯菌灵等药剂。④培育抗（耐病）品种。

5　褐胚病

5.1　为害与症状

褐胚病（Brown germ）是一种常见病害。该病症状为芽点上产生褐色斑点，病斑随着幼胚的发育不断扩展并汇集到一起，随后病组织开始变黏、腐烂。湿热法催芽容易引起褐胚病，10~38℃潮湿条件下最易发生。

5.2　病原

与该病有关的真菌有 27 种，其中曲霉菌和青霉菌最常见。

5.3　防治

①催芽时注意采取卫生措施，清除发病油棕种子。②采用干热法催芽：在 39.5℃干燥条件进行催芽。

6　白纹病

6.1　为害与症状

白纹病（White stripe）为零星发生。在亚洲，该病在有机黏土或腐殖土中种植的油棕上较常见。该病中度发生可造成减产 15%，严重发生时能导致减产约 50%[5]。黄白色细条斑分布于叶片中脉两侧。染病油棕通常约 7 个月后能恢复绿色，嫩叶恢复更快。在马来西亚，该病发生严重且常发生在2~3龄油棕上，3~5龄油棕发病最为严重，随后病害缓慢扩展。

6.2 病因

先前认为该病为遗传原因所致，一些相关研究证据也支持该观点。但是，现在普遍认为该病是营养失调所致，如缺硼以及叶片氮与钾比值较高引起的[21]，但氮与钾比值并不是一个合适的病害指标。

6.3 防治

多施用碳酸钾同时减少氮肥使用量；当穗轴钾含量较低时应施用钾肥，而叶片氮含量超过2.8%时应减少氮肥施用量[21]。

7 果穗和果实腐烂病

7.1 为害与症状

从世界范围来看，果穗和油棕果实腐烂病偶尔发生，目前对其研究较少。该病在马来西亚发生较多；在国内，该病曾普遍发生，发病最重的地区腐烂果穗高达80%，最低为7.23%[22]。在西非，一种果穗梗部腐烂的病害被称为叶基枯萎病（leaf base wilt），该病引起的腐烂发生较少，果穗弯曲后大部分油棕果都能发育。常见的症状为果穗和油棕果腐烂，症状为叶片朝地面弯曲，腋叶处的穗梗也随之弯曲，随后开始腐烂。油棕果腐病可分为干枯型和湿腐型两种症状，影响因子包括环境条件、栽培管理等。

7.2 病因

在马来西亚，果穗顶端腐烂常与Deli油棕材料有关。叶基枯萎病可能是机械原因造成。油棕果腐病是一种非侵染性生理病害，尽管果实可被各种腐生细菌和真菌（炭疽菌、镰刀菌和*Marasmius palmivorus*）侵染，但这些腐生菌对健康果实和果穗并无致病力。

7.3 防治

最好的防治方法是通过卫生措施创造不利于真菌生长的环境条件：在抽穗盛期前保证肥料和水分供应，去除腐烂的果穗；在多雨季节，防止油棕园积水；在干旱季节，要适时适量灌溉；防止过度割叶，要保留较多的有效叶片。

8 次要油棕病害

除了主要油棕病害，在苗圃中还有几种箭叶腐烂病和芽腐病（Seedling spear and bud rots），但危害不重。此外，油棕叶片常因一些次要病原菌的侵染而易呈现褪绿和坏死症状，叶面也易被附生和腐生微生物所覆盖，从而显著影响光合作用。油棕次要或很少发生的病害非常多[5,22-32]，这些病害多是根据症状特点命名的，如铜纹病、环斑病等。其中，苗期病害有炭疽病[5,22]、幼苗枯萎病、帚梗柱孢菌叶斑病、幼苗箭叶腐烂病[5,19]、伏革菌叶腐病、芽干腐病、环斑病毒病[23-26]，成龄期叶部病害有拟茎点霉叶斑病[27]、小孢拟盘多毛孢叶斑病[28]、轮枝孢菌枯萎病[29]、雀眼斑病[22]、壳斑病、枯斑病[23,30]、藻斑病、蜜环菌茎腐病[22]、基部腐烂病、基部焦腐病、茎干湿腐病[31]。果实病害有黑腐病[32]。历史上一些苗期叶部病害曾造成严重危害，但现在已问题不大。在这些病害中，煤烟病（sooty mould）较常见的次要病害，由刺盾炱属、*Brooksia*和*Ceramothyrium*属真菌引起。该病发病较轻时叶片上表面出现零星的环状病斑，下表面分布有不规则的斑块或稀疏的霉状物，发病严重时整个叶片表面几乎布满黑色霉层。另外，地衣是油棕上常见的一种附生植物，可在叶片上表面上形成灰绿色壳状体。

9 其他异常情况

油棕生长和发育受许多异常条件的影响，如贫瘠的沙质土壤、长期干旱、湿度过大或内涝、杂草丛生、土壤异常等。一种油棕生长异常情况为植株衰退症，症状表现为油棕几乎停止生长，根和箭叶生长速率降低、绿色叶片数量减少，存活的叶片仍然保持挺立、紧凑，形成一个锥形的茎干，叶片逐渐退化，褪绿，干枯，变脆。在非洲，该病被认为是遗传原因或严重缺钾和镁元素或油棕园崎岖不平所致。在亚洲，该病很少发生。

在美洲，在遭受红环腐病危害的油棕园会发生一种被称为“Choke”病或矮冠病（Dwarfed crown）的病害。该病发生时所有叶片都比正常叶小，叶片绿色、挺立，果穗萎缩、扭曲伴有大量叶片萎缩或形成皱褶。发病油棕常突然恢复健康，一簇簇正常的新叶从变形的叶片心部伸出来，使油棕树冠外观上象有两层。该病类似于“Hoja pequena”病（小叶病），但后者是指箭叶腐烂-小叶病恢复后的症状。在其他油棕种植区也有这种畸形现象。在马来西亚，“Choke”病被来描述一种类似的油棕生理性病害。

此外，水涝也可造成花序大量败育，给低龄油棕种造成严重损失，甚至有时可导致油棕死亡；雷击偶尔也会导致油棕死亡[33]。

10　油棕病害研究中存在的问题及有害生物综合防治

10.1　油棕病害研究中存在的问题

尽管不同学者对油棕病害进行了大量研究，但是依然存在一些突出的问题。首先，一些病害症状相似，致使在病害识别和命名上很容易混淆，各国语言和命名习惯的不同也容易带来病害名称上的问题。例如，环斑病（ring spot）现用于一种苗圃病害，但也曾被用于叶片斑驳病（Leaf mottle）。再如“Choke”病（一种油棕生理性病害）与“Hoja pequena”病（小叶病）均可导致“小叶”症状类似，但后者为指箭叶腐烂-小叶病恢复后的症状。此外。其次，一些病害病因复杂，需要审慎辨别症状表现、病原或病因。如油棕冠部死亡可能是发生了根部或茎部病害，通过水分和养分缺失造成的；也可能是患芽腐和箭叶腐烂，通过病原菌侵染生长点致死的。同时，一些病害病原复杂，需要分别不同病原的症状。如致死性黄化病，早期的研究认为该病可能由真菌、细菌或类病毒等引起，学界争议较大。但是，随着研究的深入，对该病的了解逐渐清晰，近年的文献更准确地使用“芽腐病”[19]或“箭叶腐烂病”（由棕榈疫霉 *Phytophthora palmivora*）代替，很少再使用“致死性黄化病”，而且后者现多用于描述植原体引起的病害。再次，一些病害在不同的国家和地区病害症状表现和发病率可能有较大差异，且环境条件的影响很重要。此外，一些“防治”措施或农业操作可能适得其反，最终反而加重了病害的发生，如除草剂的不合理施用以及修剪用的工具不注意消毒等。最后，值得注意的是尽管培育抗（耐）病品种对其他一些作物来说是一种最经济有效的措施，但对油棕而言该措施具有显著的局限性，原因是一些野生种或杂交油棕种质材料虽然抗（耐）病，但产量较低，而育种周期较长也限制了其应用。

10.2　油棕有害生物综合防治

目前，生物入侵已成为我国生物及生态安全中一个十分重要和紧迫的问题[34]。由于海南省地处热带地区，温湿度适宜，其独特的生态和环境条件特别有利于外来生物（尤其是害虫）入侵并建立种群造成严重危害，典型的案例为检疫性害虫椰心叶甲[35-38]和红棕象甲[39-40]。其他一些棕榈科植物（包括油棕）上的重大病虫害也有潜在入侵可能，如椰子织蛾，对该虫发出预警的时间为2013年年初[41]，而当年8月即发现该虫已成功入侵海南省，现已给该省海口、三亚、陵水、万宁、文昌、儋州等市县的椰子造成了严重危害[42]。在我国，尽管至今尚未出现油棕致死性黄化病、枯萎病和茎基腐病等毁灭性病害，但笔者实验室发现入侵害虫红棕象甲为害并导致油棕死亡（本实验室未发表资料），其他害虫如椰心叶甲[3]、红棕象甲[22,43]和二疣犀甲[44]等也可严重威胁油棕。因此，开展对油棕有害生物（病虫鼠害等）的系统调查、检测、建立重大病虫害预警系统、制订综合防治措施具有重要现实意义。值得关注和警惕的是一些病害（如油棕枯萎病）常通过真菌孢子进行传播，油棕种子也可能携带有线虫[5]。目前，随着油棕种质资源交流、世界贸易以及人员流动性越来越频繁，一些重要病虫害随之扩散的风险显著加大，现已有枯萎病已从非洲成功扩散到南美洲的报道[5]。目前，我国已制定了积极发展油棕和油茶等木本油料的战略规划，预计今后十年是我国油棕种质资源引种、试种的活跃期，为了防止油棕枯萎病、茎基腐病等病原以及一些可能严重危害其他植物的害虫传入，必须进行严格的检疫措施。总之，采取严格的检疫措施并系统开展油棕有害生物综合治理研究对我国油棕种植业顺利健康发展具有重要意义。

11 结语

油棕是世界上产油效率最高的植物[1,45]，发展油棕产业对于维护我国粮油安全、促进热区农民就业和增收等具有十分重要的意义[2-3]。但是由于受历史上品种适应性、配套技术、“粮油争地”等原因的显著制约，我国油棕生产发展曲折缓慢。最近，我国已把油棕纳入战略规划，油棕发展已获得国际、国家、地方等多层面的支持。但是，受 20 世纪引种油棕失败的影响，我国油棕种植业发展依然面临种种质疑、阻力和困难。因此，油棕科研工作者需要克服种种不利现实条件，扎扎实实地开展科研工作，切实做好适合我国种植的优良、高效品种的筛选工作，通过油棕种质资源圃的品种示范作用促进农业主管部门以及农民“思想解放”，推动我国油棕产业顺利发展。目前，我国发现的油棕病害有果腐病[3]、苗疫病[3,22]、拟茎点霉叶斑病[27]和小孢拟盘多毛孢叶斑病[28]等，笔者近几年的调查结果显示我国油棕病害以苗期或低龄油棕病害为主。目前，尚未发现大面积发生严重病害，但是，非洲、中南美洲以及东南亚地区已经暴发了油棕枯萎病、茎基腐病、致死性黄化病等毁灭性病害以及一些重要虫害。同时，预期我国今后十年内油棕引种、试种工作将非常活跃。因此，在引进国外油棕优良品种（种质资源）的同时，也必须开展油棕病害的病原学、流行学、生防资源以及害虫生命周期和生态学、天敌资源的系统调查工作，建立确疫情监测或调查体系，制订详细的重要病害综合防治预案和措施。最后，油棕枯萎病已从非洲成功扩散到南美洲[5]以及重要检疫害虫椰心叶甲[36-38]和红棕象甲[42-43]等入侵我国并造成严重危害的生物入侵教训告诉我们必须严格的检疫措施，防止重要病虫害（包括危害油棕的病毒和线虫）随油棕种子、中间宿主或昆虫媒介等传入我国。总之，做好入境检疫、病害调查、疫情调查体系并制订周详的重要病虫害综合防治预案对我国油棕产业顺利健康发展具有重要意义。

参考文献

[1] Singh R，Ong A M，Low E T L，*et al.* Oil palm genome sequence reveals divergence of interfertile species in old and new worlds [J]. Nature，2013a，500 (7462)：335-339.

[2] Cao H X，Sun C X，Shao H B，*et al.* Effects of low temperature and drought on the physiological and growth changes in oil palm seedlings [J]. African Journal of Biotechnology，2011，10 (14)：2630-2637.

[3] Lei X T，Cao H X. OilPalm (in Chinese) [M]. Beijing：China Agriculture Press，2013.

[4] Lin W F. Current Situation of Planting and Using of Oil Palm and Its Analysis of Developing Prospect in China (in Chinese) [M]. Beijing：China Agricultural Science and Technology Press，2010.

[5] Corley R H V，Tinker P B. TheOil Palm (Fourth Edition) [M]. Oxford：Blackwell Science Ltd，2003：391-422.

[6] Goh K J，Chiu S B，Paramanathan S. Agronomic Principles & Practices of Oil Palm Cultivation [M]. Malaysia：Agricultural Crop Trust，2011：389-481.

[7] Flood J. A review of *Fusarium* wilt of oil palm caused by *Fusarium oxysporum* f. sp. *elaeidis* [J]. Phytopathology，2006，96 (6)：660-662.

[8] Yu F Y，Zhu H，Qin W Q，*et al.* Review on basal stem rot of oil palm (in Chinese) [J]. China Tropical

Agriculture, 2009 (2): 46-48.

[9] Ntsefong G N, Ngando Ebongue G F, Paul K, *et al*. Control approaches against vascular wilt disease of *Elaeis guineensis* Jacq. caused by *Fusarium oxysporum* f. sp. *elaeidis* [J]. Journal of Biology and Life Science, 2012, 3 (1): 160-173.

[10] Tengoua F F, Bakoume C. Basalstem rot and vascular wilt, two treats for the oil palm sector in Cameroon [J]. The Planter, 2005, 81 (947): 97-105.

[11] Paterson R R M. *Ganoderma* disease of oil palm: a white rot perspective necessary for integrated control [J]. Crop Protection, 2007, 26 (9): 1369-1376.

[12] Zhu H, Yu F Y, Lv L B, *et al*. Advances of research on basal stem rot of oil palm (in Chinese) [J]. Chinese Agricultural Science Bulletin, 2008, 24 (9): 465-469.

[13] Rahamath Bivi M, Siti Noor Farhana M, Khairulmazmi A, *et al*. Control of *Ganoderma boninense*: a causal agent of basal stem rot disease in oil palm with endophyte bacteria in vitro [J]. International Journal of Agriculture & Biology, 2010, 12 (6): 833-839.

[14] Rao V, Lim C C, Chia C C, *et al*. Studies on *Ganoderma* spread and control [J]. The Planter, 2003, 79 (927): 367-383.

[15] Rees R W, Flood J, Hasan Y, *et al*. Effects of inoculums potential, shading and soil temperature on root infection of oil palm seedlings by the basal stem rot pathogen [J]. Plant Pathology, 2007, 56 (5): 862-870.

[16] Chung G F. Management of *Ganoderma* diseases in oil palm plantations [J]. The Planter, 2011, 87 (1022): 325-339.

[17] Wong L C, Bong C F J, Idris A S. *Ganoderma* species associated with basal stem rot disease of oil palm [J]. American Journal of Applied Sciences, 2012, 9 (6): 879-885.

[18] Beuther E, Wiese U, Lukacs N, *et al*. Fatal yellowing of oil palms: search for viroids and double-stranded RNA [J]. Journal of Phytopathology, 1992, 136 (4): 297-311.

[19] Torres G A, Sarria G A, Martinez G, *et al*. Bud rot caused by *Phytophthora palmivora*: a destructive emerging disease of oil palm [J]. Phytopathology, 2016, 106 (4): 320-329.

[20] Jacquemard J C. Oil Palm [M]. London: Macmillan Education, 1998.

[21] Tohiruddin L, Prabowo N E, Foster H L. Cause and correction of white stripe in oil palm [A]. International Oil Palm Conference, Indonesian Oil Palm Research Institute, Bali, 2002: 8-12.

[22] Qin W Q, Zhu H. Identification and management of pests, diseases and rats in palms (in Chinese) [M]. Beijing: China Agriculture Press, 2011.

[23] Hanold D, Randles J W. Detection of coconut cadang-cadang viroid-like sequences in oil and coconut palm and other monocotyledons in the south-west Pacific [J]. Annals of Applied Biology, 1991, 118 (1): 139-151.

[24] Morales F J, Lozano I R, Sedano M, *et al*. Partial characterization of a potyvirus infecting African oil palm in South America [J]. Journal of Phytopathology, 2002, 150 (4-5): 297-301.

[25] Vadamalai G, Hanold D, Rezaian M A, *et al*. Variants of Coconut cadang-cadang viroid isolated from an African oil palm (*Elaies guineensis* Jacq.) in Malaysia [J]. Archives of Virology, 2006, 151 (7): 1447-1456.

[26] Lozano I, Morales F J, Martinez A K, *et al*. Molecular characterization and detection of African oil palm ringspot virus [J]. Journal of Phytopathology, 2009, 158 (3): 167-172.

[27] Niu X Q, Tang Q H, Yu F Y, *et al*. Identification and biological characteristics of pathogen of leaf spot in oil palm (*Elaeis guineensis* Jacq) (in Chinese) [J]. Acta Agriculturae Jiangxi, 2011, 23 (11): 103-105.

[28] Shen H F, Zheng L, Zhang J X, *et al*. First report of *Pestalotiopsis microspora* causing leaf spot of oil palm (*Elaeis guineensis*) in China [J]. Plant Disease, 2014, 98 (10): 1429-1429.

[29] Wei A M, Huang M J, Wang J. First report of pathogen identificaion and pathogenicity assay of wilt disease

inNeodypsis decaryi and *Elaeis guineensis* (in Chinese) [J]. Guangdong Landscape Architecture, 2005, 30 (4): 33-34.

[30] Selvaraja S, Balasundram S K, Vadamalai G, *et al.* Spatial variability of orange spotting disease in oil palm [J]. Journal of Biological Sciences, 2012, 12 (4): 232-238.

[31] Chander Rao S. Some observations on stem wet rot disease of oil palm (*Elaeis guineensis*) in Andhra Pradesh, India [J]. The Planter, 1997, 73 (860): 611-614.

[32] Eziashi E I, Uma N U, Adekunle A A, *et al.* Biological control of *Ceratocystis paradoxa* causing black seed rot in oil palm sprouted seeds by *Trichoderma* species [J]. Pakistan Journal of Biological Sciences, 2006, 9 (10): 1987-1990.

[33] Teoh C H, Ng A, Prudente C, *et al.* Balancing the need for sustainable oil palm development and conservation: the lower Kinabatangan floodplains experience [M]. Pushparajah E. Strategic directions for The sustainability of the oil palm industry. Kuala Lumpur: Incorporated Society of Planters, 2001: 1-25.

[34] Wan F H, Guo J Y, Zhang F, *et al.* Research on biological invasions in China (in Chinese) [M]. Beijing: Science Press, 2009.

[35] Lu B Q, TangC, Peng Z Q, *et al.* Biological assessment in quarantine of *Asecodes hispinarum* Boucek (Hymenoptera: Eulophidae) as an imported biological control agent of *Brontispa longissima* (Gestro) (Coleoptera: Hispidae) in Hainan, China [J]. Biological Control, 2008, 45 (1): 29-35.

[36] Li C X, Huang S C, Qin W Q, *et al.* The control efficiency of parasitoids on *Brontispa longissima* (Gestro) in Sanya City (in Chinese) [J]. Chinese Journal of Tropical Crops, 2012, 33 (7): 1288-1292.

[37] Lv B Q, Jin Q A, Wen H B, *et al.* Current status of *Brontispa longissima* outbreaks and control (in Chinese) [J]. Chinese Journal of Applied Entomology, 2012, 49 (6): 1708-1715.

[38] Jin T, Jin Q A, Wen H B, *et al.* Research progress and perspective outlook of using parasitic wasps to control *Brontispa longissima* (in Chinese) [J]. Chinese Journal of Tropical Agriculture, 2012, 32 (7): 67-74.

[39] Li L, Qin W Q, Ma Z L, *et al.* Effect of temperature on the population growth of *Rhynchophorus ferrugineus* (Coleoptera: Curculionidae) on sugarcane [J]. Environmental Entomology, 2010, 39 (3): 999-1003.

[40] Wei J, Qin W Q, Ma Z L, *et al.* Occurrent status and research advvance on main control measures of *Rhynchophorus ferrugineus* (Olivier) (in Chinese) [J]. Guangdong Agricultural Sciences, 2009 (6): 110-112.

[41] Lv B Q, Yan Z, Jin Q A, *et al.* Exotic pest alert: *Opisina arenosella* (Lepidoptera: Oecophoridae) (in Chinese). Journal of Biosafety, 2013, 22 (1): 17-22.

[42] Yan W, Tao J, Liu L, *et al.* Precaution of an alien pest invasion, the coconut black-headed caterpiller, *Opisina arenosella* (in Chinese) [J]. Plant Protection, 2015, 41 (4): 212-217.

[43] Qin W Q, Yan W. Mornitoring and control of the red palm weevil (in Chinese) [M]. Beijing: China Agriculture Press, 2013.

[44] Geoffrey B O. Biology and management of palm dynastid beetles: recent advances [J]. Annual Review Entomology, 2013, 58: 353-372.

[45] Singh R, Low E T L, Ooi L C L, *et al.* The oil palm SHELL gene controls oil yield and encodes a homologue of SEEDSTICK [J]. Nature, 2013b, 500 (7462): 340-344.

橡胶树炭疽病无人机精准施药技术初探*

郑肖兰[1,3,4]**，郑行恺[2]，刘先宝[1]，李博勋[1]，时　涛[1]，黄贵修[1]***

（1. 中国热带农业科学院环境与植物保护研究所，海口　571101；2. 海南大学环境与植物保护学院，海口　570228；3. 农业部热带农林有害生物入侵监测与控制重点开放实验室，海口　571101；4. 海南省热带农业有害生物监测与控制重点实验室，海口　571101）

摘　要：鉴于目前智能化植保机具研制与精准施药技术的推广，本研究针对橡胶树炭疽病防治新型药剂“保叶清”微乳剂无人机飞防试验结果发现，在苗圃地飞行高于树冠2m，未添加飞防助剂，以3%和4%浓度的“保叶清”微乳剂进行施药校正防效分别为46.33%和55.91%；飞行高于树冠1m，添加1%飞防助剂，以3%和4%浓度的“保叶清”微乳剂进行施药校正防效分别为67.42%和73.27%；飞行高于树冠2m，添加1%飞防助剂，以3%和4%浓度的“保叶清”微乳剂进行施药校正防效分别为71.27%和80.07%。在试验剂量下对橡胶树及环境无不良影响。此外，通过对橡胶树成龄胶林（海南阳江和云南勐捧）进行无人机精准施药防治技术熟化，并使用水敏纸图像分析的方法检测，结果发现橡胶树割龄12年的胶林下层叶片雾滴粒良好、飘移损失较小、雾群分布均匀；该结果显示使用无人机进行成龄开割树防治炭疽病方法可行，且能胜任平地胶林和山地胶林的相关防治。

关键词：橡胶树；炭疽病；无人机；精准施药技术

---

* 基金项目：国家重点研发计划项目（2018YFD0201105）；海南省重点研发计划课题（ZDYF20182402）；海南省自然科学基金面上项目（317230）

** 第一作者：郑肖兰，硕士，副研究员，研究方向：植物病理学；E-mail：orchidzh@163.com

*** 通信作者：黄贵修；E-mail：hgxiu@vip.163.com

第八部分　其　他

A rapid method to quantify fungicide sensitivity in the wheat stripe rust pathogen *Puccinia striiformis* f. sp. *tritici*[*]

Peng Furong[**], Fu Yuhang, Yu Yang, Yang Yuheng, Fang Anfei, Bi Chaowei[***]
(*College of Plant Protection*, *Southwest University*, *Chongqing* 400715, *China*)

Abstract: Wheat stripe rust is one of the most important and devastating disease in wheat production. Under intense disease pressure, it is of great significance to quantify fungicide sensitivity of wheat stripe rust quickly and accurately for monitoring and controlling the disease. A rapid method based on detached leaf is proposed in this paper. The specific methods are as follows. 5−10 μg/mL uredospore suspension, prepared by Tween 20 (concentration 0. 02%) was inoculated by art−brush (the diameter of 0. 03 mm) when wheat seedlings fullyunfoldthefirstleaf. After 8 d of inoculation, 75 μg/mL 6−Benzylaminopurine and different concentrations of fungicide was added into 0. 5% water agar. to prepared drug−containing medias. The 9 cm circular filter papers were cut into 3 segments of 2. 5 cm, 4. 0 cm, 2. 5 cm along the diameter, and the 4 cm filter paper was placed in the middle of the plate. 5cm middle leaf segments of infected wheat with obvious lesions were cut and placed on a 4cm filter paper in parallel, and−the ends of the leaf segments were fixed with the remaining filter paper to ensure them adhered to the surface of the drug−containing mediums. Each dish contains 5 leaf segments of different pots. After 8 days of inoculation, the proportion of spore heap was accurately measured by Photoshop. The control effect and 50% effective concentration (EC_{50}) were calculated by EXCEL. The test results show that this method can make the fungicide better acted on *Pst* through the isolated wheat leaf segment. And the sensitivity of *Pst* to triadimefon in this way is close to that of spraying potted wheat and the control effect is relatively stable. Virulence of triadimefon to wheat stripe rust, collected in Chongqing, was determined by the above method. The results showed the frequency distribution of 81 isolates to triadimefon distributed approximate normal (W=0. 204). Therefore, the measured 0. 1453 ± 0. 0081 μg/mL can be used as a baseline for the sensitivity of *Pst* to triadimefon. And this method can be used to measure the virulence of the fungicide against wheat stripe rust indoor. Compared with the traditional living pot spray and the chemical dressing method, the method of the tablet containing medicine is more precise, and the amount of plant material is small and the repeatability is high.

Key words: *Puccinia striiformis* f. sp. *tritici*; Fungicide Sensitivity; Detached leaves; Triadimefon

---

* 基金项目：国家重点研发计划“长江流域冬小麦化肥农药减施技术集成研究与示范”（2018YFD0200500）；中央高校基本科研业务费专项资金（XDJK2017B026）

** 第一作者：彭复蓉，硕士研究生，主要从事植物真菌病害方面的研究；E-mail：1334599779@ qq. com
*** 通信作者：毕朝位，副教授，主要从事植物真菌病害及病原菌抗药性研究；E-mail：chwbi@ swu. edu. cn

Apolygus lucorum effector AI6, encoding a Glutathione peroxidase manipulates PAMP-triggered immunity

Dong Yumei*, Jing Maofeng, Shen Danyu, Zhang Meiqian, Wang Chenyang, Karani T. Nyawira, Zuo Kairan, Wu Shuwen, Dou Daolong, Xia Ai**

(*Department of plant protection*, *Nanjing Agricultural University*, *Nanjing* 210095, *China*)

Abstract: The mired bug *Apolygus lucorum*, an omnivorous species has become a major agricultural pest in china since the large-scale cultivation of Bt-cotton in recent decade. As a phloem sap insect, *A. lucorum* secretes saliva that contains many functional proteins into plant interface to perturb host cellular processes during feeding. However, the molecular functions of these secreted enzymes and effectors in host plants are still unknown. In this study, a putative salivary gland effector, *Apolygus lucorum* Cell Death Inhibitor 6 (AI6), was identified using the transcriptome database of *A. lucorum* and aphid effector analysis. AI6 encodes an active glutathione peroxidase with ability to inhibit INF1-induced cell death upon overexpression in the leaves of *Nicotiana benthamiana*. We further demonstrated that AI6 acted as a negative regulator of plant resistance by repressing PAMP-induced ROS. To ensure successful colonization in host cells, our results revealed that AI6 may induce hormone-related pathway such as jasmonic acid (JA), salicylic acid (SA) and ethylene (ET), promote insect feeding and pathogen infection. Taken together, the identification and molecular functional analysis of AI6 effector in the *A. lucorum* will provide valuable information on insect-plant interaction.

* First author: Dong Yumei; E-mail: 2017102095@njau.edu.cn

** Corresponding author: Xia Ai; E-mail: xiaai@njau.edu.cn

Effects of bio-organic fertilizer on soil microbial biomass and soil enzyme activities in roots of tobacco plants*

Liang Liuyang**, Kang Yebin***

(*College of Forestry*, *Henan University of Science and Technology*, *Luoyang* 471003, *China*)

Abstract: Tobacco is one of China's major economic crops, and its output and quality directly affect the development and benefits of the tobacco industry. Bio-organic fertilizer has the functions of improving soil structure, improving soil fertility and permeability, promoting root growth, reducing crop diseases, providing nutrients for microorganisms in soil, increasing their quantity and activity, and applying them to crops. In order to explore the application amount and application effect of tobacco-specific bio-organic fertilizer, the number of microorganisms in soil samples was detected by dilution coating plate method. The activity of invertase, urease, phosphatase and catalase in soil samples were determined by 3, 5-dinitro colorimetric method, colorimetric method, phenylpyrazine colorimetric method and titration, The results showed that 250 kg of bio-organic fertilizer, 15kg of NPK fertilizer, 15kg of heavy superphosphate, 15kg of potassium sulfate and 4-5kg of potassium nitrate per 667 m^2 could significantly increase the root microbes of tobacco roots and rhizosphere soil. The number of colonies, especially the number of colonies of bacteria and actinomycetes; significantly increased the activities of invertase, urease, phosphatase and catalase in the soil around the roots and roots of tobacco; the invertase in the rhizosphere soil The activities of urease and catalase were higher than those in the root zone. 500 kg bio-organic fertilizer per 667 m^2, heavy superphosphate weight 15kg, potassium sulfate 15kg, potassium nitrate 4-5kg, can only increase the number of fungi and actinomycetes in the soil around the roots of tobacco, as well as sucrase, urease And phosphatase activity. According to the test results, it is preliminarily determined that the suitable application rate of bio-organic fertilizer in Luoyang tobacco area is 250kg per $667m^2$.

Key words: Tobacco; Microecological preparation; Soil microorganism; Soil enzyme activity

* 基金项目：河南省烟草公司科学研究与技术开发重点项目（HYKJ201610）；河南省烟草公司洛阳市公司科学研究与技术开发项目（LYKJ201804）

** 第一作者：梁留阳，在读硕士研究生，主要从事植物免疫学研究

*** 通信作者：康业斌，教授，博士研究生导师，主要从事植物免疫学研究；E-mail：kangyb999@163.com

Dissect the rice immune system using a whole-genome sequenced mutant population

Li Guotian[1,2,3 *], Sha Gan[1], Jain Rashmi [2,3], Jenkins Jerry[4], Shu Shengqiang [3], Chern Mawsheng[2,3], Copetti Dario[7], Xing Feng[6], Xie Weibo[6], Wing Rod A[7], Schmutz Jeremy[4,5], Ronald Pamela C[2,3 **]

(1. *State Key Laboratory of Agricultural Microbiology and College of Plant Science and Technology*, *Huazhong Agricultural University*, *Wuhan* 430070, *Hubei*, *China*; 2. *Department of Plant Pathology*, *University of California*, *Davis*, *CA* 95616, *USA*; 3. *Joint BioEnergy Institute*, *Lawrence Berkeley National Laboratory*, *Berkeley*, *CA* 94720, *USA*; 4. *U. S. Department of Energy Joint Genome Institute*, *Walnut Creek*, *CA* 94598, *USA*; 5. *HudsonAlpha Institute for Biotechnology*, *Huntsville*, *AL* 35806, *USA*; 6. *National Key Laboratory of Crop Genetic Improvement*, *National Center of Plant Gene Research*, *Huazhong Agricultural University*, *Wuhan* 430070, *China*; 7. *Arizona Genomics Institute*, *School of Plant Sciences*, *University of Arizona*, *Tucson*, *AZ* 85721, *USA*)

Abstract: Rice provides food for more than half of the world's population and also serves as a model for studies of other monocotyledonous species. Here, we report the *de novo* genome sequencing and analysis of an early flowering Geng/*japonica* variety KitaakeX, which completes its life cycle in 9 weeks. Our KitaakeX sequence assembly contains 377. 6 Mb, consisting of 33 scaffolds (476 contigs) with a contig N50 of 1. 4 Mb. Complementing the assembly are detailed gene annotations of 35 594 protein coding genes. The high-quality genome of the model rice plant KitaakeX will accelerate rice functional genomics.

To further facilitate rice research, we generated and sequenced a fast-neutron-induced mutant population in KitaakeX. We sequenced 1 504 mutant lines at 45-fold coverage and identified 91 513 mutations affecting 32 307 genes, i. e., 58% of all rice genes. We detected an average of 61 mutations per line. Mutation types include single-base substitutions, deletions, insertions, inversions, translocations, and tandem duplications. We observed a high proportion of loss-of-function mutations. To facilitate public access to this genetic resource, we established an open access database called KitBase that provides access to sequence data and seed stocks. We have screened this mutant population for lines altered in plant immunity and identified multiple mutant candidates. Progress on these mutants will be presented.

* First author: Li Guotian; E-mail: li4@ mail. hzau. edu. cn

** Corresponding author: Ronald Pamela C; E-mail: pronald@ ucdavis. edu

Cultural practices impact soil microbial communities and eliminate negative effects of cucumber monoculture*

Gao Yuhan[1], Lu Xiaohong[1**], Guo Rongjun[1], Miao Zuoqing[1], Li Shidong[1]
(1. *Institute of Plant Protection*, *Chinese Academy of Agricultural Sciences*, *Beijing* 100193, *China*)

Abstract: Cucumber monoculture commonly causes reduced plant growth and yield due to the accumulation of soilborne pathogens. However, we have observed this problem disappeared in commercial greenhouses after extended long-term monoculture. To investigate possible mechanisms of this phenomenon, soil properties, microbial communities in rhizosphere soil and length of cropping cycles were analyzed from multiple greenhouses with length of cropping varying from 1 to 25 years. Long-term cucumber monoculture combined with input of organic and inorganic fertilizer was associated with the accumulation of soil organic matters, nutrients and soil salinization. Soil pH decreased and reached the lowest value of 4.80 in 7 years of cropping but bounced back afterwards. To examine microbiomes, regions of fungal internal transcription spacer (ITS) and bacterial 16S rDNAs were amplified and sequenced with soil DNAs as templates using next-generation sequencing. The abundance of microbial population was positively correlated with length of cropping due to the accumulation of soil nutrients. However, microbial diversity decreased as the cropping years increased. Salinity accumulation and acidification were main contributors for reduced fungal and bacterial diversity. Bacterial abundance of operational taxonomic units (OTUs) related to potentially antifungal bacteria *Bacillus*, *Pseudomonas* and *Streptomyces* spp. intended to be higher as the cycles of monoculture increased. Pathogen-like *Fusarium* spp. increased as cropping years increased until 7 years and then continuously decreased as the cycle of cultivation increased. Our findings suggested that cucumber monoculture increased soilborne pathogen populations in the first 7 years due to their interactions with hosts. However, as the cropping cycles continued to increase and input of organic fertilizer, the accumulation of high soil organic matter content and the enhanced beneficial microbes suppressed soilborne pathogens, and therefore helped overcoming negative effects of continuous cucumber cropping.

Key words: soil properties; soil microbiome; organic fertilizer; greenhouse cultural management

---

* Funding: National Key Research and Development Program of China (2016YFD0201000) and the Program of Modern Agricultural Industry Technology System (CARS-25-D05)

** Corresponding author: Lu Xiaohong; E-mail: Luxiaohong@caas.cn

A Nano Luciferase toolkit for studying plant disease signaling pathways

Wu Caiyun*, Xie Kabin**

(*National Key Laboratory of Crop Genetic Improvement*, *Huazhong Agricultural University*, *Wuhan* 430070, *China*)

Abstract: Bioluminescence using luciferases is an efficient and convenient approach to study gene expression and protein-protein interaction to dissect the plant disease signaling pathway. Recently, a new luciferase (NanoLuciferase, NanoLuc) was characterized from deep sea shrimp *Oplophorus gracilirostris*. Owing to its small size, bright luminescence and outstanding dynamic range, NanoLuc is being engineered to study promoter activity, protein stability, protein-protein interaction and *in vivo* imaging in animals and yeast. However, NanoLuc is not adapted in plant system yet. Here, we optimized NanoLuc and developed a NanoLuc toolkit for plant functional genomics. Our results show that the NanoLuc is 1 000 brighter than firefly luciferase and Renilla Luciferase in rice protoplast. This ultra-high sensitivity of NanoLuc enables detecting target protein at almost single cell level. We are developing a set of vectors to use NanoLuc tool for plant functional genomics, including NanoLuc-based reporter to examine promoter activity, NanoLuc tagging to study the protein stability and modifications, splitting NanoLuc assays to study protein interactions, and NanoLuc-bioluminescent transgenic plant for genetic screen. All these vectors are thoroughly tested in rice. We anticipate that our NanoLuc toolkit could be broadly used in plant functional genomics as well as a versatile tool to study plant disease signaling pathways.

Key words: tools; luciferase; NanoLuc; reporter gene assay

---

* First author: Wu Caiyun, graduate student, major in molecular plant pathology; E-mail: wucaiyunwy@ 163. com

** Corresponding author: Xie Kabin, Professor; E-mail: kabinxie@ mail. hzau. edu. cn

玉米组蛋白去乙酰化酶 HDAC 家族基因的鉴定与表达规律*

于 璐，庞 茜，张 康，曹宏哲，藏金萍，邢继红**，董金皋**
（河北省植物生理与分子病理学重点实验室，河北农业大学生命科学学院，保定 071001）

摘 要：组蛋白的乙酰化修饰是表观遗传修饰的重要组成部分，组蛋白去乙酰转移酶（HDACs）是调节染色质结构和基因表达的关键表观遗传因子，在植物生长、发育和对非生物或生物胁迫反应中发挥重要作用。现阶段在拟南芥和水稻中，已对 HDACs 的系统分析和分子功能进行了大量的研究，然而对玉米 HDAC 基因家族的系统分析尚未报道。本研究确定了玉米组蛋白去乙酰转移酶（HDAC）家族的基因信息、系统进化关系、在玉米的不同组织以及抵抗生物、非生物胁迫过程中的表达情况，为阐明玉米 HDAC 家族基因在玉米抵抗生物、非生物胁迫中的功能及机制奠定基础。笔者从 WERAM 数据库中获得了玉米 HDAC 家族 17 个成员的基因信息；根据系统进化关系，HDAC 家族基因分为 Class Ⅰ、Class Ⅱ、Class Ⅲ 、STR、HD2 五个亚家族；利用公共数据分析，通过对 HDAC 家族基因的玉米组织特异性表达分析，结果表明 GRMZM2G046824、GRMZM2G56539 和 GRMZM2G008425 在花药和花粉中高度表达，而 GRMZM2G100146 和 GRMZM2G159032 在种子和胚乳中高度表达。在玉米抵抗生物逆境（拟轮枝镰孢）和非生物逆境（高温、低温、盐胁迫等）过程中，HDAC 家族基因的表达水平呈现明显的变化规律。生物逆境方面，根据拟轮枝镰孢侵染玉米后的 HDACs 数据，笔者对 HDACs 进行了激素处理，其表达模式表明了一些该家族成员可能参与了 JA 或 SA 信号通路所引起的防御反应。在非生物逆境方面，HDACs 中成员 GRMZM2G100146、GRMZM2G172883 在热、盐和干旱处理后均表现出明显的上调表达，免疫印迹结果显示，玉米苗在不同温度下处理，植物体内的乙酰化水平也相应的变化，其中 42℃高温处理下，植物体内在 H3K9ac 和 H4K5ac 位点表现明显的乙酰化水平上升。这些研究结果将有助于揭示 HDAC 在控制玉米发育和对胁迫反应中的作用。

关键词：玉米；组蛋白；乙酰转移酶（HDAC）；基因表达；胁迫

---

* 基金项目：粮食丰产增效科技创新专项（2016YFD0300704）；国家玉米产业技术体系（CARS-02-12）

** 通信作者：董金皋；E-mail：dongjingao@ 126. com
邢继红；E-mail：xingjihong2000@ 126. com

N 端融合检测标签的 pMDC32 植物表达载体系列的构建*

左登攀**，陈相儒，赵添羽，张永亮，王　颖，王献兵，李大伟，于嘉林，韩成贵***
（中国农业大学植物病理学系，农业生物技术国家重点实验室，北京　100193）

摘　要：Gateway 植物表达载体 pMDC32 主要组成元件包括启动子、Gateway 盒（attR1-ccdB-attR2）、终止子和用于选择转基因植物的筛选标记基因。所用的启动子是组成型启动子为烟草花叶病毒（CaMV）双增强启动子（2 ×CaMV 35S）。本实验室已在表达载体 pMDC32 attR2 区 C 端插入 3 * Flag 序列的 pMDC32-MCS-3 * flag 载体，为了便于实验中目的基因 N 端融合检测标签，本文在以 pMDC32 为骨架基础上进一步构建了可在目的蛋白 N 端带有 3 * Flag、3 * HA 和 3 * Flag-3 * HA 检测标签的系列表达载体，分别命名为 pMDC32-3 * Flag-MCS、pMDC32-3 * HA-MCS 和 pMDC32-3 * Flag-3 * HA-MCS，3 种载体构建策略分别如下。

首先合成了 3 * Flag-3 * HA 基因并连在 pUC57-simple-225 载体上，通过 *Kpn* Ⅰ/*Sal* Ⅰ双酶切获得 3 * Flag-3 * HA（*Kpn*I/*Sal*I）片段，插入已用 *Kpn*I/*Sal*I 双酶切获得的 pMDC32（*Kpn* Ⅰ/*Sal* Ⅰ）线性化载体中，获得重组载体 pMDC32-3 * Flag-3 * HA-MCS，表达目的蛋白时可根据多克隆位点酶切插入基因（插入目的基因 5′端可使用 *HindIII*/*Sal*I 酶切位点，3′端可使用 *Pac*I/*Spe*I/*Sac*I）。pMDC32-3 * Flag-MCS 构建首先基于携带 3 * Flag-3×HA 基因的 pUC57-simple-225 载体 PCR 扩增 3 * Flag（*Kpn*I/*Spe*I）片段，再将 3 * Flag（*Kpn*I/*Spe*I）片段插入已用 *Kpn*I/*Spe*I 双酶切获得的 pMDC32（*Kpn*I/*Spe*I）线性化载体中，获得 pMDC32-3 * Flag-MCS 重组载体，表达目的蛋白时可根据多克隆位点酶切插入基因（插入目的基因 5′端使用 *Spe*I 酶切位点，3′端使用 *Sac*I）。pMDC32-3 * HA-MCS 构建以 pMDC32-3 * Flag-3 * HA-MCS 载体以 *BamH*I/*Bgl*II 进行双酶切后回收 pMDC32-3 * HA 的载体，自连接获得 pMDC32-3 * HA-MCS 重组载体，表达目的蛋白时可根据多克隆位点酶切插入基因（插入目的基因 5′端可使用 *Hind*III/*Sal*I 酶切位点，3′端可使用 *Pac*I/*Spe*I/*Sac*I）。

本实验对 Gateway 表达载体 pMDC32 改造后获得 N 端带有检测标签的载体，均能够在植物中进行瞬时表达目的基因及进行转基因遗传转化，并获得很好的表达效果，为重要基因功能研究及植物遗传转化获得转基因遗传材料提供了便利。

关键词：载体 pMDC32；载体改造；表达载体应用

---

* 基金项目：国家自然科学基金项目（31671995）
** 第一作者：左登攀，博士生，主要从事植物病毒与寄主的分子互作研究；E-mail：B20173190806@ cau. edu. cn
*** 通信作者：韩成贵，教授，主要从事植物病毒病害及抗病转基因作物研究；E-mail：hanchenggui@ cau. edu. cn

小麦病程相关蛋白 PR-10 的原核表达、纯化和抗血清的制备*

时　兴**，李　畅，王　颖，张宗英，李大伟，于嘉林，韩成贵***
（中国农业大学植物病理学系，农业生物技术国家重点实验室，北京　100193）

摘　要：病程相关蛋白（Pathogenesis-related proteins，PR）是一类受到生物胁迫和非生物胁迫而产生的抗病抗逆蛋白，共分为 17 个家族，其中 PR-10 家族具有核酸酶活性、抑菌活性及抗病毒的功能。

小麦黄花叶病由小麦黄花叶病毒（*Wheat yellow mosaic virus*，WYMV）侵染引起，导致小麦减产。由于受到传毒介体禾谷多黏菌、气候、栽培条件等诸多因素影响，防治困难。PR-10 蛋白已经被证实能够增强谷类作物的耐胁迫性。因此，利用转基因技术获得转 *PR*-10 基因小麦，探索是否可以实现对小麦黄花叶病的防控，具有一定的研究价值。

为检测 *PR*-10 基因的表达，本文制备了小麦 PR-10 的抗血清。首先通过 RT-PCR 的方法获得小麦病程相关蛋白 *PR*-10 基因，再将 *PR*-10 基因克隆到原核表达载体 pDB. *His*. MBP 上，转化大肠杆菌 BL21（DE3）。经过 0. 1 mmol/L 的 IPTG 诱导、超声破碎和 Ni 亲和层析柱纯化，得到质量较好总量约 4mg 的 PR-10 融合蛋白。然后将融合蛋白作为抗原免疫健康的新西兰白兔，制备并获得抗血清。最后，利用该抗血清对原核表达的融合蛋白以及在本生烟瞬时表达的小麦 PR-10 蛋白进行了 Western-blot 检测，结果显示该抗血清能与 PR-10 蛋白发生特异血清学反应。

关键词：小麦病程相关蛋白 PR-10；原核表达；抗血清；检测

致谢：感谢刘俊峰、王献兵教授和张永亮副教授对本研究的指导和帮助。

---

* 基金项目：转基因专项抗病虫小麦新品种培育（2016ZX08002001）
** 第一作者：时兴，硕士生，主要从事植物病毒检测研究；E-mail：shixing@ cau. edu. cn
*** 通信作者：韩成贵，教授，主要从事植物病毒学，抗病毒基因工程；E-mail：hanchenggui@ cau. edu. cn

表达检测带标签的 pGD 系列衍生载体的构建

胡汝检[*]，左登攀，吴占雨，王 颖，韩成贵[**]
（中国农业大学植物病理学系，农业生物技术国家重点实验室，北京 100193）

摘 要：pGD 作为一种常用的双元表达载体，在植物细胞中能够高效地表达载体上的目的蛋白。为了便于对表达的目的蛋白进行血清学检测，本文构建了 pGD 带有 N 端标签的系列载体，包括：pGD-3 * Flag-MCS、pGD-3 * HA-MCS 和 pGD-3 * Flag-3 * HA-MCS 三种表达载体。具体步骤如下，本载体以 pGD 为骨架构建，以含有合成的携带 3 * Flag-3 * HA 基因的质粒 pUC57-simple-225 为模板，通过 *Bam* HⅠ和 *Bgl* Ⅱ双酶切获得 3 * Flag（*Bam* HⅠ/*Bgl* Ⅱ）片段，插入 pGD（*Bgl*Ⅱ）线性化载体中（注：BamHⅠ与 BglⅡ为同尾酶），获得重组载体 pGD-3 * Flag-MCS。通过 *Bgl* Ⅱ和 *Hin*d Ⅲ双酶切获得 3 * HA（*Bgl*Ⅱ/*Hin*dⅢ）片段，插入 pGD（*Bgl*Ⅱ/*Hin*dⅢ）线性化载体中，获得重组载体 pGD-3 * HA-MCS 通过 *Bam* HⅠ和 *Hind* Ⅲ双酶切获得 3 * Flag-3 * HA（*Bam* HⅠ/*Hind* Ⅲ）片段，插入 pGD（*Bgl*Ⅱ/*Hin*dⅢ）线性化载体中（注：*Bam* HⅠ与 *Bgl* Ⅱ为同尾酶），获得重组载体 pGD-3 * Flag-3 * HA-MCS，分别转化大肠杆菌感受态细胞 MC1022，经 PCR 检测后选取阳性克隆提质粒，送公司测序，测序结果比对正确，说明成功获得重组载体。另外，构建了 pGD-MCS-6 * Myc 表达载体，将人工合成带有酶切位点的 6 * Myc 基因序列克隆通过 *Bam* HⅠ和 *Bgl* Ⅱ双酶切获得 6 * Myc（*Bam* HⅠ/*Bgl* Ⅱ）片段克隆到 pGD（*Bam* HⅠ）线性化载体中。以上 4 种重组表达载体已应用于实验室的瞬时表达实验中，均能够很好地表达带有标签的目的蛋白，并利用标签特异性抗体对靶标蛋白进行检测。

关键词：pGD 载体；载体改造；标签

致谢：感谢于嘉林、李大伟、王献兵教授和张永亮副教授对本研究的建议。

* 第一作者：胡汝检，硕士研究生，主要从事植物病毒与寄主的分子互作研究；E-mail：Hurujian@ cau. edu. cn
** 通信作者：韩成贵，教授，主要从事植物病毒病害及抗病转基因作物研究；E-mail：hanchenggui@ cau. edu. cn

昆明地区野生食用菌资源调查及鉴定*

刘 艳**，祝友朋，韩长志***

（西南林业大学 云南省森林灾害预警与控制重点实验室，昆明 650224）

摘 要：近些年，国内诸多学者就浙江省、西藏林芝、云南省等省（地区）开展了野生食用菌的调查研究工作，取得了较为突出的成绩。前人已经明确云南省野生食用菌有882种，占世界2 000种食用菌的44.1%，占中国966个分类单元的91.3%。云南全省120多个县市均有野生食用菌分布，且野生食用菌发生较早，年生产周期长达8个月，集中发生在6—9月，产量高质量优。云南野生贸易真菌已知有64个属207种，以担子菌中的牛肝菌属 *Boletus*、口蘑属 *Tricholoma*、蚁巢伞属 *Termitomyces*、革菌属 *Thelephora*、红菇属 *Russula*、乳菇属 *Lactarius*、枝瑚菌属 *Ramaria* 为主。然而，尚不清楚昆明市周边野生食用菌资源的分布情况，严重制约着当地野生食用菌资源的进一步保护和利用。本研究通过国内文献的梳理及实地调研相结合的方法，对昆明市周边野生食用菌资源情况进行研究，结果表明，自2000年至今，我国对于野生食用菌的研究报道逐年增多，且大多集中于形态分类、生境特征及食用、药用价值的研究；同时，对昆明周边地方及农贸市场进行充分调查，并选择周边农贸市场进行主要野生食用菌资源收集的方法，获得大量野生食用菌种类，明确野生食用菌多数长在林区森林地面，有的长在倒木、腐木、树桩或树杆的腐朽部分，有的长在腐草堆上或粪肥上，且大多在夏秋季生长，分布广泛。此外，通过对所收集的野生食用菌进行宏观和微观形态学观察、鉴定，以及分子生物学鉴定等工作，明确位于昆明周边地区的野生食用菌资源主要以蘑菇属、牛肝菌属、绒盖牛肝菌属、乳菇属、红菇属等为主，尤以担子菌纲、伞菌目、牛肝菌科居多。本研究为进一步开展昆明地区野生食用菌的DNA条形码研究打下了坚实的基础。

关键词：昆明地区；野生食用菌；资源调查；鉴定

---

* 基金项目：国家级大学生创新创业训练计划项目（项目编号：201610677004）；云南省大学生创新创业训练计划项目（项目编号：S201610677004）

** 第一作者：刘艳，本科生；E-mail：1558723914@ qq. com

*** 通信作者：韩长志，博士，副教授，研究方向：经济林木病害生物防治与真菌分子生物学；E-mail：hanchangzhi2010@163. com

核桃根际土壤微生物群落功能多样性分析[*]

官　鑫[**]，向　阳，祝友朋，韩长志[***]
（西南林业大学生物多样性保护学院　云南省森林灾害预警与控制重点实验室，昆明　650224）

摘　要：核桃根际土壤微生物群落多样性对于发挥土壤作用具有重要作用，然而，尚不明确核桃根际土壤微生物群落功能情况。因此，为了明确不同土层深度核桃根际土壤微生物群落多样性特征及其变化趋势，本研究采用 Biolog-ECO 板技术及高通量测序技术，对不同土层深度核桃根际土壤微生物群落多样性差异进行了比较。结果表明，不同土层土壤微生物群落功能多样性差异显著，土壤平均颜色变化率（AWCD）随培养时间延长而逐渐增加，不同土层土壤 AWCD 大小顺序为 20～40 cm>40～60 cm>0～20 cm；同时，土壤微生物群落 Shannon 指数的总体趋势为 20～40 cm>0～20 cm>40～60 cm。不同土层深度核桃根际土壤微生物群落均匀度指数和优势度指数之间差异不显著。不同土层深度核桃根际土壤微生物对不同碳源的利用能力存在差异，其中 40～60 cm 土层利用率最高，0～20 cm 利用率最低，糖类、氨基酸类、羧酸类和多聚物类为各深层土壤微生物的主要碳源。此外，通过主成分分析，结果表明，主成分 1 和主成分 2 分别能解释变量方差的 45.21%和 33.47%，在主成分分离中起主要贡献作用的是糖类、氨基酸类和羧酸类。该研究为进一步解析核桃根际土壤微生物与生防菌之间的关系提供重要的理论支撑。

关键词：Biolog-ECO；高通量技术；群落多样性；核桃根际土壤微生物

* 基金项目：云南省大学生创新创业训练计划项目（201610677034）；西南林业大学大学生创新创业项目（201825）

** 第一作者：官鑫，本科生；E-mail：2922057051@qq.com

*** 通信作者：韩长志，博士，副教授，研究方向：经济林木病害生物防治与真菌分子生物学；E-mail：hanchangzhi2010@163.com

马铃薯栽培种‘合作88’孤雌生殖诱导群体的倍性分析*

王洪洋**，郑英转，栾宏瑛，李灿辉***
（云南师范大学马铃薯科学研究院，昆明　650500）

摘　要：四倍体马铃薯栽培种‘合作88’是我国西南地区的主栽品种，其在品质、抗病害性、加工、种植面积等方面都具有重要地位。四倍体马铃薯存在遗传基础狭窄且复杂、难以创新等问题，严重阻碍了相关研究的深入开展。2017年对5 000朵‘合作88’花去雄后，与‘Phureja’进行杂交并创建了一个‘合作88’孤雌生殖诱导 F_1 群体（824份）。通过流式细胞仪检测、分析发现，该群体包含265份二倍体，128份三倍体，335份四倍体，22份五倍体，64份六倍体，10份八倍体材料。该结果对深入挖掘和解析‘合作88’重要农艺性状的调控机理至关重要，同时也对马铃薯种质资源创新具有重要意义。

关键词：马铃薯；合作88；孤雌生殖；倍性分析

---

* 基金项目：国家自然科学基金（31800134）
** 第一作者：王洪洋，博士，讲师，主要从事马铃薯与致病疫霉菌互作研究；E-mail：hongyang8318@126.com
*** 通信作者：李灿辉，博士，教授，主要从事马铃薯遗传育种研究；E-mail：ch2010201@163.com

适合于 2-DE 分析的水稻质膜磷酸化蛋白质富集方法的建立*

聂燕芳[1,2]，邹小桃[1,3]，王振中[1,3]，李云锋[1,3]**

（1. 华南农业大学广东省微生物信号与作物病害重点实验室，广州 510642；
2. 华南农业大学材料与能源学院，广州 510642；3. 华南农业大学农学院，广州 510642）

摘　要：细胞质膜（plasma membrane，PM）作为植物细胞与外界的屏障，在细胞壁的合成、离子转运和细胞信号转导过程等方面起着重要的作用。质膜蛋白质的磷酸化也被证实广泛参与了病原菌的信号识别等反应。开展植物细胞质膜蛋白质磷酸化变化的研究，对于了解植物和病原菌的相互识别和信号跨膜转导等具有重要意义。

采用双水相分配法和离心法对水稻叶片的细胞质膜进行了纯化，结果表明由 6.3/6.3%（*w/w*）Dextran T 500/PEG 3350 组成的双水相体系可以获得高纯度的质膜微囊，其质膜标志酶 VO_{43-}-ATPase 的相对活性达 93%以上。采用 Al（OH）$_3$-MOAC 法，对水稻质膜磷酸化蛋白质进行了富集；结果表明质膜磷酸化蛋白质的富集得率为 3.46%～4.21%。对质膜磷酸化蛋白质进行了 2-DE 分析，结果表明同一凝胶的 Pro-Q Diamond 染色和硝酸银染色的 2-DE 图谱背景清晰，蛋白质点分布均匀，没有明显的纵向拖尾和水平横纹现象；其中 Pro-Q Diamond 染色图谱中的蛋白质点数为 296 个，硝酸银染色图谱中的蛋白质点数为 316 个；表明该技术体系可以获得高分辨率和高重复性的 2-DE 图谱。采用 PDQuest 8.0 图像分析软件对 2 种染色图谱进行叠加分析，结果表明 2 种图谱中蛋白质点的吻合度高。

上述结果表明，应用双水相分配法和 Al（OH）$_3$-MOAC 法，并结合 2-DE 技术和 Pro-Q Diamond磷酸化蛋白质特异性染色技术，建立了高效的水稻叶片质膜磷酸化蛋白质的富集和 2-DE 分析技术体系。

关键词：水稻；质膜磷酸化蛋白质；富集；双向电泳

---

* 基金项目：国家自然科学基金（31671968）；广州市科技计划项目（201804010119）；广东省科技计划项目（2016A020210099）

** 通信作者：李云锋；E-mail：yunfengli@ scau. edu. cn

不同泡桐种带毒组培苗症状观察与植原体质粒分子变异分析*

孔德治**，田国忠，张文鑫，林彩丽***

（中国林业科学研究院森林生态环境与保护研究所，国家林业局森林保护学重点实验室，北京 100091）

摘 要：由植原体引起的泡桐丛枝病（Paulownia witches'-broom，PaWB）是我国重要的植物病害之一，其分布范围广，危害严重，具有重要的研究价值。本研究在通过对外植体的培养获得福建福州、四川通江、江苏南京和苏州、山西太原、陕西西安和渭南共 7 个地区 7 个品种的 12 组带毒泡桐组培苗基础上，对连续继代培养 4 个月以上的 12 组带毒泡桐组培苗的症状进行了观测比较，结果显示不同地区不同泡桐品种组培苗间发病症状存在差异。同时对采自福建福州与南平的 23 个发病泡桐植株中的植原体质粒编码的 *orf*4 基因进行 PCR 扩增和序列变异分析。结果发现 23 个发病泡桐样品中，有 3 个样品中的 *orf*4 基因序列与泡桐丛枝植原体南阳株系 pPaWBNy-1编码的 *orf*4 基因序列相同，有 17 个样品中的植原体质粒编码的 *orf*4 基因序列 3′-端存在 108bp 的核苷酸缺失，其余 3 个样品中的植原体质粒缺失 108bp 核苷酸与不缺失的 *orf*4 基因同时存在。先后 2 次对福建福州与南平发病泡桐样品采集分析发现，福建地区泡桐丛枝植原体中 pPaWBNy-1 编码的 *orf*4 基因 3′-端缺失 108bp 核苷酸的现象是稳定且普遍存在的。同时通过组培获得了一株福建的植原体质粒变异的南方泡桐，该变异株系的成功组培为下一步关于病原致病性变异和泡桐抗性差异的研究奠定基础。

关键词：泡桐丛枝病；植原体；组织培养；质粒

---

* 基金项目：国家重点研发计划子课题“泡桐丛枝植原体扩散流行的生态适应性与分子基础研究（2017YFD0600103-3）”

** 第一作者：孔德治，硕士研究生，主要从事植物病理学研究；E-mail：kdzcaf@ 163. com

*** 通信作者：林彩丽，助理研究员，主要从事森林病理学研究；E-mail：lincl@ caf. ac. cn

褪黑素诱导水稻相关基因 qRT-PCR 中内参基因的选择[*]

陈　贤[**]，赵延存，刘凤权[***]
（江苏省农业科学院植物保护研究所，南京　210014）

摘　要：目前，实时荧光定量 PCR（qRT-PCR）是基因表达分析中应用最为广泛的方法之一。选择合适的内参基因对于实验进行校正和标准化至关重要。为了筛选出稳定表达的内参基因，本研究中以褪黑素处理水稻叶片为研究对象，以褪黑素处理植株后叶片中第 0 h、3 h 及 12 h 的总 RNA 为材料，利用 qRT-PCR 技术和 3 个统计学软件讨论了 10 个常用内参基因（18*S*、25*S*、*ACT*、*β-TUB*、*eEF-1a*、*eIF-4a*、*UBC*、*UBQ-5*、*UBQ-10*、*GAPDH*）在褪黑素处理下的稳定性。其结果如下：

（1）利用 LinRegPCR 软件分析单个 PCR 反应的扩增效率（E_R）和线性相关系数 R^2 值。结果显示：各内参基因的 $R^2 \geqslant 0.999$，PCR 扩增效率接近 2.0，符合 qRT-PCR 对扩增效率的要求。

（2）利用 geNorm 软件分析内参基因的稳定性。结果显示：在 0 h，稳定性排名前三的基因分别为 *UBQ-5*> *UBQ-10*> *eEF-1a*；在 3 h，稳定性排名前三的基因分别为 *eEF-1a*>18*S*> *GAPDH*；在 12 h，稳定性排名前三的基因分别为 *GAPDH*>25*S*>18*S*。根据 M 值及最合适的候选基因配对数，最佳的组合为 18S/GAPDH。

（3）利用 Normfinder 软件分析内参基因的稳定性。结果显示：在 0 h，稳定性排名前三的基因分别为 *UBQ-10*> *eIF-4a* >*UBQ-5*；在 3 h，稳定性排名前三的基因分别为 *eIF-4a* >*UBC*> *UBQ-10*；在 12 h，稳定性排名前三的基因分别为 *eIF-4a*> *eEF-1a* >*GAPDH*。

（4）利用 BestKeeper 软件分析内参基因的稳定性。结果显示：在 0 h，稳定性排名前二的基因分别为 *UBC*>*UBQ-5*；在 3 h，稳定性排名前三的基因分别为 *β-TUB* >*UBQ5*>*UBQ10*；在 12 h，稳定性排名前二的基因分别为 *UBQ10*> *UBQ5*。

综上，*UBQ5* 和 *eIF-4a* 组合可作为分析褪黑素处理水稻过程后的内参基因。

关键词：褪黑素；水稻；内参基因

* 基金项目：江苏省自然科学基金（BK20170606）
** 第一作者：陈贤，助理研究员，从事水稻与病原细菌互作的研究；E-mail：cxbmh@126.com
*** 通信作者：刘凤权，研究员；E-mail：fqliu20011@sina.com

基于转录组测序的槟榔叶片黄化分子机理研究*

禤　哲**，车海彦，曹学仁，罗大全***
（中国热带农业科学院环境与植物保护研究所，农业部热带作物有害
生物综合治理重点实验室，海口　571101）

摘　要：槟榔（*Areca catechu* L.）作为海南省重要的经济作物，其叶片黄化现象日趋严重，是制约槟榔产业的持续健康发展的一大障碍。本研究运用转录组测序技术分析在植原体、干旱和除草剂胁迫下差异基因的表达情况，筛选出与叶片特异性黄化相关的差异基因，以探索槟榔在受到植原体侵染、干旱以及除草剂胁迫下叶片表现出特异性黄化症状的分子机理。研究结果如下：

应用 Illumina HiSeq 平台对采集到的表现出特异性黄化症状（受植原体、干旱和除草剂胁迫）的叶片、全叶黄化型叶片以及健康槟榔叶片进行转录组测序，共获得了321 402条 Unigenes。筛选出了 654 个在特异性叶片黄化组和全叶型黄化组为共表达的差异基因，440 个仅在特异性叶片黄化组中为共同显著差异表达基因。对差异基因进行 GO 注释和 KEGG 富集，发现与叶片黄化相关的差异基因主要富集在次级代谢途径、激素信号转导途径和能量代谢途径中，其中在生长素、茉莉酸和乙烯信号转导途径中 SAUR、JAZ 和 ERF1 显著上调表达可能对槟榔叶片发生黄化症状有重要的调控作用；而特异性黄化样本中的差异基因特异注释到油菜素内酯信号转导和合成途径中，说明该通路可能对槟榔叶片发生特异性黄化症状起关键的调控作用。通过对特异性黄化样本中共表达的差异基因进行次级代谢通路分析，注释到类黄酮生物合成、苯丙烷类生物合成以及二苯乙烯、二芳基庚醇和姜辣素生物合成这 3 条次级代谢途径上的基因显著上调表达，对植物着色色素的合成途径中基因的表达起着重要的调控作用，可能导致槟榔叶片在植原体、干旱和除草剂胁迫下叶片表现出相似的黄化症状。

关键词：槟榔；叶片黄化症状；转录组测序

---

* 基金项目：中国热带农业科学院基本科研业务费（1630042017023）；中国热带农业科学院环境与植物保护研究所自主选题项目（hzsjy2017007）；海南省槟榔病虫害重大科技计划项目（ZDKJ201817）

** 第一作者：禤哲，硕士，主要从事植物病害研究

*** 通信作者：罗大全，研究员，主要从事热带作物病理学研究

林下和传统种植三七的两类土壤中可培养微生物多样性比较研究*

王玉玺[1**]，李迎宾[1]，罗来鑫[1***]，徐秀兰[2***]，李健强[1]

（1. 中国农业大学植物保护学院，种子病害检验与防控北京市重点实验室，北京 100193.
2. 北京市农林科学院蔬菜研究中心，农业部华北地区园艺作物生物学与种质创制重点实验室，农业部都市农业（北方）重点实验室，北京 100097）

摘　要：三七（*Panax notoginseng*）是五加科人参属植物，具有重要的药用价值。随着市场需求量的不断增加，三七种植面积不断扩大；同时，三七生长存在严重的连作障碍，传统的新地轮作等农田种植管理方式，已使适宜三七种植的耕地面积不断减少。近年来，利用云南省丰富的林下资源开展三七原生态种植，是解决目前三七生产困境的重要方式之一。然而，关于林下种植三七土壤中的微生物区系研究鲜有报道。基于此，本研究采用可培养的方法，对林下种植三七的土壤与传统种植三七的土壤中微生物的丰度进行了初步比较分析，力图发现可能的资源微生物，同时为三七的种植栽培提供参考。

（1）材料和方法。①土壤样品采集。一年生林下土壤采集自普洱市澜沧县竹塘乡老达寨林下三七原生态种植试验基地（22°47′31.84″N；99°48′7.61″E），一年生传统土壤采集自昆明市寻甸县（25°31′3′′N，103°16′51′′E）；样品类型为表面土（0~1 cm）、根围（2 cm 左右）、根际土（根部抖落）和土表以下 20 cm 土壤 4 种。②微生物分离。取 5 g 土壤，溶于 45 mL 灭菌水中，120r/min 摇培 2 h 后进行稀释涂板。其中，真菌分离培养采用孟加拉红培养基，25℃培养；细菌分离培养采用 NA 培养基，28℃培养。定期观察并拍照、计数、纯化、提取核酸鉴定。

（2）主要结果。基于可培养的方法，对两种生境土壤中的微生物进行分离培养，主要结果如表 1 所示。①对于传统种植土壤而言，土表优势真菌为 *Gonytrichum*（120 个菌落/g）、*Penicillium*（100 个菌落/g），优势细菌为 *Paenibacillus*（18 000 CFU/g）、*Rhizobium*（6 000 CFU/g）；根围优势真菌为 *Gonytrichum*（40 个菌落/g）、*Trichoderma*（20 个菌落/g），优势细菌为 *Paenibacillus*（10 000 CFU/g）、*Bacillus*（1 400 CFU/g）；根际优势的真菌为 *Gonytrichum*（60 个菌落/g），优势细菌为 *Microbacterium*（8 000 CFU/g）、*Paenibacillus*（12 000 CFU/g）；土面以下 20 cm 优势真菌为 *Gonytrichum*（120 个菌落/g）、*Penicillium*（60 个菌落/g），优势细菌为 *Microbacterium*（2 000 CFU/g）。②对于林下种植土壤而言，土表优势真菌为 *Aspergillus*（40 个菌落/g），优势细菌为 *Bacillus*（2 000 CFU/g）、*Rhizobium*（1 000 CFU/g）；根围优势真菌为 *Aspergillus*（100 个菌落/g）、*Penicillium*（100 个菌落/g），优势细菌为 *Bacillus*（2 600 CFU/g）；根际优势真菌为 *Penicillium*（80 个菌落/g），优势细菌为 *Paenibacillu*（10 000 CFU/g）；土面以下 20 cm 优势真菌为 *Gonytrichum*（160 个菌落/g）、*Penicillium*（80 个菌落/g），优势细菌为 *Stenotrophomonas*（4 000 CFU/g）。

---

* 基金项目：云南省重大科技计划“克服三七连作障碍技术体系构建及应用”（2016ZF001）

** 第一作者：王玉玺，在读硕士研究生；E-mail：yuxiwww@ sina. com

*** 通信作者：罗来鑫，副教授；E-mail：08030@ cau. edu. cn

徐秀兰，副研究员；E-mail：xuxiulan@ nercv. org

表1　三七2种类型的种植土壤生境中优势微生物分离情况

土壤来源	优势微生物	土壤类型			
		表面土（0~1 cm）	根围（2cm）	根际	土表以下 20 cm
传统种植	真菌（个菌落/g）	*Gonytrichum*（120） *Penicillium*（100）	*Gonytrichum*（40） *Trichoderma*（20）	*Gonytrichum*（60）	*Gonytrichum*（120） *Penicillium*（60）
	细菌（CFU/g）	*Paenibacillus*（18 000） *Rhizobium*（6 000）	*Paenibacillus*（10 000） *Bacillus*（1 400）	*Microbacterium*（8 000） *Paenibacillus*（12 000）	*Microbacterium*（2 000）
林下种植	真菌（个菌落/g）	*Aspergillus*（40）	*Aspergillus*（100） *Penicillium*（100）	*Penicillium*（80）	*Gonytrichum*（160） *Penicillium*（80）
	细菌（CFU/g）	*Bacillus*（2 000） *Rhizobium*（1 000）	*Bacillus*（2 600）	*Paenibacillu*（10 000）	*Stenotrophomonas*（4 000）

（3）主要结论。①土壤优势微生物种类存在差异：传统种植4种类型土壤优势真菌种类单一，均主要为 *Gonytrichum* 和 *Penicillium*；优势细菌主要为 *Paenibacillus* 和 *Microbacterium*。林下4种类型土壤优势真菌种类较多，主要为 *Aspergillus*、*Penicillium* 和 *Gonytrichum*；优势细菌主要为 *Bacillus*、*Paenibacillu* 和 *Stenotrophomonas*。②微生物总量存在差异：总体而言，基于可培养方法分离获得的传统种植土壤生境中真菌和细菌总量较林下种植土壤生境中更多。

关键词：三七；连作障碍；林下土壤；传统种植土壤；优势菌群

微生物溯源技术研究进展*

许晓丽[1**]，罗来鑫[1]，冯建军[2,3]，章桂明[2,3***]
（1. 中国农业大学植物保护学院，北京 100193；2. 深圳海关动植物检验检疫技术中心，深圳 518045；3. 深圳市外来有害生物检测技术研发重点实验室，深圳 518045）

摘　要：微生物溯源研究有助于人们更深入地了解病原菌种群结构及其多样性的起源和进化，为病原菌的检疫、流行监测、综合防治等研究提供重要的信息资料和科学依据。在病原微生物溯源技术研究中，生物学方法、噬菌体敏感性、多位点酶电泳、等表型分析方法和扩增片段长度多态性、核糖体分型、重复序列 PCR、简单重复序列、单核苷酸多态性、多位点可变串联重复序列、多位点序列分析等分子方法均发挥了重要作用。本文根据近年国内外相关文献资料，对病原微生物常用的溯源技术的原理、优势、局限性及应用进展进行了总结，并对未来溯源研究方法的发展进行了展望。

关键词：溯源技术；表型分析；分子标记；研究进展

Research progress of technology on Microbial Source Tracking

Xu Xiaoli[1]，Luo Laixin[1]，Feng Jianjun[2,3]，Zhang Guiming[2,3]
（1. *College of Plant Protection*，*China Agricultural University*，*Beijing* 100193，*China*；2. *Animal and Plant Inspection and Quarantine Technology Center*，*Shenzhen Customs District*，*Shenzhen* 518045，*China*；3. *Shenzhen Key Laboratory of Inspection Research & Development of Alien Pests*，*Shenzhen* 518045，*China*）

Abstract：Microbial traceability studies will help people to understand more deeply of the pathogen population structure，the origin and evolution of diversity. And it can provide important information and scientific basis for the study of epidemic surveillance，comprehensive prevention and control of pathogen. Phenotypic analysis methods such as biological method，phage sensitivity，and molecular methods such as RFLP，rep-PCR，SSR，SNP，MLST play an important role in the study of the pathogenic microorganism traceability technology. Based on the relevant literature recently，this article will describe the principles of various approach techniques and their pros and cons. Finally the future of traceability research methods will be discussed.

Key words：traceability technology；phenotypic analysis；molecular marker technology；progress

---

* 基金项目：国家重点研发计划课题（2016YFF0203204）
** 第一作者：许晓丽，硕士研究生；E-mail：xuxl1123@126.com
*** 通信作者：章桂明；E-mail：zgm2001cn@163.com

微生物溯源（microbial source tracking，MST）技术最早兴起于20世纪80年代，该技术通过比较污染样品和可能污染源中微生物的差异或其他生物标记的有无来确定两者的联系，从而确定污染来源。生活在不同地区或不同宿主的的微生物群，在自然选择压力下，会产生特异的环境适应性，并将这些特性整合到基因组中遗传给后代，从而使这一地区的微生物带有其生存环境的特异性指纹图谱，如特异性的遗传特征、代谢产物等，通过分析这些呈现出多样性的表型或基因型特征，便可以达到溯源的目的。

长期以来，致病性微生物传播对人类健康、食品安全、生态安全以及经济贸易均造成了重大损失。尽管国内外已经研发出针对各类致病性微生物的快速、灵敏检测技术，但是通过致病性微生物溯源技术可以有效比较致病菌株是否一致、追踪传染源以及调查传播途径，为致病性微生物的流行学研究提供重要数据，如对进口产品携带的微生物开展溯源可以明确该微生物的来源，便于防控危险性微生物的传入以及贸易争端的解决。

微生物溯源技术随生物学尤其是遗传学和分子生物学的发展而不断提高和完善。目前用于检测比较污染样品和可能污染源中微生物差异的方法主要可分为两类：表型方法（即生物化学方法）和分子方法。表型方法是以微生物各种代谢产物、酶的生理生化反应、噬菌体的敏感性等方面为基础，来鉴定微生物种类。而分子方法是基于遗传物质为研究对象的，通过分析菌株特定的基因位点或染色体组的多态性差异鉴定菌株种类。用于细菌溯源研究的分子技术主要包括：扩增片段长度多态性（AFLP）、单核苷酸多态性（SNP）分析、多态性可变串联重复序列（MLVA）、多位点序列分析（MLST）等。本文就病原微生物溯源技术的原理、优缺点及应用进展等方面予以综述。

1　表型方法

1.1　生物学方法

生物学方法主要以微生物的形态学或生理生化特性为依据，如对微生物的培养特征、染色特性及对温度、pH值等的适应性，酶的活性等进行测定，由于形态学特征、酶等都依赖于基因的表达，具有相对稳定性，对微生物生理生化特征的比较也是对基因组的间接比较。任建国等[1]对柑橘溃疡病菌株A菌系的35项生理指标、16项生化指标进行聚类分析，将来自广西不同地区的13株菌聚成4类，揭示了A菌系存在普遍分化现象。

1.2　噬菌体敏感性

噬菌体有严格的宿主特异性，故可利用噬菌体对待检菌体进行裂解实验来进行细菌的流行病学鉴定与分型，以追查传染源。噬菌体方法最早基于对由噬菌体特异感染引起的细菌菌落中形成噬菌斑的观察，应用于伤寒沙门氏菌的分型研究中。目前，这种简单的方法已经被更先进的技术和噬菌体取代，不仅用于鉴定细菌分离株的种类，而且还用于细菌检测方法[2]。与传统的分析方法相比，该方法更具有特异性，可将同一血清型的病原菌分成若干个噬菌体型，且具有速度快的优点。然而在实际应用中，该方法要求实验室必须有一套分型所用的噬菌体，而且分型噬菌体侵染的宿主范围也有限[3]。

1.3　多位点酶电泳（MLEE）

MLEE（multilocus enzyme electrophoresis）是基于同工酶的多态性进行微生物分型的一种方法。该方法根据看家基因酶电荷性质的差异，通过电泳将由不同遗传位点编码或由同一位点不同等位基因编码的酶分开，从而达到鉴别基因型的目的。由于这些同型性可以遗传给下一代，所以通过比较足够多的这类酶，便可探究菌株之间的系统发育关系。该方法建立后，不久便被称为群体微生物学“分类金标准”，并得到广泛应用[4]。

多位点酶电泳技术的主要优点在于所需设备较简单，成本较低。但当核酸的变化不引起蛋白质改变或有些氨基酸组成改变但是电泳图谱不发生变化的基因型会被漏检，同时等位基因的选择也会影响微生物溯源的效果[5]。另一方面，MLEE 分析至少需要 10 个及以上看家基因酶以分析菌株之间的差异性，存在周期长、工作量大的问题。

2 分子方法

2.1 限制性片段长度多态性（RFLP）

RFLP 是最早被使用的分子标记技术，它是根据限制性内切酶能特异识别 DNA 分子中的特异序列，特异酶切不同个体的基因组 DNA，将其与特异性探针进行杂交，检测其与探针中同源 DNA 片段是否存在长度差异[6]。Mariana 等[7]以 *rpoB* 基因作为分子标记，通过 PCR-RFLP 技术对黄单胞菌属进行系统发育分析，发现该技术具有良好的区分性。

作为分子遗传标记，RFLP 技术可以直接反映整个基因组的遗传信息，以达到准确溯源的目的。但由于该技术需要一系列复杂的操作程序，成本高且细菌基因组小、酶切位点少、多态性不丰富等原因，也使得该技术在细菌溯源方面的应用受到限制。

2.2 扩增片段长度多态性（AFLP）

AFLP 是由荷兰科学家 Zabeau 和 Vos 于将 PCR 和 RFLP 相结合发展起来的一种 DNA 指纹技术[8]，该技术基于 PCR 技术扩增基因组 DNA 限制性片段。Khodakaramian 等[9]运用 AFLP 技术对从伊朗收集到的 57 个柑橘溃疡病菌和 16 个已知不同菌系的溃疡病菌株（A、B、C、D 和 E）进行遗传多样性分析，聚类分析发现共产生五个簇群，分别为 C 型簇群、B 型和 D 型簇群、E 型簇群、致病型 A 簇群和与其他四类差异明显的独立的一类致病型。

AFLP 本质是对整个基因组进行随机探测，不需要预知基因组的序列特征，具有低成本、高分辨率、重复性好的优点，避免了 RFLP 技术克隆、探针制备及分子杂交等一系列复杂操作，目前已在基因组作图、群体遗传学和 DNA 指纹分析中有广泛的应用[10]。

2.3 重复序列 PCR（rep-PCR）

rep-PCR 是 Versalovic 等[11]描述的一种细菌基因组指纹分析方法，其原理是利用特定的引物扩增细菌基因组中广泛分布的短重复序列，如 REP、ERIC 和 BOX，扩增位于这些序列之间的不同区域可得到不同的图谱，产物经凝胶电泳便可以分析其多态性。姚廷山[12]和魏楚丹等[13]将 rep-PCR 技术成功应用于我国柑橘溃疡病菌菌株遗传多样性的研究中，证明我国柑橘溃疡病菌基因组存在丰富的遗传多样性。漆艳香等[14]采用 REP、ERIC 和 BOX 引物对不同杧果产区的杧果细菌性黑斑病菌菌株进行 rep-PCR 分析，结果显示杧果细菌性黑斑病菌菌株有丰富的遗传多态性和较大的遗传变异。rep-PCR 耗时短、所需 DNA 量少，但该技术的分辨率取决于菌株中重复序列的数量，且 PCR 试剂、热循环及凝胶电泳条件等因素也会对结果产生影响[15]，因此该方法的可重复性较差。

2.4 简单重复序列（SSR）

SSR 又称微卫星（Microsatellite）是指基因组中以 1~6 个核苷酸为单位多次串联重复组成的长达几十个核苷酸的序列，在不同的基因组上、不同的生物个体间某段 DNA 序列重复次数不同而产生多态性。由于等位基因众多、分散性高等优点，SSR 常用于基因定位、遗传多样性分析、亲缘关系鉴定等[16-17]。SSR 两侧区域相对较为保守，可设计特异性引物进行 PCR 扩增，再经电泳便得到其长度的多态性。卢小林[6]、彭耀武[18]等将 SSR 分子标记技术应用于中国柑橘溃疡病菌的遗传多样性分析中，证明我国不同柑橘产区之间的柑橘溃疡病菌存在多态性。

2.5 单核苷酸多态性（SNP）

单核苷酸多态性（single nucleotide polymorphism，SNP）即基因组中单个核苷酸突变而引起的 DNA 序列的多态性。早期研究证实，在获得更加丰富的 SNPs 基础上可以大大增加对大肠杆菌进行分型的能力[19]。Florence Hommais [20]利用 MLST 技术研究大肠杆菌 SNP 位点以及基于 SNPs 的进化关系。作为第三代遗传分子标记，SNP 所独有的数量多、分布广、易于基因分型的特性决定了它更适合于对细菌进行地域性溯源识别。目前已发展出多种高效、稳定、经济的 SNP 分析技术，如限制酶内切和实时荧光定量 PCR、高分辨熔解曲线及微列阵分析等。

2.6　多位点可变串联重复序列（MLVA）

MLVA 是通过基因组中可变数量串联重复序列（VNTR）的特征来实现分型的分子分型技术。VNTR 是由短片段 DNA 序列串联重复组成的重复 DNA 片段，其重复的次数在不同个体间存在高度的可变性[21]。该技术可以通过一组通用基因，对所有原核生物实现分级分类，具有更为灵活、快速、分辨力强的优点[22]。Pruvost 等[23,24]对世界范围内柑橘溃疡病菌遗传多样性运用 MLVA 技术和主成分分析法进行分析，发现可分为四种类型：A 型广泛分布于世界各地；A 和 A^* 型分布在孟加拉国、印度、阿曼和巴基斯坦；A^w 型分布在印度和佛罗里达州；A^* 型分布在伊朗、阿曼、柬埔寨和泰国。并进一步证明柑橘溃疡病菌的发源地西亚是该病原菌多样性的中心。

2.7　多位点序列分析（MLST）

1998 年，Marden 等人首次提出 MLST 技术，用于分析脑膜炎奈瑟氏菌（*Neisserria meningitidis*）的种群分类[25]。其基本原理是针对持家基因设计引物进行 PCR 扩增和测序，根据得出的每株病原体各个位点等位基因数目进行序列类型鉴定及聚类分析。冯建军等[26]利用 7 个看家基因，对不同地区的西瓜细菌性果斑病菌（*Acidovorax citrulli*）进行多位点序列分型，将供试菌株分为 CC1 和 CC2 两大克隆复合体型；C. Moretti 等[27]将 MLST 与 rep-PCR 技术结合，对来自 15 个国家的 124 株橄榄节疤病原菌 *Pseudomonas savastanoi* pv. *savastanoi* 的地理起源进行研究，分析发现世界范围内的 Psav 种群是由单一的遗传谱系产生的。

该技术具有快速简便、重复性好、数据标准等优点，且不需要参考菌株对数据进行标准化，可实现实验室间数据共享及比较，越来越多的被作为能进行国际间菌株比较的常用工具。但不足之处在于需要预先知道待测微生物的基因组序列，以便推测看家基因并设计引物；此外，MLST 技术需要测序，成本较高，因此该技术的推广需要测序价格的降低。

3　总结与展望

伴随着经济全球化，我国进出口贸易呈现出快速增长的态势。与此同时，粮食、环境、生物等的安全问题也日益突出。在进出境检疫中，亟待建立有效的危险性微生物溯源技术体系。传统的表型分析方法已不能完全适应当代复杂多变的微生物进化现状，与之相比，基于微生物基因的分子方法具有更好的稳定性和可重复性，还可以开展遗传进化、追踪溯源、分子流行病学等多方面的研究，但单独的分子生物学溯源技术也需要病原菌的表征作为辅助判断的依据，因此二者应相辅相成，结合特定的微生物采取合适的溯源方法。

随着基因测序技术的快速发展、DNA 序列数据库的不断完善以及成本的不断降低，以新一代基因测序为基础的全基因组溯源方法，将为高频跨境危险性微生物风险评估、除害处理效果评估、监管政策制定等提供科学依据，也将为为双边贸易谈判、国门安全风险预警提供科技支撑。

参考文献

[1] Ren J G, Huang S L, Li Y R, *et al.* Studies on Differentiation of Strains of *Xanthomonas axonopodis* pv. *citri* in Guangxi (in Chinese) [J]. Microbiology China, 2007, 34 (2): 216-220.

[2] Łukasz Richter, Janczuk-Richter M, Niedziółka-Jönsson J, *et al.* Recent advances in bacteriophage-based methods for bacteria detection [J]. Drug Discovery Today, 2017.

[3] Su S B, Ma H X, Xu F Y. Review of application of bacteriophage in the diagnosis and therapy for bacteria in-

fection (in Chinese) [J]. Veterinary Science in China, 2011 (5): 546-550.

[4] Boerlin P. Applications of multilocus enzyme electrophoresis in medical microbiology [J]. Journal of Microbiological Methods, 1997, 28 (3): 221-231.

[5] Taylor J W, Geiser D M, Burt A, *et al.* The evolutionary biology and population genetics underlying fungal strain typing [J]. Clinical Microbiology Reviews, 1999, 12 (1): 126.

[6] Lu X L. Genetic Diversity of *Xanthomonas citri* subsp. *citri* Isolates from China on the Basis of Simple Sequence Repeat (SSR) (in Chinese) [D]. Chongqing: Chongqing University, 2013.

[7] Ferreiratonin M, Rodriguesneto J, Harakava R, *et al.* Phylogenetic analysis of *Xanthomonas* based on partial rpoB gene sequences and species differentiation by PCR-RFLP [J]. International Journal of Systematic & Evolutionary Microbiology, 2012, 62 (Pt 6): 1419.

[8] Velappam N, Sondgrass J L, Hakovirta J R. Rapid identification of pathogenic bacteria by single-enzyme amplified fragmem length polymorplfism analysis [J]. Diagnostic microbiology and infections disease, 2001, 39 (2): 77-83.

[9] Khodakaramian G, Swings J. Genetic Diversity and Pathogenicity of *Xanthomonas axonopodis* Strains Inducing Citrus Canker Disease in Iran and South Korea [J]. Indian Journal of Microbiology, 2011, 51 (2): 194-199.

[10] Kumar A, Misra P, Dube A. Amplified fragment length polymorphism: an adept technique for genome mapping, genetic differentiation, and intraspecific variation in protozoan parasites [J]. Parasitology Research, 2013, 112 (2): 457-466.

[11] Versalovic J, Koeuth T, Lupski J R. Distribution of repetitive DNA sequences in eubacteria and application to fingerprinting of bacterial genomes [J]. Nucleic Acids Research, 1991, 19 (24): 6823-6831.

[12] Yao T S, Hu J H, Tang K Z, *et al.* Primary analysis on genomic diversities of *Xanthomonas axnopodis* pv. *citri* in nine provinces of China (in Chinese) [J]. Journal of Fruit Science, 2010, 27 (5): 819-822.

[13] Wei C D, Ding D, Ye G, *et al.* Genetic diversities of *Xanthomonas citri* subsp. *citri* in Guangdong and Jiangxi Provinces (in Chinese) [J]. Journal of South China Agricultural University, 2014 (4): 71-76.

[14] Qi Y X, Zhang H, Pu J J, *et al.* Genetic diversities of *Xanthomanas campestris* pv. *mangiferaeindicae* by rep-PCR (in Chinese) [J]. Jiangsu Agricultural Sciences, 2017, 45 (6): 88-91.

[15] Ranjbar R, Karami A, Farshad S, *et al.* Typing methods used in the molecular epidemiology of microbial pathogens: a how to guide [J]. New Microbiologica, 2014, 37 (1): 1-15.

[16] Kaur S, Panesar P S, Bera M B, *et al.* Simple sequence repeat markers in genetic divergence and marker-assisted selection of rice cultivars: a review [J]. Critical Reviews in Food Science & Nutrition, 2015, 55 (1): 41-9.

[17] Végh A, Hevesi M, Éva Pǘjtli, *et al.* Characterization of *Erwinia amylovora*, strains from Hungary [J]. European Journal of Plant Pathology, 2016: 1-7.

[18] Peng Y W. Genetic polymorphism differentiation from '*Xanthomonas axonopodis* pv. *citri*' in China (in Chinese) [D]. Chongqing: Southwest University, 2014.

[19] Zhao H, Zhao H B, Qu P, *et al.* A Review of *Escherichia coli* Traceability Technology (in Chinese) [J]. Food Research and Development, 2013 (11): 123-127.

[20] Hommais F, Pereira S, Acquaviva C, *et al.* Single-nucleotide polymorphism phylotyping of *Escherichia coli* [J]. Applied & Environmental Microbiology, 2005, 71 (8): 4784-4792.

[21] Gevers D, Cohan F M, Lawrence J G, *et al.* Re-evaluating prokaryotic species [J]. Nature Reviews Microbiology, 2005, 3 (9): 733-739.

[22] Young J M, Park D C, Shearman H M, *et al.* A multilocus sequence analysis of the genus *Xanthomonas* [J]. Syst. appl. microbiol, 2008, 31 (5): 366-377.

[23] Pruvost O, Magne M, Boyer K, *et al.* A MLVA genotyping scheme for global surveillance of the citrus pathogen *Xanthomonas citri* pv. *citri* suggests a worldwide geographical expansion of a single genetic lineage [J].

Plos One, 2014, 9 (6): e98129.

[24] Pruvost O, Goodarzi T, Boyer K, *et al.* Genetic structure analysis of strains causing citrus canker in Iran reveals the presence of two different lineages of *Xanthomonas citri* pv. *citri* pathotype A* [J]. Plant Pathology, 2015, 64 (4): 776-784.

[25] Liu J H, He D, Yang Y Q, *et al.* Application of Multilocus Sequence Typing method on Pathogenic Microorganisms Typing and Identification (in Chinese) [J]. Microbiology China, 2007, 34 (6): 1188-1191.

[26] Feng J, Schuenzel E L, Li J, *et al.* Multilocus sequence typing reveals two evolutionary lineages of *Acidovorax avenae* subsp. *citrulli* [J]. Phytopathology, 2009, 99 (8): 913.

[27] Moretti C, Vinatzer B A, Onofri A, *et al.* Genetic and phenotypic diversity of Mediterranean populations of the olive knot pathogen, *Pseudomonas savastanoi* pv. *savastanoi* [J]. Plant Pathology, 2016.